The Russian Gas Matrix

The Russian Gas Matrix

How Markets are Driving Change

Edited by
JAMES HENDERSON and SIMON PIRANI

Contributors
TATIANA MITROVA
JONATHAN STERN
KATJA YAFIMAVA

Published by the Oxford University Press
for the Oxford Institute for Energy Studies
2014

OXFORD
UNIVERSITY PRESS

Great Clarendon Street, Oxford OX2 6DP
Oxford University Press is a department of the University of Oxford.
It furthers the University's objective of excellence in research, scholarship
and education by publishing worldwide in
Oxford New York
Auckland Cape Town Dar es Salaam Hong Kong Karachi
Kuala Lumpur Madrid Melbourne Mexico City Nairobi
New Delhi Shanghai Taipei Toronto
with offices in
Argentina Austria Brazil Chile Czech Republic France Greece
Guatemala Hungary Italy Japan Poland Portugal Singapore
South Korea Switzerland Thailand Turkey Ukraine Vietnam

Oxford is a registered trade mark of Oxford University Press
in the UK and in certain other countries

Published in the United States
by Oxford University Press Inc., New York

© Oxford Institute for Energy Studies 2014

The moral rights of the author have been asserted
Database right Oxford Institute for Energy Studies (maker)

First published 2014

All rights reserved. No part of this publication may be reproduced,
stored in a retrieval system, or transmitted, in any form or by any means,
without the prior permission in writing of the Oxford Institute for Energy
Studies, or as expressly permitted by law, or under terms agreed with
the appropriate reprographics rights organization. Enquiries concerning
reproduction outside the scope of the above should be sent to Oxford
Institute for Energy Studies, 57 Woodstock Road, Oxford OX2 6FA

You must not circulate this book in any other binding or cover
and you must impose this same condition on any acquirer

British Library Cataloguing in Publication Data
Data available

Library of Congress Cataloguing in Publication Data
Data available

Cover designed by Cox Design Ltd
Typeset by Davidson Publishing Solutions, Glasgow
Printed by Berforts Information Press Ltd, Witney

ISBN 978-0-19-870645-8

The contents of this book are the sole responsibility of the authors.
They do not necessarily represent the views of the
Oxford Institute for Energy Studies or any of its members.

1 3 5 7 9 10 8 6 4 2

CONTENTS

List of Maps	*vii*
List of Figures	*vii*
List of Tables	*x*
Preface	*xiii*
Acknowledgements	*xv*
Units and conversions	*xvi*
Glossary	*xvii*
Notes on contributors	*xx*

	Introduction *James Henderson and Simon Pirani*	1

Part 1: Political Economy of Russian Gas

1	The Political and Economic Importance of Gas in Russia *Tatiana Mitrova*	6
2	The Matrix: an Introduction and Analytical Framework for the Russian Gas Sector *Jonathan Stern*	39

Part 2: The Impact of Competitive Markets on Russian Gas

3	Russian Responses to Commercial Change in European Gas Markets *Jonathan Stern*	50
4	The Impact of European Regulation and Policy on Russian Gas Exports and Pipelines *Jonathan Stern*	82
5	Russia's Domestic Gas Market Development, Prices, and Transportation *James Henderson, Simon Pirani, and Katja Yafimava*	108
6	Sources of Demand in the Russian Domestic Gas Market *Simon Pirani*	155
7	CIS Gas Markets and Transit *Simon Pirani and Katja Yafimava*	181
8	Asia: a Potential New Outlet for Russian Pipeline Gas and LNG *James Henderson*	216

| 9 | Summary: the Influence of Markets on Russia's Gas Supply Strategy
Simon Pirani | 246 |

Part 3: Russia's Gas Supply Strategy

10	Sources of Russian Gas Supply *James Henderson*	252
11	Gazprom's West-Facing Supply Strategy to 2020 *Jonathan Stern and James Henderson*	258
12	The Dynamics of Gazprom's Future Strategy *James Henderson and Jonathan Stern*	286
13	Non-Gazprom Russian Producers: Finally Becoming Truly Independent? *James Henderson*	314
14	Central Asian and Caspian gas for Russia's Balance *Simon Pirani*	347
15	Summary: the Likely Balance of Russian Gas Supply *James Henderson*	368

Part 4: Conclusions

| 16 | The Changing Balance of the Russian Gas Matrix and the Role of the Russian State
James Henderson and Simon Pirani | 376 |

Appendices

Appendix 1. Statistical issues	391
Appendix 2. Russian domestic gas consumption by region	394
Appendix 3. Data on gas transmission and storage in Russia	398
Appendix 4. Proposed legal framework for third-party access to the Russian gas transport system	399
Appendix 5. Gas transportation tariff methodology	401
Appendix 6. Statistical information on the power sector	404
Appendix 7. Reserves, production, and costs at Gazprom's major production associations	410
Bibliography	412
Index	430

LIST OF MAPS

Chapter 3
Map 3.1	The Ukrainian and Yamal–Europe pipelines	71
Map 3.2	The Nord Stream pipelines	74
Map 3.3	The Blue Stream and South Stream pipelines	76

Chapter 5
Map 5.1	Gas consumption by Federal district	111
Map 5.2	The Russian gas transportation system	137

Chapter 6
Map 6.1	Fuel shares in power generation	159

Chapter 8
Map 8.1	Russia's Eastern Gas Programme development plan	230
Map 8.2	Licences and fields at Sakhalin Island	235
Map 8.3	Russia's existing and potential LNG projects for the Asian market	241

Chapter 11
Map 11.1	Yamal–Nenets fields and pipelines	261
Map 11.2	Yamal peninsula and Ob–Taz Bay fields	266

Chapter 14
Map 14.1	Central Asian and Caspian gas pipelines	350

LIST OF FIGURES

Chapter 1
Figure 1.1	Structure of Russian exports by value, 2012	7
Figure 1.2	The share of oil and gas in Federal budget revenues	8
Figure 1.3	Russia: growth in GDP and industrial value added	12
Figure 1.4	Russia: industrial production growth	13
Figure 1.5	Direct foreign investment in Russia	13
Figure 1.6	The share of oil and gas MET and export duties in Federal budget income	15

Figure 1.7	Structure of the primary energy mix of leading gas producing countries	16
Figure 1.8	Installed electricity and CHP capacity in Russia	16
Figure 1.9	Gas and coal prices in Moscow region	17
Figure 1.10	The share of gas purchases in household spending	20
Figure 1.11	Framework of strategic documents in the Russian energy sector	26

Chapter 3
Figure 3.1	Russian long-term gas contracts with OECD European countries to 2035	53
Figure 3.2	European long-term oil-linked and spot prices of gas August 2010–September 2013	58
Figure 3.3	Russian long-term export contracts with OECD European countries to 2035	60
Figure 3.4	Russian gas export prices relative to NBP	65

Chapter 5
Figure 5.1	Russia and Ukraine: GDP and gas consumption, 2008–13	116
Figure 5.2	The regulated domestic gas price in Russia	122
Figure 5.3	Comparison of Gazprom and Novatek gas sales prices	125
Figure 5.4	Comparison of domestic gas price growth and export netback parity levels	128
Figure 5.5	Split of Gazprom sales by major customer segment	129
Figure 5.6	A possible outlook for Russian domestic gas prices	131
Figure 5.7	Comparison of old and new MET rates for gas production in Russia	133
Figure 5.8	Household gas tariffs compared to industrial gas tariffs in Russia	135
Figure 5.9	Comparison of regulated domestic gas price and gas transport tariff movements	152

Chapter 6
| Figure 6.1 | Fuel shares of thermal electricity generation | 158 |

Chapter 8
| Figure 8.1 | An estimate of the Chinese gas supply and demand balance to 2030 | 223 |
| Figure 8.2 | Possible supply volumes to China as part of its multi-vector gas import policy | 227 |

| Figure 8.3 | Russian gas export forecast from the Energy Strategy to 2030 | 231 |
| Figure 8.4 | Estimates of breakeven gas prices for Russian gas to Asia | 239 |

Chapter 10
| Figure 10.1 | The main elements of Russian gas supply since 1991 | 253 |

Chapter 11
| Figure 11.1 | Gazprom projections of gas production to 2030 | 270 |
| Figure 11.2 | Gazprom's LNG spot sales since 2005 | 277 |

Chapter 12
Figure 12.1	Breakdown of Gazprom's gas revenues, 2008–12	288
Figure 12.2	Breakdown of Gazprom's total overall revenues, 2008–12	289
Figure 12.3	Comparison of Gazprom and Novatek EBITDA margins, 2007–12	292
Figure 12.4	Gazprom's free cash flow, 2007–12	295
Figure 12.5	Gazprom's potential eastern gas production	301

Chapter 13
Figure 13.1	Average Novatek gas price to domestic consumers compared to average Gazprom price	317
Figure 13.2	Share of Independent producers in Russian gas supply	318
Figure 13.3	Share of Novatek revenues from gas and liquids sales	321
Figure 13.4	Estimate of potential Novatek gas output	326
Figure 13.5	Share of end users in Novatek's gas sales	327
Figure 13.6	Split of Novatek's gas sales in 2012	328
Figure 13.7	Rosneft's planned gas sales portfolio in 2020	331
Figure 13.8	Rosneft's potential gas output to 2020	333
Figure 13.9	Lukoil gas production estimate to 2025	336
Figure 13.10	Potential Gazprom Neft gas production	339
Figure 13.11	Potential non-Gazprom gas production to 2025	342

Chapter 14
| Figure 14.1 | Estimates of costs of gas delivered to Moscow region in 2020 | 360 |

Chapter 16
Figure 16.1	Gazprom's gas export price to Europe	368
Figure 16.2	The potential oversupply of Russian gas	384

LIST OF TABLES

Chapter 1
Table 1.1	Oil and gas in Russian GDP, budget revenues, and total exports in 2011–12	8
Table 1.2	Russia in international rankings	14
Table 1.3	MET rate for gas (up to 1 July 2015)	29
Table 1.4	Competing interests in the gas industry	36

Chapter 2
Table 2.1	The Russian gas matrix: major building blocks	41

Chapter 3
Table 3.1	Gazprom's exports to Europe, 2005–13	51
Table 3.2	Russian long-term contract gas prices to European countries, 2010–13	61
Table 3.3	Russian gas export pipeline capacity to Europe, 2013–20	80

Chapter 4
Table 4.1	Volumes of Russian gas and EU borders crossed in 2011	85

Chapter 5
Table 5.1	Russian domestic gas consumption	110
Table 5.2	Gas demand growth trends	112
Table 5.3	Russia's gas demand outlook	114
Table 5.4	Gas contracts signed by Independent gas producers since 2009	126
Table 5.5	Gazprom's transport of its own and third-party gas, 2008–12	144
Table 5.6	Average levels of transportation tariffs charged by Gazprom to third parties	147
Table 5.7	Transportation tariffs charged by Gazprom to its own transmission companies	150

Chapter 6
Table 6.1	Gas for power and district heating: overview	156
Table 6.2	Thermal, nuclear, and hydro power	157
Table 6.3	Electricity demand growth and GDP growth	160
Table 6.4	Gas consumption in industry	173
Table 6.5	Prilled urea to Europe: Evrokhim cost estimates, 2010	174
Table 6.6	The cost of Evrokhim's own gas and its purchases	175

Chapter 7
Table 7.1	Supplies of gas to CIS countries	182
Table 7.2	Ukraine's gas imports	186
Table 7.3	Gas consumption in Ukraine	192
Table 7.4	Gas prices in Ukraine	193
Table 7.5	Belarus gas consumption, imports, and prices	198
Table 7.6	Moldova gas consumption, imports, prices, and tariffs	203

Chapter 8
Table 8.1	Russia's eastern gas reserves and resources	217

Chapter 11
Table 11.1	Gazprom gas reserves and production	262
Table 11.2	Gazprom production projections, 2010–20	263
Table 11.3	Changes in estimates of peak production dates for Gazprom fields	270

Chapter 12
Table 12.1	Increase in Gazprom's production and transmission costs	290
Table 12.2	Gazprom's total capital expenditures	292
Table 12.3	Gazprom's investments in the gas sector, 2013–30	294

Chapter 14
Table 14.1	Gas exports from the Central Asian and Caspian region	348
Table 14.2	Prices of Central Asian gas exports, 2005–9	351
Table 14.3	Gazprom purchase prices in Central Asia and Caspian, and sales prices	354
Table 14.4	Russian gas demand in areas close to infrastructure links with Central Asia	361

Table 14.5	Central Asian gas exports to China and Ukraine: prices compared	363
Table 14.6	Russian companies upstream in Central Asia and the Caspian	365

Appendix

Table A1.1	Statistical measurements of Russian gas consumption	393
Table A2.1	Russian gas consumption: regional breakdown	395
Table A3.1	Gazprom's high-pressure Gas Transmission Network 31 December 2012	398
Table A6.1	Territorial structure of fuel consumption for thermal power generation	404
Table A6.2	Gas and coal prices for power stations, 2000–11	405
Table A6.3	Gas and coal prices for power stations, 2011	406
Table A6.4	Total new electricity generating capacity commissioned, 2005–11	407
Table A6.5	Timing of new power generation capacity installation	408
Table A6.6	Commissioning plans by generating companies, 2012–18	409
Table A7.1	Remaining proven and probable reserves at Gazprom's major production associations (31 December 2012)	410
Table A7.2	Production and costs of production at Gazprom's major production associations	411

PREFACE

The world of natural gas has become increasingly complex over the past decade. This has manifested itself in the accelerating number of timely and relevant books and research papers we have published since our inception as a gas programme in 2003.

Natural gas, with its relatively low energy density and hence high cost of transportation and storage, lent itself to the 'old world' model of long-term contracts, destination clauses, and 'captive' end consumers. This business model suited Russia and Gazprom well. Through the take or pay mechanism the buyer took the 'market risk' and the seller (through price linkage to oil) took the 'price risk'. In the early to mid-2000s, when European gas demand growth seemed assured and oil prices were expected to remain robust, the risk incurred by both parties seemed low. In this environment Russia continued to expand its 'matrix' of **markets**: Russian domestic, the FSU, Europe, and the beginnings of Asian LNG (Sakhalin-2), and **supply**: Gazprom production dominated by its West Siberian fields, Russian Independents (which could be controlled through access to transportation infrastructure), and Central Asian imports. Gazprom's central role in the matrix was bolstered by Russian political support, enabling it to maintain its role in the domestic and the FSU 'sphere of influence' markets.

During the 2000s, a coincidence of factors imposed significant external stress on the largely regionally circumscribed gas markets and evoked (in our view) an irreversible evolutionary response, one that we discussed in our 2012 book, *The Pricing of Internationally Traded Gas*. These factors included the growth of LNG development ambitions by Qatar and others, the establishment of gas trading hubs in continental Europe after years of pro-competition EU policy, the unforeseen 'Black Swan' of US shale gas (and the prospect of LNG exports), and the financial crisis and subsequent economic stagnation (and gas demand reduction) experienced in Europe, Russia, and the FSU in particular. While our 2012 book on international gas pricing concluded by postulating what such changes might mean for an arbitrage-related price linkage (but not convergence) between regional world trading hubs, this book assesses in detail the impact of these changes on Russia.

Since Russia is the single most important actor in global gas, such changes in the world's gas markets and in the behaviour patterns of its other key players prompt several questions: How will Russia fare as its 'old world' system in Europe is irrevocably changed? Can the prior system of centralized political control effectively operate in this more complex world where even the Russian domestic market is open to competition? Will Russia secure a share of the fast-growing Asian gas market and if so who will lead the charge?

These and other issues are addressed in this book. The editors and chapter authors are widely acknowledged experts in their fields and the creative process, which was superbly managed by editors James Henderson and Simon Pirani, has produced a work which is very much more than the sum of its component parts.

The Natural Gas Research Programme at the Oxford Institute for Energy Studies is justly proud of its Russia and CIS research capability and I am delighted to see this book added to the canon of publications produced by our programme in a field where academic literature is unfortunately relatively sparse.

Howard Rogers
Director
Natural Gas Research Programme
Oxford Institute for Energy Studies
Oxford, 2014

ACKNOWLEDGEMENTS

We express our warmest thanks are to our fellow authors: Jonathan Stern, Katja Yafimava, and Tatiana Mitrova. Working with a comparatively small team enabled us to draw on a large body of experience, while integrating our arguments in a narrative in a way that is not usually possible in multi-author works. Thanks, too, to Howard Rogers, who took part in our editorial discussions; to colleagues on the Natural Gas Research Programme from whose collaboration we benefited; and to others at the Oxford Institute for Energy Studies who have commented on our work, both at a seminar in November 2013 and at other times.

We also owe a debt of gratitude to many colleagues, friends, and acquaintances who are involved in one way or another with the Russian gas sector, discussions with whom are invaluable.

Thanks are due to those who did editorial work on the book, including Laura El-Katiri; David Sansom, who prepared the maps; Catherine Gaunt, our copy editor; Christopher Riches, who compiled the index; and Evelyn Sword and Margaret Walker for typesetting. We also thank Kate Teasdale, the administrator at the OIES, and Jo Ilott, the administrator of the Natural Gas Research Programme.

James Henderson and Simon Pirani
January 2014

UNITS AND CONVERSIONS

Volumes of gas are stated in this book in billions of cubic metres (Bcm). For some reserves figures, trillions of cubic metres (Tcm) are used. (1 trillion = 1,000 billion.) In energy terms, 1 Bcm = 0.9 million tonnes of oil equivalent (mmtoe).

Production is usually stated in billions of cubic metres/year (Bcm/year). The conversion to the units usually used in North America is: 1 billion cubic feet per day (1 BCF/d) = 10.34 Bcm/year. 1 cubic metre = 35.3146667 cubic feet.

There are some references to production of liquefied natural gas (LNG), measured in millions of tonnes (mt). 1mt of LNG = 1.36 Bcm of natural gas.

Gas condensate is measured in millions of tonnes of oil equivalent (mmtoe), or millions of barrels of oil equivalent. 1 mmtoe = 7.33 mmboe.

Gas prices are quoted in US dollars per thousand cubic metres ($/mcm). The conversion to the units usually used in the European gas market is: $1/million British thermal units ($1/MMBtu) = $36.20/mcm.

Some prices are quoted in rubles per thousand cubic metres (RR/mcm). In January 2014, the exchange rate was $1 = RR 33.09.

The reader should bear in mind that gas volumes are measured in Russia and Europe according to different criteria. In Russia, gas is usually measured at 2 °C, and has 37,046 kJ of energy content and 8850 kJ of gross calorific value (GCV) per cubic metre. In Europe, gas is usually measured in 'standard cubic metres', at 15 °C, with 40,000 kJ of energy content and 9393 kJ of net calorific value (NCV). As a result, 1 Bcm of gas as measured in Europe is equivalent to 1.0797 Bcm of gas as measured in Russia. However, since conversion factors differ between organizations and many publications are not specific about the heating and calorific values of the data they publish, readers should be careful when comparing the units in this book with those published by other sources.

GLOSSARY

ACQ	Annual contract quantity
ADR	American Depository Receipt
APBE	Agency for Forecasting Energy Balances
BCF/d	billion cubic feet per day
Bcm	Billions of cubic metres
Bcm/year	billions of cubic metres/year
CBM	Coal bed methane
CCGT	Combined cycle gas turbines
CDAs	Capacity delivery agreements
CHP	Combined heat and power
CIS	Commonwealth of Independent States
DG COMP	European Commission's Competition Directorate
EBITDA	Earnings before interest tax and depreciation
EBRD	European Bank for Reconstruction and Development
ECJ	European Court of Justice
ECT	Energy Charter Treaty
EDF	Électricité de France
EE system	Entry/exit system
EIA	US Energy Information Administration
EnCT	Energy Community Treaty
EOR	Enhanced Oil Recovery
ERI RAS	Russian Academy of Sciences
ERI	Energy Research Institute
ES-2030	Energy Strategy to 2030.
ESPO	the East Siberia-Pacific Ocean
ETS	Electronic Trading System
FAS	Federal Antimonopoly Service
FEC	Federal Energy Commission
FSU	Former Soviet Union
FTS	Federal Tariff Service
GDF	Gaz de France
GMT	Gazprom Marketing and Trading
GTS	Gas Transmission System
IEA	International Energy Agency
IOC	International Oil Company
IP	Interconnection point

IPO	Initial public offering
JCC	Japan Crude Cocktail
kpbd	thousand barrels per day
LNG	Liquefied natural gas
LPG	Liquefied petroleum gas
LTA	Long-term agreements
LTSC	Long-term supply contracts
mcm	thousand cubic metres
MET	Mineral Extraction Tax
MICEX	Moscow International Commodities and Currency Exchange
mmbls	million barrels
mmboe	millions of barrels of oil equivalent
mmbpd	millions of barrels per day
MMBTu	million British thermal units
mmtoe	million tonnes of oil equivalent
MOEK	Moscow United Energy Company
MRET	Mineral Resources Extraction Tax
mt	million tonnes
mtpa	millions of tonnes per annum
NPT	Nadym Pur Taz
OAO	Open joint stock company
OGKs	Wholesale generating companies
OIES	Oxford Institute for Energy Studies
PCI	Projects of Common Interest
PP	Point-to-point
PSA	Production Sharing Agreement
RAB	regulated asset base
RR	Russian rubles
SES	Single Economic Space
SKV	Sakhalin–Khabarovsk–Vladivostok (pipeline)
SPIMEX	St Petersburg International Mercantile Exchange
TAG	Trans Austria Gasleitung (pipeline)
TAP	Trans-Adriatic Pipeline
Tcm	Trillions of cubic metres
TENS	Trans-European Networks
TGK	Territorial generating company
ToP	Take-or-pay
TPA	Third-party access
TsDU TEK	Central Dispatching Unit of the Fuel and Energy Complex

UGSS	Unified Gas Supply System
VAT	Value added tax
VTPs	Virtual trading points or hubs
WEO	World Energy Outlook

NOTES ON CONTRIBUTORS

James Henderson has been analysing the Russian oil and gas industry for more than 20 years, including as Head of Equity Research at Renaissance Capital in Moscow (1997–9), Head of Russia at Lambert Energy Advisory (2006–13), and currently as Senior Research Fellow at the Oxford Institute for Energy Studies (OIES). He is author of *Non-Gazprom Gas Producers in Russia* (OIES/OUP, 2010) and of many publications on Russian energy issues.

Tatiana Mitrova is Head of Oil and Gas Research at the Energy Research Institute of the Russian Academy of Sciences. She joined the Institute in 1998 and is a recognized commentator on Russian energy issues and the author of many publications on Russian and international energy issues. She is an expert on the EU–Russia Gas Advisory Council. She also teaches at the Gubkin Oil and Gas University and the Higher School of Economics in Moscow.

Simon Pirani is Senior Research Fellow on the OIES Natural Gas Research Programme. Before joining the Institute in 2007 he wrote about Russia and Ukraine as a journalist and historian. He is the editor of *Russian and CIS Gas Markets and their Impact on Europe* (OIES/OUP, 2009) and author of many publications on Russian, Ukrainian, and Central Asian gas issues.

Jonathan Stern has been writing about Soviet, and then Russian, gas issues, for more than 30 years, and is the author of *The Future of Russian Gas and Gazprom* (OIES/OUP, 2005). His recent publications include (as editor) *The Pricing of Internationally Traded Gas* (OIES/OUP, 2012). He founded the OIES Natural Gas Research Programme in 2003 and is now its Chairman and a Senior Research Fellow. Since 2011 he has been the EU speaker for the EU–Russia Gas Advisory Council.

Katja Yafimava is Senior Research Fellow on the OIES Natural Gas Research Programme and an expert adviser to the EU–Russia Gas Advisory Council. She is the author of *The Transit Dimension of EU Energy Security: Russian gas transit across Ukraine, Belarus, and Moldova* (OIES/OUP, 2011). She joined the OIES in 2006 after spells at Shell and the Energy Charter Secretariat.

INTRODUCTION

James Henderson and Simon Pirani

The dramatic changes in international natural gas markets since the 2008 economic crisis have impacted powerfully on Russia, one of the largest gas producers and consumers. It is not only that the pre-crisis certainty of steadily rising demand in the main markets for Russian gas came to an end. Pricing trends and regulatory regimes in Europe (Russia's main export market) have changed dramatically; CIS markets that had been relatively predictable have been disrupted by recession and price disputes with Russia; and change in the huge Russian domestic market is being driven by price reform.[1] Whereas Gazprom (Russia's dominant gas company) was formerly a near-monopoly supplier to domestic customers, the company now faces increasingly aggressive competition from other producers, not only for domestic market share but also for access to Asian markets, which are seen as crucial to the future of Russian gas.

Because gas is seen as a strategic, not just an economic, asset, it is government that has taken and will take key decisions about the industry's future – on export strategy, domestic market reform, and models of upstream development and taxation. This book, following on from previous work by the Natural Gas Research Programme at the Oxford Institute for Energy Studies,[2] therefore considers political as well as economic and energy factors in the Russian gas sector's development.

The 'matrix' of this book's title is an analytical framework designed to capture the relationship between the sources of demand for Russian gas

---

[1] The political crisis in Ukraine, and Russian military action in Crimea, took place in February–March 2014 when this book was at an advanced stage of production. This is certainly the most serious crisis in Russo-Ukrainian relations, and probably the most serious crisis in Russia's relationship with western Europe, since the fall of the Soviet Union. Chapter 7, on CIS markets, reflects trends and scenarios that were evident to us at the time of final revision, in January 2014, but not the most recent developments. For a more recent assessment, see: Pirani S. *et al.* 'What the Ukraine Crisis Means for Gas Markets', Oxford Energy Comment, OIES, March 2014, www.oxfordenergy.org/2014/03/what-the-ukrainian-crisis-means-for-gas-markets/.

[2] This book directly follows two other books published as outcomes of the programme's work that covered the production of, and markets for, Russian gas – Jonathan Stern's *The Future of Russian Gas and Gazprom* (Oxford: OIES/OUP, 2005) and Simon Pirani's *Russian and CIS Gas Markets and their Impact on Europe* (Oxford: OIES/OUP, 2009) – as well as other relevant books, working papers, and articles.

(Europe, the CIS importers, the domestic market, and, for the future, Asia) and the sources of supply (Gazprom's gas, that produced by its domestic competitors, and imports from Central Asia). A unifying theme is that whereas in the past the sector's development was driven largely from the supply elements of this matrix, over the past five years the demand elements have come to the fore.

By way of introduction to this study of the Russian gas industry in 2014, it is instructive briefly to review the past 35 years of its history.[3] As the USSR entered its final decade in the 1980s, its gas industry mirrored the centralized structure of the whole economy. A powerful central organ, the Ministry of Gas, dominated production through its control of all upstream operations, and exerted influence over all the dominion states of the Union via its control of the vast pipeline network established to supply them. Imports from Central Asia were called upon to supplement the vast resources of the Volga Urals and West Siberia, and gas was offered to demand centres across Russia, into Soviet states such as Ukraine, Belarus, and Moldova, and on to Eastern Europe. By the 1980s exports to Western Europe were firmly established and were being ramped up, comprising a vital part of the supply and demand balance and providing much-needed finance for the Soviet treasury.

The collapse of the Soviet Union in 1991 brought significant change to the economies of the former Union states. In Russia, it generated dramatic change in the oil sector, with a collapse in production, break-up and privatization of the industry, and the introduction of foreign oil majors and service companies to assist in the recovery process. But the gas industry experienced no such dramatic change. In the spirit of the times, Viktor Chernomyrdin, the Soviet Union's last minister of gas, corporatized the ministry and made it a joint stock company, OAO Gazprom. In many other respects, the sector changed very little. In the first decade of the post-Soviet era Gazprom accounted for 90–95 per cent of Russian gas production, controlled the distribution of the remaining 5 per cent (essentially, associated gas production from the Russian oil companies), and managed imports from Central Asia to satisfy its own need to balance the overall gas system. Turkmenistan, Uzbekistan, and Kazakhstan had only one route – via Russia – to export their gas, and their oil export options were also limited. Since hydrocarbons remained a key source of budget revenues, at least for Kazakhstan and Turkmenistan, the gas trade reinforced their economic dependence on Russia.

---

[3] Recent important publications include Thane Gustafson's *Wheel of Fortune: the battle for oil and power in Russia* (Harvard: Harvard University Press, 2012), which gives an overview of the post-Soviet history of Russian oil, and Per Hogselius's *Red Gas* (Basingstoke: Palgrave Macmillan, 2013), a study of the evolution of Soviet gas exports to western Europe.

Until 2000, therefore, any talk of a Russian gas matrix would have been somewhat irrelevant. There were three main markets – Russia, the FSU, and Europe – supplied by one dominant producer, Gazprom, which used two small alternative sources of supply to balance the system. However, as the fallout from the Asian and Russian debt crises of 1997–8 gave way to economic recovery in the 2000s, so the balance of Russia's supply sources began to change. Gas consumption in all three of Gazprom's markets began to rise quite rapidly, thanks to economic growth in Russia and the FSU and to the switch to gas-fired power in Europe.

Questions began to be asked about Gazprom's ability to meet this rising demand, and the company started to assess its options for increasing supply. Given its poor financial state at the end of the 1990s, it was keen to defer major investment in new fields for as long as possible. It therefore turned to the cheap alternative of increased third-party production in Russia and extra imports from Central Asia, both of which could be acquired at the equivalent of very low Russian domestic prices. The supply side of the Russian gas matrix began to develop, with companies such as Novatek and Itera offering 'independent' gas supply alongside the oil companies' gas output, freeing up Gazprom production for export to the lucrative European market. At the same time, rising imports from Central Asia were being used to supply the markets of the FSU, in particular Ukraine, again allowing Gazprom to sell its own gas to more profit-generating customers. The matrix of three main markets and three main sources of supply had been established – albeit with the supply still dominated and controlled by one main player, Gazprom. In 2006–7, all the sources of demand were expanding, and many of the questions facing the gas sector were framed in terms of how Gazprom would ensure they were adequately supplied. Then came the world financial and economic crisis of 2008 and the resulting steep fall in demand.

This is the point at which the chapters in this book pick up the story. Its central thesis is that over the past five years the balance of this matrix has been fundamentally altered, and that as we consider the future it is probable that the new trends which have been established – declining Gazprom influence, increasing Independent production in Russia, and a shift in Central Asian exports away from Russia towards China – are likely to continue. These trends have been catalysed by forces in a global gas market that has become increasingly competitive and progressively interconnected, meaning that the Russian gas industry and the Russian government have had to start to respond to challenges that they have not previously faced. Their initial reaction has in itself, we would argue, displayed many of the characteristics of a market at work, even if the players themselves have been reluctant to acknowledge this fact.

The book is structured in four parts. The first part sets the scene. Chapter 1 describes the political and economic context in which the Russian gas sector operates; Chapter 2 introduces the gas matrix in more detail.

The second part analyses developments in the markets for Russian gas. Russia's export trade to Europe is the subject of two chapters: Chapter 3, which discusses the changing pattern of Russian exports, the substantial changes to gas pricing and market structure, and the development of pipeline infrastructure, while Chapter 4 examines how European policy and regulation has impacted Russia's exports. The Russian domestic market is the subject of the next two chapters: Chapter 5 on the evolution of market structure, gas pricing, and pipeline regulation, and Chapter 6 on the main drivers of demand. Chapter 7 explains the changes in CIS markets – Ukraine, Belarus, Moldova and the Caucasus – for Russian gas, together with the transit disputes between Russia and some of those states. Chapter 8 discusses the potential for Russian gas exports to China and other Asian markets. Chapter 9 draws together our arguments on the demand side.

The third part of the book discusses Russia's gas supply strategy. Chapter 10 provides an overview and introduction. Chapter 11 discusses Gazprom's west Siberian and Yamal peninsula production assets, which remain the main source of west-facing supply (in other words, for European Russia and other western export markets). Chapter 12 discusses Gazprom's future, in terms of its plans to open up Asian markets and of its corporate and financial strategy in a wider sense. The remaining chapters focus on the other suppliers of gas to the Russian matrix: Chapter 13, on the non-Gazprom Russian producers, and Chapter 14 on the Central Asian and Caspian producers. Chapter 15 draws together our views of gas supply.

The fourth part of the book, Chapter 16, presents our conclusions.

Part 1

Political Economy of Russian Gas

CHAPTER 1

THE POLITICAL AND ECONOMIC IMPORTANCE OF GAS IN RUSSIA

Tatiana Mitrova

The macroeconomic background

Although it seems to be a commonplace, oil and gas really are the main drivers of the Russian economy. Russia is a world leader in crude oil and natural gas production and exports. Between 2000 and 2010 exports rose dynamically: oil by 70 per cent; gas by 15 per cent. In the same period, annual production of oil rose by more than half and exceeded 500 million tonnes per year, while annual production of gas rose by 10 per cent. Revenues from oil and gas exports equated to more than a quarter of Russia's gross domestic product (GDP), and amounted to a third of the national budget.

During the last two decades, the Russian economy has become increasingly dependent on commodities exports (particularly hydrocarbons), despite numerous statements about the need to reduce dependence and the setting of targets. While, as mentioned above, the role of energy export revenues in the Russian economy keeps growing, the manufacturing sector's share keeps falling – reaching 4.8 per cent in 2010. Fuel exports as a proportion of total exports have risen from 43 per cent in 1996 to 64 per cent in 2010, and to 70 per cent in 2012 (Figure 1.1), while the share taken by manufactured goods in total exports fell from 26 per cent in 2005 to 14 per cent in 2012. At the same time the composition of imports changed: the share taken by manufactured products increased from 45 per cent to a peak of 79 per cent before the financial crisis, falling back to 69 per cent in 2010.[1]

Ways in which Russia's dependence on global fuel prices can be reduced, while increasing the contribution of the domestic market and of high-value-added products (engineering, chemical, and other products) to economic growth, have been discussed since the early 2000s. But it cannot be said that these aims have been achieved. The structure of the Russian economy has not changed significantly since 2006; the effect of the announced 'modernization' is modest, if not invisible, and the share of

---

[1] Covi, G. 'A case study of an advanced Dutch disease: The Russian oil', IMEF Ca Foscari University of Venice, May 2013.

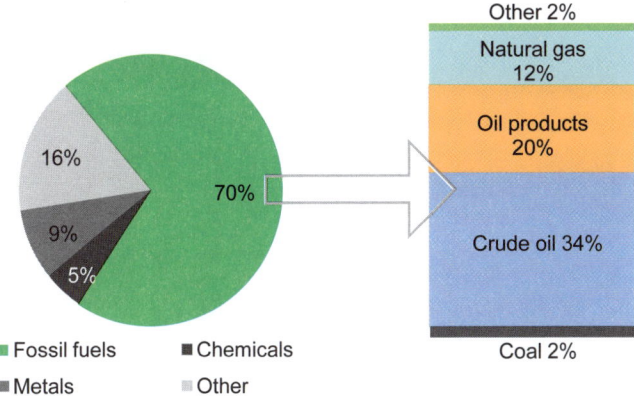

Figure 1.1: Structure of Russian exports by value, 2012
Source: Russian Federation Custom Statistical Yearbook 2012.

high-tech industry has not grown at all.[2] In fact, according to statistics compiled by Rosstat, the national statistics agency, the share taken by high-technology industries in Russia's GDP fell from 1.14 per cent in 2003 to 1.04 per cent in 2012. Such an insignificant level of high-tech activity is unable to make any impact on an economy entirely focused on raw materials.

Moreover, Russia's economy is demonstrating increasingly strong signs of the Dutch disease.[3] In 1994, the share of the oil and gas sector's proceeds in the national budget was below 2 per cent, by 2012 it had reached about 50 per cent (Figure 1.2). This leap has been driven by the oil price hike since 2004. Indeed, Russia's GDP growth is largely explained by rising international prices of oil, gas, and other raw materials, rather than by the successful development of other industries. Against the background of stagnation in other industries, the oil and gas sector has become the main source of growth both for GDP and for budget revenues (see Table 1.1), and this trend shows no sign of slowing[4] – although the share taken by the oil and gas sector in Russia's GDP (18.6 per cent in 2012) is still quite low compared to services (51 per cent in 2012).

---

[2] van der Marel E. 'Beyond Dutch Disease: When Deteriorating Rule of Law affects Russian Trade in High-Tech Goods and Services with Advanced Economies', London School of Economics, July 2012, pp. 1–26.
[3] Dülger F., Lopcu K., Burgaç A., and Balli, E. 'Is Natural Resource-Rich Russia Suffering from the Dutch Disease?', International Conference on Eurasian Economies 2012, pp. 54–9; Covi G. 'A case study of an advanced Dutch disease', pp. 1–27.
[4] Novak A. 'Prioritety gosudarstvennoi politiki v rossiiskoi neftegazovoi otrasli', Natsional'nyi Neftegazovyi Kongress. 19 March 2013.

8   *The Russian Gas Matrix*

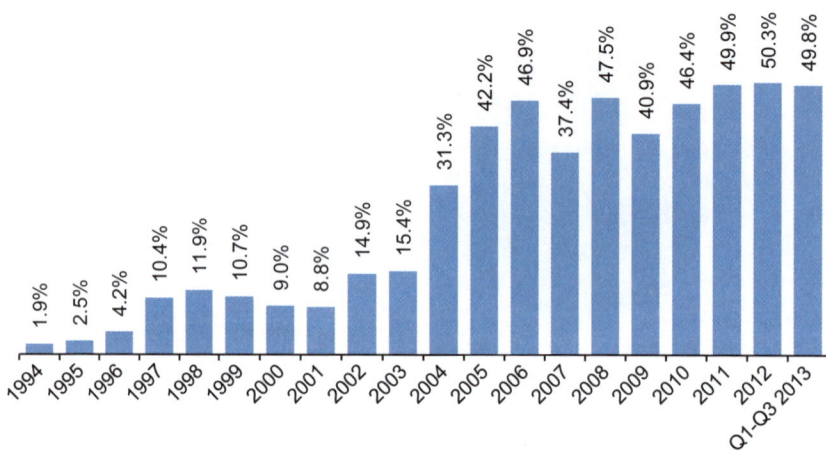

Figure 1.2: The share of oil and gas in Federal budget revenues
Source: www.roskazna.ru/reports/fb.html, Ministry of Finance.

Table 1.1: Oil and gas in Russian GDP, budget revenues, and total exports in 2011–12

	Oil	*Gas*
GDP	15.4	3.2
Budget revenue	36%	5%
Export	34%	12%

Sources: Russian Federation Ministry of Finance, Customs Service, The World Bank Database http://data.worldbank.org/indicator/NY.GDP.PETR.RT.ZS.

The significance of oil and gas sector revenues for the entire Russian economy can hardly be overestimated. Oil and gas sector activity has a tremendous multiplicative effect, as it creates a hefty domestic demand for other industries' products and ensures provision of the infrastructure development required for economic growth. Proceeds from hydrocarbon exports have an impact on the financial resources of manufacturers and service providers and, therefore, on business activity in the country and, thereby, on its economic development prospects.[5]

These revenues play an even more crucial role in meeting such budget expenditures as allocations for military and social purposes, and hence in

---
[5] Vliianie vneshnikh tsen na otsenku perspektiv razvitiia ekonomiki Rossii, *Voprosy Ekonomiki*, 2012, No.4, pp. 84–96.

the maintenance of the country's social stability and integrity. The government is now trying to implement the president's electoral assurances, which require greater budget expenditures, and its main hopes are placed on the oil and gas sector. Thus *a direct significance for national security is attributed to all events in this key economic sector*. Special emphasis is hence laid on the oil and gas industry by government and the country's top authorities, together with the desire to exert maximum control over it.

However, in recent years we have arrived at an impasse. On one hand, external challenges to the oil and gas sector's further sustainable development are increasing: there is a growing likelihood of a fall in world prices and in the volumes of Russia's hydrocarbon exports. On the other hand, the national economy itself is gradually sliding into a recession, and the authorities have no tools to stop this, other than their traditional recourse to oil and gas.[6]

In this context *the situation in international markets* gains critical significance. A comparison may be made with the Soviet Union in the mid-1980s, when changes in the international oil markets led to a collapse in the USSR's export revenues – which became one of the reasons for the country's demise. Now, changes in the international oil and gas markets and, in particular, uncertainties associated with the rapid development of shale oil and gas production, are creating new risks for the sustainability of Russia's economy.

First and foremost, there is a threat of a potential decrease, or even just of a stabilization, in oil prices. The price of Urals oil is the most influential factor for Russia's economy, not only because oil exports account for more than a third of federal budget revenues (more than 36 per cent in 2012) but also because gas prices in Europe are related to the oil price. Each 10 per cent rise in world oil prices can increase Russia's GDP growth rate by 0.75 percentage points. The role of gas is less significant, though not insubstantial. In 2012 it accounted for slightly over 5 per cent of federal budget revenues. Each 10 per cent rise in the European price of Russian gas can increase Russia's GDP growth rate by 0.21 per cent.[7] However, as of late 2013 oil prices have actually stabilized, and some researchers even predict that they will fall. And given their high level, they are no longer supporting economic growth in Russia. The break-even oil price, necessary to balance the Russian budget, is really high: in 2013 the official target was $97/bbl.[8]

---

[6] Gurvich E. 'Dolgosrochnie perspektivi Rossiiskoi ekonomiki', *Ekonomicheskaia politika*, No.3, 2013, pp. 7–32.

[7] Malakhov V.A. 'Otsenka zavisimosti VVP i sprosa na energonositeli ot udorozhaniia topliva I energii na vnutrennem I vneshnem rynke', ['Assessment of dependence of the GDP and demand for energy resources on the growth of fuel and energy prices in internal and international markets'], TEK Rossii, No. 1, 2012, pp. 16–24.

[8] 'Neft' i gaz prinesut v biudzhet Rossii v 2013 godu bol'she zaplanirovannogo', 17 September 2013. http://lenta.ru/news/2013/09/17/dohod/.

If world oil prices fall, economic resilience and prospects of maintaining social stability will be questioned. This scenario is a direct threat to the Russian authorities.

Another threat is a potential reduction in hydrocarbons export volumes. Favourable transformations in the world energy sector, and particularly in hydrocarbons markets – such as the growth of production from shale – represent a major risk for Russia, and the Russian energy sector has not previously been faced with such difficult conditions. The gravity of the risk is manifested in the stagnation in oil and gas production and exports: thus gas exports in 2013 are at the level of 2000. Recent studies[9] show that owing to the lack of development of an institutional framework, an outdated tax system, low competition, and low investment efficiency Russia will be the most sensitive of all the largest energy producers to fluctuations in global hydrocarbon markets.

The serious risks facing Russia, which have arisen from the transformation of global energy markets, include declining oil and gas exports and export revenues, relative to planned official indicators; this may lead to slower-than-expected GDP growth and to deterioration of the main parameters of the Russian energy sector. Falling revenues from exports of gas and, especially, oil could considerably reduce the contribution they make to GDP.[10] The powerful multiplicative effects typical of these sectors, together with lower inflows of foreign capital, could magnify the impact of decreasing export revenues and slow down economic development. A preliminary assessment of the effect of these factors on economic growth indicates a slowdown of one percentage point each year, due to decreased energy exports.[11]

As far as export volumes are concerned, at least with regard to oil exports, in addition to external factors there are also *very serious internal limitations*. As early as in 2006, OECD researchers warned that hydrocarbons production stagnation was highly probable, citing as the main reasons governmental regulation of the sector (which has become more pronounced in subsequent years); an unfriendly tax regime; restrictions on foreign investment; and a generally unfavourable business environment.[12]

The Russian oil and gas sector is now approaching the exhaustion of capacities created in Soviet times. The greater part of Russian production is

---

[9] Makarov, A., Grigoriev, L., and Mitrova, T. (eds.), 'Global and Russian Energy Outlook up to 2040', ERI RAS – ACRF, 2013, p.110.

[10] Gurvich, E. and Prilipskii, I. 'Kak obespechit vneshniuiu ustoichivost' rossiiskoi ekonomiki', *Voprosi Ekonomiki*, No. 9, 2013, pp. 4–39.

[11] Makarov A., Mitrova T., and Malakhov V., 'Prognoz mirovoi energetiki i sledstviia dlia Rossii', op. cit., pp. 34–51.

[12] Ahrend R. and Tompson, W., 'Realising the Oil Supply Potential of the CIS: The Impact of Institutions and Policies', OECD Working Paper. No ECO/WKP (2006)12.

based on discoveries made in the Soviet era: 90 per cent of oil output, and an even higher proportion of gas output, is from fields discovered before 1998. In order to sustain production, Russia needs to develop new provinces, such as Eastern Siberia and the Arctic offshore; to develop its vast unconventional reserves; and to apply more efficient production techniques to existing fields. But such development is not competitive under the current tax regime.

For example, Russia has huge potential for Enhanced Oil Recovery (EOR). According to Lukoil,[13] the use of best practices would provide an additional 4 billion tonnes of oil reserves without the need to build new infrastructure, but this would require the use of technologies that are not economically justified under the current tax regime. The Mineral Extraction Tax (MET) and export tax require much lower break-even costs than those currently applying to all significant sources of new supply ($30–40/bbl, compared to the costs of new sources of around $70–220/bbl). This destroys incentives to increase oil production. At the same time, the government is not prepared to replace the existing system with profit-based taxation, claiming that it is much more difficult to administer and that a transitional period from the existing system to a new one could be dangerous for the country. Therefore all exemptions and adjustments are made on a case-by-case basis, giving some temporary relief, but not solving the problem. If the government prefers short-term revenues to longer-term sustainability, and does not make the appropriate changes to the taxation regime, there is a danger of a rapid oil production decline by the end of this decade, with a corresponding slowdown of GDP and decline of its budget incomes.

Of no less significance are *domestic macroeconomic challenges*. A period of rapid economic recovery, from 2000 to 2008, ended with the onset of the global economic crisis in 2008. *Since then, the Russian authorities have been struggling with an increasingly evident economic slowdown* (Figure 1.3). In 2013 Russian economic performance turned out to be much weaker than expected, despite high hydrocarbon prices. The annual growth rate slowed to 1.4 per cent in the first half of 2013, compared to 4.5 per cent in the first half of 2012, due to a slowdown in consumption, stalled investment demand, and a continuing weak external environment.[14] The economics ministry downgraded its 2013 GDP growth estimate from 3.6 per cent to 2.4 per cent in its baseline, or 'moderately optimistic', scenario. This scenario assumes 'active government policy aimed at improving the investment climate, competitiveness and economic efficiency', which seems now, as of

---

[13] 'Global Trends in Oil and Gas Markets to 2025', Moscow: Lukoil, 2013, p. 49, www.lukoil.com/materials/doc/documents/Global_trends_to_2025.pdf.

[14] Russia Economic Report. 'Structural Challenges To Growth Become Binding', The World Bank, No. 30, September 2013.

12   *The Russian Gas Matrix*

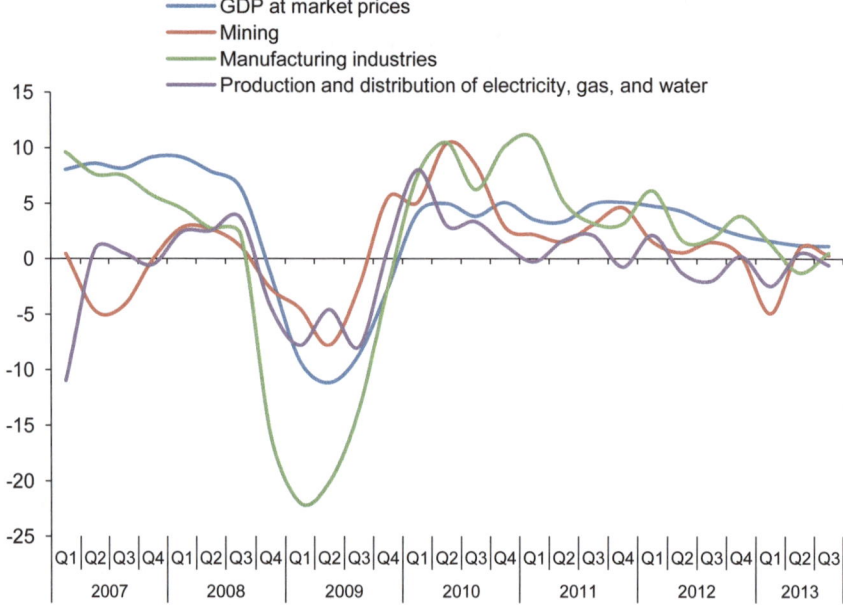

Figure 1.3: Russia: growth in GDP and industrial value added (as a proportion of the corresponding quarter of the previous year)
Source: Rosstat.

late 2013, rather questionable. The ministry's estimate of growth in its 'conservative' scenario – in fact, a 'muddling through' scenario – is just 1.7 per cent.[15] The Russian government is strongly concerned by such visible decline.

Russian industry is obviously entering recession. Industrial production is in decline: this is already visible in the statistics. Starting from 2013, there have been signs of stagnation in industrial output – to zero in the first quarter of 2013 (Figure 1.4). The growth of capital investment in industrial assets fell from 14–16 per cent in 2012 to below zero in summer 2013. Producers started to close plants – which could mean increasing unemployment and social tensions – saying that the current situation is even worse than that in 2008. Furthermore, the lack of significant institutional reforms aimed at solving key problems – ranging from reducing the numerous pressures on businesses from the authorities and barriers to business, to reducing the scope of involvement of governmental and quasi-governmental companies, and developing competition – gives no grounds to hope for a fast self-sustained recovery.

---

[15] Minekonomrazvitiia snizilo prognoz po rostu VVP na 2013 god do 1,8%. 26 August 2013, www.rbc.ru/rbcfreenews/20130826160715.shtml; Kuvshinnikova O. Rossiia gotovitsia k desiati toshchim godam, 7 November 2013. www.vedomosti.ru/finance/news/18435801/rossiya-gotovitsya-k-desyati-toschim-godam?full#cut.

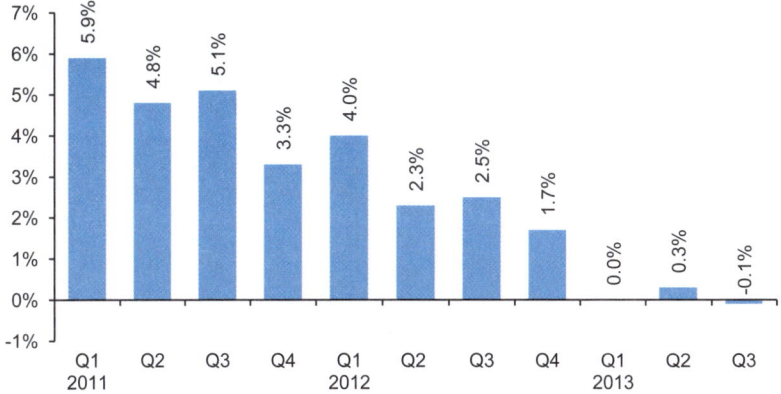

Figure 1.4: Russia: industrial production growth (compared to the same quarter of the previous year)
Source: Rosstat.

Weakness in domestic demand is reflected in subdued investments. The value of fixed-capital investments rose by just 0.1 per cent in the first quarter of 2013, compared to 15.5 per cent growth in the same period of 2012. Russian business confidence also looks weak as research by the World Bank, among others, has shown.[16]

Moreover, Russia continues to be far from the most attractive market for foreign investment (Figure 1.5). Survey data suggests that Russia is still perceived as being prone to serious problems relating to corruption and bureaucratic interference, which contribute to the costs and risks of doing business (see Table 1.2). Doubts persist about the respect for contracts and private sector property rights – factors which discourage investment.

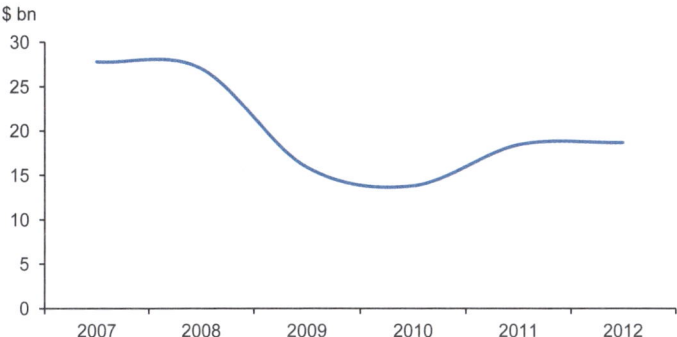

Figure 1.5: Direct foreign investment in Russia, $ billion/year
Source: Rosstat.

---
[16] Russia Economic Report.

Table 1.2: Russia in international rankings

Date, index	Organization	Russia's ranking	Total countries rated	Details
2013 Index of Economic Freedom	Heritage Foundation	139	161	Russia is in the Mostly Unfree group. The Repressed group starts from the 145th ranking.
2013 Ranking of Countries for Mining Investment: 'Where Not to Invest'	Behre Dolbear	25	25	
2012 Corruption Perception Index	Transparency International	133	174	Russia shares its rating position with Iran, among others.
2013 Doing Business Report (overall rank)	World Bank	112	185	In 2012 President Putin instructed the government to take measures to raise Russia's ranking to the 50th position in 2015 and to the 20th position in 2018. However the 'Doing Business' rating may be biased and dependent on insignificant factors. In this connection a group of independent experts set up by the World Bank, including Sergei Guriev, former head of the New Economics School, recommended discontinuing it.
2013 Doing Business Report: Protecting Investors	World Bank	117	185	Russia took the 60th position in 2007. Since then it has been declining.
Report, 28 June 2013	Standard & Poors	BBB (Lower medium grade)	-	The BBB rating has been confirmed since December 2008. Russia seeks to boost its credit rating by no less than two steps to A– by 2016 and another level to A by 2020, according to a government plan approved in March.
Report, 11 June 2013	Fitch	BBB (Lower medium grade)	-	The BBB rating has been confirmed since February 2009.
Report, 27 March 2013	Moody's	Baa1 (equivalent to BBB+)		The BBB rating has been confirmed since July 2008.

Sources: Heritage Foundation, Behre Dolbear, Transparency International Index, World Bank, Standard & Poors, Fitch, Moody's.

The role of gas in the Russian economy

The role of gas in Russia's economy differs significantly from that of oil. Oil accounts for the largest share of budget revenues – 36 per cent in 2012 – because of high export duties and a high tax on production. Gas's share is far less important – only 5 per cent – as both export duties and production tax are considerably lower than those on oil (see Figure 1.6).

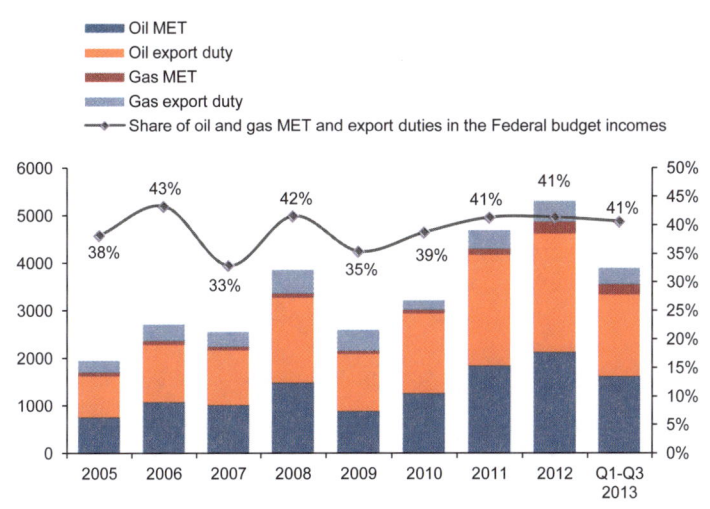

Figure 1.6: The share of oil and gas MET and export duties in Federal budget income

Source: www.roskazna.ru/federalnogo-byudzheta-rf/yi/.

On the other hand, gas is distinguished from oil by its growth potential. Oil exports are predicted to fall (this is seen even in official documents, for example, the *Economic Strategy of the Russian Federation for the Period up to 2030* and the *Master Plan for the Oil Industry for the Period up to 2030*) due to the stabilization (at best) or even decline of production, against a background of growing domestic demand for refined products. But the gas industry faces no such limitations. If the international market situation is favourable and gas exports are competitive, Russia is potentially capable of increasing gas production by almost 70 per cent (see the *Master Plan for Development of the Gas Industry for the Period to 2030*). This creates substantial expectations for the gas industry, which it may not be able to live up to.

There are, however, more fundamental reasons for the fact that gas – despite its notably smaller contribution to the national budget – is so important for Russia's economy. These are: its predominant place in Russia's own energy sector, its role as a domestic political tool, and its use as an instrument of foreign policy.

The role of gas in the energy balance

To start with, *gas is the basis of Russia's energy sector*: it accounts for more than 54 per cent of primary energy consumption. This is one of the highest figures in the world, even among the main gas producing countries (see Figure 1.7).

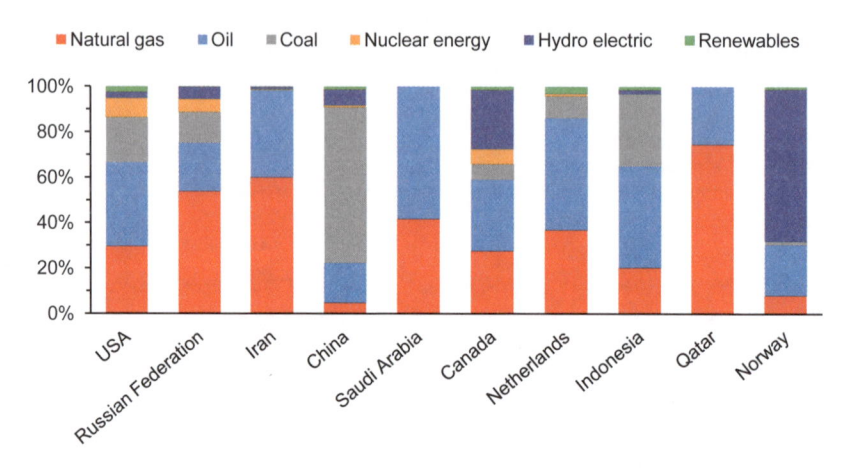

Figure 1.7: Structure of the primary energy mix of leading gas-producing countries
Source: BP Statistical Review of World Energy 2013, London: BP.

Gas is the main fuel for electric power generation in Russia. Gas-fired plants account for 44 per cent of installed capacity in electricity generation (Figure 1.8) and provide for 70 per cent of the output from thermal plants.

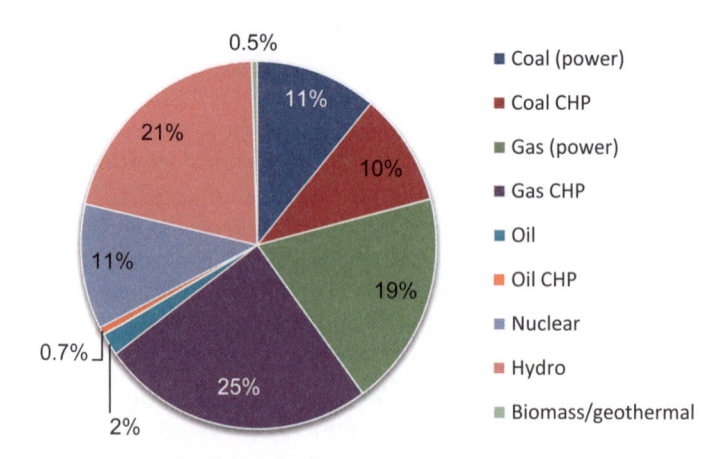

Figure 1.8: Installed electricity and CHP capacity in Russia
Source: IEA World Energy Outlook 2011.

During the last two decades the share of gas in Russia's primary energy consumption has increased continuously, from 43 per cent in 1991 to 54 per cent in 2009. Russia's domestic gas consumption, measured by volume, rebounded almost to the levels of the early 1990s, while domestic oil and coal demand remain at just over half of those levels. This gas demand growth was encouraged by domestic pricing policies that kept gas prices low, in large part to manage the social and industrial impact of the post-Soviet recession, while those for coal and oil were liberalized. As a result, gas squeezed coal out of thermal power plants. Gas has the advantage of being the fuel which is in place across large parts of European Russia, where most Russian industrial and residential demand is concentrated. Where price differences are marginal, gas remains the preferred fuel for new equipment in industry and power generation because of its flexibility and environmental performance.

In order to limit the rise in gas consumption, promote inter-fuel competition (first of all with coal), and support the necessary supply-side investments in new gas production, from 2000 the government started to develop gas price growth policies. But even substantial gas price increases in 2007–13 could not limit the increasing role of gas in the energy balance. The main coal producing assets are located in Eastern Siberia, and transporting coal pushes up the delivered cost substantially. In fact coal prices were almost increasing in line with gas prices (Figure 1.9). With incremental transportation costs by rail of at least $30/tonne, the price of steam coal to industry or power plants in the heart of European Russia rises to $80–90/tonne.

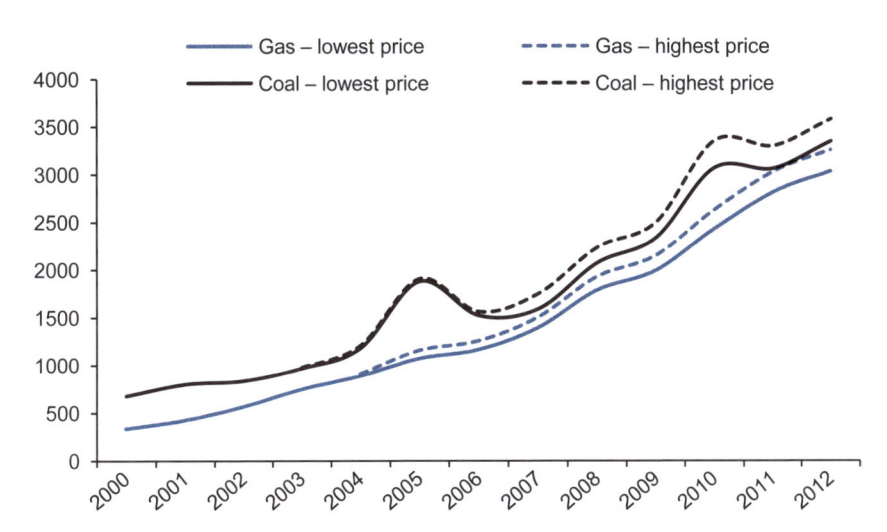

Figure 1.9: Gas and coal prices in Moscow region, rubles/tonne of coal equivalent
Source: Rosstat, ERI RAS.

At these price levels, the benefits of choosing coal over natural gas are not evident, particularly in the power sector, where coal-fired power generation achieves a lower conversion efficiency than natural gas. According to estimates by the International Energy Agency (IEA): at a price of $85/tonne for steam coal in European Russia, gas prices would have to rise to $7.5/MmBtu before coal-fired power would be competitive.[17] (See also Chapter 6 on the sources of gas demand in Russia.)

As a result of this price imbalance, gas plays an excessively large role in energy supplies in central and southern Russia. Gas's share of boiler and furnace fuel consumption exceeds 95 per cent in some regions (Belgorod, Briansk and Orel, Mordovia, Penza, and almost all of the North Caucasus Federal District). In order to reach domestic consumers, gas is supplied through three major corridors and is transported 2000 km (on average) to consumers. This is the main reason why the government is so concerned about high dependence on this fuel and why, for more than a decade, it sought to reduce gas's share in the primary energy mix.

Reducing the role of gas in the domestic economy is a major strategic issue, highlighted in the government's *Energy Strategy to 2030*, which states that promoting a 'rational energy balance' means achieving a:

> ... reduction in the share of gas in the structure of domestic energy consumption and [an] increase in the share of non-fuel energy in the structure of the fuel and energy balance.[18]

The target set in 2009, and published in the *Energy Strategy*, is to reduce the share of gas in the fuel mix from 54 per cent to between 46 per cent and 47 per cent in 2030. It is supposed to be achieved in three stages:

First stage (2011–15):
- Development and introduction of the economic mechanism of effective inter-fuel competition for replaceable energy carriers (gas/coal); bringing the ratio of gas to coal prices in the domestic market to 1.8–2.2.
- Reduction in the share of gas in the fuel and energy balance to 51–52 per cent.
- State support for thermal coal-fired power generation development (tax stimulus, preferential lending, and other regulatory support); state support and direct funding for nuclear power generation development.

---

[17] World Energy Outlook 2011, IEA, Paris, 2011.
[18] 'Energy Strategy of Russia for the Period Up to 2030', Ministry of Energy of the Russian Federation, Moscow, 2010, p. 158.

- Renewable energy development: creation of an institutional base, tax stimulus, and introduction of guaranteed access to electric grids for the power plants operating on renewable energy.
- An increase in non-fuel energy's share of the energy balance to 11–12 per cent.

Second stage (2015–20):
- Effective inter-fuel competition based on advanced monitoring of gas and coal prices, and reduction of gas's share of the energy balance to 48–49 per cent.
- State support for nuclear and coal-fired power generation development.

Third stage (by 2030):
- Development of renewable and non-hydrocarbon energy based on the introduction of advanced technologies and use of public–private partnership mechanisms.
- Bringing the ratio of gas to coal prices in the domestic market to 2.5:2.8.
- Reduction in the share of gas in the energy balance to 46–47 per cent, and an increase in the share of non-fuel energy to 13–14 per cent.[19]

This plan demonstrates quite clearly that gas is the key factor in Russian energy sector development.

Gas as a domestic political tool

In addition to its vital role in energy supply and in ensuring energy security, gas has an important social function – not just in economically and socially disadvantaged regions, but in the country as a whole. In fact, *Russian gas is an important domestic political tool*, used in a number of ways:

- *Keeping energy costs as a share of household budgets low, by cross subsidization at the expense of industrial consumers.* Since the Soviet period, guaranteed power and water supplies have been seen as an inalienable and basic right for all citizens. More than 20 years of development of the market economy have not been able to break down this perception among the population and it is a major potential cause of social unrest. In order to avoid such unrest, both the federal and local Russian authorities prefer to subsidize domestic consumers at the expense of large-scale industry – whose gas price is several times higher than that facing the country's household consumers. While increasing gas prices for industrial

---

[19] *World Energy Outlook 2011*, p. 265.

consumers have, in the last decade, raised gas's share of their costs above that of their foreign competitors, the situation with the Russian population is quite different: the share of household spending on gas, as a proportion of total spending, is still lower than that of their foreign counterparts, and in recent years has stabilized (Figure 1.10).

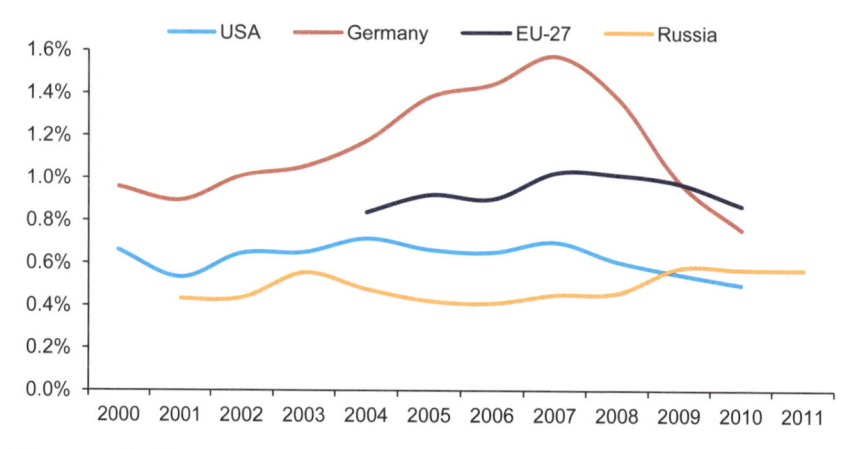

Figure 1.10: The share of gas purchases in household spending
Sources: IEA, Rosstat, Eurostat, EIA US DOE, US BEA.

- *Acting as a social stabilizer and providing improved living standards in remote areas, by means of regional gasification.* From the standpoint of economic efficiency, these projects look very doubtful: laying pipes to remote villages is often considerably more expensive than the delivery of liquefied petroleum gas (LPG) or the development of forms of decentralized energy supply. However, the key argument is that 'the central government takes care of rural residents', and low-cost, easy-to-use gas is the main element of this 'care'. Interestingly, while that entire workload has previously been carried exclusively by Gazprom, with the involvement of other companies (Novatek, Rosneft) in regional gas trading, the responsibility for gasification has increasingly shifted towards them (although there has been no legal consummation of this deal between the companies and the government).

- *Subsidization of individual depressed regions.* Gas tariffs are set in such a way that consumers in regions near production sites pay a significantly higher specific tariff for transportation than those in remote regions – for example, the unstable republics of North Caucasus or the strategic Kaliningrad exclave. Considering the high level of non-payments in the Caucasian republics, it is hard to view supply to these regions as anything other than a form of 'loyalty payment' from the centre.

- *Gas's new role in the Russian fiscal system as, apparently, the only realistic source of additional incomes for the state.* Oil taxes cannot be increased without undermining production volumes (which would not be in the state's interest), while other branches of industry are in decline and can hardly provide additional support for the budget. However, election promises have to be fulfilled and budget spending tends to grow, so gas is becoming the 'cash source of last resort'. There has already been a long and painful discussion on MET growth, in which the gas industry is regarded as a 'milch cow' – even though the government realizes that the industry would have difficulty in being able to provide the hoped-for income from new, extremely expensive projects; these, therefore, apply for tax breaks and 'special exemptions'.

- *Gazprom's use as a government 'treasury'.* It is quite common for Gazprom (and perhaps Novatek and Rosneft in the future) to finance 'projects of state importance'. The most recent examples are the construction of Sochi Olympics facilities funded and supervised by Gazprom, and the extremely expensive Sakhalin–Khabarovsk–Vladivostok gas pipeline built by Gazprom in preparation for the Asia–Pacific Economic Cooperation Summit in Vladivostok in September 2012 – to say nothing of alleged financial support for election campaigns.

- *The gas industry's generation of a strong multiplicative effect for producing regions.* This effect makes some projects, which are questionable from an economic point of view, very attractive for local governors. Shtokman is a well known example: local government continues to advocate project development – due to the prospect of gas supply to Murmansk region, which could underpin gas-fired power generation and heat supply development – even though the markets are extremely unfavourable for it. Another example is the vigorous promotion of the Yamal LNG project, due to the synergy expected for shipbuilding yards and Northern Sea Route development. Similar interest in developing gas projects was shown by the governments of Yakutia, Sakhalin, and other regions.

- *The gas industry's influence on other industrial sectors* is very strong. The gas sector is traditionally a prime source of demand for steel works and pipe manufacturers. Recently, cooperation with car manufacturers has been developing actively, aimed at the promotion of natural gas vehicles.

- *Relations between the gas industry and the environmental agenda, which are nearly absent in Russia.* There is no doubt that greater gas use would be the most realistic and affordable option for Russia to reduce carbon dioxide emissions, if the country joined the post-Kyoto process. However, Russia does not participate in the second commitment period of the Kyoto Protocol, and the development of a post-Kyoto agreement remains at an initial stage. Thus, the prospects of 'green' energy development in Russia remain unclear.

- And last but not least – *the use of gas to subsidize vested interests of loyal groups close to the country's leadership*. While historically this was a single group, associated with the Gazprom management and subcontracting companies, in recent years it has been joined by the group of Gennadii Timchenko (Novatek) and the Rotenberg brothers, whose companies receive contracts to build the largest pipelines.[20]

This extensive list of the gas industry's domestic functions provides a good explanation of the particular attention that the Russian political leadership pays to it. But in addition, gas is the most important (already almost the only) and effective instrument of foreign policy.

Gas as an instrument of Russian foreign policy

Gas is one of the main tools of Russia's integration into global trade, and especially of Russia's economic relationship with the EU. Gas provides 12 per cent of all Russian export revenues (see Figure 1.1). The development of various joint ventures and direct investments, and of major foreign projects, helps to implant Russian business more deeply into global economic relations, gradually making Russia a fully-fledged participant in the global economic system. In fact, gas is one sector of the global economy in which Russia possesses vast expertise and competitive advantages.

These economic considerations and the available assets shape geopolitical consequences in many ways. Gas supplies, or implementation of gas projects, serve as a tool to preserve Russia's geopolitical impact in certain regions (for example, the CIS, Eastern Europe, and the Balkans). In the post-Soviet area gas is perhaps the main instrument of integration, allowing Russia to exercise its influence over CIS countries (albeit by means of cost-ineffective measures such as expensive gas imports from Central Asia and Azerbaijan or gas supplies to Belarus at reduced prices). Another

---

[20] 'Koroli goszakaza: kto osvoit $20 mlrd', 19 March 2013, *Forbes*, http://m.forbes.ru/article.php?id=235724; Zagorodskii, A. 'Neudivitel'no, chto pobezhdaiut imenno te, kto naibolee blizok k sisteme raspredeleniia', *Kommersant*, 2 March 2012, www.kommersant.ru/doc/1884562; Serov, M., Mordiushenko, O., Solodovnikova, A. 'Gennadii Timchenko saditsia na trubu', *Kommersant*, 21 August 2013, www.kommersant.ru/doc/2260037; Gennadiy Timchenko, 'U menya net planov v Rossiu', *Kommersant*, 16 April 2013. www.kommersant.ru/doc/2172126; Melnikov K. and Mordiushenko O. 'U nas tolko odin biznes-plan – byt' gotovym v luboi situatsii', *Kommersant*, 18 September 2012, www.kommersant.ru/doc/2024432.

striking example is that of the negotiations on the construction of the South Stream pipeline. For each country whose territory it will traverse there will be new investments, employment, and additional tax revenues, not to mention the transit payments. During an economic downturn, these arrangements look very attractive, and allow Russia gradually to incorporate the Balkans into its sphere of economic (and geopolitical) influence.

It is possible to consider Russia's desire to diversify its supply routes from this perspective. The goal of reducing dependence on the European market, and increasing gas sales to Asian customers, announced in the early 2000s, has become increasingly prominent in recent years (though not currently embodied in specific agreements). This is partly in line with economically justified aims of monetizing gas reserves in Eastern Siberia and the Far East, expanding economic ties and enhancing integration with the Asia–Pacific region, and partly with a view to expanding relations, and maybe even building strategic alliances, with Asian countries (mainly China) in opposition to the USA and Europe.

These uses of gas by the Russian government – particularly, for example, in the Ukrainian transit crises or in inflated prices for the Baltic States – often lead to it being described as a geopolitical 'weapon'. Rather than 'weapon' (a means of destruction or punishment) a more appropriate term would be that gas is an instrument of pressure, of attempts to keep other countries in the Russian sphere of influence. Gas was, and still is, the most important element in negotiations not only with Ukraine and Belarus, but with practically all countries bordering Russia. Moreover, Russia usually links gas agreements with other aspects of cooperation (membership in regional integration organizations promoted by Russia, the Customs Union, the Black Sea Fleet, arms supplies, etc.)

Sometimes Russia proposes purely geopolitical projects that, while lacking any economic sense, showcase its ambitions and capabilities, primarily to the USA (for example, the supply of gas via a pipeline across North Korea to South Korea, or participation in construction of a Turkmenistan–Afghanistan–Pakistan–India pipeline). In practice, however, these initiatives are limited to declarations, without reaching the stage of practical preparation and investment. However, it is important for the Russian government to possess such an instrument for international negotiations and for strengthening its authority.

Institutions and corporate entities in the gas sector

State regulation

Given the importance of the Russian energy sector to the economy, most key decisions about energy policy and regulation are taken at the highest levels of government. This is particularly true for the oil and gas sector: control over oil and gas revenues is of colossal importance for the authorities. Moreover, analyses of the decision-making process in the gas industry in particular show that key decisions are usually made directly by President Putin, who regards this sphere as strategically critical and follows its operation in detail. Below this level, multiple ministries and other executive offices work on the development of energy sector policy proposals and aspects of policy making. The energy sector's regulatory functions are distributed among the following authorities:

- The *Energy Ministry*, which has primary responsibility for the energy sector. It develops investment programmes, authorizes energy projects, and takes the lead on day-to-day regulation and supervision of the energy sector, while also overseeing energy policy and having responsibility for developing strategy for the whole energy sector and for particular energy industries, including gas. From September 2012 the Ministry approves all export contracts signed by Gazprom (including any contract renegotiations).

- The *Ministry of Economic Development*, which has influence over the regulation of tariffs and energy sector reforms and deals with general energy regulation issues in the framework of economic planning and development. It also coordinates energy and energy efficiency policies with the overall economic development priorities.

- The *Federal Tariff Service (FTS)*, is a federal executive body responsible for tariff regulation that reports directly to the government. The FTS regulates tariffs of natural monopolies, in particular in the electricity, oil, and gas sectors. It sets transportation, transmission, and other regulated tariffs, including wholesale tariffs for natural gas destined for industrial and power sector use, and tariffs to residential and municipal customers. The FTS was established in its current form in 2004, replacing the former Federal Energy Commission.

- *The Federal Anti-Monopoly Service (FAS)* is a federal-level executive governmental body, responsible for competition policy, which controls the execution of competition laws and related laws.[21]

- The *Ministry of Natural Resources*, which issues field licences, regulates upstream activities, monitors compliance with licence agreements, and levies fines for violations of environmental regulations.

---

[21] The FAS was established by the Decree of the President of Russia No. 314 on 9 March 2004; 'Russia Country Profile', EBRD. 2008, www.ebrd.com/downloads/legal/irc/countries/russia.pdf.

- The *Finance Ministry*, which is responsible for tax policy for the energy sector, and oversees fiscal policy, a critical component of the investment climate.

- There are two additional competing bodies which deal with the implementation of state policy in the energy sector: the *Presidential Commission on the strategy of the fuel and energy complex development and environmental security*, founded in June 2012, is chaired by the President, with Igor Sechin as an executive secretary; and the *Governmental Commission of the Russian Federation on the fuel and energy complex*, established under the auspices of the Prime Minister in February 2013, which deals with operational issues and problems in the energy sector.

Energy policy framework

The *Energy Strategy to 2030* is the main document that provides a detailed overarching framework of long-term policy priorities for the energy sector. In addition to the promotion of a 'rational energy balance', discussed above, the main targets of the gas sector with regard to exports are:

- to maintain its position in Europe while diversifying energy supplies and reducing dependence on European customers and to diversify export markets, primarily to the Asian market. The target is to increase the Asian markets' share to 26–27 per cent of total energy exports (and up to 20 per cent of gas exports) by 2030; and
- diversification of the product structure of exports (with the share of LNG in gas exports reaching 15 per cent by 2030).

The *Energy Strategy* is supplemented and, in some cases, modified by so-called *General Schemes* (*Master Plans*) for the oil, gas, and coal sectors, and a similar document for the power sector, adopted in 2008 and then amended in 2010. There are also several specific *Conceptions* and *Programmes*, such as the Eastern Gas Programme (see Figure 1.11).

Other strategic documents correspond to the *Energy Strategy*; their priorities are efficiency, security, and reliability. For example, Russia's pledge to the Copenhagen Accord is a 15–25 per cent reduction in emissions by 2020 compared to a 1990 baseline. Another aim, to reduce Russia's energy intensity by 40 per cent by 2020 compared to 2007, is much more ambitious. This target was announced by (the then President) Dmitrii Medvedev in 2008 and its achievement would have substantial implications for energy use. Another target adopted by the Russian authorities for 2020 is to increase the share of renewable energy resources in the electricity mix to 4.5 per cent. Many of the environmental policies come from the *Climate Doctrine Action Plan* adopted by the

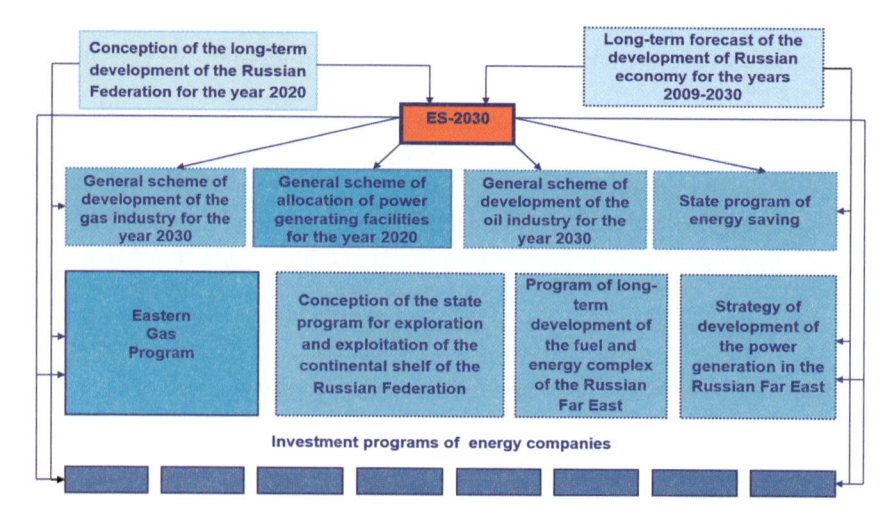

Figure 1.11: Framework of strategic documents in the Russian energy sector
Source: Institute of the Energy Strategy.

Government of Russia in April 2011. This plan sets out a range of measures for different sectors of the Russian economy and includes economic instruments for limiting greenhouse gas emissions in industry and power generation.

Access to subsoil

The legal framework with respect to the use of subsoil resources in Russia is established by the Federal Law 'On Subsoil Resources' of 21 February 1992. According to the Subsoil Law, geological surveys and exploration and extraction of minerals (including oil and gas) can be performed under a licence for subsoil use. This licence certifies the subsoil user's right to perform certain activities on a certain part of subsoil within a limited period of time, subject to compliance with the licensing conditions (usually established by a 'licence agreement', an integral part of the licence). As a general rule, a licence for subsoil use is granted based on the results of an auction or tender. The Subsoil Law provides for significant limitations in relation to the granting of licences for subsoil use with respect to areas of subsoil considered to be of 'federal significance', including areas of subsoil:

- containing extractable oil reserves of 70 million tonnes or more;
- containing natural gas reserves of 50 Bcm or more;
- located in internal waters, territorial seas, or on the continental shelf of the Russian Federation.[22]

---

[22] Tax and Legal Guide to the Russian Oil & Gas Sector. Deloitte, 2012.

For areas of subsoil considered to be of 'federal significance', a licence may be granted only to a Russian legal entity. Upon holding an auction/tender for the right to use such an area of subsoil, the government may also place restrictions on the participation of Russian legal entities which are owned by foreign investors in whole or in part. For areas of subsoil located entirely or partly on the continental shelf, a licence may be granted only to a Russian legal entity having experience of no less than five years in working on the continental shelf, and in which the Russian Federation directly or indirectly holds more than 50 per cent of shares. In practice, this means that these licences are granted only to state-owned oil and gas companies (such as Gazprom and Rosneft) or, in some cases, to joint ventures with these companies (provided that the Russian Federation retains more than 50 per cent of shares in the venture).[23]

Not only upstream investments, but all foreign capital investments in the gas sector (in exploration, production, transmission, wholesale supply and export) are included in the list of business activities 'which have strategic value for the defence of the state and national security support'.[24] This regulation de facto requires the president's personal permission for any deal involving foreign partners. It effectively restricts international cooperation in this sphere to mega-projects with state-controlled companies.

Russian legislation also provides for Production Sharing Agreements (PSAs) between the Russian Federation and investors (including foreign legal entities). Under a PSA, the state grants an investor exclusive rights to explore and extract subsoil resources in a specified subsoil area; production is shared between the investor and the state. The investor performs work related to exploration and extraction of subsoil resources at its own expense and risk, either itself or through a third-party contractor (operator of the PSA). The PSA regime does not release the investor and operator from general Russian licensing and regulatory requirements. The right of the investor to work on a certain area of subsoil under a PSA should be confirmed by the licence for subsoil use.

PSAs were of the utmost interest in Russia in the 1990s, but since 2000 no new PSAs have been signed in the oil and gas sector. There are only three active PSAs (Sakhalin-1, Sakhalin-2 and the Khar'iaga project), which were concluded before the current Federal Law 'On PSAs' entered into force in 1996, and as such are 'grandfathered'. Government officials have made several public announcements, stating that the Russian Federation would not enter into new PSAs.[25]

---

[23] Ibid.

[24] Federal Law of 29 April 2008 N 57-FZ on 'On the Procedure of Foreign Investments into Economic Organizations of Strategic Importance for the Defense of the State and National Security Support'; 'Russia Country Profile'.

[25] Tax and Legal Guide to the Russian Oil & Gas Sector.

Pricing policy

During the 1990s and up to 2006, Russia had very low regulated domestic gas prices which were far below prices on international markets. These low prices stimulated very rapid gas demand growth but did not provide for an adequate cash flow for producers, resulting in underinvestment. At the same time, the depletion of the Soviet inheritance of low-cost gas was reaching the point at which it was impossible to postpone major new investments. Higher gas prices became necessary to support the higher costs of gas production and transportation involved in developing the next generation of gas fields.[26]

In 2006, a new policy was adopted on gas price increases,[27] aimed at bringing domestic prices to 'parity' with export prices (less transportation and excise duty) by 2011. Annual price increases of 15–25 per cent were implemented, and as a result average final industrial wholesale gas prices increased by more than 4.5 times in the last decade. But the initial target date to reach netback parity, 2011, was postponed to 2014–15, and most recently to 2018 or even later. Indeed, it is questionable whether the target of netback parity, conceived in an era of lower oil prices, will remain the formal objective of gas pricing policy. (This issue is discussed in detail in Chapter 5 ('Domestic gas prices and market reform', page 117).

Gas taxation in the Russian Federation

For many years the state's tax take from gas was much lower than that from oil, with export tax being no higher than 30 per cent and the MET being about one-tenth of that for oil on an energy equivalent basis. This was one side of the 1990s deal between state and industry; the other side was the extremely low regulated domestic gas prices. However, along with the recent gas price increases the situation is changing: facing a budget deficit and potential decline in oil revenues, in 2011 the government announced significant increases in MET for gas starting in 2012, indicating that the government is aiming to take most of the benefit of price growth through taxes.

Companies conducting their business in the Russian gas sector must pay any and all standard taxes imposed on Russian companies, together with a number of specific sectoral taxes (including MET, subsurface use tax, and export duties) as listed here:[28]

Standard taxes. The main ones are: value added tax (VAT) and corporate income tax. In Russia VAT is rated at 18 per cent; this rate is charged on any and all goods, works, and services. Exclusions relevant to the gas sector

---

[26] Pirani, S. (ed.), *Russian and CIS Gas Markets and their Impact on Europe* (Oxford: OIES/OUP, 2009).

[27] Governmental Decree No.333, May 2007.

[28] 'Obzor nalogovogo rezhima v neftegazovoi otrasli Rossii', Ernst&Young, 2009.

include: services in connection with gas transportation outside Russia, gas transit services, and services in connection with the transportation of gas imported to Russia for processing.[29] Under a law passed in 2013, all operations in connection with the disposal of goods abroad in the course of raw hydrocarbons extraction at offshore fields are also exempt from VAT.[30] The corporate income tax standard rate in Russia is 20 per cent.

MET. The MET rate has been the subject of fierce discussion in government for some time, and various fundamentally different proposals have been made. As of November 2013, it has been decided to levy MET at gradually rising levels, with the difference between the rates for Gazprom and for independent producers narrowing (see Table 1.3).

Table 1.3: MET rate for gas, rubles/mcm (up to 1 July 2015)

	1 July 2013 to 31 December 2013	*1 January 2014 to 31 December 2014*	*From 1 January 2015*
Gazprom	622	700	788
Independent producers	402	471	552

Source: Tax Code of the Russian Federation, Part 2, Section 8, Clause 26.

Furthermore certain categories of recovered gas are MET zero rated. They are:

- associated gas;
- gas pumped into the reservoir in the course of condensate recovery;
- gas fields in subsurface areas of the Yamal peninsula used solely for producing LNG, up to 250 Bcm recovered, provided the term of reserves development do not last for more than 12 years;

On 30 September 2013 a new act was passed (Federal Act 263),[31] under which, from 1 July 2014, a new MET formula shall apply:

$$S=(Si*Erf*Kc)+Tr$$

where

Si = the initial rate fixed at the amount of 35 rubles/mcm;

Erf = the basic value for a reference fuel unit;

---

[29] Tax Code of the Russian Federation, Clause 164.
[30] Federal Act 268 dd. 30 September 2013.
[31] Federal Act 263 dd. 30 September 2013; 'Gosduma smeshala neft's gazom', *Kommersant*, 23 September 2013.

Kc = *a coefficient characterizing the degree of complexity of natural gas and (or) gas condensate from the hydrocarbon deposits;*

Tr = *gas transportation costs.*

The aforementioned coefficients in their turn depend on more than a dozen parameters – the oil price; oil export duty rate; currency exchange rates; gas price for the domestic (with reference to netback parity) and export markets; transportation costs; the amount of gas supplied on the domestic market and abroad; geographic position; level of gas reservoir depth and depletion; and other parameters. So far it is not clear how companies will calculate all this data or how the tax authorities will double check it. Moreover, it should be noted that hardly a month had passed after the adoption of the new MET schedule before the government started discussing possible amendments to the formula. (This issue is further discussed in Chapter 5 ('Domestic gas prices and market reform', page 117.)

Subsurface Use Tax. This includes:[32]

- single payments for subsurface use in certain circumstances specified in the licence;
- regular payments for subsurface use;
- dues for taking part in a tender (public sale).

The payment amounts are fixed in accordance with the provisions of the licence. The regular payments for subsurface use are fixed and depend on economic and geographical environment, subsurface area dimensions, type of mineral deposits, duration of work, state of the territory, geological exploration, and risk level.

Export Duty of 30 per cent, on natural gas exports outside the Customs Union (Russia, Belarus, Kazakhstan), paid on the realized export price.[33] Exemptions are currently in place for: gas exported via the Blue Stream pipeline across the Black Sea to Turkey;[34] some gas exports to neighbouring countries; and, thus far, for all LNG export projects. (LNG exports are not charged any export duty, and according to the rules of the WTO, which Russia joined recently, this can not be changed, as the tariffs should be frozen.)

Taxation of PSAs. Under PSAs special tax treatment is applied. In the course of the agreement being implemented, the investor makes single payments for subsurface use in relation to circumstances arising which are specified therein and in the licence (bonuses); annual payments for

---

[32] Federal Law 'On Subsurface Resources', Clause 39.
[33] Decree No. 754 adopted by the Government of the Russian Federation on 30 August 2013.
[34] World Energy Outlook 2011.

agreements on marine body area and sea bed areas; regular payments for subsurface use (rentals); reimbursement of the state's expenses in connection with exploration; and compensation for damage caused by works implementation.[35]

Changes in corporate structure

Russia's gas sector has some structural and institutional specifics that stem from the Soviet-era command economy. Its institutional structure seemed to have been frozen for a long period, which started in early 1990 when the Soviet-era gas ministry was turned into Gazprom, a corporation. In 1999 Gazprom was deemed the owner and operator of the Unified Gas Supply System (UGSS) under the law 'On Gas Supply in the Russian Federation',[36] which at that time facilitated continuity in gas supply. Gazprom was required by law to supply pre-negotiated volumes of gas to customers at regulated prices, regardless of profitability. Additional gas could be purchased from Gazprom or independent producers at higher prices. During the 1990s, the gas sector was practically exempted from restructuring. The whole sector was politically defined as a 'natural monopoly', even though the law stipulated that gas transmission and distribution network owners were obliged to provide non-discriminatory third-party access to free capacity, according to procedures determined by the government.[37]

During the early 2000s, some independent gas producers emerged, but compared to Gazprom their role was negligible. Some of the vertically integrated oil companies attempted to commercialize their gas production, but as Gazprom controlled the access to the transportation system, they did not succeed. Some, such as Lukoil, ended up selling gas to Gazprom at the wellhead, while others, such as TNK-BP and Surgutneftegaz, utilized most of their associated gas for their own needs. Gazprom's dominance was unquestionable.

In 2002–6 a major campaign for gas market liberalization was initiated by the Ministry of Economic Development. Various governmental bodies prepared nearly a dozen competing 'market concepts'. But in 2006 a very clear signal came from the president that these discussions should be stopped, as Gazprom's strategic and geopolitical role had to take precedence

---

[35] Clause 13 of Federal Act N 89-FZ 'About Production Sharing Agreements' as amended 19 May 2010.
[36] The Federal Law of the Russian Federation No. 69-FZ, 31 March 1999, 'On gas supply in the Russian Federation'.
[37] Regulation 'On the Provisions for Access of Independent Enterprises to the Gas Transportation System of JSC Gazprom' approved by Resolution No. 858 of the RF Government, 14 July 1997.

over all commercial arguments. Since 2006 new legislative initiatives have included the encoding in law of the gas export monopoly, and the consolidation of state majority ownership of Gazprom.[38]

Discussions on market liberalization have recently started again, much more modestly. In fact the government is not developing any regulatory framework to unbundle Gazprom, which means that that question is not on the agenda at least for the next few years, bearing in mind the long period of time that would be needed to implement such a regulation. The government is frightened by the prospect of a transitional period when something might go wrong – and these fears are understandable, taking into account the huge economic and political role of gas.

However, Gazprom's performance in the second half of the 2000s was becoming more and more disappointing, even though it was enjoying high sales volumes and high prices until the 2008 crisis. Falling export volumes and revenues in 2009 were the last straw for the government, which realized that Gazprom needed incentives in order to improve its efficiency. There was already, in Novatek, a candidate to play the role of a competitor. With the acquisition of a shareholding by Gennadii Timchenko (the founder of the oil trader Gunvor and a friend of President Putin) Novatek gained a strong administrative resource and lobbying power. It then started its crusade for market share. In 2012–13 another new player, Rosneft, came to the market. After leaving his position as vice-president responsible for the energy sector in the government in May 2012, Igor Sechin returned to Rosneft, and immediately started developing its gas business, recruiting senior managers from Gazprom and launching ambitious projects. So, after a decade of steady increases in the non-Gazprom producers' role, in the post-crisis period, a real breakthrough occurred.

In 2006 Gazprom achieved what it had aimed at: higher regulated prices. But ironically, particularly because of the economic crisis, there were unintended consequences: those higher prices enabled the non-Gazprom producers to compete for customers. (See Chapter 13 for a detailed account.)

In 2009, Gazprom had to absorb the bulk of domestic demand reduction, while the independents used the opportunity to enhance their position. In 2013, with the Russian gas market still oversupplied, Rosneft and Novatek are competing more and more aggressively with Gazprom, proposing discounts to the regulated price. They were successful in 'cherry-picking' Gazprom's best industrial customers. It seems, nevertheless, that a mechanism of 'market division' and government control was applied – that is, the largest power generating companies and

---

[38] Belyi A. 'Trends of Russia's Gas Sector Regulation', Fourth Annual Conference on Competition and Regulation in Network Industries, Brussels, 25 November 2011.

metallurgical plants could not simply change their suppliers without consulting government.

The gas market structure is now more reminiscent of the oil industry, with its regional monopolies or oligopolies. Non-Gazprom production is increasing, while market consolidation and concentration is taking place. After Rosneft's acquisition of TNK-BP and Itera in 2013, there are just three large non-Gazprom producers left on the market: Novatek, Rosneft, and Lukoil. Rosneft, which has contracts to supply more gas than it is able to produce, apparently aims to become a 'consolidator' of oil companies' gas output. There is now an oligopoly (and regional monopolies), instead of the former monopoly. The government prefers this, and is applying the good old 'system of blocks and checks', and 'division of markets' between groups of interest.

In the next few years, competition on the domestic market will increase. A large number of power sector and industrial consumers renewed their five-year contracts in 2012–13, and the majority of industrial customers now have contracts. So further changes in market shares are likely to happen through sales to residential and smaller business customers, which is much more complicated and considerably less profitable. Rosneft and Novatek are already moving to the regions: in Cheliabinsk, Sverdlovsk, and Kostroma they are the dominant suppliers, not Gazprom. But the regulatory framework is still designed only for Gazprom, and there is a huge legal 'grey zone'. The gas distribution business is in a regulatory vacuum; there are complicated schemes of ownership and operation, many vested interests, and gas supply chains are configured in numerous ways. Here the independents will have to deal more and more with non-payments and the consequences of the communal sector reform. How regional monopolies are to be regulated is still unclear, but there are already signals from government that dominant regional suppliers will be made responsible for regional gasification programmes and gas distribution infrastructure development. (These issues are further discussed in Chapters 5, 6, 12 and 13.)

Access to export

Before 2006, Gazprom enjoyed a de facto monopoly over gas exports, but not a legal one. From the mid-1990s, this monopoly was challenged, first, by independent companies such as Itera that began to export gas to CIS countries; and then, in the 2000s, by Eural Trans Gas, which started exports to Europe. Partly in response to these developments, and also as a consequence of the January 2006 Ukrainian gas transit crisis, the 2006 law 'On Gas Exports' gave Gazprom a legal monopoly. According to the concept of a 'single export channel', even LNG production from Yamal should be sold by Gazprom Export under an agency agreement.

As growth potential and profitability on the domestic market are limited, both Novatek and Rosneft are keen to get access to export markets. Obviously, the pipeline export monopoly is a 'sacred cow' that the government is not ready to remove in any case, so the independents have focused their appetites on LNG projects, principally Yamal LNG (Novatek) and Sakhalin-1 (Rosneft and Exxon Mobil).

In November 2012 Novatek for the first time proposed that LNG supplies be excluded from Gazprom's export monopoly, on the grounds that this would help to develop LNG projects and secure a strong position for Russia in world LNG trade. The main government bodies responsible indicated their approval, and in October 2013 the government approved amendments to the law on gas export drafted by the energy ministry. Though this law is far from a real LNG export liberalization, as it imposes strict requirements for the companies and projects able to export, it is nevertheless an important step towards more liberal market conditions.

Competing political and commercial agendas

In the history of the Russian gas industry's development, 1991–2009 could be described as the period of a 'traditional contract', that worked as follows:

- subsidization of the economy through low gas prices (cost containment to ensure the social stability and competitiveness of Russian industry), and in exchange;
- low taxes on gas production;
- the carrying of social and other unforeseen financial burdens by Gazprom;
- the protection by the state of Gazprom's interests in Russia and abroad, including its exclusive export rights and ownership of the UGSS, while export revenues offset losses in the domestic market; and
- a monopolistic market structure with one dominant player, Gazprom, and a number of much weaker independent gas producers, enjoying higher prices but with no access to export markets.

However, in recent years a 'new deal' has started to emerge, whose shape is still not entirely clear, but the pillars of which will include:

- prices which have already reached a level that ensures profitability on the domestic market, and the concept of a phased transition to equal profitability with European netback offers higher prices in the distant future;
- taxes which should be increased to remove the bulk of rents received by gas companies due to the increase of gas prices;

- the gradual redistribution of exclusive export rights among several major players (Gazprom, Rosneft, Novatek), but in return the non-Gazprom producers will be forced to take on certain social obligations; and
- a gradual reduction in the dominance of Gazprom in the domestic market, although market concentration remains high due to the consolidation of assets in Novatek and Rosneft.

However, this is only a very general outline of the 'new deal'. The Russian gas industry has become hostage to a large number of government interests that are non-industrial and often even non-economic. The main groups of conflicting interests are shown in the Table 1.4. The most important question is: which of these conflicting interests will be the most potent?

Depending on the outcome of the interaction of these contradicting interests, a number of strategic choices will need to be made for the gas sector. They include:

- How much gas should be produced (and transported)? All the official strategic documents (for example, the *Energy Strategy* or *General Scheme of Gas Industry Development*) assume intense growth of production volumes, reaching 1 trillion cubic metres/year by 2030 in some scenarios – which is definitely a challenge in the low-demand growth environment. The investment plans of both Gazprom and the independents imply production levels significantly higher than all projected domestic demand and potential exports. Such a volume-oriented strategy could work only for low-cost projects – but none of the greenfield projects or new pipelines are low-cost. But lower production volumes would mean rejecting some 'political' projects, which might not be acceptable for the government.

- On export policy: should Russia focus on the pricing principle, protecting the ideology and renegotiating long-term pipeline contracts under the energy ministry's control, or should Gazprom get more freedom and flexibility and finally change its attitude towards spot indexation?

- Should Russia go for the new remote and expensive giant fields, or stimulate development of the satellite and smaller fields in traditional areas with existing infrastructure ('scale vs efficiency')? If gas projects are still regarded as tools to improve regional performance or solve social problems, then the abandonment of Shtokman, the Power of Siberia pipeline, or the Vladivostok LNG project would be impossible. But focusing on smaller fields and projects necessarily demands reform of the market structure.

Table 1.4: Competing interests in the gas industry

The government's political interests	*Companies' commercial interests*
Geopolitical and geostrategic considerations of Russian government underlie a potential confrontation with the EU over gas exports; the mega-projects bypassing Ukraine; projects in eastern Russia that are not commercially viable; costly and risky LNG projects, economically dubious projects of cooperation in the CIS (including gas imports from Central Asia and Azerbaijan) in order to retain and strengthen Russia's political influence. Promotion of the '**social responsibility' policy for the gas companies on the domestic market** (including financing of different federal and regional projects, gasification, etc.) in order to prevent social tensions.	There is a purely commercial goal – **to improve oil and gas companies' performance**, thus boosting economic growth – that often contradicts the geopolitical objectives of the state. Nearly all the projects supported by the state are value-killers for gas companies. A key question is: will the state go ahead if these projects destroy value? Can this situation persist, or will market forces oblige the Russian government to change its attitude towards the gas sector?
Government's fiscal policy targets. Taking into account growing social expenditures, the government seeks additional budget incomes. It requires greater export revenues and MET payments. Gas price growth is regarded as a potential source of additional tax base.	**Additional taxation will constrain gas production growth and undermine the commercial rationale of nearly all new projects.** Export duty is increasingly undermining the competitiveness of Russian pipeline gas abroad.
The goal of **increasing Russia's energy efficiency** supports the trend towards high gas prices. There is huge bureaucratic activity in energy efficiency promotion, and a special law was adopted, but energy intensity continued to grow in the last few years (despite aggressive price growth). The goal of **maintaining internal social stability through price subsidies** for the population and the implementation of large-scale (though not always economically viable) gasification. The goal of **supporting domestic industrial output and economic growth** by curbing the growth of gas prices (and thus of electricity prices).	**Growing production and transportation costs for all gas producers** require higher prices on the domestic market and contradict the paternalistic state policy of cross-subsidization.

- Gas pricing: should it be regulated at the rate of inflation, or according to the netback principle, or some other system? If equal profitability is the choice, then exactly which netback should be applied – from Gazprom's border prices or from the European spot markets? The Union of Energy Consumers and other organizations of industrialists are now arguing that energy prices should be frozen. The explanation is very simple: their sales are dropping (for external reasons) and their margins are shrinking due to electricity and gas price growth. They have very strong incentives to put up a fight, and they have direct access to President Putin and the government to argue their position. Stagnating industrial output will most probably become their strongest argument.

- The pricing issue is linked to the question of transportation tariffs. At the end of the day, will there be price regulation, or just transportation tariff regulation? The principles of inclusion of capital costs in the tariff, and the non-transparency of operating costs, lead to unnecessary increases in transportation tariffs, which are undermining non-Gazprom producers' supply economics. As of late 2013, they can not reach customers to the west of Moscow. With the further development of the UGSS – if, for example, the Eastern Programme investments are included in transportation tariffs – these tariffs might become prohibitive. So if the government prefers to focus on the pipeline construction projects, it would mean that the independents will not be competitive, or that domestic prices would have to be raised substantially in order to pay for this inefficient construction.

- Growing taxation is another dimension of the prices issue. If the government pressures the producers too strongly, it will undermine their investment programmes, while going for numerous tax breaks means that the Government will not be able to collect all the necessary budget revenues. Usually in such a situation, choices are made in favour of short-term benefits, rather than taking into account longer-term consequences.

The importance of the gas industry for Russia economically and politically, now and in the future, can not be overestimated. Gas is not only the backbone of the Russian energy sector, but is also one of the most powerful tools of domestic and foreign policy. The Russian gas industry is undergoing some very fundamental, albeit slow, changes, with increasing competition both domestically and abroad. External forces are starting to create imbalances in the sector, which means that the status quo may not be able to continue.

The gas industry is indeed at a crossroads. Choices should be made in the near future between geopolitical goals and economic efficiency; between long-term export volume maximization and obtaining the highest possible prices; between additional tax incomes and sustainable production volumes; between market relationships and price regulation and subsidization. At the end of the day, the choice is between market-oriented industry development with a purely economic rationale, or the maintenance of 'strategic' considerations as the main priority, achieved through state guidance.

The future path of the Russian energy sector's development, and even the evolution of its whole economy, depends on how these issues are resolved. It should be stressed that the decision will mainly be made by the Russian political leadership.

CHAPTER 2

THE MATRIX: AN INTRODUCTION AND ANALYTICAL FRAMEWORK FOR THE RUSSIAN GAS SECTOR

Jonathan Stern

Introduction

The biggest problem facing any study of Russian gas is how to approach such a vast and multi-faceted sector, with many different inputs and outputs. The geographical extent of the industry – both domestically and externally – means that analysis of the different supply sources and markets is extremely complicated. The size and complexity of the Russian gas sector does not lead easily to comparisons with gas sectors in other countries. Depending on the metric adopted, Russia has one of the two largest gas sectors in the world, as well as being the world's largest exporter. While gas demand is larger – and in recent years production has also been larger – in the USA, the latter does not have the complex international dimensions of the Russian sector. And while Canada shares similar gas geography, in the sense that the majority of production has to be brought long distances from fields to centres of consumption, this is not on the scale of the Russian industry.

Previous work by this author[1] suggested that the Russian gas balance should be approached as a matrix with three major supply sources (Gazprom production, non-Gazprom Russian production, and Central Asian imports), and three major markets (Russian demand, exports to CIS countries, and exports to Europe). With the start of exports to Asia in the 2010s, an additional element was added on the demand side, bringing with it the complexities of the global LNG trade in which Russia is a small but growing participant.

This chapter provides an introduction to the Russian gas sector, using the matrix as an analytical framework through which the relative size of its major components can be viewed. Without such a context, there is a tendency to concentrate on a single component – in the case of non-Russian commentators this is usually exports to Europe – and relate all other developments to that element. This leads inevitably to misleading analysis and conclusions.

---

[1] Stern, J., 'The Russian Gas Balance to 2015: difficult years ahead', in Pirani, S. (ed.), *Russian and CIS Gas Markets and Their Impact on Europe* (Oxford: OIES/OUP, 2009), pp. 54–92.

One of the major themes of this book is that all elements of the Russian gas matrix have been subject to change at an accelerated rate over the past decade, but particularly since 2008. During the five years prior to 2008, booming economic growth and energy demand – reflected in oil (and therefore international gas) prices – had created concerns as to whether Russian gas production could keep up with demand in its major markets. This in turn had led to the start of liberalization of Russian domestic prices, and moved export prices to former Soviet countries towards international levels, allowing market forces to enter these gas markets for the first time in their history. The 2008–9 global recession caused demand to fall substantially in all of the markets for Russian gas, accelerating the impact of price signals and competition in the country's domestic market. Meanwhile in Europe, the fall in demand which coincided with excess LNG supply – not needed in the USA due to the shale gas revolution – and liberalization of gas transportation in EU countries, caused spot gas prices to fall to much lower levels than oil-linked long-term contract prices, establishing market hubs as price discovery locations.

These developments led to another major theme – and the overall approach – of this book, which is that the sector needs to be viewed from the perspective of changing market conditions, rather than availability of supplies, which has been the traditional approach of previous work.[2] The emphasis on changing market conditions leads to questions of Russian responses – both corporate and governmental – as gas markets move from surplus to shortage and back again. Because of the extremely large scale and high costs of production and transportation projects – both domestic and export – which characterize the Russian gas sector, this commodity cycle, with its accompanying price changes, leads to particular complexity for investment decision making.

The evolution of the matrix 2002–12

Table 2.1 shows the overall shape of the matrix, with its major supply and market building blocks, to give the reader an impression of the overall size and relative importance of the different elements.[3] Looking back a decade from the early 2010s, certain aspects of the evolution of the matrix are striking. Perhaps most obvious is that, while production rose very strongly in the 2000s, peaking in 2008 and still slightly below that peak in 2012,

---

[2] Stern, J., *The Future of Russian Gas and Gazprom* (Oxford: OIES/OUP, 2005); Pirani, S. ed., *Russian and CIS gas markets and their impact on Europe* (Oxford: OIES/OUP, 2009).

[3] Smaller (but nevertheless important) elements, not shown in Table 2.1, include storage injection and withdrawal, and deliveries to customers outside the high pressure pipeline network (UGSS).

the share of production shifted decisively in favour of non-Gazprom producers which rose from more than 11 per cent in 2002 to nearly 25 per cent in 2012. Elsewhere on the supply side, Central Asian imports roughly doubled between 2002 and 2008 but had declined by more than 50 per cent four years later.

Table 2.1: The Russian gas matrix: major building blocks (Bcm)

	2002	*2008*	*2012*
Sources of supply			
Gazprom production	522	550	487
Non-Gazprom production	73	114	169
Central Asian imports	34	68	29
Total	**629**	**732**	**685**
Markets			
Russian gas demand (UGSS)	412	462	465
Exports to CIS countries	89	89	63
Exports to Europe (physical Russian gas)	129	159	139
LNG Exports to Asia	0	0	14
Total	**630**	**710**	**681**

Note. The differences between the supply and market totals are accounted for by gas required to run the high pressure pipeline network, changes in net storage levels, and losses.

Sources: Rosstat, Gazprom, Russian Ministry of Energy, OIES research.

On the market side of the matrix, domestic demand increased relatively rapidly up to 2007 and the onset of the global recession, surpassing that level again in 2011. Exports to Europe increased strongly in the mid-2000s, peaking in 2008; they declined significantly following the recession, from which most EU countries had not recovered by 2012. Even more dramatically, exports to CIS countries declined by nearly one-third from the levels of the 2000s. Although the fall in deliveries to Ukraine was responsible for two-thirds of this decline, the loss of markets in the Caucasus was also important. LNG exports to Asia, which started in 2010, did not come close to compensating for market losses in Europe and CIS, but they added an extra dimension to the matrix which could become of much greater significance, particularly in the 2020s.

The structure of the book: how markets are driving change

Parts 2 and 3 of this book look closely at the individual elements of the matrix, addressing the complexities and revealing the detail behind the numerical generalizations in Table 2.1. As noted above, we believe that changes in market conditions, and in particular price dynamics, are the drivers of overall Russian gas development and these are the subject of Chapters 3–9. Chapters 3 and 4 deal with exports to Europe, the focus of most non-Russian commentary on Russian gas. Russia exported gas to 25 European countries in 2012 (Chapter 3) and although three countries accounted for more than half the total volumes, there are separate contracts and complex bilateral relationships with each country. In addition, Table 2.1 does not include the very considerable volumes of gas which do not originate in Russia but which are sold in Europe by Gazprom. The European market has traditionally been by far the largest source of Gazprom's revenue and, despite a decline in volumes during 2008–12, this remained the case.

In the post-2008 period there have been serious challenges to Gazprom's support of traditional oil-linked pricing, as markets have become more competitive and hub-based pricing more prevalent, and as the EU's Third Energy Package unfolds across the continent (Chapters 3 and 4). Part of these reforms have serious implications for Gazprom's proposed new pipelines, as the Gas Target Model and Network Codes are set to place restrictions on the share of capacity that can be reserved by a single company. The question of whether the worsening European market outlook in the early 2010s acted as a catalyst for changes in the Russian domestic gas market, and Gazprom's export policy towards CIS countries and Asia, is a proposition that this book will examine. But one of the general themes of the book is that supply and demand – and particularly pricing – developments in all the markets for Russian gas are driving change in the matrix.

Analysis of the sectoral and regional dimensions of domestic gas demand (Chapter 6) in a country as large as Russia is hugely challenging. For example: in 2012, the volume of gas used by the Russian pipeline network for fuel and technical purposes (40.9 Bcm) was greater than *total* national consumption in all but 12 countries in the world. This gives an indication of the challenge posed by just this element. Developments in pricing and regulation have had very substantial impacts on the domestic market since the mid-2000s. With very strong growth of volumes sold by non-Gazprom companies, there has been a dramatic shift towards market-related pricing and competition (Chapter 5) which is an illustration of the approach which we are taking in this book.

Over the past decade, Russia exported gas to seven CIS countries (Chapter 7) and although volumes are dominated by Ukraine and Belarus,

Russian political and pricing relationships have led to substantial changes in exports over the past five years. The breakdown in relations with Ukraine, which led to serious interruptions of deliveries to Europe in 2009, gave renewed impetus to initiatives from Gazprom and the Russian government to build new transit-avoidance pipelines to deliver gas to Europe via existing and new Nord Stream, South Stream, and Yamal–Europe corridors; despite declining export volumes in the post-2008 period. The geopolitical implications and motivations of these pipelines for Russia, Europe, and CIS countries, have been widely debated (Chapter 4).

Up to 2010, the markets for Russian gas were overwhelmingly 'west-facing', towards European Russia, CIS countries, and Europe, and therefore supply and transportation needed to be developed with this geographical orientation. In 2012, liquefied natural gas (LNG) from Sakhalin in the Russian Far East was delivered to three Asian countries and, given the combination of vast quantities of stranded gas in Eastern Siberia and the Far East with rapidly growing demand in (especially) China and other Asian countries, there are substantial prospects for expansion of both pipeline gas and LNG exports to this region. New markets, and therefore new supply and transportation systems, seem likely to be increasingly 'east-facing' (pipeline gas to China – and possibly Korea) and global (through Atlantic and Pacific-based LNG exports). In the early 2010s, such projects had become a strong aspiration but not yet a reality, and although they will command considerable attention over the next decade (Chapter 8), progress will require much greater success in concluding export contracts than has been seen thus far. If for any reason the anticipated surge of Russian gas exports to Asia does not happen, or is delayed until later in the 2020s, there will be significant consequences for both Asian markets (needing LNG from elsewhere) and Gazprom (which will remain much more heavily dependent on other gas markets for revenues).

On the supply side of the gas sector – the subject of Part 3 – the Gazprom production figures conceal a complex combination of declining Soviet-era assets (Chapter 10) and new field and pipeline developments (Chapters 11 and 12). The start of production from Gazprom's Bovanenkovo field on the Yamal peninsula in 2012 created the means to access the next generation of gas fields, which could potentially replace existing Nadym Pur Taz production. But costs of production and transportation from these new fields place Gazprom at a considerable competitive disadvantage to other producers. Non-Gazprom producers (Chapter 13) have a range of different, and very dynamic, production aspirations. This is not simply about differences in corporate strategies, numbers and location of fields, but also highlights a crucial distinction between gas produced with liquids ('wet gas') and non-associated gas ('dry gas'), which has substantial commercial implications, and also additional complexities such as gas gathering and processing. Since liquids command a much higher price than gas, wet gas is generally much more profitable than dry gas.

Just as it is difficult to generalize about other supply components, so taking Central Asian countries (Chapter 14) as a group conceals very substantial differences: Turkmenistan has abundant gas reserves in relation to domestic demand, while Uzbekistan and Kazakhstan have larger domestic requirements, and Russian companies have joint ventures to produce gas that they hope to market themselves. But as in other regions, the post-2008 period has seen a very substantial change, with China becoming a major – and very soon the overwhelmingly dominant – export market for Central Asian gas. As prices charged by Central Asian countries increased to European netback levels, so Gazprom reduced its purchases to relatively low levels. Indeed the main reason for continued Gazprom gas purchases may be related more to foreign policy rather than gas supply. Meanwhile Chinese purchases of Central Asian (principally Turkmen) gas could reach 65 Bcm later this decade, although the oil-linked price mechanism may pose problems for Chinese buyers.

Changing dynamics and inter-linkages between the components of the matrix

The concluding section (Part 4) explores the inter-linkages between the market and supply sides of the matrix, and between the different elements, both historical and future. While it is difficult to make generalizations even about individual components, let alone the direction of the sector as a whole, the issue of connectivity between the different elements of the matrix involves even greater complexity. Much previous commentary has made a direct connection between Russian (and Gazprom) production trends and the availability of gas for export to Europe. This may have some validity during periods of very severe weather, which reduces production (due to very low Siberian temperatures) and increases demand in Russia and CIS countries. But on an annualized basis, there is no simple connection between how much gas Russia (or Gazprom) is producing and how much it is exporting or could export (and to which markets). Because of its size, a relatively small percentage increase or decrease in domestic gas demand is substantially more important than a similar percentage change in exports. By contrast, Central Asian imports were extremely important up to the mid-2010s, but the introduction of market-related pricing, and the arrival of large scale non-Gazprom production, has greatly reduced the attractiveness of this gas, and entering the 2010s it appears to have been given the role of 'swing supply' – supply which can be called upon when needed, but not taken if demand falls or cheaper alternatives are available.

The decline of the traditional utility model and the rise of market-based relations

The matrix is intended to make sense of growing complexities and accelerating change in the markets supplied by Russian gas. All of these markets are moving away from the traditional monopoly/monopsony utility model, in the direction of competition, liberalization, and market-based relations, especially in relation to pricing.[4] Increased prices have created feedbacks on the demand side – in the general direction of demand reduction – which post-2008 reduced the need for additional supply, and accelerated competition between producers. The extent to which these trends will continue over the next decade is not at all certain, but it seems clear that the monopoly/monopsony model will not be restored. Russian gas, and Gazprom, therefore faces a much more competitive future in all of its markets.

The biggest change throughout the matrix over the past decade has been the rise of market-based relations – in other words, changes in supply/demand and pricing in different markets have a much more important and immediate impact on flows from the supply side. This operates at the level of sub-components of the matrix – that is, at the level of specific sectors of the Russian domestic market and exports to (and imports from) individual countries. Since the mid-2000s, and particularly post-2008, progress towards market-related pricing has accelerated, and this has caused substantial changes on both the supply and demand sides of the matrix that seem likely to intensify in the future.

The position of Gazprom and the role of government

The role of government – and specifically the role of the president – remains crucial to strategy and decision-making in the Russian gas sector. The state is the majority owner of Gazprom and keeps the gas sector under close scrutiny. The late 2000s and early 2010s have shown a progressive willingness – and even perhaps desire – to reduce Gazprom's dominance of the sector. But at the same time, the government's domestic and international geopolitical ambitions in relation to gas are an important driving force behind the large new investments noted above, which will have a very substantial impact on Gazprom's financial position.

These changes are difficult to capture directly within the matrix framework but are an important theme of this book. Whereas, up to 2000, the Russian gas sector could be said to amount to Gazprom plus a few minor players, the past decade and particularly the post-2008 period has

---

[4] The monopoly/monopsony model refers to an industry structure where a single company dominates the entire gas chain, and/or one company dominates production and another gas purchase and sale of gas.

seen other powerful players emerge both in the upstream and the domestic market – specifically, the rise of competitors such as Novatek and Rosneft, which have large gas reserves, rapidly increasing production, and specific LNG export projects supported by the political leadership. While Gazprom remains by far the largest single player on both the market and supply side of the matrix, by 2012 other players accounted for 25 per cent of production and domestic consumption, and their market shares were growing significantly year on year (Chapters 5 and 11). This has been made possible by liberalization of transportation to facilitate third-party access, and the apparent determination of the political elite to allow lower cost non-Gazprom gas to reach customers on payment of a tariff for use of the network (Chapter 5).

This degree of erosion of Gazprom's dominance, which was by no means predictable a decade ago, has raised questions relating to the fundamental restructuring of the company, and potential 'unbundling' of Gazprom's transportation network into a separate company under different ownership. While in 2013 such ideas still seemed unlikely to be implemented in the near future, the fact that they could even be a subject for discussion shows the need to be open-minded about structural change later this decade.

Gazprom has retained its monopoly in export markets, but the 2010s have seen an intensification of debate at the highest levels of government as to whether this should continue. By the end of 2013, LNG projects had been exempted from the monopoly, on a case-by-case basis. This means that Gazprom's export dominance is increasingly being challenged in ways which few would have foreseen even five years ago.

All of this increases the complexity of Gazprom's future development strategy, previously assumed to cover the entirety of Russian gas production, demand, and exports. As other players have entered the sector and exploited attractive commercial opportunities, so Gazprom's 'business model' of the 1990s and early 2000s (cross-subsidizing deliveries to loss-making domestic and CIS customers with profitable deliveries to Europe) no longer operates. As prices charged to domestic and CIS customers increased in the late 2000s, these markets became increasingly profitable and (with the exception of a small number of protected household customers and those who do not pay their bills) no longer require subsidies.[5]

Gazprom is required both by state and private investors to increase its market capitalization and make profits. But there is a very clear conflict between the agendas of those groups of stakeholders, which is at the heart of Gazprom's financial dilemma. Increasingly the state wants (and arguably

---

[5] By 'no longer require subsidies' it is meant that prices cover the cost of production and delivery to the customer.

needs) to increase tax revenues from the gas sector (which could include dividends, since the state is a 50 per cent shareholder in Gazprom) while private sector shareholders want dividends and capital growth. During the latter part of the 2000s, when gas prices were rising but taxation was not, both sets of stakeholders could be satisfied. But in the 2010s much of this landscape changed: the state wants higher taxes, and appears to be restricting regulated gas price increases (charged by Gazprom) to no more than the rate of inflation. Gazprom's position is further complicated by an increasing number of very large gas projects with investment requirements of $20–50 billion, targeting both domestic and international markets, which appear at best doubtfully profitable and at worst potentially loss-making. With the company's cost base increasing much more rapidly than those of its competitors (because of its Yamal peninsula and East Siberian/ Far Eastern operations) the rationale for such projects appears increasingly questionable, and the consequences for Gazprom's profitability increasingly problematic. The fact that government policy is the ultimate determinant of Gazprom's decision making highlights the dilemmas and complexities in which the company finds itself.

The main aim of this book is to explain and explore the complexities of the different elements of the matrix and to show how combinations of these elements can result in a range of different outcomes for the markets supplied with Russian gas. The ways in which changing internal and external dynamics are impacting Russia's position in global gas markets will also be examined. We believe these dynamics will be important not only for those directly involved in the Russian gas sector, but for the gas sector worldwide.

Part 2

The Impact of Competitive Markets on Russian Gas

CHAPTER 3

RUSHIAN RESPONSES TO COMMERCIAL CHANGE IN EUROPEAN GAS MARKETS

Jonathan Stern

In volume terms, Russian gas sales to European customers reached their highest-ever level in 2005–8 (annual average of 168.4 Bcm). In 2009–12 they fell back significantly (to an annual average of 155.5 Bcm). A key cause of the change was the economic recession and its impact on European energy demand, but other factors were at work, including changes in the structure of European energy markets, and Gazprom's reaction to pricing trends. We have divided the discussion of Russian exports to Europe into two chapters: this chapter deals primarily with the commercial and logistical aspects of Russian exports and Chapter 4 deals mainly with the regulatory, political, and security issues. The conclusions relating to both chapters can be found at the end of Chapter 4.

Russian gas exports to Europe since the mid-2000s

The Soviet Union, and from 1992 the Russian Federation, have been exporting gas to the eastern part of Europe since the Second World War, and to the western part since 1967. However, in contrast to the near-constant growth of Russian gas exports to Europe up to the early 2000s, volumes peaked in 2007–8 and declined thereafter, returning to near peak levels only in 2013.[6] Table 3.1 shows Gazprom's exports to 25 European countries since 2005. These sales are of considerable financial importance to Gazprom, comprising just over half of the company's revenues over the past decade, although only 27–31 per cent of its sales by volume.[7]

An important statistical anomaly in Table 3.1 is that the 'Grand Total' is the volume that Gazprom delivers to European customers, which is not the same as volumes of gas delivered under long-term contracts to those customers (which we believe to be the gas that physically leaves Russia).

---

[6] Previous OIES publications (Stern, J.P., *The Future of Russian Gas and Gazprom* (Oxford: OIES/OUP, 2005), Chapter 3) focused on exports to Europe in the 1990s and early 2000s.)

[7] Figures are for 2002–3 and 2011–12, revenue figures are net of all taxes, ibid. Table 3.4, p.128; Management Report OAO Gazprom (2012), p. 35.

Table 3.1: Gazprom's exports to Europe, 2005–13 (Bcm)*

	2005	2006	2007	2008	2009	2010	2011	2012	2013**
Austria	6.8	6.6	5.4	5.8	5.4	5.6	5.4	5.4	5.2
Belgium	2	3.2	4.3	3.4	0.5	0.5	0	0	0
Estonia	1.3	0.7	0.9	0.6	0.8	0.4	0.7	0.6	0.7
Finland	4.5	4.9	4.7	4.8	4.4	4.8	4.2	3.7	3.6
France	13.2	10	10.1	10.4	8.9	8.3	8.5	8.2	8.2
Germany	36	34.4	34.5	37.9	33.5	35.3	34.1	34	40.2
Greece	2.4	2.7	3.1	2.8	2.1	2.1	2.9	2.5	2.6
Italy	22	22.1	22	22.4	19.1	13.1	17.1	15.1	25.3
Latvia	1.4	1.4	1	0.7	1.1	0.7	1.2	1.1	1.1
Lithuania	2.8	2.8	3.4	2.8	2.5	2.8	3.2	3.1	2.7
Netherlands	4.1	4.7	5.5	5.3	4.3	4.3	4.5	2.9	2.1
Switzerland	0.4	0.4	0.4	0.3	0.3	0.3	0.3	0.3	0.4
Turkey	18	19.9	23.4	23.8	20	18	26	27	26.6
UK	3.8	8.7	15.2	7.7	11.9	10.7	12.9	11.7	12.5
Sub-Total	**118.7**	**122.5**	**133.9**	**128.7**	**114.8**	**106.9**	**121**	**115.6**	**131.7**
Bosnia and Herzegovina	0.4	0.4	0.3	0.3	0.2	0.2	0.3	0.3	0.2
Bulgaria	2.6	2.7	2.8	2.9	2.2	2.3	2.5	2.5	2.8
Croatia	1.2	1.1	1.1	1.2	1.1	1.1	0	0	0
Czech Republic	7.4	7.4	7.2	7.9	7	9	8.2	8.3	7.3
Hungary	9	8.8	7.5	8.9	7.6	6.9	6.3	5.3	6
Macedonia	0.1	0.1	0.1	0.1	0.1	0.1	0.1	0.1	0.04
Poland	7	7.7	7	7.9	9	11.8	10.3	13.1	9.8
Romania	5	5.5	4.5	4.2	2.5	2.6	3.2	2.5	1.2
Serbia	2	2.1	2.1	2.2	1.7	2.1	2.1	1.9	1.2
Slovakia	7.5	7	6.2	6.2	5.4	5.8	5.9	4.3	5.4
Slovenia	0.7	0.7	0.6	0.6	0.5	0.5	0.5	0.5	0.5
Other countries	0	0.4	0.5	0.6	1.2	2.1	1.3	1.4	n/a
Sub-Total	**42.9**	**43.9**	**39.9**	**43**	**38.5**	**44.5**	**40.7**	**40.2**	**34.4**
Grand Total	**161.6**	**166.4**	**173.8**	**171.7**	**153.3**	**151.4**	**161.7**	**155.8**	**177.1**
Deliveries under long-term contracts			158.8	168.5	142.8	138.6	150.3	139.9	166.1

* These data differ from those provided in the original source because they show exports to the Baltic countries as 'European' exports whereas Gazprom counts them as 'exports to former Soviet Union (FSU) countries'.

** 2013 data are preliminary and are not consistent with previous years.

Sources: *Gazprom in Figures 2005–9*, p. 56; *Gazprom in Figures 2008–12*, p. 63; 'UK, Italy, Germany had biggest Russian gas import growth in 2013; Turkey led in H2', *Interfax Russian Oil and Gas Weekly*, 16–22 January 2014, p. 46..

52   *The Russian Gas Matrix*

The difference is the volume which Gazprom sources from elsewhere, by means of swaps and trading. For example, most of Gazprom's exports to the UK are in fact gas sourced from elsewhere.[8] This is not necessarily because the company does not have sufficient physical gas available – although during very cold winter periods this may be the case (see Chapter 4) – but because transportation distances and constraints within Europe mean that it is cheaper and easier to source gas from elsewhere.

While there is no completely consistent pattern in volume exports to individual countries, the trend is a general increase up to 2008 followed by a decline thereafter. By the early 2010s, Russian exports were significantly below the levels seen in the 2005–8 period. To some extent the data is misleading due to lack of temperature correction, which can account for substantial year-to-year differences, but the trend is clear: of 26 European countries importing Russian gas in the period 2005–12, two (Belgium and Croatia) have ceased imports entirely, 19 have reduced their imports during this period, and only five (Greece, Turkey, Poland, Czech Republic, and the UK) are importing larger volumes. In 2013 exports reached their highest levels since 2008 of around 166 Bcm partly due to a protracted spell of unusually cold weather in the early spring, and partly connected with price reductions discussed below.[9]

Figure 3.1 shows the profile of major Russian long-term export contracts to Europe up to expiry in 2035.[10] Contracted volumes peak during 2012–14 at around 180 Bcm/year and then decline gradually to around 140 Bcm/year until the late 2020s when they drop off substantially.

The outlook for European gas demand and imports

In historical perspective, the year 2005 may be seen as the end of the 'golden age of gas' in Europe.[11] For around three decades from the mid-1970s

---

[8]   The company's 2007 Annual Report (61) states that, 'In December 2007, Russian natural gas started being supplied to the UK market through the BBL gas pipeline'. It is highly unlikely that this gas is physically sourced from Siberia – although there is no way to be certain. One source suggests that out of 11.7 Bcm sold to the UK in 2012, 8.1 Bcm was Gazprom's gas and 3.6 Bcm was acquired from other companies (*Russian Energy Monthly*, July 2013, p. 22).

[9]   The figure of 163 Bcm is a preliminary Gazprom figure and does not include sales to the Baltic states of 4–5 Bcm, 'On Key Preliminary Results 2013', www.gazprom.com/press/miller-journal/ 14 January 2013.

[10] They do not include the Baltic States, Romania, Bulgaria, and former Yugoslav republics to which Gazprom delivered 12.5–13.0 Bcm in 2012.

[11] The 'golden age of gas' is a phrase coined by the International Energy Agency America (IEA) (*World Energy Outlook 2011: Are We Entering a Golden Age of Gas?*, Paris: IEA/OECD) to denote the possibility of a much larger share of gas in global energy demand following the shale gas revolution in North America. But it is used here simply to refer to an era of rapid market expansion.

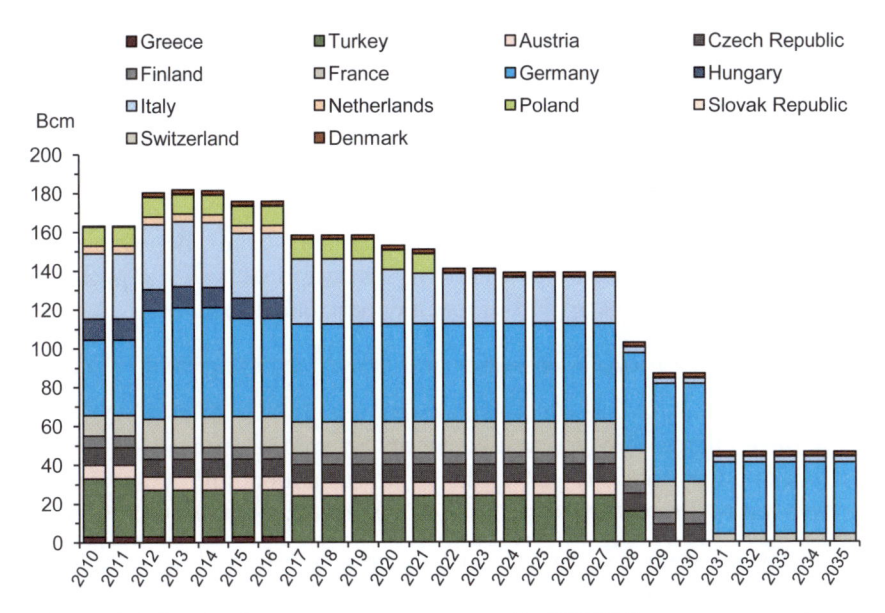

Figure 3.1: Russian long-term gas contracts with OECD European countries to 2035
Contracts with Baltic and some south-east European countries are not included.

Source: research by the Energy Research Institute of the Russian Academy of Sciences (ERI RAS) based on data from Cedigaz and Nexant.

to the mid-2000s, European gas demand increased steadily, as did imports of (former Soviet now) Russian gas. The reasons for this form an important part of European energy history: from the mid-1970s, gas progressively replaced oil products in the stationary energy balances of most countries, first in north-west Europe and then in central and southern Europe. In the major European markets, substitution of gas for oil products had largely been completed by the end of the 1990s.[12] Some sectors in individual countries (for example power generation in Poland and Germany which continued to favour coal), and some regions (notably the Balkans and south-east Europe where usage of oil products continued even in power generation), were resistant to this general trend.[13] During the 1990s and 2000s, power

---
[12] For a statistical analysis of these trends see the appendices to Stern, J.P., 'Is there a rationale for the continuing link to oil product prices in Continental European long-term gas contracts?', Working Paper NG19, OIES, 1 April, 2007, and Stern, J.P. 'Continental European long-term gas contracts: is a transition away from oil product-linked pricing inevitable and imminent?', Working Paper NG34, OIES, 1 September, 2009.

[13] For details of gas demand development in Europe up to the late 2000s see Honoré, A. *European Natural Gas Demand, Supply and Pricing: Cycles, Seasons and the Impact of LNG Price Arbitrage* (Oxford: OIES/OUP, 2010), Chapter 1.

generation in the form of combined cycle gas turbines (CCGTs) substantially increased gas demand, but by the middle of the decade, rates of growth had begun to slow. After 2005, demand plateaued and with the 2008 recession declined; by 2012 European gas demand had fallen to the level of 2002 – ten years of growth had been lost – and at mid-2013 this decline showed few signs of being reversed.[14]

The reasons for this decline are complex and cannot be explored in detail here.[15] There are at least four major components of the problems which gas has encountered in European energy balances:

- recession in Eurozone countries and slower than expected economic recovery;

- a huge increase in renewable energy in many countries – greatly facilitated by subsidies – in order to meet national and EU targets in relation to those sources and also to carbon reduction targets;

- a substantial increase in cheap imported coal – largely from the USA where it was displaced by shale gas production – made possible by the very low carbon prices produced by the EU Emissions Trading Scheme;

- high gas prices – especially for gas sold under long-term contracts with oil-linked prices – created by the post-2008 surge in oil prices to above $100/bbl.

These developments could be considered a short-term discontinuity caused by a coincidence of unusual events (recession, subsidized renewables, and cheap coal), or could be heralding the start of a secular decline of gas in European energy balances. An assessment of European gas supply and demand trends over the next decade is beyond the scope of this chapter; many different views and projections have been advanced.[16] For our purposes, the most important issue is the difference between European and OECD projections, and Gazprom's view of how European gas demand and the continent's need for Russian gas is likely to unfold over the next decade.

---

[14] In terms of temperature-corrected gas demand and excluding Turkey, which has been a rapidly growing market throughout the 2000s, the picture is even worse. *Natural Gas Information* (2012 Edition), Paris: IEA/OECD, Table 4, V8–9; *IEA Monthly Gas Survey*, January 2013, Paris: IEA/OECD, Table 1.1, 3.

[15] For more details see Honoré, A., 'Economic Recession and Natural Gas Demand in Europe: what happened in 2008–2010?', Working Paper NG47, OIES, 25 January, 2011 and Honoré, A. *The Outlook for Natural Gas Demand in Europe* (Oxford: OIES/OUP, 2014 forthcoming).

[16] For a range of estimates see *Sovremennie stsenarii razvitiia mirovoi energetiki: resultaty issledovanii 2009–12 gg*, Institut Energetiki i Finansov [Institute of Energy and Finance], 2012.

The most widely quoted scenarios of European gas demand are those of the International Energy Agency (IEA)'s World Energy Outlook (WEO). The 2012 edition of the WEO projected OECD European gas demand as falling from 569 Bcm in 2010 to 550 Bcm, before recovering to 585 Bcm in 2020.[17] The Agency's 2013 WEO, was substantially more pessimistic, with the corresponding demand scenario reaching just 537 Bcm in 2020 and not returning to the 2010 level until 2025.[18] Despite this, Gazprom management continued to be optimistic, with its CEO claiming that 'Europe needs more gas and ... we are ready to supply as much as Europe wants'.[19]

However, even if these demand projections are directionally correct, this may not affect the volume of Russian gas which Europe will need to import. With declining European conventional gas production, and no significant unconventional gas production likely before 2020 (and perhaps substantially later), Europe will become increasingly dependent on imported supplies.[20] As far as pipeline gas is concerned, Norwegian supply has probably peaked and will plateau, but will not decline significantly until after 2020.[21] North African gas exports appear to have peaked, with Egyptian supplies in decline, and although transportation capacity exists (this could increase Algerian and Libyan exports), there is a lack of available gas due to delays in field development, rapidly rising domestic demand, and political turbulence.[22] A maximum of 10 Bcm of Southern Corridor gas from Azerbaijan will start to be delivered at the end of the 2010s (see Chapters 4

---

[17] New Policies Scenario, *World Energy Outlook 2012*, Paris: IEA/OECD, Table 4.2, p. 128. For details of compatibility between Russian and international gas data see Units and Conversions (page xvi).

[18] *World Energy Outlook 2013*, Paris: IEA/OECD, Table 3.2, p. 103.

[19] 'Role of pipeline gas in Europe to grow: LNG overestimated – Miller', *Interfax Russia & CIS Oil and Gas Weekly*, 30 May–5 June 2013, pp. 29–30.

[20] Despite huge publicity about the prospects for shale gas in Europe, no substantial research study from any source sees significant quantities of unconventional (including shale) gas being produced in Europe prior to 2020, and many do not believe this picture will change substantially even by 2030. See *World Energy Outlook 2013*, Figure 3.6, p. 118; *Unconventional Gas: Potential Energy Market Impacts in the European Union*, JRC Scientific and Policy Reports, European Commission Joint Research Centre, Luxembourg: European Union, 2012; Gény, F. 'Can Shale Gas be a Game-Changer for European Gas Markets?', Working Paper NG46, OIES, 1 December, 2010; BP Energy Outlook 2030, (London: BP, 2014), p. 55.

[21] We are treating Norway as an exporter to Europe rather than as an indigenous supplier. On its production outlook, see Johnsen, E. 'Norwegian gas update', paper presented at the FLAME Conference, Amsterdam, March 2013.

[22] This was the conclusion of Fattouh, B. and Stern, J.P. (eds.) *Natural Gas Markets in the Middle East and North Africa* (Oxford: OIES/OUP, 2011), Chapters 1–4, written before the 'Arab Spring' political turbulence which created considerably more uncertainty about future North African exports.

and 14). If any pipeline gas from the Eastern Mediterranean reaches Europe, the volumes and timeframe are likely to be similar.[23] This leaves Europe in the 2010s with two major sources of incremental gas – Russia, and LNG from a potentially large number of different regions, including North America. However, as we saw in the period after the Fukushima nuclear disaster in March 2011, and the progressive closure of Japanese nuclear power stations, LNG supplies apparently destined for Europe can very quickly disappear if demand and prices in Asia increase substantially.[24]

Changing European utility structures and competition

To explain why Gazprom's European export outlook has changed, it is necessary to say a few words about the changing European utility landscape at the beginning of the twenty-first century. The post-war organization of European gas industries was that, with rare exceptions such as Germany, each country had a dominant – usually state-owned – utility which controlled virtually the entire gas market. The commercial strategy of these utilities was to segment their customer base, depending on the ability of the customers to access alternative fuels (and hence the relative value of gas for each customer group), and to price differentially between (and sometimes within) classes of customer, confident that without access to alternative gas supply at transparent prices, their customer base was essentially captive.

The long-term gas contracts which the utilities signed with all major suppliers, domestic and foreign (such as Gazprom), reflected this relatively simple commercial model, but included sufficient flexibility to allow adaptation if and when market fundamentals changed. For the first several decades of European gas trade, they were largely successful in this task, greatly assisted by the fact that the dominant companies had a significant measure of control over market fundamentals, because they were mostly monopsony buyers and monopoly sellers to a customer base whose only alternative to buying their gas was to use a different fuel.

The dominant price mechanism in European long-term gas contracts was the netback market value principle, the origins of which can be traced back to the early 1960s.[25] According to this principle, the price paid by the

---

[23] Darbouche et al. 'East Mediterranean Gas: what kind of a game-changer?', Working Paper NG71, OIES, 19 December, 2012.

[24] In 2012, European LNG imports fell by 27% compared with 2011. Asian LNG imports rose by more than 9%, Japanese imports by 11%. 'The LNG Industry in 2012', Brussels: GIIGNL, 2013, p. 8.

[25] For details of this pricing structure and its historical importance in European gas markets see Stern, J.P. 'The Pricing of Gas in International Trade – An Historical Survey', in Stern, J.P. (ed.), *The Pricing of Internationally Traded Gas* (Oxford: OIES/OUP, pp. 40–84, 2012), especially pp.54–9.

gas company to the foreign or domestic gas producer, at the border or the beach, is negotiated on the basis of the weighted average value of the gas in competition with other fuels, adjusted to allow for transportation and storage costs from the beach or the border and any taxes on gas. In continental Europe the competitive fuels were largely oil products – gas oil and (heavy or light) fuel oil.

As gas expanded its market share, so the logic of the oil-linked price mechanism which had been established in long-term contracts began to disappear and, beginning in the 1990s, the pricing of internationally traded (and domestically produced) gas moved increasingly out of line with market fundamentals. However, this did not cause major problems because the commercial model of continental European gas utilities (described above) allowed them substantially to control national gas markets irrespective of these market fundamentals.[26]

But this control began to break down in the second half of the 2000s, when European energy regulation and competition law – sometimes reinforced, but often opposed, by national governments – created increasing momentum towards effective third-party access, ownership unbundling, and regulatory oversight. These developments, combined with the elimination of destination clauses (see below), completely transformed the regulatory and market context in which existing contracts were operating. Of fundamental importance were: the arrival of workable third-party access, and the emergence of hubs with transparent prices which could be readily accessed by any customer via the internet. By the end of the decade, most consumers in the largest EU gas markets increasingly had a credible choice of suppliers, and competition was spreading across north-west Europe under the twin influences of national regulators, and the network codes required by the EU Third Energy Package, which were aimed at liberalizing transportation across Europe and promoting the role of market hubs for price formation and trading.[27] (See Chapter 4.)

Other developments also had a significant impact on the long-term contracts of European gas utilities. The shale gas revolution collapsed North American gas prices from levels in excess of $10/MMBtu in 2008, to $2–4/MMBtu for most of the period since 2009.[28] In anticipation of gas shortages and high prices, nearly 200 Bcm of regasification capacity had

---

[26] For a definition and discussion of the economic and market fundamentals of gas pricing see *The Pricing of Internationally Traded Gas*, pp. 486–9.

[27] For a discussion on the evolution of European market hubs see Heather, P. 'Continental European Gas Hubs: are they fit for purpose?' Working Paper NG63, OIES, 14 June 2012.

[28] For details see Foss, M.M. 'Natural Gas Pricing in North America', in *The Pricing of Internationally Traded Gas*, 85–144.

58 *The Russian Gas Matrix*

been built in North America during the 2000s, with LNG supplies arranged to fill it. By 2009, those supplies were no longer needed in North America and large volumes of LNG became available for Europe. This exerted significant downward pressure on spot prices as oil prices began to march upwards beyond $100/bbl, while recession collapsed European (energy and) gas demand.

For European utilities this represented a 'perfect storm' of commercial problems: progressive loss of monopoly, surplus supply, falling demand, and sharply increasing long-term contract prices (because of the increase in oil prices). In 2009, European hub prices fell significantly, to levels as much as 50 per cent below oil-linked contract prices, and aside from short periods in particularly cold winters, have averaged 25–33 per cent below oil-linked prices since then (Figure 3.2).

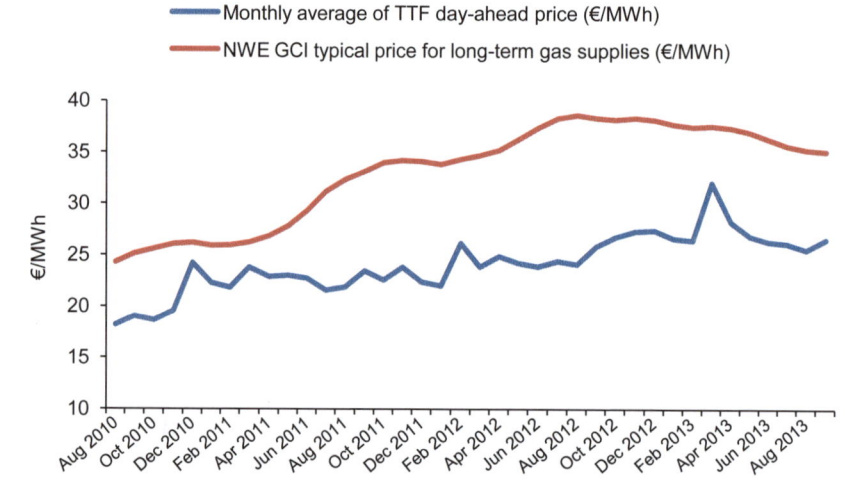

Figure 3.2: European long-term oil-linked and spot prices of gas (monthly averages), August 2010–September 2013

Note: TTF is the Dutch gas hub. The NWE GCI (north-west European Gas Contract Indicator) is a price, calculated by Platts news agency, in a typical 'pure oil-linked' contract formula.

Source: *Platts European Gas Daily Monthly Averages* for respective months.

Contractual – volume and price – problems

One of the major problems of writing about prices in long-term European gas contracts – whether with Gazprom or any other seller – is the degree of confidentiality of contracts, which prevents any detailed assessment which can be independently confirmed or verified. This should be borne in mind when reading the rest of this section.

When European demand crashed post-2008, many importing companies struggled to meet their contractual minimum take-or-pay (ToP) volumes. The impact of this on Gazprom was of special significance, because of the size and centrality of its supplies to the European gas market. Traditional take-or-pay levels in Russian long-term contracts are 85 per cent of annual contract quantity (ACQ).[29] Figure 3.3 shows the extent of purchasers' failure to meet ToP during 2008–10, which resulted in renegotiations between them and Gazprom. At the beginning of 2010, it was widely reported that a number of companies had demanded both reductions in contractual take-or-pay volumes, and reductions in prices.[30] As a result, Gazprom agreed with many of its customers that minimum ToP quantities would be reduced to 70 per cent of ACQ, and would be paid for at the contract price, but that any volumes taken in excess of minimum ToP would be sold at hub-based prices for three years beginning in October 2009.[31] Because we do not know how many customers received these concessions, Figure 3.3 shows an illustration of ToP commitments at both 85 per cent and 70 per cent; we believe that the actual level of ToP is at one of these levels.

Although Gazprom sold nearly 9 Bcm more gas to its European customers in contract year 2009/10 compared with 2008/09, the company's customers incurred take-or-pay liabilities (against a level of 85 per cent of ACQ) of 5 Bcm in 2009 and around 10 Bcm in 2010.[32] However the reasons were different: in 2009 the take-or-pay shortfall was spread across a number of companies, while in 2009/10 it was concentrated on ENI and Edison (Italy) and Botas (Turkey). The sharp drop in 2010 imports with

---

[29] 'Take-or-pay' clauses require the buyer to take an annual minimum volume of gas, or to pay for that volume whether or not it is taken. The post-2008 period is probably the first time in history that buyers had to pay for substantial volumes of gas which they were unable to take; for details see Stern, J. and Rogers, H. 'The Transition to Hub-Based Gas Pricing in Continental Europe', op. cit.

[30] These included E.ON, Wingas, Botas, Eni, RWE and Econgas. 'Europe rethinking contracts with Gazprom', *Interfax Russia & CIS Oil and Gas Weekly*, 4–10 March 2010, pp. 4–5.

[31] Anecdotal evidence suggests that these concessions were given to buyers in north-west and some in Central Eastern Europe. 'Gazprom agrees to sell a portion of gas delivery to Ruhrgas at spot prices', *Interfax Russia and CIS Oil and Gas Weekly*, 18–24 February, 2010, p. 20; Anton Doroshev, 'Gazprom adjusts gas pricing to defend market share', *Reuters*, 19 February 2010. http://uk.reuters.com/article/idUKLDE61I1M320100219?pageNumber=2&virtualBrandChannel=11700&sp=true.

[32] These volumes are for 'contract years' which in European gas contracts are usually 1 October–30 September, and are thus different from the calendar year data in Table 3.1 and Figure 3.3.

60   *The Russian Gas Matrix*

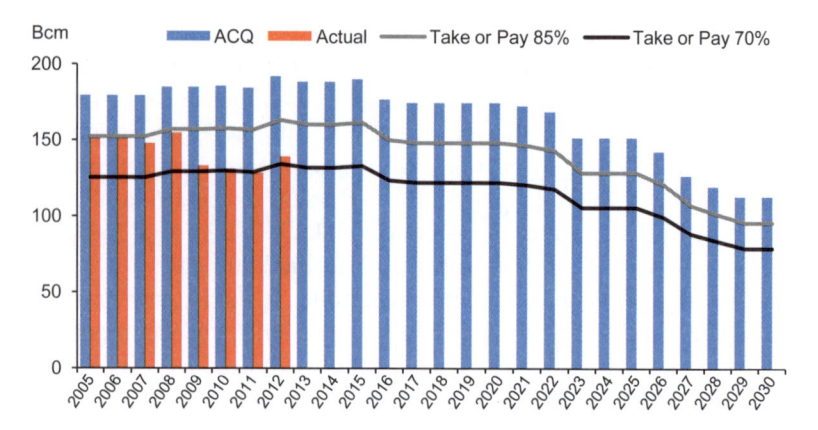

Figure 3.3: Russian long-term export contracts with OECD European countries to 2035: annual contract quantity and take-or-pay levels

Note: Contracts with Baltic and some south-east European countries are not included.

Source: Research by the Energy Research Institute of the Russian Academy of Sciences (ERI RAS) based on data from Cedigaz and Nexant.

recovery the following year is clearly visible in Table 3.1.[33] Figure 3.3 shows that, measured against a 70 per cent ToP level, in aggregate Gazprom's customers took their minimum quantities in 2011–12. By contrast, Table 3.1 shows that Russian export volumes in 2011 recovered principally because of significant increases in exports to Turkey (due to an increase in gas demand of 20 per cent compared with the previous year) and Italy (due to the loss of Libyan supplies for most of that year). While Italian imports fell back again in 2012 (although not to the same extent as in 2010), Turkish gas demand, and hence imports from Russia, continued to increase. Turkey became by far the most important non-CIS market for Russian gas after Germany. Preliminary data for 2013 (Table 3.1) show that exports were boosted to near-record levels by substantial increases in sales to Italy and Germany – the former due to continued problems with North African supplies and the latter affected by the drop in Norwegian deliveries and the need to refill storage – deliveries to other countries show no substantial changes.

---

[33] Turkish payments for gas not taken in 2010 exceeded $750m. 'Botas pays Azerbaijan, Russia, Iran over $1.6bn for unpurchased gas', *Interfax Russia & CIS Oil and Gas Weekly*, 6–12 September 2012, p. 57. ENI's take-or-pay shortfall was 5 Bcm in in 2011 and nearly 9 Bcm in 2010, 'Gazprom agrees to change terms for gas supplies to Italy', *Interfax Russia and CIS Oil and Gas Weekly*, 1–7 March 2012, p. 62. ENI's take-or-pay payments for the period 2009–11 exceeded $1.5bn (although not all of this amount was paid to Gazprom), 'Eni wants out of take-or-pay clauses in gas supply contracts', *Interfax Russia and CIS Oil and Gas Weekly*, 11–17 October 2012, p. 48.

Table 3.2: Russian long-term contract gas prices to European countries, 2010–13 ($/mcm)

	2010	2011	2012	2013
Germany	270	379	353	366
Switzerland	296	400	333	378
Italy	331	410	438	399
Denmark	0	480	394	382
Austria	305	387	394	402
France	306	399	398	404
Netherlands	308	366	346	400
Finland	273	358	373	367
Poland	331	420	433	429
Romania	325	390	424	387
Bulgaria	311	356	435	394
Slovenia	312	377	400	396
Hungary	350	383	416	418
Czech Republic	326	419	500	400
Slovakia	371	333	428	438
Serbia	341	432	405	386
Bosnia and Herzegovina	339	429	500	421
Macedonia	381	462	558	493
Turkey	326	381	416	382
Greece	359	414	475	469
Average*	**308.1**	**398.8**	**421**	**387**
Gazprom Average sales price**	**305**	**383**	**402**	

* Arithmetic average of prices for the year

** As quoted in Gazprom Annual Reports: 2012, p.77; 2011, p. 75. (converted in the source at average exchange rates of RR24–25 = $1, significantly lower than rates of RR29–30 from *Gazprom in Figures* 2008–12, p. 4.)

Sources: *Interfax Russia & CIS Oil and Gas Weekly*, 'OPAL decision, South Stream pipeline talks delayed', 6–12 March 2014, pp. 28–30.

European gas price data are confidential and cannot be independently verified, as was mentioned earlier. The data in Table 3.2 are believed to originate from Gazprom Export, because it is difficult to see where else journalists could have obtained such detailed country by country assessments.[34] The final two rows of the table show an arithmetic average of the figures for individual countries, and Gazprom's average received price (net of excise tax and customs duties), published in its Annual Reports, which should be a volume weighted average price. The data have a number of interesting features. In general, southern and (to a lesser extent) central Europe pay higher prices than do northern and western Europe. It has been hypothesized that this is due to the dominance of Russian gas in southern and central European markets and therefore the lack of any competitive pressure for Gazprom to reduce prices. However, this generalization does not hold true for all countries for all years.

It might be expected that countries further away from Russia in southern Europe would be required to pay more than their central and north European counterparts, due to additional transportation charges. In addition, the netback market pricing methodology (outlined above) ties the gas price to the prices of competing fuels. Countries that are still principally replacing oil products with imported gas – that is, some Balkan and other south European countries – will therefore not be in a position to resist demands for 100 per cent oil-based prices. With crude oil prices above $100/bbl, these will be extremely high. This is particularly the case for countries which are completely dependent on Russian gas.[35] Another possible explanation for price differences is that companies which enter into litigation with Gazprom to achieve lower prices (for example PGNiG in Poland in 2010–11 and RWE in the Czech Republic) are penalized with higher prices, whereas those who cooperate with Gazprom in projects such as South Stream (some companies in Hungary and Turkey) are rewarded with lower prices.[36] Such assumptions give rise to widespread claims that Russian gas pricing is 'political', a subject to which we return in Chapter 4.

However, following the events noted above, in respect of competitive and structural change in European gas markets since the late 2000s, gas pricing in north-western Europe, and some other countries, became increasingly competitive and determined at hubs. This caused very

---

[34] However, as noted above this is impossible to verify.

[35] However, there are exceptions: Table 3.1 shows that Croatia ceased importing Russian gas in 2011 when it received a more competitive offer from the Italian company ENI.

[36] 'Gazprom prices high for litigious in 2011, lower for South Stream Partners', *Interfax Russia and CIS Oil and Gas Weekly*, 14–20 June 2012, p. 47.

substantial problems, to which we now turn. It is very clear from Table 3.1 that by 2012 Russian gas exports remained around 10 per cent below the pre-recession highs of 2007/8. But despite this loss of market share, Gazprom remained unapologetic about its commercial strategy of defending price levels, if necessary at the expense of volumes. It pointed out that in 2012 its receipts from deliveries to 'far abroad' (in other words, non-CIS) customers exceeded $49bn, compared with the $50.8bn which it earned in 2008 – a 3.5 per cent drop in revenues, compared with a drop in volumes of more than 9 per cent.[37]

In 2009, Gazprom agreed that volumes taken in excess of minimum take or pay would be moved to hub-based prices for three years, as was noted above. The reasoning behind this was that by 2012, the recession would be over and pricing would have returned to normal – in other words, hub prices would return to oil-linked contract levels as gas demand recovered after the recession (and surplus LNG supplies were absorbed by fast-growing Asian economies).[38] As shown in Figure 3.2 this did not happen, but Gazprom remained insistent that it 'has no plans to abandon oil indexation', despite its acknowledgement that Norway, its major competitor on the European gas market, 'is actively moving to 100 per cent gas-indexed pricing which contributes to the expansion of its customer base'.[39] In a paper published in January 2013, Sergey Komlev, Head of Gazprom Export's contract structuring and pricing department, published an analytical defence of oil indexation.[40] However, as Figure 3.2 shows during 2010–13, with the exception of a few days in cold winter periods, not only did hub prices remain significantly below oil-linked contract levels, but (particularly in north-west Europe) hub prices became the dominant gas price mechanism.[41]

Gazprom's support of oil-indexed gas prices has been endorsed by Vladimir Putin – as both president and prime minister. Moreover, the 2013 Moscow Declaration of the Gas Exporting Countries Forum, over which

---

[37] Data from *Gazprom in Figures 2008–2012*, p. 66.

[38] 'Gazprom's Miller sees demand for gas slackening in Europe', *Interfax Russia & CIS Oil and Gas Weekly*, 10–16 June 2010 pp. 4–7.

[39] Gazprom Annual Report, 2012, pp. 76–7.

[40] The paper was published on line and is no longer available. For a commentary see Stern, J.P. and Rogers, H. 'The Transition to Hub-Based Pricing in Continental Europe: A Response to Sergei Komlev of Gazprom Export', OIES, 12 February 2013.

[41] The 2012 International Gas Union price survey (IGU) shows that in that year, 72% of gas sold in north west European countries – which represent 50% of total European gas demand – was sold at hub-based prices. *Wholesale Gas Price Survey 2013 Edition – a global review of price formation mechanisms 2005–2012*, International Gas Union, June 2013.

President Putin presided, emphasized that the Forum will:

> ... continue to support gas pricing based on oil/oil products indexation to ensure fair prices and stable development of natural gas resources.[42]

It may therefore be the case that oil indexation should be seen not solely as a Gazprom position but as a government and presidential position that Gazprom is required to implement.

But despite its public stance on oil indexation, starting in 2012 Gazprom began to agree a different type of price mechanism with its customers: the new mechanism retained the oil index but reduced the base price (the P0) in the pricing formula by 7–10 per cent – the effect of this was to bring the Russian price much closer to hub levels. A further mechanism was added which guaranteed buyers a limit on their exposure to hub pricing: at the end of the (one or two year) price period, if the price paid by the buyer under the new P0 + oil indexation formula exceeded the hub price by more than a defined percentage (reported anecdotally to be in the range of 5–15 per cent), the buyer would receive a 'rebate' reflecting the difference. In 2012, Gazprom set aside 114bn RR ($3.8bn) for rebates, and stated that the equivalent figure could be 'somewhat more' in 2013.[43] Gazprom could then claim that it had not compromised on the principle of retaining oil-indexed prices; and although buyers agreed to pay an uncompetitive price for their gas, they had a guarantee that a significant part of any difference between oil-indexed and hub prices would be repaid to them at the end of the period. This means that, from 2012 onwards, the prices in Table 3.2 are misleading unless the details of the rebate mechanism for individual countries are known. Figure 3.4 shows that, as a result of this new mechanism, from the beginning of 2013 Gazprom's prices fell from a level of more than 30 per cent above NBP in 2010, to less than 5 per cent above it and, despite its public stance (referred to above) the company can therefore be said to have adjusted to hub prices.[44]

International price arbitrations

A major reason for the change in pricing at the start of 2012 was that Gazprom's major customers, desperate to stem their losses, had entered

---

[42] Gas Exporting Countries Forum, Moscow Declaration of the GECF, The Second Gas Summit of the Heads of State and Government of GECF Countries, 1 July, 2013, Moscow, Para 5.

[43] According to Gazpom CFO Kruglov, 'Retroactive payments', *Interfax Russia and CIS Oil and Gas Weekly*, 27 June–3 July, 2013, pp. 62–3.

[44] This is not necessarily the case for all European markets, but it certainly applies to all the major customers of Russian gas.

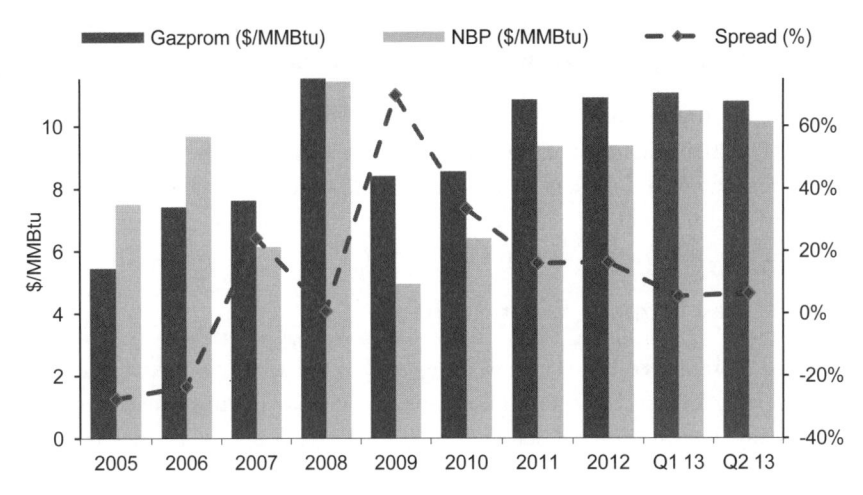

Figure 3.4 Russian gas export prices relative to NBP (month ahead)
Source: *European Gas Special*, Société Générale Cross Asset Research, 27 November 2013.

into international arbitration proceedings on the price terms of their long-term contracts. Over the approximately 40 year history of long-term European gas contracts, arbitrations had been extremely rare events. However, starting in the second half of the 2000s, they became much more frequent.[45] In 2010, the Italian company Edison initiated arbitral proceedings against the Gazprom/ENI joint venture Promgaz, although the case was subsequently settled. During 2011, Erdgas Import Salzburg, E.ON, and PGNiG announced that they had commenced arbitration proceedings with Gazprom, although agreement was reached on a new price mechanism (the P0 reduction combined with rebate mechanism described above) and proceedings were discontinued.[46]

In 2013, Gazprom's major arbitration case with the German company RWE, relating to the pricing of gas deliveries to the Czech Republic during 2010–12, reached a final judgement. These proceedings were extremely important in terms of both volumes and, potentially, in establishing a

---

[45] This applies not only to Gazprom but particularly to the Algerian company Sonatrach. As with all commercial issues related to European gas, it was the practice that even the existence of arbitral proceedings – let alone the result – was confidential. In the 2010s, more information about these proceedings has become available.

[46] 'Gazprom, E.ON revise contract retroactively from Q4 2010', *Interfax Gas Daily*, 3 July 2012, p. 5. In the Polish case, the price reduction was reported as 10%, resulting in a rebate of at least $420m for 2011. 'Gazprom, PGNiG agree gas-price adjustment, terminate arbitration proceedings', *Interfax Russia and CIS Oil and Gas Weekly*, 1–7 November 2012, p. 39.

benchmark for future litigation on oil-indexed gas pricing.[47] The judgement appeared to satisfy both parties, with RWE claiming that:

> ... the tribunal awarded RWE a reimbursement for payments made since May 2010 and adjusted the purchase price formula of the contract by also introducing an element of gas market indexation, which according to the arbitral tribunal, reflects the relevant conditions on the gas market at the time of the price revision in May 2010.[48]

By contrast Gazprom Export stated that:

> ... most of [RWE's] claims were rejected by the arbitration. As a result, the pricing formula has been adjusted in a certain way, but this adjustment is far away from the original demands made by RWE Supply & Trading CZ.[49]

The only conclusion that can be drawn from these statements is that the tribunal found that a significant proportion of the gas price should be determined by market (hub) prices.

Gazprom's price arbitration with Lithuania was ongoing at the time of writing, although there were discussions about whether this could be settled in the context of a new long-term contract (with a new price formula) starting in 2015.[50] Meanwhile Edison, which had settled similar proceedings in 2011, again filed for arbitration over prices in 2013 in the immediate aftermath of the RWE decision.[51]

---

[47] There are unconfirmed reports that the Tribunal awarded RWE a sum of €1.6bn. Note that the RWE price arbitration was additional to the arbitration on take-or-pay volumes (for the same contract) which was won by RWE with an award of €500m, although, at the time of writing, Gazprom was seeking to appeal the judgement. 'Gazprom unable to recover $500m from RWE under take or pay contract'; 'Court upholds ruling on Gazprom's take-or-pay claim against RWE, Russian company to appeal', *Interfax Russian & CIS Oil and Gas Weekly*, 18–25 October 2012, p. 58 and 11–17 July 2013, p. 32.

[48] 'Arbitration court rules in favour of RWE on price revision of its long-term gas supply contract with Gazprom', RWE Press Release, 27 June 2013.

[49] 'Gazprom Export on arbitration ruling on price review with RWE Supply & Trading CZ', Gazprom Export Press Release, 28 June 2013.

[50] 'Lietuvos Dujos H1 net profits rise 1.4%', *Interfax Russia and CIS Oil and Gas Weekly*, 25–31 July 2013, pp. 32–3. The price arbitration is in addition to another arbitral case which commenced in 2011, concerning the Lithuanian government's decision to unbundle Lietuvos Dujos's network assets into separate transmission (Amber Grid) and distribution businesses. Both of these proceedings are separate from the arbitration on heat tariffs at the Kaunas CHP plant.

[51] Anna Shirayevskaya and Tara Patel, 'Gazprom faces arbitration with Edison after RWE ruling', *Business Week*, 30 July 2013. www.businessweek.com/news/2013-07-30/gazprom-faces-arbitration-with-edison-after-ruling-on-rwe-price.

The 2012 EU competition investigation and Gazprom's export monopoly

Highly relevant to Gazprom's price renegotiations and arbitrations – although not *directly* connected with them – is the investigation by the European Commission's Competition Directorate (DG COMP) into Gazprom sales to eight Central and East European countries. One year after unannounced inspections of corporate records, DG COMP announced in September 2012 that it was opening proceedings against Gazprom for possible abuse of a dominant position on three grounds. First, Gazprom may have divided gas markets by hindering the free flow of gas across EU member states. Second, Gazprom may have prevented the diversification of supply of gas. Finally, Gazprom may have imposed unfair prices on its customers by linking the price of gas to oil prices.[52] In October 2013, DG COMP made it known that it was preparing a 'statement of objections' against Gazprom, but gave no indication as to how widespread the charges, and what the proposed remedies, might be.[53]

The case may have been somewhat over-hyped as 'the anti-trust clash of the decade', with the possibility of intransigent European Competition authorities and an equally intransigent Kremlin provoking a major crisis in EU–Russia energy and natural gas relations.[54] This is a possible scenario but not the most likely one. It is more likely that a DG COMP judgement against Gazprom would be followed by an appeal to the European Court of Justice (ECJ), which could take several years to resolve. However, any DG COMP finding that oil-linked prices are anti-competitive – even if an appeal against it were to be heard at the European Court of Justice – would severely damage what remains of Gazprom's ability to insist on retaining this mechanism, at least for sales to EU countries. Moreover, potential for confrontation remains. This was highlighted by the General Director of Gazprom Export, who made a connection between the investigation and the European price issue, asking:

> ... could this be an attempt to use political methods to secure a reduction in Russian prices by artificial means?[55]

---

[52] 'Antitrust: Commission opens proceedings against Gazprom', EU Commission Press Release, 4 September 2012. http://europa.eu/rapid/press-release_IP-12-937_en.htm.

[53] 'EC preparing anti-trust charges against Gazprom', *Platts European Gas Daily*, 4 October 2013, p. 1.

[54] For details and scenarios see Riley, A. 'Commission v. Gazprom: the antitrust clash of the decade?' CEPS Policy Brief No. 285, 2012 and Sartori, N. 'The European Commission vs. Gazprom: an issue of fair competition or a foreign policy quarrel?', IAI Working Papers 13 March, 2013.

[55] 'Role of pipeline gas in Europe to grow: LNG overestimated, *Interfax Russia and CIS Oil and Gas Weekly*, 30 May–5 June 2013, pp. 29–34.

This is not the first EU Competition investigation into Gazprom contracts.[56] In the early 2000s, DG COMP opened an investigation into 'territorial sales restrictions' (more commonly known as 'destination clauses') in long-term gas contracts. At that time it was made abundantly clear that these were hard-core restrictions of competition, and illegal under European law. The relevance for the current investigation is that similar restrictions appear to be present in Gazprom's long-term contracts with the Baltic States, and should have been removed when those countries became EU member states in 2007.[57] If they are present, this represents a serious problem for Gazprom, which can hardly plead ignorance of the legal situation.

However, the immediate reaction of the Putin administration to the DG COMP investigation was to issue an executive order stating that Russian commercial entities such as Gazprom:

> ... should supply information on their activities [...] upon request from the authorities and agencies of foreign countries ... only subject to prior consent of a respective federal executive body authorised by the Russian Government.[58]

In European circles, this order was generally interpreted as a direct instruction to Gazprom not to cooperate with the DG COMP investigation, although it is probably more correct to see it as a traditional Putin response, standing up for Russian interests against 'foreign intimidation'.

Questions have been raised as to how the competition authorities could construe linking the price of gas to oil prices as 'unfair', given that this had been the principal way of pricing gas in Europe for several decades prior to 2008. Part of the answer appears to be that for countries which had a choice of suppliers and could choose which gas to purchase, oil price linkage was not unfair. But for countries that were entirely dependent on Russian gas, a refusal to sell on terms other than oil linkage could potentially – under competition law – be considered abuse of a dominant position. This is reinforced by the export monopoly, which means that countries dependent on Russian gas have no alternative to buying from Gazprom.

In 2006, the legal status of Gazprom – or to be specific, the company's subsidiary Gazprom Export – in relation to exports, changed. Previously, Gazprom had a de facto, but not a legal, monopoly of gas exports. However,

---

[56] However, in that instance the investigation also involved contracts with other gas suppliers (notably Algeria), whereas this case involves only Gazprom contracts. For details see Stern, J.P. *The Future of Russian Gas and Gazprom*, op. cit., pp. 132–4.

[57] 'Gazprom allows Baltic customers to trade gas with third countries – paper', *Interfax Russia and CIS Oil and Gas Weekly*, 24–30 October 2013, p. 52.

[58] Executive Order 'On Measures to Protect Russian Federation Interests in Russian Legal Entities' *Foreign Economic Activities*. http://eng.kremlin.ru/acts/4401.

Russian Responses to Commercial Change in European Gas Markets 69

in the mid-1990s, companies such as Itera began to export gas to CIS countries and in the mid-2000s, Gazprom's de facto monopoly of exports to Europe was breached by Eural Transgas. Partly in response to these developments, and partly in response to the January 2006 Ukraine crisis, the 2006 Law on Gas Exports gave Gazprom a legal monopoly on exports.[59] Until 2012 there was no serious challenge to the export monopoly, aside from Novatek's 10 year contract to sell 2 Bcm/year of gas to the German company EnBW. Initially this gas will come from non-Russian sources, but Novatek made clear that it plans to

> ... begin [the] initial process to create a consumer market for LNG and natural gas in Europe ... [and] establish a market position when legislative changes are eventually made to exports of natural gas [from Russia].[60]

In relation to LNG, Gazprom's export monopoly was removed at the end of 2013, which would allow Novatek to supply EnBW with gas from the Yamal LNG project (Chapter 13).[61] The legislation allows two categories of companies to export LNG:[62]

- companies developing deposits of federal importance whose licence states that as of 1 January 2013 they are required to build LNG plants or send gas to be liquefied at other LNG plants;
- companies (and their subsidiaries) in which the state owns a stake of more than 50 per cent and which are developing offshore fields to produce LNG or gas produced under production-sharing agreements.

From this, it is clear that there is no blanket authorization for LNG exports and that relaxation of the monopoly over pipeline gas exports to Europe is not on the agenda. On the other hand, few would agree with the General Director of Gazprom Export that:

---

[59] *Federal'nyi zakon Rossiskoi Federatsii ot 18 Iuliia 2006 g. 117-F3, Ob Eksporte Gaza*. To be precise: Article 3 of the law confers an exclusive right to export gas on the owner of the Unified Gas Supply System and its subsidiary companies, currently Gazprom and Gazprom Export.

[60] 'Novatek, EnBW sign long term supply deal', *Platts European Gas Daily*, 15 August 2012, p. 4.

[61] 'Russia to liberalise exports soon – Putin', *Interfax Russia and CIS Oil and Gas Weekly*, 3–9 October, 2013 pp. 26–7. Although at the time of writing it was not clear whether this would take the form of a blanket – or a case by case – exemption for LNG.

[62] 'Russian government approves bill on LNG export liberalisation', *Interfax Russia and CIS Oil and Gas Weekly*, 24–30 October pp. 12–13.

... a unified export channel will be maintained not only throughout the life span of the current generation, but throughout the life spans of our children, grandchildren and great grandchildren.[63]

President Putin himself was far less definite and, when prime minister, did not rule out relaxation of the export monopoly in the future.[64] But his Executive Order in response to the DG COMP investigation formalizes the Ministry of Energy's approval of negotiations and renegotiations of long-term contracts and makes the Ministry part of the 'single export channel'. It also implies that, even should some type of unbundling of Gazprom's network be contemplated in the future (see 'Gas transmission and storage', page 136), the government itself might maintain a single export channel.

The transit diversification and avoidance pipelines

The break-up of the Soviet Union caused very substantial problems for the Russian gas industry. Not only were important assets suddenly located in different sovereign states, but what had been intra-industry transfers of gas and money needed to be commercialized. This caused immense problems, the consequences of which continued to be felt through the first two decades of the post-Soviet era, and some remained unresolved into the 2010s. The main transit corridor, carrying around 80 per cent of Russian gas to Europe, was through Ukraine; Moldova (an extension of the Ukrainian system to south-east Europe and Turkey) and Belarus (transiting gas to Poland and the Baltic States) were also important (see Chapter 7).

Yamal–Europe

The commercial and political difficulties in the gas relationships between Russia and the three western CIS transit countries have been extensively documented in previous OIES research.[65] As these problems unfolded in the 1990s so Gazprom began to consider gas delivery systems to Europe which would lessen its dependence on Ukraine. The first was the Yamal–Europe pipeline (Map 3.1) running from Torzhok in Russia across Belarus

---

[63] 'Role of pipeline gas in Europe to grow: LNG overestimated, *Interfax Russia and CIS Oil and Gas Weekly*, 30 May–5 June 2013, pp. 29–34.
[64] 'Russia might liberalise gas exports in future, Putin', *Interfax Russia and CIS Oil and Gas Weekly*, 6–14 October 2011, p. 49.
[65] Yafimava, K. *The Transit Dimension of EU Energy Security: Russian Gas Transit Across Ukraine, Belarus, and Moldova* (Oxford: OIES/OUP. 2011), Chapters 6–8; Stern, J.P. *The Future of Russian Gas and Gazprom*, op. cit., Chapter 2.

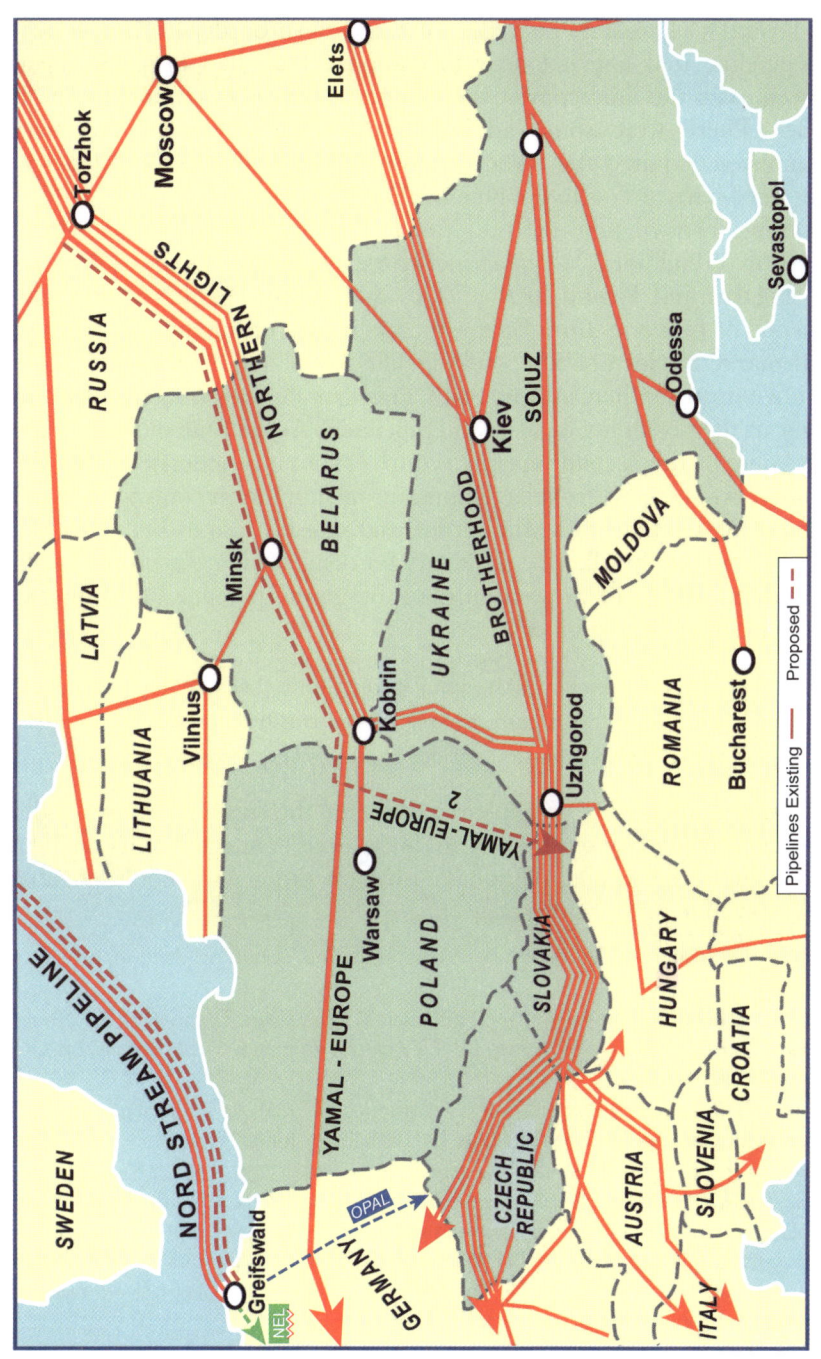

Russian Responses to Commercial Change in European Gas Markets 71

Map 3.1: The Ukrainian and Yamal–Europe pipelines
Source: Oxford Institute for Energy Studies.

and Poland to Frankfurt on Oder on the eastern border of Germany.[66] The pipeline took much longer to complete than had originally been foreseen, with the final compressor station becoming operational only in 2006.[67] There were some additional problems when the European Commission required the owner of the Polish section (Europol Gaz) to grant third-party access to the pipeline.[68]

During construction of the Yamal–Europe line, there had been discussion of building a second string. However, due to regular gas disputes with Belarus and Poland in the 1990s and 2000s, Poland's contractual reduction of deliveries from Gazprom, and its lack of intention to increase gas demand, the idea of a second pipeline was abandoned.[69] It therefore came as a surprise when, in April 2013, President Putin instructed Gazprom to look at the possibility of a second pipeline.[70] At the time of writing the exact route of this second pipeline is unclear; but it is understood that the intention is to connect to Velke Kapusany, the long-term contract delivery point on the Ukraine–Slovakia border, and possibly also to Beregovo, the delivery point on the Ukraine–Hungary border, allowing Gazprom to fulfil its contracts to both of those countries through this pipeline. According to Gazprom's CEO, the project:

> … does not require any high capital expenditures [because the first line was built with] an infrastructure reserve for the construction of a second line [and hence the construction period] would be very, very short.[71]

However, in late 2013 it appeared that even the feasibility study for the project was being blocked by Poland, for reasons connected to its political relationship with Ukraine. Without Polish cooperation it will not be possible for Gazprom to progress this project.

---

[66] For more technical details see www.gazprom.com/about/production/projects/pipelines/yamal-evropa/.

[67] For more details of the history see Yafimava, K. *The Transit Dimension of EU Energy Security*, op. cit., pp. 218–20; Stern, J.P. *The Future of Russian Gas and Gazprom*, op. cit., pp. 118–20.

[68] 'Commission requests Poland to stop violation of EU rules on internal gas market', European Commission Press Release, IP/10/945, 14 July 2010.

[69] Stern, J.P. *The Future of Russian Gas and Gazprom*, op.cit., pp. 120–2; Yafimava, K. *The Transit Dimension of EU Energy Security*, op. cit., pp. 218–49. Negotiations with Poland were also not helped by the controversy over the ownership of Europol Gaz (the company which owns the Polish section of the line). 'Gazprom, PGNiG unable to agree on appraisal of 4% in Europol Gaz, *Interfax Russia and CIS Oil and Gas Weekly*, 5–11 July 2012, p. 54.

[70] www.gazprom.com/about/production/projects/pipelines/yamal-evropa-2/.

[71] 'Gazprom sees agreement with Poland giving edge to Yamal–Europe 2 project', *Interfax Russia and CIS Oil and Gas Weekly*, 9–15 May 2013, p. 41.

Yamal–Europe is a transit-diversification pipeline which Gazprom expected would have the effect of demonstrating to Ukraine that, unless it changed its behaviour, it would lose lucrative gas transit business. However, not only did this seem to have little impact on Ukrainian policy, but periodic Belarusian transit crises caused Moscow to conclude that diversifying transit between Ukraine and Belarus was insufficient to solve its problems, and that a means of avoiding transit through both of these countries was required.[72]

Nord Stream

Because transit security through Ukraine and Belarus had remained poor throughout the post-Soviet period, Gazprom periodically considered the possibility of a direct pipeline from Russia to Germany across the Baltic Sea. In 1997 a joint venture was established between Gazprom and the Finnish company Neste to carry out a feasibility study of the offshore part of what had been known as North Transgas; this then became known as the North European Pipeline, before finally becoming the Nord Stream pipeline.[73] Construction began in April 2010, in the aftermath of the 2009 Russia–Ukraine gas crisis. The original 1990s concept was that the Shtokman field in the Barents Sea should provide gas for Nord Stream in addition to an LNG export terminal at Murmansk (see Chapter 11).[74] During the 2000s it became clear that a combination of delays to that project – principally due to uncertainties surrounding foreign participation and the commercial viability of LNG exports – meant that if Nord Stream were to begin operating in the early 2010s, then its gas supply would need to be sourced from elsewhere. The Nord Stream consortium – of Gazprom, E.ON Ruhrgas, and Wintershall (subsequently joined by GDF Suez and Gasunie) – was established in 2005 to build two pipelines each with a capacity of 27.5 Bcm/year (Map 3.2).[75] The first Nord Stream pipeline went into operation in November 2011, followed by the second line one year later, supplied with gas from Western Siberia. While the two pipelines

---

[72] See Yafimava, K., 'The 2007 Russia–Belarus gas agreement', Oxford Energy Comment, OIES, January, 2007 and Yafimava, K., 'The June 2010 Russian–Belarusian gas transit dispute: a surprise that was to be expected', Working Paper NG43, OIES, 1 July, 2010 for details of Belarus crises.

[73] For some early history of the pipeline see Stern, J.P. *The Future of Russian Gas and Gazprom*, op. cit., pp.120–2; Yafimava, K. *The Transit Dimension of EU Energy Security*, op. cit., pp. 94–6.

[74] For details of the Shtokman project and LNG exports see Chapters 8 and 10.

[75] A great deal of historical, technical, and corporate information on Nord Stream can be found on the company's website www.nord-stream.com/. Following the completion of the first two lines, claims have been made that the capacity of the system is in fact closer to 60 Bcm/year.

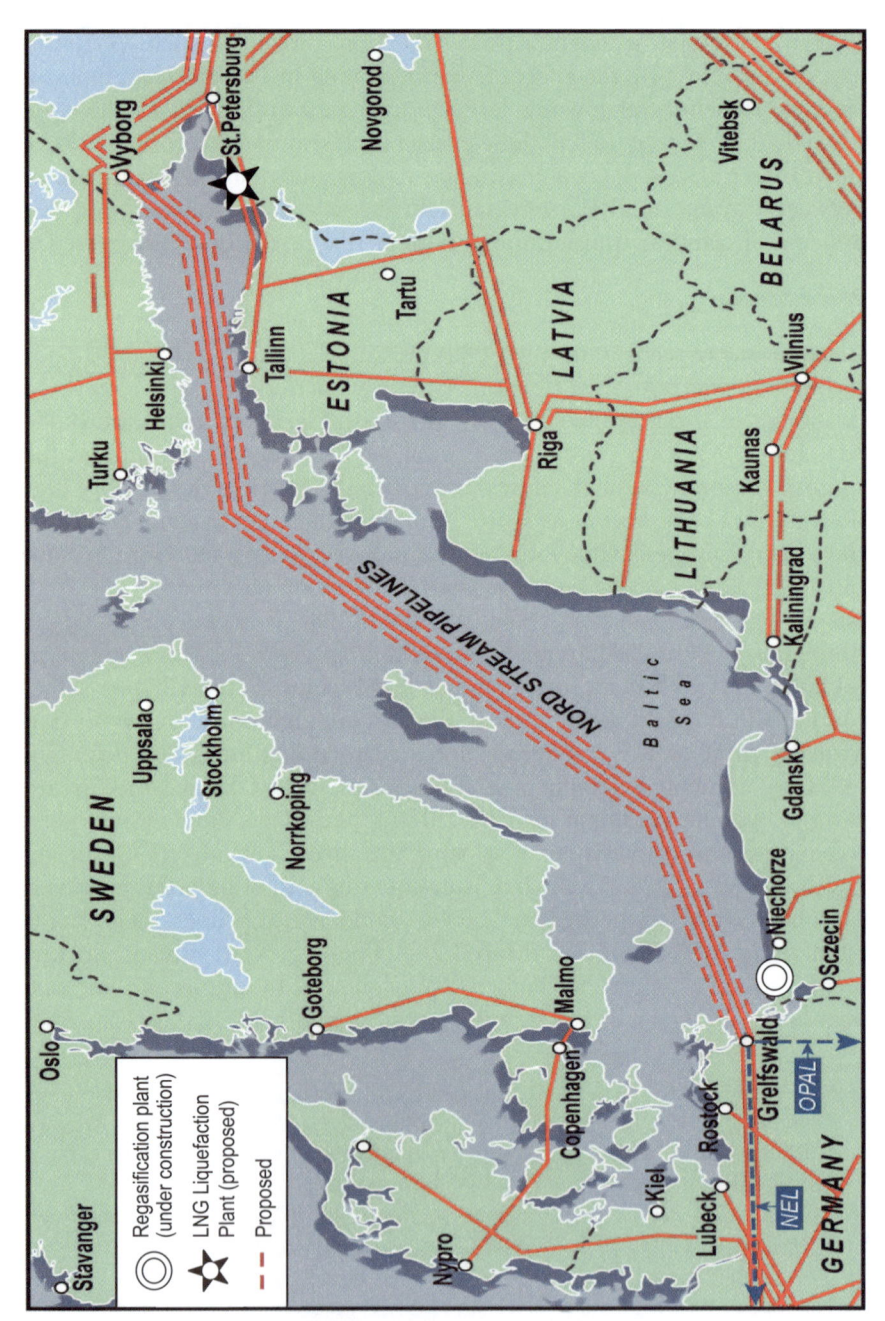

Map 3.2: The Nord Stream pipelines
Source: Oxford Institute for Energy Studies.

have a total capacity of 55 Bcm/year, delays and problems in utilizing the capacity of the onshore pipelines (NEL and OPAL) (see Chapter 4) have meant that until late-2013, capacity utilization remained at relatively low levels. In 2012, Gazprom proposed building two additional lines, Nord Stream 3 and 4, which would double the capacity of this corridor to 110 Bcm/year, with project documentation showing a preliminary timeline for both pipelines to be commissioned by the end of 2018.[76]

Blue Stream and South Stream

Security concerns in Turkey (one of Gazprom's newest and potentially biggest markets) caused by transit problems through Ukraine led to the construction of the Blue Stream pipeline across the Black Sea (Map 3.3).[77] This was a technically challenging project due to significant water depth (more than 2000 metres) and difficult corrosion and seabed conditions. Gas supplies through the pipeline commenced in 2003. Although Blue Stream carried only relatively small volumes for many years, by 2012 these had risen to 14.7 Bcm out of a capacity of 16 Bcm.

For a few years after it was commissioned there were discussions about building additional strings of Blue Stream, but in 2006 Gazprom and the Italian company ENI announced a joint venture to create the much larger South Stream pipeline system across the Black Sea.[78] Since then, EDF and Wintershall have joined the offshore section of South Stream. The offshore section is technically challenging – four pipelines, 930 km in length, to be laid from Anapa on the Russian Black Sea coast to Varna in Bulgaria, in water depths up to 2250 metres (Map 3.3). Originally the project was planned to be two lines with a capacity of 31 Bcm/year but following the January 2009 Russia–Ukraine crisis, it was expanded to four lines and 63 Bcm/year. The current schedule is that gas will flow through the first pipeline by the end of 2015, and full capacity will be reached by the end of 2017. Gazprom announced in November 2012 that a final investment

---

[76] There have been suggestions that one of these lines might be built directly to the UK, although this seems unnecessary given the large volume of existing transportation capacity between the UK and the Continent. Nord Stream AG, *Nord Stream Extension: project information document (PID)*, March 2013 www.nord-stream.com/extension/.

[77] For more details see www.gazprom.com/about/production/projects/pipelines/blue-stream/.

[78] A memorandum of understanding was signed the following year. A great deal of historical, technical, and corporate information on South Stream can be found on the Gazprom website www.gazprom.com/about/production/projects/pipelines/south-stream/, the South Stream offshore website www.south-stream-offshore.com/about-us/, and the South Stream onshore website www.south-stream.info/en/pipeline/.

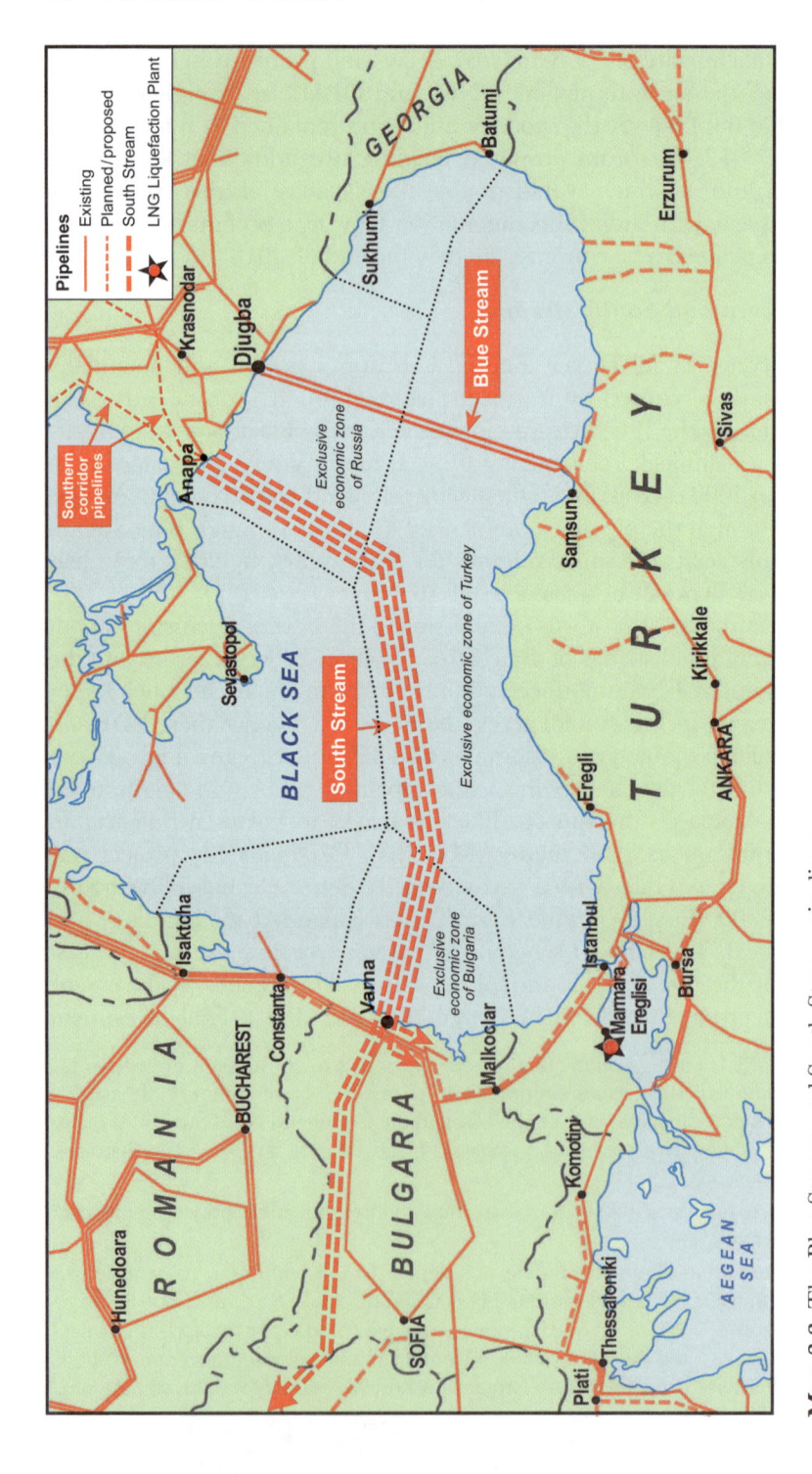

Map 3.3: The Blue Stream and South Stream pipelines
Source: Oxford Institute for Energy Studies.

decision had been taken on the offshore section, and in 2013 tenders were issued for the steel and the laybarge(s) for the four pipelines and Gazprom Export signed a ship or pay agreement for the entire 63 Bcm/year capacity.[79]

As for the onshore European section of South Stream, during 2008–10 Russia signed intergovernmental agreements with seven European countries for the onshore section(s).[80] Although indicative route maps can be found on the company's website, at the time of writing no onshore route has been finally decided. Initial plans to build a southern leg to Greece and a northern leg to Austria had been abandoned due to lack of demand and an unfriendly regulatory environment (respectively). The onshore section of the project across Europe is facing significant regulatory challenges due to the implementation of the Third Energy Package and the Network Codes (see Chapter 4).

Much scepticism was expressed in Europe about the likelihood of South Stream being built and, as mentioned above, the project still has hurdles to surmount. However, construction of what the company is calling its 'Southern Corridor' project, comprising 2500 km of pipeline and 10 compressor stations to bring West Siberian gas to the Black Sea to supply South Stream, started in 2013.[81] The cost of this corridor is estimated at $17bn and these preparations, combined with tenders for equipment and construction of the offshore section, confirm that South Stream will go ahead, although there may still be uncertainty about the timetable of construction and gas flow.[82]

Motivations for transit avoidance pipelines

All of the transit avoidance pipelines are owned roughly 50 per cent (usually plus one share) by Gazprom and 50 per cent by a consortium of importing European utilities.[83] These pipelines are mainly replacing gas transportation

---

[79] Gazprom Press Releases, 'FID taken for South Stream offshore pipeline', 12 November 2012; 'South Stream Transport B.V. shareholders confirm terms and conditions of South Stream's offshore section construction', 4 October 2013.

[80] Hungary, Bulgaria, Serbia, Greece, Slovenia, Croatia, and Macedonia.

[81] Not to be confused with Southern Corridor pipeline from Central Asian/Caspian and Middle East countries to Europe. For details and maps see www.gazprom.com/about/production/projects/pipelines/southern-corridor/.

[82] Construction of the Bulgarian section of the line began in October 2013. 'Gazprom mulls raising 100 bln rubles with infrastructure bonds – paper', *Interfax Russian & CIS Oil and Gas Weekly*, 18–24 July 2013, p. 23; 'South Stream construction starts in Bulgaria', Gazprom Press Release, 31 October 2013.

[83] The partners in the pipelines are: Blue Stream: Gazprom 50%, ENI 50%; Nord Stream: Gazprom 51% E.ON 15.5%, Wintershall Holding 15.5%, GDF Suez 9%, Gasunie 9%; South Stream: Gazprom 50%, ENI 20%, EDF 15%, Wintershall 15%. The ownership of Europol Gas (Polish section of the Yamal–Europe pipeline) is 48% Gazprom, 48% PGNiG and 4% GazTrading, but in 2013 – following the death of the owner of GazTrading – the other two partners were negotiating to purchase its shareholding. 'PGNiG, Gazprom to discuss Yamal stakes', *Platts European Gas Daily*, 8 October 2013, p. 4.

via Ukraine and will therefore neither serve new markets nor provide for additional sales to existing markets. Moreover, there is uncertainty as to how rapidly demand for Russian gas in Europe will grow. All these circumstances have led to substantial discussion as to whether these ventures will be profitable or rather what the financial community refers to as 'value-destroying'. Research has shown that Nord Stream should earn a positive rate of return, but that the profitability of South Stream is considerably more doubtful.[84]

This raises the question as to why foreign partners are willing to invest in these projects, to which one answer is that foreign companies find it difficult to resist an invitation from Gazprom (and the Russian government) to invest in a project, due to their keenness to invest in upstream gas projects in Russia. For Nord Stream, the crucial element which persuaded E.ON Ruhrgas and Wintershall (the original investors) to participate was equity ownership in the Iuzhno-Russkoe gas field in Western Siberia; the only time that Gazprom has allowed foreign investment in a large Cenomanian gas field.[85] In addition, it is Gazprom which takes the major tariff risk in these pipelines, having signed ship-or-pay agreements with their owners for 100 per cent of their capacity.[86]

The transit-avoidance pipelines, in particular Nord Stream which has not involved any additional sales of gas to Europe since the first line was completed at the end of 2011 (see Table 3.3) have been an extremely costly exercise for Gazprom. In addition, as Mikhail Korchemkin has shown, even with new pipelines such as Nord Stream 1 and 2 in operation, Gazprom is still required to honour its ship-or-pay commitments in its existing transportation contracts with Slovakia and the Czech Republic, hence paying twice for transportation.[87] While Gazprom claims that 'at least a quarter of South Stream pipeline's capacity will be filled with gas

---

[84] For Nord Stream see Chyong, C.K., Noel, P., and Reiner, D.M., 'The Economics of the Nord Stream Pipeline System', University of Cambridge, EPRG Working Paper 1026, 2010; also 'Nord Stream internal rate of return 10% – sources', *Interfax Russian & CIS Oil and Gas Weekly,* 4–10 July 2013, p. 41.

[85] The importance of this field is that it produces large volumes of gas from shallow horizons; all other joint ventures with foreign companies involve fields with some degree of difficulty involving location, gas composition or depth. Both German companies, through asset swaps, have acquired a 25% share in Sevmorneftegaz which owns the licence to the Iuzhno-Russkoe field with 834 Bcm of proven and probable reserves and 180 Bcm of possible reserves (www.gazprom.com/about/production/projects/deposits/yrm/).

[86] 'Gazprom signs agreement on South Stream gas transportation', *Interfax Russia and CIS Oil and Gas Weekly*, 26 September–2 October 2013, p. 36.

[87] 'In 2012, the new export route to Waidhaus was four times more expensive than the old one', *East European Gas Analysis*, 1 June 2013.

under new contracts', this seems implausible, as finding an additional 15–16 Bcm/year of gas demand in south and south-eastern Europe does not seem possible, unless new contracts with Turkey can be concluded.[88] This means that the principal – if not the sole – purpose of South Stream (and of the majority of the rest of the transit-avoidance pipeline capacity) is to take gas currently flowing through Ukraine and transport it to the same customers via alternative routes. The cost can be justified only by the view that continued dependence on Ukraine is considered economically unattractive and/or posing an unacceptable security risk (see Chapter 4).

While the transit-avoidance pipelines are clearly projects which are dominated, in a corporate sense, by Gazprom, there is a very clear inference that President Putin takes a strong personal interest in their development and may be considered a key (and probably *the* key) political driving force behind them. Over the past decade, Putin appears to have taken a very public and personal role in encouraging – to the point of publicly ordering – Gazprom to create and accelerate the construction of Nord Stream, South Stream, and most recently Yamal–Europe 2.[89] The question therefore arises as to why Putin feels the need to play this role. For some, the answer is straightforward and is connected with widespread allegations that the political elite receive corrupt payments via equipment and service contracts for these large pipelines.[90] However, there are other major reasons, such as the determination of the Russian government to avoid dependence on Ukraine, and the importance of oil and gas supplies to the maintenance of Russia's geopolitical standing – to which we return in Chapter 4.

Export capacity utilization and contractual commitments

Table 3.3 provides an illustration of likely export capacity to Europe up to 2020 and compares this with average annual take-or-pay volumes under long-term contracts during this period. Up to 2015 it is assumed that no new capacity will be built (with the start of South Stream deliveries planned for the end of that year), and no significant investment in the Ukrainian network will mean that operational capacity will fall to 100 Bcm/year. In the minimum 2020 variant, it is assumed that four strings of South Stream will be built, but no other new capacity; and the operational capacity of the Ukrainian network will decline to 90 Bcm – or 50 Bcm if a decision is taken to close capacity. In the maximum 2020 variant it is assumed that all

---

[88] 'Construction of South Stream pipeline officially underway', *Interfax Russia and CIS Oil and Gas Weekly*, 6–12 December 2012, pp. 5–7.

[89] The same can also be said of LNG projects and Eastern Gas Programme pipelines. See Chapters 8 and 12.

[90] See the section on reform and governance in Chapter 12.

80 The Russian Gas Matrix

proposed pipelines – four South Stream, two additional Nord Stream, and another string of Yamal–Europe (via Belarus) – are built. The minimum and maximum capacity figures are then compared with two levels of exports: take-or-pay levels of 70 per cent and 85 per cent of annual contract quantity (ACQ) and 100 per cent of ACQ, the latter being significantly above the levels of physical delivery of Russian gas to Europe during 2009–12 (see Table 3.1).

Table 3.3: Russian gas export pipeline capacity to Europe, 2013–20 (Bcm/year)

Pipeline capacity	2013	2015	2020 Minimum	2020 Maximum
To Finland	5	5	5	5
Through				
Belarus	48	48	48	63
Ukraine*	120	100	50	90
Blue Stream	16	16	16	16
Nord Stream	55	55	55	110
South Stream	0	0	32	63
Total	**244**	**224**	**237**	**347**
Surplus capacity at ToP level**	81–110	63–91	84–109	199–224
Surplus capacity at ACQ level**	52	34	58	173

* Useable Ukrainian capacity is estimated at 120 Bcm/year and, without substantial investment, is likely to decline during the 2010s. In 2013 the Ukrainian prime minister suggested that if Russian utilization fell below a certain level, two lines could be closed leaving only 50 Bcm of capacity.

** ACQ and ToP levels of 70–85 per cent are based on data from Figure 3.3 and therefore somewhat over-stated because not all contracts are included.

Source: adapted from Pirani, S. (ed.), *Russian and CIS gas markets and their impact on Europe* (Oxford: OIES/OUP, 2009), Table 12.8, p. 427.

Table 3.3 makes clear that, even with no further pipeline capacity availability in 2015, under most circumstances Gazprom will have substantial surplus export capacity throughout the decade.[91] The only situation in which Gazprom might find itself somewhat pressed would be if in 2020 it had built only South Stream, and Ukraine had carried through

---

[91] See Chapter 11 for a discussion of Gazprom's gas availability over this period.

the suggestion of its prime minister that it might close some unused capacity, leaving only 50 Bcm available.[92] However, this would imply a European import requirement of well above 180 Bcm/year which, while possible, is not likely. But if, on the other hand, Gazprom has four South Stream lines in operation by 2020 – which currently is the likely outcome as long as it is allowed to use the full capacity of these lines on-land (which as we shall see in Chapter 4 is by no means certain) – then, as long as 50 Bcm of Ukrainian capacity remains available, and even without additional Nord Stream and Yamal–Europe lines, it should be able to cope comfortably with increased European requirements. Only if (a) Ukrainian capacity is completely shut down, or Gazprom chooses not to use it, and (b) European demand for Russian gas rises above 200 Bcm/year, could the additional Nord Stream 3 and 4 and/or Yamal–Europe 2 capacity be needed.

Therefore, from a political perspective, during the 2010s Gazprom becomes progressively less dependent on Ukrainian transit capacity to Europe. By 2020 it can reduce transit volumes through Ukraine to very low levels. Unless it has reason to increase exports to around 200 Bcm/year, which market conditions currently do not suggest will be required, it could completely abandon Ukraine as a transit route for European exports. Indeed Aleksei Miller, Gazprom CEO, observed that:

> ... when South Stream reaches its design capacity, Gazprom's transit risks for supply to its European consumers will be brought down to almost zero.[93]

But in two important respects this analysis is much too simple: first it assumes that if Gazprom builds new pipelines on EU territory, it can then reserve 100 per cent of their capacity for Russian gas. Second, it shows only *annual* capacity figures, whereas it is becoming increasingly important to know how much Russian gas can be delivered in the winter season – particularly on days of high demand (in Russia, CIS countries, and Europe). Discussion of these issues and overall conclusions on Russian gas exports to Europe can be found in Chapter 4.

---

[92] 'Ukraine may stop buying Russian gas due to high price – Azarov', *Interfax Russia and CIS Oil and Gas Weekly*, 19–25 September 2013, pp. 45–7.
[93] Gazprom Annual Report, 2012, p. 5.

CHAPTER 4

THE IMPACT OF EUROPEAN REGULATION AND POLICY ON RUSSIAN GAS EXPORTS AND PIPELINES

Jonathan Stern

This chapter, a companion to Chapter 3, discusses the impact on Russia's gas exports to Europe of the changing legal and regulatory environment, and of broader political, economic, and geopolitical issues. First, by way of introduction, we briefly outline the evolution up to the mid-2000s of EU–Russia energy relations. The first section then shows the importance of European regulation for Russian gas exports and pipelines to Europe. This is followed by a section dealing with European perceptions and realities of the commercial, political, geopolitical, and security dimensions of Russian gas. The final section summarizes and draws conclusions for both this and the previous chapter.

The 1994 Agreement on Partnership and Cooperation formed the basis for EU–Russia institutional cooperation in the energy sector, and this agreement led to the establishment of the EU–Russia Energy Dialogue (also known as the 'Prodi Initiative') in 2000. The Dialogue process was reinforced by the Trans-European Networks (TENS) initiative which had already led to the Yamal–Europe pipeline, the Shtokman field, and (the North European pipeline which was the forerunner of) Nord Stream being designated as 'priority projects' eligible for funding under TENS.

While little progress has been made towards creating a new Partnership and Cooperation Agreement, in March 2013 the two sides signed an EU–Russia Energy Cooperation Roadmap to 2050, with sections dealing with all major energy sources.[1] The gas section set out tasks to be completed by specified dates, with the major recommendation for 2020 being the mitigation of supply–demand, infrastructure/regulatory, and political risks for both sides in EU–Russia gas trade.

EU policy and regulation to promote liberalization and competition in natural gas markets is a multi-faceted subject, and its unfolding and implementation over the past two decades has been (to say the least) a long and winding road. Although the future direction of travel has been set, much of the detail has yet to be finalized. In the 1990s, the Russian Federation was preoccupied with establishing a post-Soviet order in relation to gas, and the

---

[1] EU, *Roadmap: EU–Russia Energy Cooperation Roadmap Until 2050*, The Coordinators of the EU–Russia Energy Dialogue, March 2013.

EU was struggling to achieve first steps towards liberalization of the gas industry, which finally culminated in the First Gas Directive of 1998.[2] During this period, Russian interest in EU-related pipeline regulation focused almost entirely on the Energy Charter Treaty and its Transit Protocol.[3]

By the time that the Second Gas Directive was passed in 2003, the new Gazprom management under Aleksei Miller had begun to pay more attention to the process, and condemned the EU liberalization initiative in general terms as an attack on long-term gas sales contracts which would be damaging to EU energy security.[4] Russian doubts about the liberalization process were reinforced by the 2005 auctions for expanded capacity in the Trans Austria Gasleitung (TAG) pipeline through Austria, which resulted in 149 parties (the majority of which had no access to gas supply) being allocated capacity on a pro rata basis; this was followed by a lottery which resulted in 29 capacity winners.[5] Although this very early experience of capacity auctions has never been repeated, it seriously damaged the credibility of the initial process for Gazprom, and signalled the start of growing problems between the EU and Russia on the regulation of gas transportation; problems which have become increasingly complex with the implementation of the Third Energy Package in the 2010s.

The impact on Russian gas exports of EU policy and regulation

The Third Energy Package and the Gas Target Model

Following the European Commission's adoption in 2009 of the Third Energy Package, which included both the Third Gas Directive and Regulation 715

---

[2] Directive 98/30/EC of 22 June, 1998 *Concerning Common Rules for the Internal Market in Natural Gas*. For details of the tortuous route to passing this Directive in the 1990s see Stern, J.P., *Competition and Liberalization in European Gas Markets: a diversity of models* (London: RIIA, 1998), Chapter 4.

[3] For details see Yafimava, K., *The Transit Dimension of EU Energy Security: Russian Gas Transit Across Ukraine, Belarus, and Moldova* (Oxford: OIES/OUP, 2011) especially Chapter 9.

[4] Directive 2003/55/EC of the European Parliament and of the Council, 26 June 2003, concerning common rules for the internal market in natural gas, Official Journal L 176, 15.7.2003, 57. For details of the passing of this Directive in the 2000s, see Haase, N., *European Gas Market Liberalisation* (Groningen: Energy Delta Institute, 2009), Chapters 6–8. For Gazprom comments in the early 2000s see Stern, J.P., *The Future of Russian Gas and Gazprom* (Oxford: OIES/OUP, 2005), pp. 130–2.

[5] For a complete account of this episode see Konoplyanik, A.A., 'Pravovye aspekty protseduri nediskriminatsionnogo konkurentnogo dostupa k svobodnym moshchnostiam transportirovki (DEKH, TAG i ECG)', *Neftegaz, Energetika i Zakonodatel'stvo*, vypusk 8, 2009, pp. 142–56.

on access to gas networks, both the Russian government and Gazprom started to pay close attention to the liberalization process. It became increasingly clear that, notwithstanding reservations from many parties (both European and non-European) this was an unstoppable process with very important consequences for all gas stakeholders (some of which have been described in Chapter 3 in relation to commercial and structural changes in the industry).[6]

The most important elements of the Third Directive and Regulation 715 are 'unbundling' (the separation of gas supply from transportation services) and the new (national but especially) EU network codes, created to regulate cross border transportation. The principles or 'vision' of how this new system would work was set out in a Gas Target Model.[7] European gas transportation has moved from a point-to-point (PP) system with delivery points largely at the borders of countries, to an entry/exit (EE) system with capacity bookings at interconnection points (IPs) between different entry and exit zones. Delivery of gas will be largely at virtual trading points (VTPs or hubs). Transportation capacity at IPs will be auctioned for up to 15 years ahead in annual and monthly tranches.

The transition to this new system raised much greater problems for Gazprom than for other suppliers of gas to the European market for three reasons: the size of the volumes of Russian gas requiring delivery to European customers under long-term contracts; the duration of those contracts, some of which do not expire until the 2030s; and the number of borders which these volumes need to cross, in east–west and east–south directions. Table 4.1 shows that in 2011, almost three-quarters of Russian gas delivered to EU countries crossed more than one EU border, while 52 Bcm crossed three or more borders. An additional challenge – again different to the vast majority of other suppliers – is Gazprom's ambition to build major new pipelines to (and through) Europe at the same time as new regulations governing transportation are being devised.

---

[6] Directive 2009/73/EC of the European Parliament and of the Council of 13 July 2009 concerning common rules for the internal market in natural gas and repealing Directive 2003/55/EC. *Official Journal* L 211/94, 14.8.2009. Regulation (EC) 715 on conditions for access to the natural gas transmission networks and repealing Regulation (EC) 2005/1775, OJ L 211/36.

[7] For much greater detail on the Gas Target Model, EU transportation regulation and network codes, and Russian problems with this process, see Yafimava, K., 'The Third EU Package for Gas and the Gas Target Model: major contentious issues inside and outside the EU', Working Paper NG75, OIES, April, 2013. The next few paragraphs are a brief summary of the arguments made there.

Table 4.1: Volumes of Russian gas and EU borders crossed in 2011*

Number of borders crossed to reach a border delivery point	Volumes (Bcm/)y
1	26.1
2	29.7
3	43.3
4	8.6

* Figures are necessarily approximate since it is impossible to know the exact route of gas flows.

Source: Yafimava, K., 'The Third EU Package for Gas and the Gas Target Model: major contentious issues inside and outside the EU', Working Paper NG75, OIES, April 2013, Table 2, p. 35.

European long-term gas import contracts are of two types: covering gas supply and transportation capacity. When European gas utilities were merchant transmission and sales companies – companies that controlled gas supply and owned/operated transmission networks – both contracts were signed with the same company. The unbundling of transportation networks from gas supply meant that suppliers such as Gazprom began to deal with two separate companies, increasingly in separate ownership. While this is administratively more complex, a more serious problem is that in some cases the two contracts are of different durations, with the length of the supply contract exceeding that of the transportation contract. In addition, the new system means that capacity can no longer be booked on a PP basis, but will be auctioned across a number of IPs, and instead of being delivered at the border, delivery points will be moved to VTPs. These changes to the provisions of existing long-term transportation contracts have potentially important repercussions for Gazprom supply contracts:[8]

- the problem of 'contractual mismatch', in other words, that a transportation contract with a customer may be shorter than a gas supply contract, raising the possibility that Gazprom will have obligations to deliver gas to a customer, but may lack the transportation capacity to allow it to do so;

---

[8] For existing contracts, there is no requirement to change the delivery point. However, if and when those contracts are extended or renewed, the delivery point will need to be moved to the VTP. From the Russian side, there is particular sensitivity that such changes may place Gazprom at a disadvantage in price negotiations.

- the problem of a 'missing (transportation) link' such that, with its obligations to deliver large volumes of gas across a number of European borders (now entry/exit zones), a shortage of capacity at one single IP, or a failure by Gazprom to obtain sufficient capacity from an individual auction, could prevent delivery of gas to customers beyond that point.

The most important issues which arise for Gazprom relate to the potential loss of, and difficulties in securing, capacity, and potentially increased costs of transportation. The conclusion of analysis by Katja Yafimava[9] on these issues is that the contractual risks to Gazprom which arise from moving to the new EU transportation regime are low, but are by no means negligible in terms of potential financial exposure:

> [A] mechanism whereby Gazprom would receive legally-binding assurances that under an entry exit system it will be able to secure sufficient capacity for delivering under its existing supply contracts at a cost not higher than under a previous point to point system, needs to be established independently of the magnitude of risk. Development of a special procedure of coordinated allocation of capacity along a route of IPs, which could ensure against a missing capacity link at any individual IP, could form part of such assurances.[10]

The general problem of the new EU regulatory framework is one of uncertainty, with the four main network codes not due to be finalized until the end of 2014, leaving another eight codes still outstanding, and the main emphasis being on existing (rather than incremental or new) capacity.[11] Because, at the time of writing, so much of this framework is still evolving – and will almost certainly continue to evolve even after the four main network codes have been finalized at the end of 2014 – it is impossible to give definite answers to detailed questions about future regulation, particularly in relation to new large-scale pipelines such as South Stream, Nord Stream 3 and 4, and Yamal–Europe 2. However, by 2013, it was

---

[9] Yafimava, K. 'The Third EU Package for Gas and the Gas Target Model', op. cit., p. 54.
[10] ibid.
[11] See Yafimava, K. 'The Third EU Package for Gas and the Gas Target Model', op.cit., for a discussion of 'new' versus 'incremental' capacity. At the time of writing only the Capacity Allocation code had been finalized, with Balancing, Interoperability and Tariffs (crucial for calculation of transportation costs) still outstanding. For an explanation of incremental capacity see CEER, *CEER Blueprint on Incremental Capacity*, Council of European Energy Regulators, Brussels: CEER, 23 May, 2013. By the end of 2013 an amendment to the CAM network code on incremental and new capacity had been proposed.

already clear that Gazprom would not be able to control 100 per cent of the capacity in new pipelines, as it had not applied for an 'exemption' from third-party access under Article 36 of the Third Gas Directive. In addition, none of the Russian pipelines had been accepted in the first list of 'Projects of Common Interest' (PCI), which would have made them eligible for fast-track permitting.[12]

The OPAL pipeline episode provides an illustration of the difficulties which Gazprom has already faced in respect of new pipelines on the territory of the EU. The Nord Stream pipeline has two onshore extensions – NEL and OPAL (see Map 3.2) – both of which applied for exemptions from third-party access rules: NEL was refused, and OPAL was granted an exemption by the German Federal Network Agency for 100 per cent of capacity.[13] But the European Commission, which has to approve all exemptions, reduced this to 50 per cent of capacity, unless Gazprom agreed to a 3 Bcm gas release programme in the Czech Republic to allow other suppliers to challenge its dominant position.[14] Gazprom declined to do so, arguing that the decision to deny it the full capacity in OPAL was illogical because no other supplier had – or is currently foreseen to have – gas available at Greifswald to deliver into the pipeline. At mid-2013, this situation had still to be resolved but technical tests, which allowed more than 50 per cent of the capacity to be used during extremely cold weather in March 2013, proved a useful demonstration of the pipeline's capability.[15] By the end of 2013, it was reported (but not yet confirmed) that agreement had been reached, with Gazprom being allowed to book 50 per cent of firm capacity and bid for the remaining capacity without the need for a gas release programme.[16] The OPAL episode is held up by Gazprom as an example of what it believes to be discrimination against its participation in European gas markets.

---

[12] See Yafimava, K. 'The Third EU Package for Gas and the Gas Target Model', op.cit., pp. 45–51, for additional explanation of: regulatory treatment of new capacity, the 2011 EU Infrastructure Regulation, PCI, and exemptions. If a project has already taken its final investment decision – which is the case for South Stream – then it is not eligible for an exemption, since one of the eligibility conditions is that a project will not be built unless it is exempted from TPA. For information on PCI projects see http://ec.europa.eu/energy/infrastructure/pci/pci_en.htm.

[13] 'Federal Network Agency grants partial exemption for the OPAL pipeline', Press Release Federal Network Agency, Bonn, 25 February, 2009.

[14] 'South Stream set to cede 10% of capacity', *Interfax Gas Daily*, 22 July 2013, p. 4.

[15] 'Europe lucky that Nord Stream tested under heightened pressure during cold snap – Gazprom', *Interfax Russia and CIS Oil and Gas Weekly*, 25–30 April, 2013, p. 45.

[16] 'Russia, EU agree terms of OPAL pipe usage', *Platts European Gas Daily*, 18 September 2013, p.3. At the time of writing there had been no official confirmation of the agreement.

88   *The Russian Gas Matrix*

The Russian side sees Article 11 of the Third Gas Directive (generally known as the 'Gazprom clause') as another major example of such discrimination in relation to:

> [...] circumstances that would result in a person or persons from a third country or third countries acquiring control of a transmission system or a transmission system operator (TSO). The regulatory authority shall refuse the certification [of such persons as a TSO] if it has not been demonstrated that the entity concerned complies with the requirements of Article 9 [i.e. is itself an unbundled utility] ... that granting certification will not put at risk the security of energy supply of the Member State and the Community.[17]

This clause is believed to have been inserted at the insistence of, particularly, central/eastern European and Baltic member states determined to prevent Gazprom taking ownership in their networks.

In general, while Russian politicians remain sceptical as to whether the Third Energy Package will deliver the benefits claimed by the European Union, their main complaint is what they see as the EU's 'unilateral approach'. According to Prime Minister Medvedev:[18]

> We have never said [the Third Package] is bad. It is the European Union's business how to regulate energy flows within the Community. But it would be desirable not to impose unilateral approaches on other partners. Their arguments need to be heard. We are not trying to meddle in the European Union's jurisdiction, but we believe some things should be discussed with us, as we are important suppliers of energy resources to Europe.

In 2011, on the initiative of the European Commission president and the Russian energy minister, the EU–Russia Gas Advisory Council was created in order to 'reduce risks and exploit opportunities in EU–Russia gas cooperation'.[19]

Access to EU customers, the demise of the trading house system, and Gazprom's downstream investments

The consequences of the Third Energy Package are by no means wholly negative for Gazprom. Liberalization of access to transmission and

---

[17] Article 11 of Directive 2009/73/EC (the Third Gas Directive), op. cit.

[18] 'Medvedev warns EU against imposing unilateral approaches in implementing Third Energy Package', *Interfax Russia and CIS Oil and Gas Weekly*, 21–7 March 2013, pp. 20–3.

[19] EU–Russia Gas Advisory Council – rules of procedure, October 2011. For more information on the work of the Council see http://ec.europa.eu/energy/international/russia/dialogue/dialogue_en.htm.

distribution networks means that the company is free to supply gas directly to customers throughout the EU. This amounts to an opportunity for Gazprom to fulfil a long-held ambition to develop direct sales to customers, independent of the major European utilities which for so long acted as 'gatekeepers', preventing competitors from reaching their captive customers. Gazprom's trading house policy, which began in 1990 with the creation of the Wingas and WIEH joint ventures (with the German company Wintershall, a subsidiary of BASF), had been intended to remedy this problem.[20] However, despite the creation of joint ventures in many countries, the success of Wingas in the German market was not replicated elsewhere.[21] After 20 years, the annual direct sales volumes to final customers European countries (excluding Wingas) did not exceed 4 Bcm. There were also allegations that trading houses served as vehicles for corrupt transactions, and in 2010 a decision was taken to phase out such intermediaries.[22]

Recognition that the strategy had run its course came in 2012 when, following agreement on an asset swap with Wintershall, it was decided that Gazprom would become the sole owner of Wingas.[23] This was the logical conclusion of gas market liberalization in Germany, where gas marketing no longer provides significant profits for a company such as Wintershall (as it had done under the previous monopoly structure) while it is still valuable to Gazprom in marketing its own gas in a vertically integrated structure. By the late 2000s, in place of its trading house strategy, Gazprom had established energy and gas trading companies such as Gazprom Germania and Gazprom Marketing and Trading.

While Gazprom devoted considerable attention to potential downstream investments in European utility industries – specifically transmission and

---

[20] For the history of Gazprom's joint ventures and trading houses in the 1990s and early 2000s see Stern, J.P. *The Future of Russian Gas and Gazprom*, pp. 111–14.

[21] A list of Gazprom's European sales subsidiaries can be found in *OAO Gazprom, IFRS Consolidated Financial Statements*, 31 December 2012, pp. 54–5. In 2012, the joint ventures with Wintershall contributed nearly 60% of the sales revenues from these companies.

[22] 'Gazprom eyes shedding middlemen in exports', *Interfax Russia and CIS Oil and Gas Weekly*, 25 February–3 March 2010, p. 20. Over the next two years, Gazprom bought out of GDF Suez's interest in Fragaz, OMV's interest in GWH, ENI's interest in Promgas, and the elimination of Overgaz when the Bulgarian long-term contract was renewed in 2012. 'EC greenlights Gazprom buying stake in dealer Promgas fromn ENI', 'Gazprom signs 10-year direct gas supply contract with Bulgaria', *Interfax Russia and CIS Oil and Gas Weekly*, 8–14 December 2011, p. 57, 15–21 November 2012, p. 52.

[23] The swap will not be completed until 2014, for details see *Management Report OAO Gazprom 2012*, p. 12.

distribution utilities and power plants – in general these did not work out, due to a combination of domestic political opposition and unattractive commercial prospects. Aside from Wingas and investments in the Baltic countries (see below), the only major downstream physical assets owned by Gazprom in EU countries are storage facilities.

In addition to the transit-avoidance pipelines to assure security of transportation, Gazprom has been faced with another security problem in Europe. Physical security of Russian gas supply has always been a factor in trade with Europe given the severity of winter weather in Russia, particularly in Siberia where most gas is produced. Russia's main storages were built in Ukraine during the Soviet era, to provide very large additional gas supplies to Europe, either during the cold winter months of peak European demand, when it is impossible to pump sufficient additional volumes from Siberia, or at times when insufficient gas was available because of high domestic demand. There are periodic reminders of this problem, notably in early 2012 when, due to severe weather in Russia, exports to Europe were constrained and the leadership prioritized domestic demand over exports.[24] Also in February 2013, when nine European countries experienced cuts in Russian gas deliveries during a cold spell affecting Russia, the western CIS, and European countries.[25] Shortfalls in Russian gas supplies to Europe not only mean lost revenues for Gazprom, but failure to deliver can also give rise to penalty payments to European customers. For these reasons, Gazprom is extremely concerned to ensure such shortfalls are minimized, and this is that reason for creating the company's own storage capacity in Europe.

With the problems encountered in the Russia–Ukraine relationship in the 2000s, Gazprom ceased using the Ukrainian facilities, and began to replace them with storages in European countries. By 2013, Gazprom owned (or jointly owned) European storages in Austria (Haidach), Germany (Rehden and Katharina), and Serbia (Banatski Dvor), with new facilities under construction in the Netherlands (Bergermeer) and Germany (Etzel and Jemgum). These totalled 4.5 Bcm of working gas storage in Europe, with plans to expand to 5 Bcm in 2015, and a target of increasing this capacity to 5 per cent of annual exports

---

[24] 'Putin says Gazprom's primary task is to supply domestic demand as cold causes drop in supplies to Europe', *Interfax Russia and CIS Oil and Gas Weekly*, 2–8 February 2012, pp. 6–9. See also Heather, P. and Henderson, J., 'Lessons from the February 2012 European gas "crisis"', Energy Comment, OIES, April, 2012.

[25] Chazan, G., 'Russian reduces gas supplies to Europe'; Chazan, G., 'Italy hit by shortage of Russian gas', *Financial Times*, 4–5 February 2012 and 7 February 2012; see also Westphal, K., 'Security of Gas Supply: four political challenges under the spotlight', SWC Comments 2012/C, June, Stiftung Wissenschaft und Politik.

(in the range of 8–10 Bcm) by 2030.[26]

Gazprom has also considered buying or building electricity assets in Europe. The rationale for such purchases would be that Gazprom could supply gas to power stations and might be able, either alone or in combination with European utilities (by controlling the price of the gas), to sell electricity at competitive prices. Agreements of this sort were discussed with EDF, RWE, and the German state of Bavaria, but they never materialized, presumably because Gazprom could not see sufficient value in such propositions.[27]

Security, Politics and Geopolitics

Contrary to popular perceptions, Russian gas deliveries to Europe (however measured) declined in the 2007–12 period, as Table 3.1 showed. Nevertheless, many countries in Europe – particularly in the central, southeast, and Baltic region – remain largely or wholly dependent on Russian gas.[28] The historical and political legacy of the Soviet era has created an understandable desire in these countries to reduce – or even eliminate – this dependence, reinforced by a conviction that they are subject to commercial discrimination in relation to the prices charged by Gazprom. Nevertheless, despite these concerns, until the end of the 2000s – nearly 20 years after the break-up of the Soviet Union and independence of that country's socialist allies in Europe – comparatively little progress had been made towards diversification away from Russian supplies. The January 2009 Russia–Ukraine crisis highlighted these problems, with several

---

[26] Gazprom Annual Report 2012, pp. 62–3. Details of underground storages can be found in *Gazprom in Figures* 2008–12, pp. 50–2; and also see 'Underground Gas Storage Facilities'. About Gazprom, Gazprom website. www.gazprom.com/about/production/transportation/underground-storage/. The target can be found in *Gazprom Loan Notes 2012*, p.98. The company also leases some additional capacity in the UK from Vitol, *Gazprom Loan Notes 2013*, p. 143.

[27] Gazprom Press Releases: 'Gazprom and EDF sign a Cooperation Agreement on gas power generation in Europe', 22 June 2012; 'Gazprom and RWE agree upon Memorandum of Understanding', 14 July 2011; 'Gazprom and Bavaria sign Roadmap for cooperation in power generation and gas supply', 21 December 2011.

[28] In northern Europe the three Baltic countries (Estonia, Latvia, and Lithuania) and Finland are wholly dependent on Russian gas. Five central and south-eastern Europe countries (Slovakia, Bulgaria, Serbia, Macedonia, and Bosnia/Herzegovina) are overwhelmingly dependent on Russian gas, and Russia provides another four (Poland, Romania, Czech Republic, and Hungary) with the vast majority of their gas imports.

south-east European countries suffering serious hardship, and created concern throughout the Continent.

It is not an overstatement to say that the January 2009 Russia–Ukraine crisis was the most serious gas security event and one of the most (if not *the* most) serious energy security events ever experienced in Europe. The details of this event, its causes and its results, can be found elsewhere,[29] but it gave rise to a new Regulation on Gas Security which extended the powers of the EU in this area far beyond the previous (2004) Directive.[30] In the wake of the crisis, a range of EU-funded measures was introduced. These included: reverse-flow capability to be installed on all cross-border pipelines, greater interconnection between countries (particularly in south-east Europe and in projects improving the ability to flow gas from north to south), additional storage, and new LNG terminals for isolated markets heavily dependent on Russian gas. Many of these security measures – in particular the interconnections and new LNG terminals – promise to bring significantly greater commercial opportunities for previously isolated markets to import non-Russian gas, and therefore have a relevance for the future that is beyond their original rationale.

'Political' gas pricing – the cases of the Baltic States and Turkey

In the minds of many European politicians, media commentators, and electorates, the principal concern about Russian gas supplies is the political threat which they appear to carry. While the pricing of Russian gas in the large, established European markets is becoming more market-related (see Chapter 3), in smaller markets further away from evolving north-west and central European hubs, it is very often referred to as 'political'. This is generally believed to be the origin of the 2012 DG COMP investigation which originated in a complaint from Lithuania.[31] The term 'political pricing' is generally meant to relate to decisions by Gazprom – supported, or perhaps ordered, by the Russian government – to tie gas prices to decisions on gas infrastructure or investments, or to other non-gas bilateral issues between the Russian Federation and the country in question.

---

[29] See Chapter 5 and Pirani, S., Stern, J.P., and Yafimava, K., 'The Russo-Ukrainian gas dispute of January 2009: a comprehensive assessment', Working Paper NG27, OIES, February, 2009.

[30] Regulation (EU) No 994/2010 of the European Parliament and of the Council of 20 October 2010 - *concerning measures to safeguard security of gas supply* and repealing Council Directive 2004/67/EC. For a detailed account of this Regulation see Pozsgai, P., 'The Evolution of EU security of gas supply policy', *EDI Quarterly*, Vol. 4, No.3, 2012, pp. 35–7.

[31] 'Lithuania files complaint against Gazprom with European Commission', *Interfax Russia and CIS Oil and Gas Weekly*, 10–26 January 2011, pp. 28–9.

Nicolò Sartori[32] reflects a widespread European perception that:

> Kremlin-directed Gazprom has often been seen to use energy supplies as a foreign policy weapon ... Gazprom's strategy [has been] to divide the EU into two parts ... bilateral preferential relations with key Western European customers to whom it gives ... cheaper gas prices and better contractual conditions ... unfair business practices resulting in heavy dependence and higher gas prices in Eastern Europe in order to preserve Russia's economic leverage and exert political influence on the region. This approach would help explain gas price differentials – hardly attributable to market forces – between Russia-friendly powerhouses such as Germany and weaker 'Russia sceptic' countries such as Lithuania.[33]

But while such views are widespread in Europe they should be questioned, not simply because of regular government denials that it uses either energy or Gazprom as a weapon,[34] but because so much of Russian behaviour appears designed to extract maximum revenues rather than political concessions. Nevertheless, many European governments believe they have been 'punished' and charged higher prices for refusing to agree to Russian gas and non-gas related proposals, particularly in relation to infrastructure and that this is reflected by the variations in prices in Table 3.2.[35] For reasons of space, we deal here briefly with the cases of the Baltic States and Turkey, at opposite ends of Europe both geographically and in terms of the size of their gas markets.

The traditional method of long-term contract gas pricing in Europe – netback market pricing based (mainly) on oil products – was, and particularly in south-east and Mediterranean Europe remains, the major price formation mechanism, despite the arrival of hub-based pricing post-2008 (see Chapter 3). However, European buyers have only been able to persuade their suppliers to move to hub pricing if they can demonstrate that their market has become genuinely competitive with their customers being able

---

[32] Sartori, N., 'The European Commission vs. Gazprom: an issue of fair competition or a foreign policy quarrel?' IAI Working Papers 13 March 2013, p.3.

[33] Ibid p. 9. In a similar vein see Lough, J., 'Russia's Energy Diplomacy', Chatham House Briefing Paper, 2011 and Hedlund, S., 'Russia as a neighbourhood energy bully', *Russian Analytical Digest*, No. 100, July 2011, pp. 2–5.

[34] For Medvedev on this when he was president see, 'Russian oil exporters interested in refinery privatisation – Shmatko', *Interfax Russia and CIS Oil and Gas Weekly*, 2–8 December 2010, pp. 5–7; for Putin's comments on Gazprom see, 'Gov't not "meddling" in Gazprom affairs – Putin', *Interfax Russia and CIS Oil and Gas Weekly*, 2–8 September 2010, p. 32.

[35] 'Gazprom prices higher for litigious in 2011, lower for South Stream partners', *Interfax Russia and CIS Oil and Gas Weekly*, 14–20 June 2012, pp. 47–9.

to access gas at prices set at liquid hubs.[36] At the time of writing, in markets such as those of the Baltic States and Turkey, liberalization and competition were very much 'work in progress', although their supply situations were completely different. Baltic gas markets remained monopolized by Russia as their sole source of supply, while in Turkey, although Russian gas remained the dominant source of supply, this was supplemented by pipeline supplies from Iran and Azerbaijan, and LNG from a variety of sources.

Only in the Baltic States and Finland did Gazprom manage to obtain ownership of significant shares in gas utilities and other downstream gas-consuming assets.[37] After 2008 (which as noted above is the date when oil prices rose above $100/bbl, with oil-linked gas prices following the trend), prices charged by Gazprom to Baltic countries rose substantially and complaints were increasingly heard – particularly from Lithuania – that they were being charged the highest prices in Europe for Russian gas, eventually leading to the formal complaint to DG COMP which resulted in the 2012 proceedings against Gazprom (see Chapter 3). In addition, the Lithuanian state entered into two international arbitrations with Gazprom alleging discrimination against the Lithuanian gas market, and in 2012 further proceedings against Gazprom alleging a $1.9 bn overpayment for gas.[38]

The Baltic gas markets are small: Table 3.1 shows that in 2012 the three countries (Estonia, Latvia, and Lithuania) together imported 4.8 Bcm (3.1 per cent of total Russian exports in that year) and only Lithuanian imports are significantly in excess of 1 Bcm/year. Moreover they are geographically isolated markets requiring significant investment to serve either by pipeline or LNG. It is significant that, more than 20 years after the break-up of the USSR and despite an apparent consensus that they had been exploited by Gazprom, the Baltic States had failed to arrange supplies of non-Russian gas. Part of the answer may be that for at least 15 years after independence, Baltic countries continued to pay prices which were lower than those paid by other European importers. Grigas notes that prices

---

[36] Presentations by Statoil have shown that sales at hub prices are confined to liberalized markets. www.statoil.com/en/InvestorCentre/Presentations/2013Conferences/Downloads/Create%20value%20from%20a%20strong%20gas%20position%20by%20EVP%20Eldar%20Saetre.pdf.

[37] For an account of how Gazprom obtained these shares see Grigas, A., 'The Gas Relationship between the Baltic States and Russia: politics and commercial realities', Working Paper NG 67, OIES, October, 2012 and Grigas, A., *The Politics of Energy and Memory between the Baltic States and Russia* (Burlington: Ashgate, 2013), Chapter 6.

[38] Grigas, A. 'The Gas Relationship between the Baltic States and Russia: politics and commercial realities', p. 15. At the time of writing, these proceedings had yet to be concluded. Lithuania has also moved ahead with unbundling the ownership of its gas industry, a development widely perceived to be aimed at reducing Gazprom's ownership.

did not reach European levels until 2008 – this was perhaps not unconnected with the accession of these countries to EU and NATO membership during 2005–7.[39]

It is uncertain whether any other supplier would have found it sufficiently profitable to deliver gas to their markets given their small demand and distance from supply sources, had these countries not been part of the Soviet Union at the time. This may account for the fact that – despite the political concerns about Russian gas supplies – it has not thus far proved possible to bring alternative gas supplies to the region. This will change when the LNG terminal under construction in Lithuania is completed around 2015.[40] However, the inability of the three countries to cooperate with each other (and possibly also with Finland) to build a single LNG terminal serving the entire region has also delayed the development of alternative gas supplies. Noel et al.[41] estimate that a security of supply levy of 9–10 per cent will be necessary to cover the cost of development of national LNG terminals. The size of this levy may depend on the price of LNG supplies and it is uncertain how this will compare with the price of Russian gas.

In contrast to the Baltic States, over the past 15 years Turkey has become one of the largest – and certainly the fastest growing – gas markets in Europe. In 2012 it accounted for 17 per cent of Gazprom's exports to Europe, and was its second largest market (Table 3.1).[42] Anxiety about the level of dependence on Soviet, and then Russian, gas supplies, which began in 1987, was exacerbated by the difficulties of transit through Ukraine in the 1990s, causing Turkey to focus on diversification of supplies.[43] As a result, LNG imports began from Algeria (starting in 1994) and Nigeria (1999); and pipeline gas imports began from Iran (2002) and Azerbaijan (2007).[44] However, the 16 Bcm/year contract for supplies through the Blue Stream pipeline starting in 2003 (see Chapter 3) confirmed that Gazprom would remain the country's major supplier.

---

[39] Ibid. pp. 10–13.

[40] Which also happens to be when the current long-term contract with Gazprom expires. 'Lietuvos Dujos H1 net profits rise 1.4%', *Interfax Russia and CIS Oil and Gas Weekly*, 25–31 July 2013, p. 32.

[41] Noel, P., Findlater, S., and Chyong, C.K. (2012). 'The Cost of Improving Supply Security in the Baltic States', Electricity Policy Research Group, University of Cambridge, EPRG Working Paper 1203.

[42] In 2012, Turkey consumed 46.3 Bcm (compared with 17.4 Bcm 10 years previously), a figure exceeded only by Germany, Italy and the UK. *BP Statistical Review of World Energy 2013* (London: BP, 2013), p. 23.

[43] For the history of Russia–Turkey gas trade in the 1990s and early 2000s, see Stern, J.P. *The Future of Russian Gas and Gazprom*, op. cit., pp. 122–5.

[44] In 2012, 4.1 Bcm of the 7.7 Bcm of Turkish LNG imports came from Algeria. The remainder were split between Norway, Qatar, Egypt, and Nigeria. *BP Statistical Review of World Energy 2013*, p. 28.

The pricing issue has caused periodic crises in Russia–Turkey gas relations. The gas trade, originally set up during the Soviet era on a barter basis, with gas being exchanged for commodities and construction services, was converted to a traditional netback market price basis in the mid-1990s. But as deliveries were due to begin, it became clear that the Turkish side had over-estimated its demand requirements and the contract was renegotiated with a much lower take-or-pay level and a downward revision of prices.[45] Following the economic crisis and recession of 2008, Turkish gas demand fell (although not as severely and for a much shorter period than was the case in most European countries) and (as shown in Table 3.1 and discussed in Chapter 3) purchases fell below take-or-pay levels in 2009–10.

This caused a major price debate between Gazprom and Botas which had two important consequences. First, the failure of Botas to renew the original 1987 contract which expired in 2011; as far as this author knows, this is the only long-term Russian gas export contract which has not been extended on expiry, although the volume was redistributed among other Turkish importers, some of which are Gazprom subsidiaries.[46] Second, the agreement of Turkey to allow the construction of the South Stream pipeline within its exclusive economic zone in the Black Sea, as a result of which a lower price was apparently agreed which allowed a reduction in its annual gas bill of $1bn.[47] A similar price reduction of 11 per cent was reported in the contract with Bulgaria following agreement to build South Stream on that country's territory.[48]

Overall, looking at developments in European gas security and diversification in terms of cold war geographical categorization in the first two decades of the post-Soviet era, it has been the pattern that gas markets in west European countries have become more competitive and supply-diverse, while former socialist and communist countries have made much slower progress towards diversification away from Russian supplies.[49] The main explanations for the latter's lack of diversification are that timorous and disorganized governments have been threatened with political

---

[45] Ulchenko, N., 'What is so special about Russian-Turkish economic relations', *Russian Analytical Digest*, No. 125, 25 March, 2013, pp. 5–9.

[46] 'Gazprom assigns 6 Bcm of gas to 4 private Turkish companies instead of Botas', *Interfax Russia and CIS Oil and Gas Weekly*, 9–15 August 2012, p. 36.

[47] 'Russia, Turkey agree on long-term gas supplies until 2025, support for South Stream', *Interfax Russia and CIS Oil and Gas Weekly*, 22–8 December 2011, pp. 5–9. In late 2012, the Russian price was quoted as $418/mcm, compared with Iranian gas at $423/mcm and Azeri imports at $282/mcm. 'Socar, Turkey's Cig Enerji create company for gas sales in Turkey', *Interfax Russia and CIS Oil and Gas Weekly*, 1–7 November 2012, p. 49.

[48] 'Gazprom signs South Stream protocol with Bulgaria, amends gas supply contract', *Interfax Russia and CIS Oil and Gas Weekly*, 23–9 August 2012, p. 36.

[49] Exceptions are Croatia, and to some extent Hungary and the Czech Republic.

and/or military consequences by aggressive Russian governments, or that corrupt energy and industrial elites have been bribed by Moscow-financed oligarchs acting as agents of the Russian government or Gazprom.[50] Whatever the truth of these suggestions, the reality of Russian gas pricing is that it has maximized its commercial position in countries without alternative gas supplies, which is exactly what would be expected from a revenue-maximizing discriminating monopolist. But to describe such behaviour as 'political' is analytically problematic.[51]

The classic netback market pricing formula (see Chapter 3) is consistent with charging very different prices to different countries, depending on their location and the fuels which compete with gas in their energy balances, and commercial confidentiality prevents detailed comparative analysis of contracts. However, two other criteria have also impacted Gazprom's pricing policy. First, the willingness of buyers to allow Gazprom (or other Russian companies) ownership of transportation assets.[52] This has particularly been the case in former Soviet countries (see Chapter 7) but also in Europe in relation to South Stream. Second, membership of international organizations seen to be anathema to Russian political and security interests, such as the EU and NATO (and, by contrast, rejection of membership of Russian-dominated organizations such as the CIS) has had significant impacts on prices paid for Russian gas. Again this has applied principally to former Soviet states but also to the Baltic countries. Only in these respects is it completely clear that Russian gas pricing in European (as opposed to CIS) countries could be described as 'political', rather than the more general sense in which that term is applied.

Diversification away from Russian gas supplies: the Southern Corridor

European concern about lack of diversification and over-dependence on Russian gas has been a feature of the Soviet and post-Soviet periods. Starting in 2000, a major element of EU external gas policy has been to create a 'Southern Corridor' bringing gas from a variety of Caspian, Central Asian, and Middle East countries as a means of diversifying European gas supplies

---

[50] See Grigas, A. *The Politics of Energy and Memory*, op. cit., Chapter 6 for events in the Baltic countries.

[51] For a discussion of discriminating monopoly gas pricing, see Allsopp, C. and Stern, J., 'The Future of Gas: what are the analytical issues related to pricing', in Stern, J.P. (ed.), *The Pricing of Internationally Traded Gas* (Oxford: OIES/OUP, 2012), pp. 10–39.

[52] The first example of this was in the oil sector in the Baltic countries, where Russian oil companies were prepared to build extremely expensive new pipelines and export terminals and abandon Soviet-era assets which they were barred from acquiring. See Grigas, *The Politics of Energy and Memory*, op. cit., Chapters 3 and 4.

and reducing the continent's dependence on Russia.[53] From the start of this initiative it was clear that Turkey would be the central transit country (and an important gas market) for the proposed major pipelines: Nabucco, TAP, and ITGI.[54] The eventual decision by the Turkish and Azeri governments to build their own Trans-Anatolian (TANAP) pipeline as far as the western Turkish border, and the subsequent decision of the Shah Deniz partners to select the Trans-Adriatic (TAP) pipeline to deliver the initial 10 Bcm/year of Azeri gas starting in 2019, confirmed that, for at least the rest of the 2010s, very little Caspian gas would be available for Europe.[55] A key factor contributing to the lack of available gas was that little progress was made towards the construction of a trans-Caspian gas pipeline which could enable large-scale Central Asian gas supplies – principally from Turkmenistan – to reach Europe.[56] While the relatively small volume agreed is clearly disappointing in comparison to the notion of a 'corridor' carrying 90–100 Bcm/year of gas (which was the general expectation during the 2000s) it is at least a start, and certainly does not deserve Gazprom's view that it is 'just about enough gas for a barbecue'.[57]

---

[53] For the principles see EU, *External Energy Relations – from principles to action*, Commission of the European Communities, Communication from the Commission to the European Council, COM (2006) 590 final, Brussels 12 October, 2006. The initiative culminated in negotiations between the EU, Azerbaijan, and Turkmenistan to build a trans-Caspian pipeline system. The specific policy spoke of opening 'the Southern Gas Corridor – a supply route for roughly 10–20% of EU estimated gas demand by 2020', EU, 'The EI Energy Policy: Engaging with Partners beyond Our Borders', European Commission, Communication from the Commission to the European Parliament, the Council, the European Economic and Social Committee and the Committee of the Regions, on security of energy supply and international cooperation –COM(2011) 539 final, Brussels, 7 September, 2011, Section 1.2, 5.

[54] For an overview of these projects see Giamouridis and Paleoyannis (2011). Giamourdis, A. and Paleoyannis, S., 'Security of Gas Supply in South Eastern Europe: potential contribution of planned pipelines', LNG and storage', Working Paper NG52, OIES, July 2011.

[55] For an overview of the TANAP and TAP projects see www.tanap.com/en/ and www.trans-adriatic-pipeline.com/ Despite contracts having been signed with European buyers, some believe that Turkey could take a significant proportion of the 10 Bcm of Shah Deniz phase two gas destined for Europe; but also that significant supplies of gas from Iraqi Kurdistan will be available to Turkey starting in the mid-2010s.

[56] Despite a mandate to negotiate a legally binding treaty to build a trans Caspian Pipeline system having been signed between the EU, Azerbaijan, and Turkmenistan. 'EU starts negotiations on Caspian pipeline to bring gas to Europe', European Commission Press Release, IP11/10/23, 12 September 2011.

[57] Chazan, G. 'Decision time for BP-led group on route of Caspian gas pipeline', *Financial Times*, 9 June 2013. www.ft.com/cms/s/0/754e27ba-cec5-11e2-ae25-00144feab7de.html?siteedition=uk#axzz2gMHEy7Sl.

From the announcement of the South Stream pipeline in 2006, a major element of European commentary, including from the highest levels of the European Commission, was that the pipeline was a 'bluff' by Russia, solely designed to prevent Southern Corridor pipelines – and specifically Nabucco – from going ahead, and that as soon as Nabucco started construction South Stream would be abandoned. Not only have such judgements proved to be wide of the mark (see Chapter 3), but they always appeared to misunderstand the principal rationale behind South Stream – which was diversification away from Ukraine – and the fact that the Russian side had a far greater familiarity with most of the potential Southern Corridor suppliers than did most European stakeholders. Thus EU confidence that South Stream would never materialize was matched by Russian confidence that the Southern Corridor would never be built, at least not on the scale anticipated by Brussels. Part of this Russian confidence may have stemmed from its legal position that a pipeline across the Caspian could not be built without the agreement of all the littoral states.[58] However, Europeans were unimpressed by Gazprom's purchase of gas from Azerbaijan, which it clearly did not need, seemingly with the sole objective of reducing supply available to the Southern Corridor.[59]

Geopolitics and the gas 'weapon'

While both sides – but especially Russia – have engaged in competitive rhetoric about each other's project, the reality is that neither South Stream nor the Southern Corridor, in the form they are likely to be implemented, will substantially change the geopolitics of gas in Europe, unless European gas demand resumes a strongly upward trajectory. As far as security and diversification of supplies are concerned, European commentary has largely ignored the problematic gas delivery track record of Caspian suppliers, which had caused particular problems in Turkey (but also in Greece). Despite cordial political relations, Turkey's natural gas trade with Iran has been especially difficult – it has included an international arbitration over prices which found generally in favour of the Turkish side – in terms of

---

[58] This is one legal interpretation. Another is that it would be legally possible for Turkmenistan and Azerbaijan, with adjoining territorial waters, to build a bilateral pipeline across the sea. Neither Russia nor Iran are likely to agree to such a pipeline. Iran has had its own agenda for wishing to deny European access to Central Asian gas. Russia's position refers to the Declaration of the 2007 Five Nations Tehran Summit that any significant Caspian development requires consensus. 'Russia has questions to ask EU about Trans-Caspian gas pipeline project – Russia EU envoy', *Interfax Russia and CIS Oil and Gas Weekly*, 15–21 September 2011, p. 9.

[59] For more detail on Gazprom purchases of Azeri gas see Chapter 14.

periodic interruptions of deliveries.[60] In every case, Gazprom stepped in and provided additional volumes to Turkey at short notice, with no suggestion that the prices charged were higher than those in long-term contracts, despite the relatively desperate situation in which Turkey found itself.

The extent to which Gazprom's transit-avoidance strategy should be considered 'political' is analytically problematic. The issues are analogous to those discussed above in relation to its pricing policy. The transit-avoidance strategy represents a Russian refusal to expose its gas exports to economic or political transit risks – which in Russian eyes is amply justified by its experience in Ukraine during the post-Soviet era which culminated in the January 2009 crisis. However, critics have viewed the transit-avoidance pipelines as political in the sense of 'isolating' the countries which it avoids, thereby allowing Gazprom to exert greater pressure on them.[61]

The Russian desire to establish transit-avoidance pipelines has clearly been sufficient for Gazprom to provide substantial financial inducements to those countries and companies whose cooperation is required.[62] The reasons why the Russian side is prepared to offer such strong incentives are both politico-economic and geopolitical. Russia has clearly been determined to eliminate dependence on 'unreliable' countries – specifically Ukraine – where transit of gas to Europe provides the possibility of political and economic leverage: political in the sense that the transit country is confident that Europeans will always blame Russia for any gas delivery problems whether or not the Russian side is objectively at fault; economic in the sense that Russia will be the bigger financial loser out of any such events. The January 2009 crisis consolidated both of these conclusions in Russian minds and therefore confirmed the justification for transit-avoidance pipelines.

---

[60] Kinnander, E., 'The Iran–Turkey gas relationship: politically successful, commercially problematic?', Working Paper NG 38, OIES, January, 2010 gives details of Iran–Turkey gas supply problems up to 2009. But these seem to have worsened since the start of Syrian instability in early 2012, 'Gazprom supplying extra gas via Blue Stream to help Turkey for the fourth day in a row', *Interfax Russia and CIS Oil and Gas Weekly*, 18–25 October, p. 58. They also seem to have caused similar problems for Azeri-Turkish supplies, 'Gazprom boosts gas supply to Turkey 60% due to pipeline explosion', *Interfax Russia and CIS Oil and Gas Weekly*, 4–10 October 2012, pp. 57–8.

[61] In the case of the Nord Stream this referred to Ukraine, Poland, and the Baltic Countries. Helm, D., *The Russian dimension and Europe's external energy policy*, September, 2007 www.dieterhelm.co.uk/node/655.

[62] Specifically, as noted above, E.ON and Wintershall being given shares in the Iuzhno-Russkoe field in exchange for participation in Nord Stream, and Botas receiving price reductions in return for Turkish acceptance of South Stream in its offshore economic zone.

The strongest geopolitical element is the fact that the energy sector provides Russia with its most important international economic visibility and relevance in Europe. Specifically it provides the Russian president with his only major focus for economic dialogue with European (and potentially also Asian) leaders. Without major new gas (and oil) projects, the geopolitical significance of Russia – and hence the standing of both the country and the president in international negotiations – would be much reduced. It is this, rather than any 'weapon' which might arise from exports to Europe (or Asia), which is the most important geopolitical element in Russian gas supplies. It is very difficult to see how Russia could successfully impose its political/military will by depriving Europe of gas, and any such action would be likely to impose serious economic and political damage on Russia. This analysis, therefore, agrees with Per Hogselius's[63] study of Soviet gas exports which concluded that 'many analysts have exaggerated the role of Soviet and Russian gas as an energy weapon'. But the more widely held, if perhaps somewhat overstated view, is that of Marshall Goldman:

> Gazprom and by extension the Russian government are already beginning to enjoy a power over their European neighbours far beyond the dreams of the former Romanov czars or the Communist Party general secretaries.[64]

Thus despite the lack of supporting evidence – or perhaps by the attribution of the term 'political' to actions which should simply be viewed as commercial – the belief that Russia uses energy, and specifically gas, as a political and geopolitical weapon is deeply ingrained in European political and energy consciousness and in conventional wisdom. So much so, that it is doubtful that this mindset can be changed for many years to come.

Summary and conclusions: the logic of Gazprom's European export strategy[65]

Commercial and market strategy

Up to the 2007–8 financial crisis and recession, European gas market development was highly beneficial for Gazprom with increasing volumes and prices. By contrast, during 2009–12, commercial developments in European gas markets, in particular the decline in demand and the arrival

---

[63] Hogselius, P., *Red Gas: Russia and the Origins of European Energy Dependence* (Basingstoke: Palgrave Macmillan, 2013), p. 222.
[64] Goldman, M.I., *Oilopoly: Putin, Power and the Rise of the New Russia* (London: OneWorld Books, 2008), p. 3.
[65] This section serves as the conclusion of Chapters 3 and 4.

of liberalization, competition, and hub-based pricing, were almost uniformly negative. Strong public resistance by Gazprom and the Russian government to hub-based pricing has tended to obscure a growing willingness to adapt long-term contract prices to market levels by a combination of reductions in the base price and annual rebates to customers. Take-or-pay conditions have also been relaxed somewhat, and buyers have been allowed to roll supplies over into future years without penalty. During 2008–12 these adjustments were not sufficient to prevent loss of export volumes in many countries, due to a combination of reduced demand, cheap coal, increased renewable energy supplies, competitors offering cheaper gas, and determined efforts by countries heavily or wholly dependent on Russia to diversify their imports. Recovery of deliveries in 2013 was partly due to exceptionally cold spring weather across Europe, and partly to the fact that Gazprom's new price mechanism made its gas far more competitive with other supply sources.

Despite its opposition to changes in European gas price formation, there is no reason to believe that Gazprom cannot compete effectively by pricing its long-term contract portfolio in relation to hub prices. Gazprom has subsidiary companies in major European gas markets, with many hundreds of staff engaged in the trading of a wide variety of energy products. However, Gazprom would need to choose between selling a certain volume or maintaining a certain price of gas at the major European hubs. This is in essence the 'OPEC dilemma' about whether to defend price or defend volume sales. But rather than being part of any 'gas-OPEC' organization, Russia is such a large player (around 25 per cent of total European gas demand) that, except for periods of time when the European market is substantially over-supplied, it could operate such a strategy on its own without needing to cooperate with other suppliers. Particularly when the supply/demand balance is tight, Gazprom's potential power over prices in European spot markets will be considerable.

Transit-avoidance pipelines – regulatory and cost problems

The transit-avoidance pipelines not only involve Gazprom in very considerable investment costs, but are likely to encounter substantial problems due to new European regulation under the Third Energy Package. Major problems have already been encountered with the on-land extensions of the Nord Stream pipeline, specifically the OPAL pipeline through Germany to the Czech Republic. These problems seem likely to be repeated on a larger scale for South Stream, which Gazprom seems determined to build by the end of 2015, despite the fact that the relevant EU regulation for new pipeline capacity is not yet in place, and no exemption from Third Package rules has been (or seems likely to be) applied for. Lack of assurance on tariffs and capacity availability involves significant financial risks given a total South Stream investment which could exceed €40bn, including the cost of pipelines in Russia and Europe and the Black Sea crossing. Similar considerations will apply to Nord Stream 3 and 4 and Yamal–Europe 2, should these be built.

But why does Gazprom – and the Russian government – want or need such a large export capacity to Europe; and why does it believe that it can increase exports to 200 Bcm in the future? There are various explanations for such behaviour in the face of falling European gas demand. The main one is Gazprom's – and perhaps more importantly President Putin's – determination to bypass Ukraine and create sufficient 'transit arbitrage', in other words, the need to create sufficient transit alternatives that no single country will be able to obtain leverage over Russian gas exports to Europe. Another relates to the Gazprom – and general Russian – view that European efforts to decarbonize energy balances are an expensive fashion requiring huge subsidies which cannot be afforded, and which will be abandoned in favour of much greater gas use. A third explanation is that the Russian leadership elites need Gazprom to build these new pipelines because this is how they are able to obtain corrupt payments by means of equipment and construction contracts.

Regulatory, political, and security responses to Russian gas

Russia is not in a good position to resist decisions on energy and gas markets taken in Brussels by 28 EU member states, but a pragmatic solution to problems and uncertainties of Russian gas transportation caused by the Third Energy Package and the gas network codes needs to be found. Equally problematic and long-lasting have been European reactions to Russian gas from a political and security standpoint. There are major political problems between Brussels and Moscow, and between the latter and individual member states, which have prevented the signing of a new EU–Russia Partnership and Cooperation Agreement. While at an institutional level the debate is about whether the EU wants a 'strategic partnership' with Russia on energy and natural gas, or just a 'commercial partnership', at a political level many EU member states – particularly those which are former Soviet countries and former Soviet allies in Europe – do not want any kind of partnership with Russia, and want to phase out dependence on Russian gas which they see as politically threatening. However, these states have made little progress towards reducing their dependence on Russian gas in the post-Soviet period, and the extent to which they will do so in the future is unclear.

The January 2009 Ukraine crisis was disastrous for the reputations of Russia and Gazprom – whether deserved or not – and much of Europe's subsequent gas security regulation, as well as political and media attitudes towards Russian gas, continue to be shaped by this event. But Gazprom's determination – encouraged and indeed apparently ordered by government and specifically President Putin – to build increasing numbers of new pipelines to avoid transit through Ukraine, has not improved European perceptions of Russian gas security. These pipelines are seen by some as an attempt to forestall diversification away from dependence on Russian gas supplies, and by others as an attempt to isolate and intimidate countries which oppose Gazprom or Kremlin policies.

Gazprom's monopoly of LNG exports was removed at the end of 2013 (see Chapter 3), but there is no suggestion that the company will lose its pipeline export monopoly. But even if it did, it is unlikely that this would have any significant impact on the political sensitivity of Russian gas in Europe. That political sensitivity is likely to remain, despite the fact that European countries which have been historically dependent on Russia seem set to import at least some portion of their requirements from alternative sources over the next decade. Particularly in those countries with historical/political experience of Soviet domination, the notion that imports of Russian gas provide the Kremlin with an 'energy weapon', operated by Gazprom, which it will not hesitate to use to exert political influence, remains deeply entrenched. Through this lens, all actions by Gazprom – specifically price increases which may have logical commercial explanations – will be interpreted as 'political'. The irony of this interpretation is that Gazprom has consistently maximized its financial position by refusing to compromise on pricing (until forced to do so either by international arbitrators or competing gas suppliers), which is the classic behaviour of a 'discriminating monopolist'.

The logic of Gazprom's European export strategy

In the face of so much commercial and regulatory complexity and financial risk, the question needs to be asked: is Gazprom's commercial behaviour since the late 2000s – in relation to pricing and transit-avoidance pipelines – *necessarily* wrong? This judgement depends on a view of future European gas demand and supply. From 2008 to 2009, European demand was falling faster than European supply but over the period 2009–11 that trend reversed, the supply/demand balance tightened and spot prices rose substantially (albeit still well below the levels of oil-indexed prices).[66] While the outlook for European gas demand for the rest of the 2010s is relatively gloomy, perhaps not returning to 2010 levels until the early 2020s (see Chapter 3), the supply outlook may be no less difficult.

---

[66] Cedigaz (2012). *Natural Gas in the World: 2012 Edition*, Rueil Malmaison, Table 22, p.44 and Table 59, p.139.

So does Gazprom have the correct – or at least not necessarily the wrong – export strategy for the 2010s and early 2020s in Europe? The answer to that question depends on whether Russian gas can in the future maintain and increase the 2013 volume levels of exports above 160 Bcm/year, which it lost in many European countries in the 2009–12 period.[67] This in turn entails a difficult calculation, for each market, of gas demand, import requirements, and the role of Gazprom's pricing policy in these variables. Such an analysis would need to look at supply and demand substitutes and their costs, as well as the likely evolution of national carbon reduction policies. Thus the logic of Gazprom's export strategy in Europe very much depends on an overall view of future European gas supply/demand balance and hence its import requirements, in the context of regional and global carbon, energy, and gas developments. One view is that – due to carbon reduction – gas is in terminal decline as a fuel in European energy balances and nothing can reverse that situation. This interpretation would suggest that Gazprom's strategy of charging the highest possible price – short of going to arbitration with its customers (and possibly losing) – has been the correct one. Another is that a wave of cheap North American LNG supplies, driven by shale gas development and supplemented by similar development in Europe, will progressively threaten Gazprom's dominance of European gas markets.

But Gazprom commentary (and to a lesser extent that of the Russian government) on the future of European gas suggests an opposite interpretation that:[68]

- the post-2008 decline in gas demand is a temporary phenomenon;

- a combination of increasing demand and falling domestic production will give rise to a very substantial additional gas import requirement prior to 2020 and beyond;

- North African pipeline gas supplies are in decline and will not recover;

- LNG will not be available in large quantities for Europe, due to high Asian demand;

- US LNG exports will be limited by rising domestic production costs of shale gas;

---

[67] We stress the term 'volume exports' here, rather than 'market share', because for countries where gas demand will fall in the future, Gazprom's market share may remain stable (or even increase) but its volume exports may nevertheless decline.

[68] 'European politicians yet to realize that shale gas production unprofitable – Gazprom Export expert', *Interfax Russia and CIS Oil and Gas Weekly*, 17–23 October 2013, pp. 45–6.

- European shale gas production will be of marginal importance, as will other sources of pipeline gas, such as the Southern Corridor.

This combination of events would leave only Russian gas in a position to fulfil European import requirements. If this interpretation is correct, then the European market will have no choice other than to import larger volumes of Russian gas.

In late 2013, this interpretation of future European and global gas trends was not a mainstream projection, but, even assuming it to be correct, it does not explain Russian determination to build such large, and extremely expensive, pipeline export capacities. Table 3.3 shows that in 2013, Russian nameplate gas export capacity was already substantially in excess of 200 Bcm/year and, with a slight dip in the middle of the decade (due to an assumption of falling capacity in the Ukrainian network), will remain so through the rest of the decade even if only half of the planned South Stream capacity is built.[69] Russian determination to phase out *substantial* dependence on the Ukrainian network had largely been achieved by 2013 (following the completion of two Nord Stream pipelines). Construction of additional pipelines in order to achieve complete independence from Ukrainian transit seems symbolic, very high cost, and extremely risky given (as has already been seen in the case of the OPAL pipeline) that new EU regulation may prevent Gazprom from using the full capacity of on-land pipeline extensions of its export lines. These pipelines are likely to be 'value-destroying' for Gazprom (to use financial analysts' terminology), with minimal security of demand advantages and substantial financial and regulatory risks. In this situation, a more logical Russian strategy would be to delay all new pipeline investments until the European business and regulatory environment becomes clearer.

Justification for building additional export capacity requires certainty that Russian gas exports would reach (and probably exceed) 200 Bcm/year by 2020, or that the entire Ukrainian capacity would become unusable because of: commercial disputes over Russian gas prices or Ukrainian financial (transportation tariff) demands, security concerns, or a decision by Ukraine to close some of its transit capacity. However, should Gazprom and the Russian government build some part of the huge additional capacity shown in the maximum 2020 scenario in Table 3.3, there will be significant financial pressure to ensure that gas is flowing through it, and Gazprom's incentive to withhold gas supplies from Europe – for commercial or political reasons – will be much reduced. The construction of Russian

---

[69] 'Nameplate' refers to the design capacity of the pipelines; actual useable capacity is likely to be lower although this has already been taken into account in the Ukrainian case.

export pipelines in addition to what had already been built in 2013, therefore, seems positive for Europe from both a commercial and future gas availability perspective. However, it is problematic from the perspective of those who believe that increased Russian gas imports constitute a political and security threat which will increase Moscow's ability to use gas as an 'energy weapon'. While we do not believe this to be the case, it is a widely held political and media view across Europe which seems likely to endure throughout, and probably beyond, the 2010s.

In conclusion, Russian gas exports to Europe will remain extremely important for at least the next decade, and probably up to 2030. But for the rest of this decade, volumes seem unlikely to expand greatly from the 150–170 Bcm/year levels of the 2005–13 period (Table 3.1) unless there are commercial or regulatory changes which somehow make gas much more attractive in the power generation sectors of the major European gas markets, and/or global gas demand and pricing continues to deprive Europe of LNG supplies up to 2020. However, even if we assume that more low-carbon energy sources are used, and that the importance of gas in European energy balances diminishes, this will not necessarily mean that the importance of Russian gas will diminish. Indeed, even if its export volumes do not increase significantly, Russia could become an even more dominant player in a shrinking European gas market. During the 2010s and early 2020s, the only substantial new source of gas likely to challenge Russian gas in Europe will be LNG. The volumes and prices of LNG available to Europe – and how Gazprom will choose to respond to these potential challenges – will therefore play a major role in determining the future of Russian gas in Europe. In turn, LNG pricing and availability for Europe will be determined principally by demand and prices in Asia, and supply and prices in North America. Thus Russian gas exports to Europe will be affected not only by Russian and European, but also global, gas market dynamics.

CHAPTER 5

RUSSIA'S DOMESTIC GAS MARKET DEVELOPMENT, PRICES, AND TRANSPORTATION

James Henderson, Simon Pirani, and Katja Yafimava

The changes in the Russian domestic gas market since the mid-2000s, and its further evolution up to 2020, are covered in this chapter and the next. There are two interrelated themes: (i) changes in the volume of consumption, and in the way that gas is consumed in the economy and society; (ii) the move from a quasi-monopoly of sales, transport, and distribution by Gazprom, towards a more liberalized (but not completely liberalized) market.

The changes in the volume of gas consumption can be considered in three time periods. In the first period, from the late 1990s until 2007, gas consumption rose each year (by 2–2.5 per cent per year), driven largely by GDP growth of 5–8 per cent per year. In the second period, from 2007 to 2012, consumption fell sharply as a result of the 2008–9 economic crisis, only returning to the 2007 level in 2011, but falling again in 2012. Gas demand continued to be determined mainly by the level of economic activity, but there were signs that price, competition among suppliers, and some efficiency measures, also began to play a role. In the third period, from 2012 onwards, it seems likely that demand will rise only gradually: the question we pose is whether it will rise at all.

The changes in gas market structure can also be divided up, very roughly, in similar time periods. Laws providing for reform of the gas market were adopted in the late 1990s and early 2000s, but only in 2006–7 did the government take significant steps to implement them – in the first place, by adopting a new pricing strategy, aimed at bringing prices to European netback levels. This pricing policy has been accompanied by other market reforms. While plans to break Gazprom into production, transportation, and trading divisions have been constantly postponed to the more distant future, from 2009 the government has compelled Gazprom to provide pipeline access to other substantial gas producers – in the first place Rosneft, the state-owned national oil company, and Novatek, the largest independent natural gas producer, but also other oil companies. These non-Gazprom producers now account for a substantial share of supply. They have moved from selling to customers prepared to pay a small premium to regulated prices to competing for customers at or below regulated prices. Between now and 2020, this trend in market forces will play an increasing role in continuing to re-shape the Russian gas sector.

This chapter views these changes in the domestic market from three different angles: Consumption trends: overview, Domestic gas prices and market reform, and Transportation tariffs and pipeline regulation.[1]

Consumption trends: overview

Total gas consumption in Russia, as measured by Rosstat, the national statistics agency, is shown in Table 5.1. (The Rosstat statistics are the most comprehensive of a number of data sets; the differences between these are explained in Appendix 1.) A breakdown of consumption between the Federal Administrative Districts is shown in Map 5.1; district-by-district statistical information from Rosstat appears in Appendix 2.

The average annual rate of consumption growth – shown in the top line of Table 5.1 – was 2.3 per cent in 2000–6. In 2007–11 it slowed to 0.3 per cent, in other words, almost to a standstill.[2] Between 2011 and 2012, total consumption fell substantially, from 473 Bcm to 465.4 Bcm, confirming the weakness of the recovery in gas demand, according to information published by Gazprom.[3] In 2013, consumption rebounded to a level slightly higher than in 2011, according to a statement from the Energy Ministry.[4]

The main reasons for the sluggish nature of the recovery of gas demand from the 2008–9 crisis are (1) lower-than-expected demand, and some efficiency gains, in the power sector, which accounts for more than half the total gas consumed; and (2) the limited demand recovery in manufacturing industry. Table 5.2 shows the results of a comparison, conducted at the

---

[1] These sections are authored by Simon Pirani, James Henderson, and Katja Yafimava respectively.
[2] A methodological anomaly in the table makes the post-2007 look slightly less dramatic than it is. Since 2007, Rosstat began to include in total consumption volumes, under 'other losses', 4–9 Bcm of gas consumed in the production process (including associated gas that is re-injected). Most methodologies would exclude these volumes and reduce the total volume of consumption correspondingly.
[3] Gazprom, *Management Report 2012*, p. 17. Gazprom's and Rosstat's measures of total Russian annual consumption have been very close (no more than 2 Bcm divergence) for the last five years. See Appendix 2, Table A2.1.
[4] Deputy Energy Minister Kirill Molodtsov stated that in 2013 consumption rose by 1.9% to 432 Bcm. This figure presumably excludes gas consumed for technical purposes (40–50 Bcm/year in recent years). At the time of writing, no statistical information was available. Itar-Tass, 'Potreblenie gaza v Rossii v 2013 godu', 19 November 2013 http://itar-tass.com/ekonomika/768460.

Table 5.1: Russian domestic gas consumption

	2000	2001	2002	2003	2004	2005	2006	2007	2008	2009	2010	2011
Total domestic consumption (Bcm)	**398.5**	**406.4**	**412.5**	**426.4**	**434.9**	**440.7**	**456.3**	**466.7**	**462.0**	**432.6**	**462.1**	**472.8**
Technical use and losses	**47.7**	**45.7**	**47.7**	**51.5**	**51.1**	**49.6**	**52.0**	**49.4**	**51.0**	**36.3**	**43.9**	**45.8**
Other losses*	0.0	0.0	0.0	0.0	0.0	0.0	0.0	4.2	4.8	8.4	6.8	6.2
Energy sector, including:	**222.0**	**228.1**	**229.6**	**233.5**	**238.4**	**242.6**	**248.6**	**253.4**	**250.9**	**238.2**	**253.4**	**256.9**
Power stations	151.6	156.0	157.0	159.8	164.8	169.3	174.3	180.4	181.7	169.4	183.6	189.5
Centralized boilers	70.4	72.1	72.6	73.7	73.6	73.3	74.3	73.0	69.3	68.8	69.8	67.3
Consumption as fuel, including:	**110.1**	**113.9**	**116.7**	**121.3**	**125.0**	**127.6**	**134.2**	**136.0**	**131.8**	**127.8**	**133.3**	**139.6**
Mining and extractive sector	10.8	10.5	10.9	11.5	12.2	12.6	12.6	12.4	12.5	11.6	12.1	12.6
Manufacture and processing sectors	31.8	32.5	32.9	33.7	34.9	35.7	38.0	37.6	35.4	33.4	35.1	36.8
Used in gas processing plants (ex. dry gas output)	8.2	8.7	8.7	9.4	10.1	10.6	11.0	10.7	10.9	9.8	11.1	11.4
Construction	0.4	0.3	0.3	0.3	0.3	0.3	0.3	0.3	0.3	0.3	0.3	0.5
Agriculture	0.6	0.5	0.5	0.6	0.5	0.5	0.5	0.6	0.6	0.8	0.7	0.8
Transport and communications	1.6	1.6	1.6	1.7	1.9	1.8	2.0	1.9	2.0	1.7	2.2	2.2
Other economic sectors	16.7	18.8	19.4	20.3	21.2	22.1	24.7	28.4	23.6	23.1	24.0	25.5
Residential use	40.1	41.0	42.3	44.0	43.9	43.9	45.0	44.3	46.4	47.2	47.8	49.7
Feedstock and other non-fuel uses	**18.6**	**18.8**	**18.5**	**20.2**	**20.3**	**21.0**	**21.5**	**23.7**	**23.4**	**21.7**	**24.7**	**24.4**

* Including reinjection of associated gas, consumption at gas fields and in LNG production

Note: for some comments on the categories used, see Appendix 1. Some columns may not add up due to rounding.

Source: Rosstat.

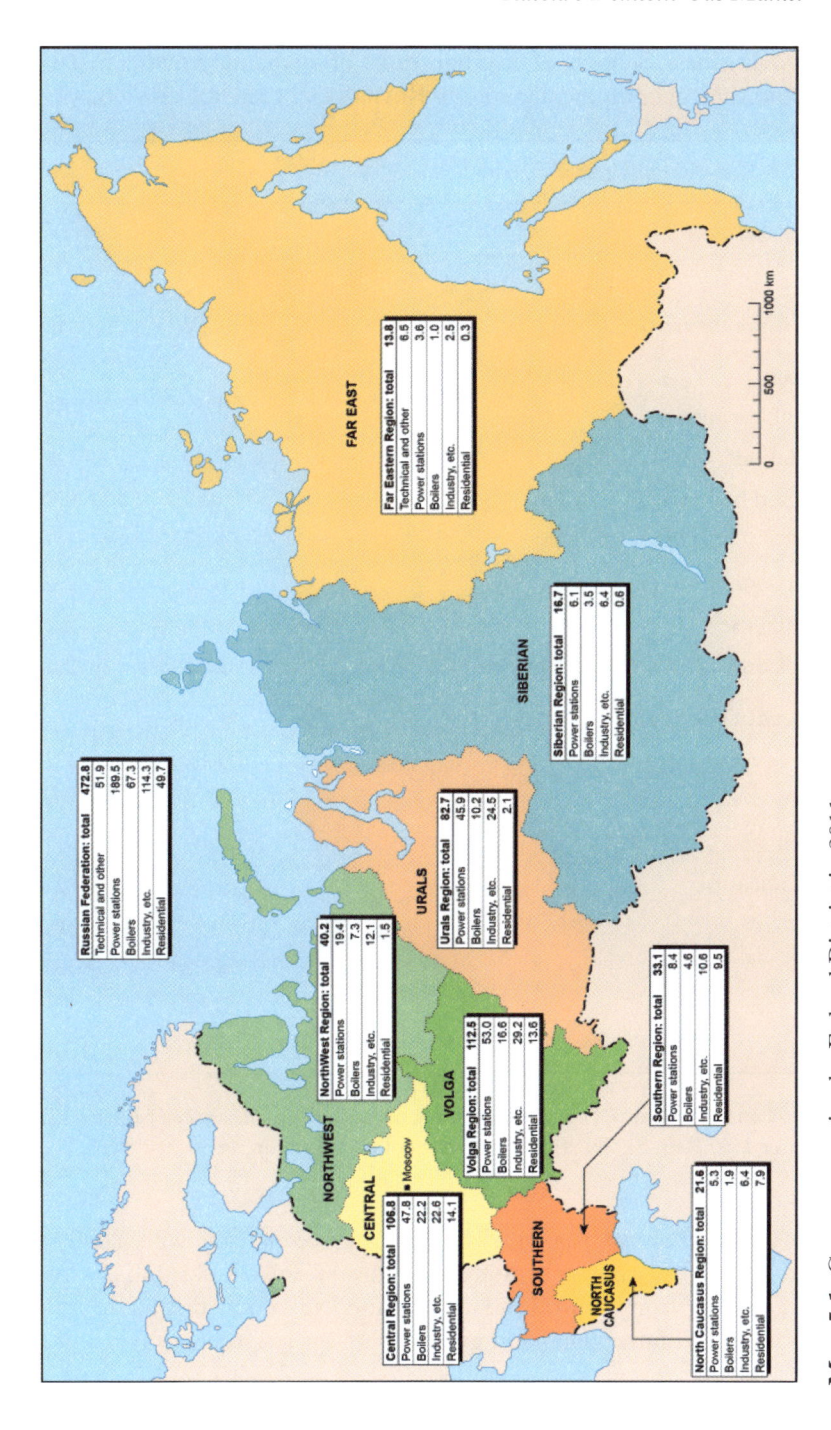

Map 5.1: Gas consumption by Federal District in 2011
Source: Agency for Forecasting Energy Balances

Skolkovo Business School, of average rates of demand growth: first by sector, and then geographically among Russia's eight federal districts.

Table 5.2: Gas demand growth trends

Demand growth rates (%) annual average	2000–6	2007–11
Total		
Russian domestic market	2.3	0.3
By sector		
Technical use and losses	1.42	–3.54
Power stations	2.36	1.42
Centralized boilers	0.9	–0.94
Industry	3.17	1.35
Other economic sectors	6.52	2.67
Residential use	1.97	2.25
Feedstock and other non-fuel uses	2.45	2.65
By Federal District		
Central	1.89	1.22
North-West	3.18	1.91
Southern	2.09	0.36
North Caucasus	1.88	–0.8
Volga	1.47	0.7
Urals	4.22	1.02
Siberia	3.65	2.24
Far East	1.81	13.49

Source: Tatiana Mitrova, 'Vnutrennyi spros na gaz', [*Domestic demand for gas: a time of transition*], Skolkovo Business School presentation, February 2013.

The analysis[5] shows that, in 2007–11, by comparison with 2000–6, demand growth from power stations and industry slowed substantially, and demand growth from centralized boilers used in urban district heating systems turned negative. Demand growth from the gas industry's technical

---

[5] Tatiana Mitrova, 'Vnutrennyi spros na gaz: perekhodnyi vozrast', Skolkovo Business School presentation, February 2013.

use and losses also turned negative. Among sectors that consume substantial volumes, only residential use saw any acceleration of demand growth. In regional terms, the sharpest slowdowns in demand growth were in the Southern, Volga, and Urals federal districts, all of which have industrial concentrations that consume large volumes of gas. The Skolkovo analysis also showed a sharp concentration of demand growth. Four sectors in four regions – oil production at Sakhalin; oil production, power, and residential use in Tiumen'; the power sector in Moscow; and the power sector in St Petersburg and Leningradskaia region – accounted for 68 per cent of demand growth in 2004–10, with their share of demand growth rising to 85 per cent in 2010. Without these sectors, average annual demand growth in 2004–10 would have been 0.4 per cent.

The consensus among economists and industry analysts is that in the next few years, annual average demand growth will continue to be very slow. Published estimates put demand growth to 2020 at 1.12 per cent (Skolkovo Business School), 1–1.2 per cent (Gazprombank), or even close to zero (Sberbank's estimate of only 1 Bcm of incremental annual demand by 2016). The Ministry of Economic Development considers that annual gas demand growth will 'not exceed 1–1.5 per cent'. An alternative view of demand growth is taken by Gazprom. While its 'low' scenario is in the same range as other forecasts, and assumes an extra 20 Bcm of extra demand by 2020/22 (an implied annual growth rate of 0.6 per cent), its 'high' scenario is substantially higher than the consensus, and assumes an extra 85 Bcm of demand (an implied annual growth rate of 1.92 per cent). The government's Energy Strategy, published in November 2009, also projected relatively high growth rates. The projections are shown in Table 5.3.

Gas demand prospects

It will be argued in this chapter, and the next, that gas demand growth will remain slow for the remainder of this decade, and could even be zero, or negative. The incremental gas required by the Russian market by 2020 is unlikely to be more than 40 Bcm. There are four groups of reasons for this:

First, *the nature of the Russian economic recovery*. The main driver of the economic recovery has been oil production, and the economic benefits of the exports of oil, gas, and other commodities. Manufacturing, which is more energy intensive, has recovered much more slowly. In 2012–13, the recovery faltered, with growth rates falling from 3.4 per cent (2012); projections of the 2013 growth rate are less than 2.5 per cent.[6] Weaker

---

[6] See Chapter 1; also, e.g. World Bank, *Russian Economic Report* no. 30: *Structural Challenges to Growth Become Binding*, Washington, September 2013, pp. 23–6; Pomelnikova, Maria et al. *Russia Country Report*, no. 2 (Raiffeisen Research, October, 2013), p. 3.

Table 5.3: Russia's gas demand outlook

		2012	*Consumption forecasts: volumes (Bcm)* 2015	2020	*Demand growth: annual avg rate*
Ministry of Econ. Devp't	Total (low scenario)	465		503.52	1
	Total (high scenario)	465		523.82	1.5
Skolkovo business school	Total, excl. tech gas	428		468	1.12
	Power sector	188		201	0.88
	Boilers	72		64	−1.36
	Industry	79		94	2.17
	Residential	75		86	1.78
Gazprombank research	Total (low scenario)	465		503.52	1
	Total (high scenario)	465		511.56	1.2
Sberbank	Total	407	407**		0
Novatek	Total	472*	479	512	0.9
		2008–13/15	2013/15–2020/22	2020/22–2030	
Energy Strategy (Nov 2009)	High	520***	560	640	1.49
	Low	475	540	600	1.68
Gazprom (February 2013)	High	460	545	600	1.92
	Low	460	480	500	0.60

* Novatek figure is for 2011, not 2012.

** For 2016, Sberbank's analysts projected consumption of 408 Bcm, implying a 0.06% demand rate during 2012–16.

*** Consumption in 2012, registered by Rosstat as 465.4 Bcm, showed this 2009 projection to be a gross overestimate.

Note. For Gazprombank, volumes extrapolated from growth rates by author; for Sberbank, Novatek, Energy Strategy, and Gazprom, growth rates calculated from volume projections. For Ministry of Economic Development, volumes extrapolated from growth rates by author (using Gazprom's estimate of volume for 2012 as starting point).

Sources: Skolkovo presentation, 'Vnutrennyi spros na gaz: perekhodnyi vozrast' February 2013; Novatek presentation, 'Russia's natural gas frontiers', September 2012; Sberbank Investment Research, 'Russian Oil and Gas – An Unexpected Journey', January 2013, p. 26; Gazprom presentation, 'Gas for the future: Gazprom Investor Day', 8 February 2013.

demand – the main factor that constrained economic growth – has been attributed by government and international institutions alike to Russia's persistent high dependence on oil and gas exports and commodity price volatility. Likewise, both national and international bodies point to the sluggish progress of the real economy. A recent report from the World Bank highlighted 'gradually deteriorating industrial performance, and the manufacture of tradables in particular'; until 2013 the effect of this deterioration was masked by growth in construction, financial services, transport, and communication, but these too have now slowed. Investment is being boosted by major state-supported infrastructure projects, such as the Sochi Olympics, the conclusion of which may again emphasize structural weaknesses.[7] All this suggests that (i) GDP growth will, during the rest of this decade, be much slower than in the pre-crisis period; and (ii) energy-intensive sectors may continue to grow more slowly, further eroding energy demand.

Second, *the unusual demographic situation in Russia*. In most countries, steady population growth tends to drive energy demand upwards; in Russia, the population has fallen throughout the post-Soviet period and is expected to continue to fall in the coming years, notwithstanding a slowing in the decline of the birth rate.[8] This effectively removes one of the underlying factors in rising energy demand that is at work in most nations.

Third, *the effect of the increase in regulated gas prices and other gas market reforms*. Due to the consistent increase in regulated prices in recent years, prices paid by power and industrial sector customers, who account for most gas purchases, are now above cost-recovery levels. Producers are competing on price, with non-Gazprom producers selling at a discount to the regulated price. In the power sector, higher gas prices are narrowing electricity producers' margins. In short, power sector and industrial consumers are increasingly sensitive to the costs of the gas they purchase. Investment has fallen because of the recession, but wherever it is made and plant is renewed or replaced, energy savings follow. On the other hand, the imbalance between prices for industrial customers and for households diminishes the impact of prices on consumption levels.

Fourth are *sector-specific issues*, which are discussed in Chapter 6. These include investment and efficiency gains in electricity generation; reform of the heat sector; changes in gas-intensive industries; municipal service reform; gasification; and so on.

---

[7] Ministerstvo ekonomicheskogo razvitiia, *Prognoz dolgosrochnogo sotsial'no-ekonomicheskogo razvitiia RF pa period do 2030 goda* (Moscow, March 2013); World Bank, *Russian Economic Report* no. 30, pp. 3–7.

[8] See, for example, the *UNDP Human Development Report for Russia*, 2013, Russian edition (*Ustoichivoe razvitie: vyzovy Rio*), pp. 26 and 60.

The potential for gas consumption actually to fall in Russia is well illustrated with reference to Ukraine. A substantial reduction in gas consumption in Ukraine since 2007 (see Chapter 7) has resulted from a combination of an industrial recession, similar to that suffered by Russia, and gas prices that, for industrial consumers, are between two and three times higher than those in Russia. The reduction is in the order of one-fifth, and may be as much as one-quarter, of total Ukrainian gas demand. While there are important differences between the two economies – in the first place, Ukraine's power sector is fuelled predominantly by coal and nuclear power, and Russia's by gas – the similarities, particularly in the urban and industrial infrastructure inherited from the Soviet Union, are at least as significant. This suggests that not only slow-to-zero demand growth, but even negative demand growth, should be considered as a possibility.

Figure 5.1 compares changes in the level of GDP, and of gas consumption, in Russia and Ukraine between 2008 and 2013. In Russia, GDP had more or less regained its 2008 level by 2012; gas consumption, having exceeded that level, more or less returned to it. In Ukraine, by contrast, since 2011 GDP has stabilized at about 6–7 per cent below its 2008 level, but gas consumption fell to almost 20 per cent below that level in 2012, and was heading still lower in 2013.

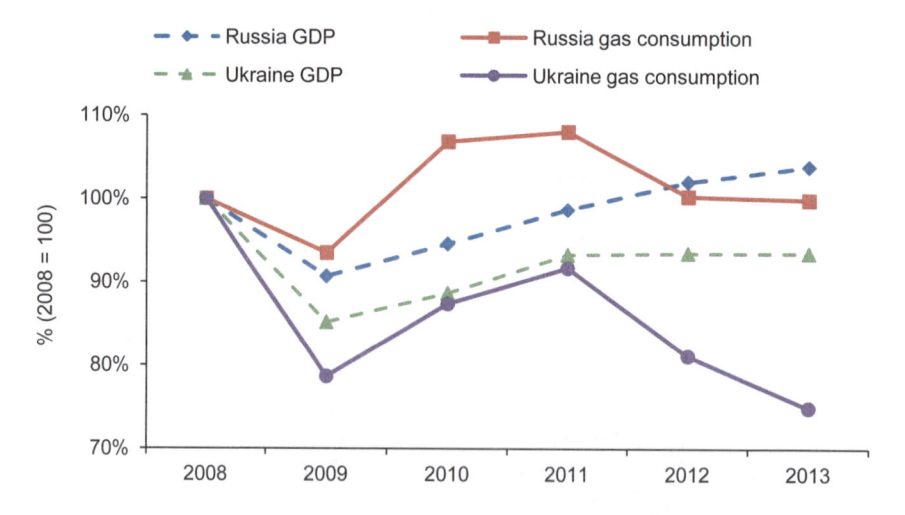

Figure 5.1: Russia and Ukraine: GDP and gas consumption, 2008–13, % (2008 = 100)
Note: 2013 figures are author's projections, based on 9 months' statistics.
Sources: World Bank (GDP), Russian Ministry of Energy and Ukraine Ministry of Energy and Coal (gas consumption).

The trends shown in Figure 5.1 suggest that, in Ukraine, energy-saving and fuel switching, as well as the effect of the recession on economic activity, contributed to the fall in gas consumption. The question for Russia is whether, as retail gas prices continue to rise, it will experience more of these same effects. The fall in gas consumption levels in 2012 suggests that it may do.

Russia's prices are not likely to rise as rapidly as Ukraine's have, as we argue below. But even if Russia's consumption fell by half as much as Ukraine's has (in other words, by about one-eighth) that would take it from the pre-crisis level of around 460 Bcm/year to just above 400 Bcm/year, further exacerbating the competition between suppliers. We do not project this outcome, but we draw attention to this, a possible scenario, to show the importance of demand-side changes.

Domestic gas prices and market reform

Since the mid-2000s the Russian domestic market has moved from a near-monopoly by Gazprom to a market in which several suppliers are competing actively for industrial and power sector customers. The volumes sold domestically by non-Gazprom producers in 2012 (150 Bcm) amounted to more than one-third of the total of gas consumed in Russia (424.5 Bcm, that is: 465.4 Bcm total consumption, minus 40.9 Bcm Gazprom own use).[9] A key driver of reform has been the increase in regulated gas price and the current debate about fundamental changes to the domestic pricing system overall.

There has been a regulated natural gas price in Russia throughout the post-Soviet era. It was introduced to exert some state control over an industry with one quasi-monopoly player (Gazprom) that had (and still has) a huge influence on the country's economy. Gas has consistently accounted for around 50 per cent of Russia's total primary energy consumption[10] and

---

[9] Gazprom and the Russian Ministry of Energy report a number of different figures on which these calculations are based. In its Databook for 2012 (rounded to the nearest Bcm), Gazprom reports total internal gas consumption in Russia of 466 Bcm, of which 41 Bcm is for internal use in the transport business. This leaves 425 Bcm of consumption. Of this, Gazprom Group gas accounts for 275 Bcm, suggesting that 150 Bcm is supplied by third parties. The Ministry of Energy reports that total Russian production in 2012 was 655 Bcm, of which Gazprom produced 487 Bcm, implying Independent production of 168 Bcm. The difference between this figure and domestic sales is accounted for largely by Sakhalin output, where 9 Bcm is re-injected at Sakhalin 1 and around 7 Bcm of non-Gazprom output is exported as LNG.

[10] US Energy Intelligence Agency Data sourced at www.eia.gov/ on 16 July 2013.

Gazprom provides around 10 per cent of Russia's GDP.[11] It also plays a vital political role in providing electricity, heat, and fuel to the population. As a result Gazprom's domestic sales price remains under state control and is fixed on a regular basis by the Federal Tariff Service (FTS), while third-party producers (often known as Independents) are allowed to price their gas as they see fit, although the regulated price has always provided a benchmark. Different prices are set for industrial (including power) consumers and households. The latter pay a lower price, as they continue to be subsidized for political reasons, but the industrial tariff is regarded as the key number as it accounts for more than 80 per cent of overall sales.

Government control of tariffs has meant that, apart from one brief period in the mid-1990s, Gazprom has been forced to set prices for its domestic customers at well below the 'global market' rate, as defined by the netback price of gas exports to Europe. This netback is calculated by subtracting transport costs and export taxes from the sales price for Russian gas in Europe to provide an 'effective price' in Moscow that is still more than double the domestic industrial gas price at the same location. This implied domestic market discount has historically been justified by three key factors. Firstly, Gazprom had an effective monopoly over supply and sales of gas in Russia and therefore needed to be regulated in order to avoid exploitation of domestic consumers. Secondly, it had also been given the exclusive right to access export prices as the monopoly marketer of Russian gas to non-CIS countries, with high revenues from export sales offsetting low prices in the domestic market. Finally, low domestic prices were also justified by the fact that Gazprom's gas production was being generated from Soviet-era fields with very low marginal costs, since most capital expenditure had already been completed. As a result, the benefits of Soviet investment were being reserved for Russian consumers rather than Gazprom's shareholders.

However, following the financial crisis of 1998–9 and the devaluation of the ruble, the domestic gas price fell sharply in real terms, to a low of $12/mcm in 2000, less than a quarter of the $50–60/mcm reached in 1995–6. In 2000 the Russian industrial gas price was at a level of only 12 per cent of the export netback price, compared with the parity achieved in 1995. Although the FTS announced a series of 20–25 per cent price increases in 2000–4, by 2005 this gap had only narrowed to 42 per cent. This meant that Gazprom was once again not only selling domestic gas at a large discount to global prices, but it was also providing a significant implied subsidy to its domestic customers, estimated at up to $2.3 billion/year,[12] because it was selling gas to them at below the long-run marginal

---

[11] *Foreign Affairs*, 6 May 2012, 'Putin's Gazprom Problem',

[12] OECD *Economic Survey: Russian Federation* (Paris: OECD, 2004), p. 135.

cost of its asset base.[13] This subsidy hampered Gazprom's ability to generate any profit from its domestic operations, and undermined Russia's attempts to fulfil its agreed obligations to achieve WTO accession.

Pressure on the regulated price system was also being generated by the changing structure of the domestic gas market, in particular via the emergence of new gas suppliers. 'Independent' gas producers such as Novatek and Itera, as well oil companies such as Lukoil and Rosneft, had begun to take an increasing interest in producing gas and even marketing it to domestic customers, and, despite the difficulty of gaining access to the trunk pipeline system owned by Gazprom, had gained a 15 per cent share of the gas market by 2005.[14] The fact that these new players could price gas outside the state-controlled structure brought into being a two-tier market, with Gazprom forced to sell at a low regulated price and non-Gazprom players able to charge whatever higher price consumers would bear. In November 2003 the OECD identified the markup as being just under 32 per cent.[15] This two-tier market became increasingly important as Gazprom, with the blessing of the Russian government, started to limit the amount of gas sold at regulated prices through its control of the allocation of gas in the annual 'Gas Balance'.[16] As a result, from 2007, consumers were allocated a fixed amount of gas that they could purchase at low

---

[13] The concept of Gazprom providing an implied subsidy to its domestic customers is a difficult one. The definition offered in the OECD 2004 *Economic Survey: Russian Federation* is that the domestic gas price subsidy is the difference between Gazprom's long-run marginal cost of production, estimated at US$34–35/mcm in 2004, and the effective industrial tariff at the time of US$26/mcm. However, the OECD itself acknowledges the difficulty in defining both the long-run marginal cost and therefore the subsidy, given the uncertainties around Gazprom's actual costs, the allowances which need to be made for future capital investment, and the lack of transparency around pricing in a non-competitive market. As a result, the notion of a 'gas subsidy' is a relatively loose one, defined in general terms as selling gas at a price which fails to cover the costs of production and delivery to market, as well as some allowance for capex to replace fixed assets. This 'subsidy', though, is not the same as the effective discount offered by Gazprom to domestic customers as defined by the difference between the domestic and export gas prices.

[14] Henderson, J. *Non-Gazprom Gas Producers in Russia* (Oxford: OIES, 2010).

[15] OECD *Economic Survey: Russian Federation*, 2004.

[16] The Gas Balance is fixed on an annual basis in a negotiation between Gazprom and the Ministry of Energy and provides a 1 to 5 year framework for gas supply in Russia, based on current demand estimates. Historically it allowed Gazprom to limit the access of third parties into the Russian gas system, as it had effective control over who would or would not be allocated space on the supply side of the Balance. However, Gazprom's ability to manipulate the Balance in its favour has been diminished as the Independents have increased their share of production and signed long-term contracts with domestic consumers.

regulated prices, and would then have to buy any extra gas they needed at a premium price, either from the Independents or from Gazprom itself.[17] Nevertheless, despite this allocation mechanism, around 75 per cent of the market remained under regulated prices, giving Gazprom a guaranteed underlying demand for its gas, albeit at low prices.

A third pressure on domestic prices – Gazprom's changing production profile – emerged in the mid-2000s. Gazprom was gradually becoming more dependent on new higher-cost fields as its cheaper Soviet legacy assets went into decline and as a result the company had an increasing need to generate profits for investment, while it was continuing to lose money in its major market, Russia. Although these losses were to an extent compensated by its export sales to Europe – where the company was selling only one-third of its volumes but generating between one-half and two-thirds of its revenues – it was becoming increasingly clear that Russia and Gazprom should no longer be so reliant on the export market, but should attempt to bring domestic prices to a level where they could sustain new investment. The government was also influenced by the need to increase energy efficiency during Russia's economic recovery, underpinned by a steadily rising oil price, which by the mid-2000s was pushing domestic gas demand up rapidly. As a result, in November 2006, President Putin and the Russian government announced a new pricing strategy, with a target for gas prices in Russia to reach European levels on a netback basis.[18] This target was confirmed by Government Decree in May 2007[19] as being a goal of 'netback parity' with export prices to Europe by 2011, importantly based on the assumption of an oil price of $50–55/barrel.[20] The decree also confirmed Gazprom's legal right to sell gas at higher prices to new customers as of July 2007, and it fixed the volumes of regulated gas sales at that date and authorized extra gas to be sold at a reducing premium to the regulated price in the four and a half years to the end of 2011. The plan was that, as the regulated price increased, the premium for new purchases would decline until all prices would be equalized by 2011.

At the same time one of the first building blocks for a liberalized market was also being put in place via the formation of a Gas Exchange, organized by Gazprom subsidiary Mezhregiongaz. Created in November 2006, the exchange was initially available for the trading of 10 Bcm/year of gas (5 Bcm by Gazprom and 5 Bcm by independent suppliers), although this

---

[17] Russian Government Directive No.333, 28 May 2007.
[18] 'Gas price in Russia, EU to get closer – Putin', *Interfax*, 24 November 2006, Moscow.
[19] Russian Government Decree No.333, 28 May 2007.
[20] Burgansky, A. *Stand and Deliver: Oil and Gas Yearbook 2010* (Moscow: Renaissance Capital, 2010).

was increased to 15 Bcm (7.5 + 7.5) in January 2008.[21] However, trading was halted in January 2009, mainly due to Gazprom's concerns over the potential for dumping gas during the 2008–9 economic crisis. But nevertheless the exchange (also now known as the Electronic Trading System (ETS)) provided further evidence that industrial consumers would be prepared to pay an average premium of up to 38 per cent over the regulated gas price.[22] Discussion about re-opening the ETS began in 2010, amid apparent enthusiasm from Gazprom and the Energy Ministry, and culminated in a draft resolution on a new trading system being sent to the government in April 2011.[23] However, there have been repeated delays in passing the legislation required to restart the Gas Exchange,[24] with the key issue being the exact functionality of the exchange. Gazprom has pushed for a market based on the trading of physical gas only (as in 2007–8), while the Russian government is also keen to also see the introduction of a more advanced futures market. Furthermore, concern has been expressed by Igor Sechin, CEO of Rosneft, about the potential for price fixing in any exchange mechanism.[25] Nevertheless, in April 2012 the government passed another resolution to allow Gazprom to sell 15 Bcm/year on commodities exchanges, rising to 17.5 Bcm, as long as this does not exceed the amount also sold by independent producers. During 2013 the Head of the St Petersburg International Mercantile Exchange (SPIMEX) expressed the belief that full gas exchange trading might begin by 2014. However, a final decision has been delayed once more, as Gazprom attempts to secure a place for its own trading system, run by Mezhregiongaz, alongside any independent exchange.[26]

Despite this setback for exchange trading, from 2006 regulated prices continued to be increased towards President Putin's export netback target. As Figure 5.2 shows, they rose by around 15 per cent per annum every year, with the exception of 2009 during the economic crisis.

However, despite these increases, the target of netback parity has remained elusive and indeed has become increasingly irrelevant, because one of the key components of the netback calculation, Gazprom's oil-

---

[21] Ibid., p. 182.
[22] Henderson, J. *Non-Gazprom Gas producers in Russia*, op.cit., p.225.
[23] 'Energy Ministry submits draft legislation on electronic trading exchange', *Interfax*, 21 April 2011, Moscow.
[24] 'Government rejects latest version of draft resolution on gas exchange', *Interfax*, 24 May 2011, Moscow.
[25] 'Electronic gas trading floor has to be independent of producers', *Interfax*, 6 September 2011, Moscow.
[26] 'Gas exchange trading might begin end-2013, probably 2014 – SPIMEX chief', *Interfax*, 6 March 2013, Moscow.

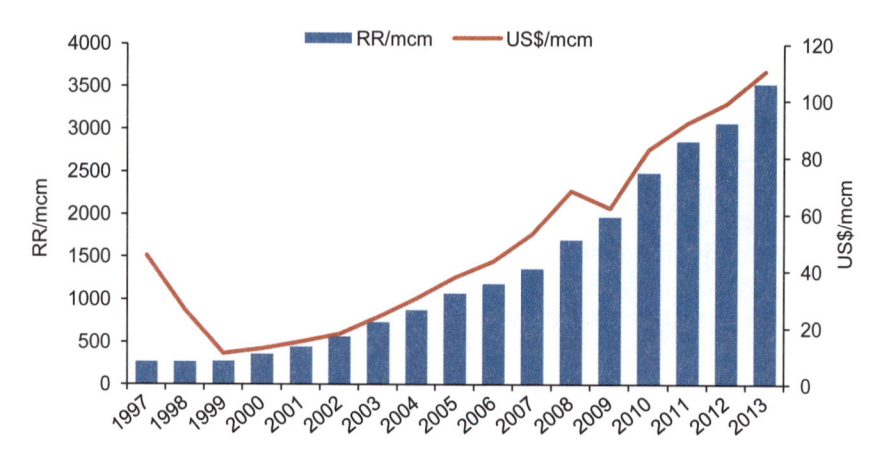

Figure 5.2: The regulated domestic gas price in Russia
Source: Federal Tariff Service of the Russian Federation, Energy Intelligence Group.

linked gas sales price to Europe, has doubled since the original target was set in 2006, pushing the netback parity to approximately $230/mcm. As early as 2010 the Russian government adopted Resolution No. 1205, which extended the transition period until 2014, with regulated prices for all consumers (excluding the residential sector) to continue being set at a discount to European netback during the 2011–13 period. The FTS approved a new formula under which wholesale prices for Gazprom's gas sold domestically (excluding the residential sector) would be set on a 'regulated price plus' basis from 2012, as long as the gas was either (a) sold under new contracts for supplies beginning after July 2007 or (b) sold under existing contracts, but where the volumes exceeded those agreed in the original contracts. Despite this alteration, however, the domestic industrial gas price in 2013 remained at only around 50 per cent of the export netback level. It should be no surprise that the parity target date, initially having been set at 2011, and then over the course of 2010–12 pushed back to 2014, then 2015, and finally to 2017/18, was by the start of 2013 effectively being ignored. Even Prime Minister Dmitrii Medvedev conceded that 'we all need to stop and think to what extent direct netback parity is optimal for operating on our [the Russian] market'.[27]

This 'thinking process' is now underway and involves not only a discussion about the relevance of netback parity, but also a debate about the correct level of regulated price increases and the interaction between

---

[27] 'Russia could depart from netback parity for gas prices – Medvedev', *Interfax*, 26 June 2013, Moscow.

regulated and 'market' prices. Following the economic crisis of 2008–9, the FTS in January 2010 introduced a standard 15 per cent per annum increase in the regulated price. It suggested that this rise should be expected in the future if netback parity were to be achieved. There was a further 15 per cent rise in January 2011, but the imminence of parliamentary and presidential elections over the winter of 2011/12 resulted in a decision to delay the 2012 increase to July, after the elections, with Deputy Economic Development Minister Andrei Klepach also reconfirming 15 per cent price rises for July 2013 and 2014 and the likelihood of a 14.6–15 per cent rise in 2015.[28] It was anticipated that the domestic industrial gas price would reach $155/mcm by 2015, still below netback parity but at least high enough to underpin Gazprom's new gas field developments on the Yamal peninsula, which are estimated to have a breakeven cost of around $120–140/mcm.[29] An additional nuance to the regulated pricing structure was introduced from January 2013, when the FTS was mandated to increase or decrease the regulated tariff on a quarterly basis by up to 3 per cent, depending on the movement of a basket of oil products relative to its base price on 1 January 2013. As a result, the regulated price fell by 3 per cent on 1 April 2013[30] but increased by the same amount on 1 August 2013,[31] providing some link to gas export prices on a regular basis.

Domestic gas prices: catalysts for a rethink

Two important events in 2012–13 have caused the Russian government to review its stance on domestic gas prices. The first, and perhaps the most immediately relevant, has been the slowdown in the Russian economy, which caused industrial growth to fall to almost zero in June 2013[32] and GDP growth to fall from 4.3 per cent in 2011 to 3.4 per cent in 2012, and 2.4 per cent in 2013 (projected by the Ministry of Economic Development).[33] This declining trend prompted President Putin to announce in June 2013 that:

---

[28] 'Govt approve price hikes of 15% in 2013–14, 14.6–15% in 2015 – Klepach', *Interfax*, 27 April 2012, Moscow.

[29] Consensus estimates from Wood Mackenzie Consultants and various investment banks and commentators, also confirmed in conversation with Gazprom in June 2013.

[30] 'Tariff regulator cuts Gazprom's prices for wholesale industrial users by 3%', *Interfax*, 1 April 2013, Moscow.

[31] 'Regulator confirms 3% Gazprom wholesale gas price rise for industry on Aug. 1', *Interfax*, 15 July 2013, Moscow.

[32] 'Russia: Industry in stagnation', Raiffeissen Research, 16 July 2013.

[33] 'Russia slashes 2013 growth forecast', *Financial Times*, 11 April 2013, Moscow.

... rapid tariff growth has become a separate and significant factor in unwinding inflation ... and a real factor behind economic deceleration, growth in costs and losses in the competitiveness of our producers. Obviously, tariffs can no longer grow at their previous rates.[34]

On this occasion Putin was referring not only to gas tariffs but also to electricity and heat tariffs, both of which are in their turn linked to gas prices for power and district heating companies.

The second key catalyst for a rethink on pricing policy came in 2012, when the price at which independent producers were prepared to sell their gas to end-consumers fell below Gazprom's average price for the first time in the post-Soviet era. This implied that the regulated gas price had reached a level at which companies such as Novatek were prepared to offer discounts to customers in order to win market share. It also demonstrated that there is an implicit oversupply of gas in Russia, driven not only by the completion of the development of Gazprom's huge Bovanenkovo field on the Yamal peninsula, but also by the increased availability of gas from non-Gazprom producers who regard their reserves as commercial at current prices. The result has been a complete reversal of the historic trend for consumers to prefer purchases of regulated price gas over those of new gas, with the implication that buyers would always turn to Gazprom first. From 2012 the preference for consumers has been to buy new gas first, at a price below the regulated price, before turning to Gazprom for gas that it is forced to sell at a higher regulated price. Figure 5.3 shows the development of this trend, with Novatek's sales price to end-consumers being 16 per cent higher than Gazprom's average price in 2007 but 2 per cent below it in 2012.

This discounting by the Independents has been reflected in the contracts that have been signed with new consumers during 2012 and 2013. Table 5.4 details the major deals signed since 2009, and demonstrates that companies including Rosneft, Novatek, Lukoil, and TNK-BP (now owned by Rosneft) have been able to tempt a significant portion of industrial and power customers away from Gazprom. They are offering prices reportedly 5–10 per cent below the regulated price to secure sales agreements lasting for up to 25 years.[35] It is clear from the contracts that have been signed that consumers of all sorts are actively seeking out the most cost-competitive gas supply. Even the power company Mosenergo, a Gazprom subsidiary, has signed a three-year gas supply contract with Novatek. In total more than

---

[34] 'Restrictions on monopoly tariffs might cause complications in energy sector', *Interfax*, 22 June 2013, St Petersburg.

[35] Henderson, J. 'Evolution in the Russian Gas Market: The Competition for Customers', Working Paper NG 73, OIES, January, 2013.

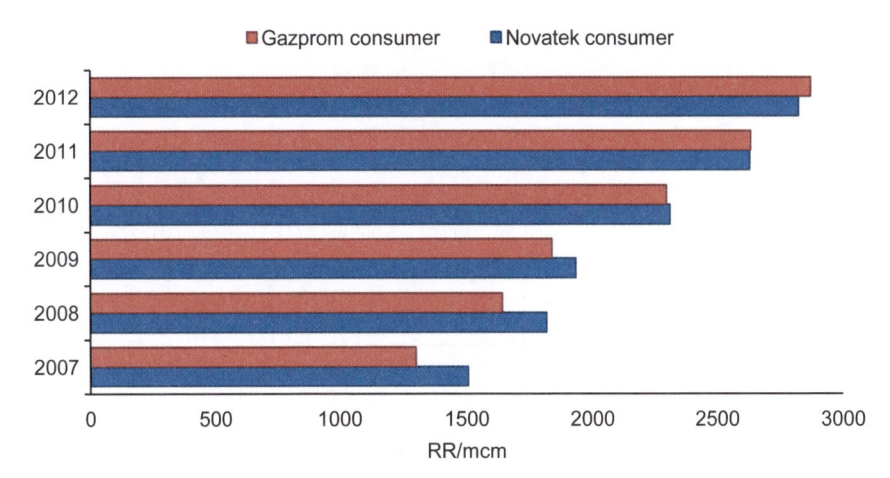

Figure 5.3: Comparison of Gazprom and Novatek gas sales prices
Source: Gazprom and Novatek IFRS Financial Statements.

1250 Bcm of gas has now been contracted for sale by Independent producers on a long-term basis, and indeed so many contracts were signed in 2012–13 that concern has been voiced about the ability of some producers to meet their obligations (see Chapter 13). Nevertheless the overriding conclusion is that real competition has now started to emerge in the Russian gas market.

It is not just price that is encouraging this change, but also the ability of both consumers and producers to have confidence in the availability of long-term transport arrangements. Third-party access to the trunk pipeline system is now effectively guaranteed for producers who can demonstrate that they have a valid sales contract and can deliver gas of suitable quality into the system. This has been demonstrated in a number of cases tested in the Russian courts, even though the pipeline regulatory regime still provides neither non-discriminatory access nor a satisfactory framework for investment. Nevertheless the implementation of third-party access laws has developed to the extent that the non-Gazprom producers can sell gas profitably, while paying a meaningful transport tariff. (See the section 'Transportation tariffs and the evolution of pipeline regulation', page 136.)

A further conclusion is that the Russian domestic market provides another example of the potential oversupply of gas, with Gazprom forced to rein in its production plans, and announce delays to some of its main field developments, in the face of reduced demand (see Chapter 11). One of the key drivers of this trend is the increasing competition from Independent producers who have been able to tempt customers away from Gazprom with lower-cost gas.

Table 5.4: Gas contracts signed by Independent gas producers since 2009

Seller	Buyer	Volume (Bcm)	Duration (year)s	Announced	Comment
Novatek	Inter RAO	7.7	6	Nov. 2009	Direct sales to Inter RAO for 2010-2015
Novatek	OGK-1	57	6	Nov. 2009	Sales to Inter RAO subsidiary for 2010-2015
Novatek	MMK	50	10	June 2012	c.5 Bcm/year+ from 2013
Lukoil	E.On Russia	2.24	10	June 2012	Supply to Yaivinskaya GRES
Novatek	Mechel	17	11.5	July 2012	Part of new marketing business in Cheliabinsk
Novatek	E.ON Russia	150	15	Aug. 2012	Four contracts to supply E.ON power assets in Russia
Novatek	Fortum	30	15	Aug. 2012	Contract to supply Niaganskaia GRES
Novatek	Uralchem	8	5	Sep. 2012	1.6 Bcm/year from 2013–17
Rosneft	E.ON Russia	5	3	Sep. 2012	Supply to Surgutskaia GRES
Rosneft	Fortum	13	5	Sep 2012	Supply to power plants
SurgutNG	E.ON Russia	n.a.	3	Sep. 2012	Supply to Surgutskaia GRES
Novatek	Severstal	12	5	Oct 2012	Contract to run from Jan 2013
Rosneft	Inter RAO	875	25	Nov. 2012	Supply to Inter RAO power assets from 2016–40
Novatek	Mosenergo	27	3	Dec. 2012	Supply to Gazprom power subsidiary in Moscow
TNK-BP	TGK-5	est. 15 Bcm	17	Dec. 2012	Supply to IES subsidiary from 2013–30
TNK-BP	TGK-7	est. 20 Bcm	15	Dec. 2012	Supply to IES subsidiary from 2015–30
TNK-BP	TGK-9	est. 20 Bcm	17	Dec. 2012	Supply to IES subsidiary from 2013–30
Novatek	E.ON Russia	est. 70 Bcm	13	Sep. 2013	Supply to E.ON's Surgut power plant, 2014–27
Rosneft	Fortum	8.3	6	Sep. 2013	Additional supplies to Fortum assets in Russia
Rosneft	Enel	NA	11	Sep. 2013	Gas supply to OGK 5 from 2014–25

Source: Company press releases.

It appears that, despite the continuing presence of a regulated domestic gas tariff, consumers and Independent producers have effectively established a 'quasi-market' price at a figure below this level. Indeed Economic Development Minister Andrei Belousov has gone so far as to suggest that if prices were liberalized now they would actually fall rather than rise.[36] In other words, the regulated price has become a ceiling rather than a floor. However, Gazprom's continuing dominance of domestic supply means that consumers cannot take full advantage of this situation, as they are essentially still obliged to purchase most of their supply at regulated prices from the state-controlled gas company. As a result, they have been actively lobbying for either an end to regulated prices or at least for a reduction in price growth.[37]

As a result of these two trends (declining industrial growth and the ability of Independents to supply lower-cost gas) the Russian government has been forced to reconsider its plans for the future direction of the regulated gas price. Firstly, in April 2013 the Economic Development Ministry announced that it would be recalculating natural gas price increases for 2013–16,[38] and although by June it had concluded that it was too late to change the outcome for 2013 (a 15 per cent increase introduced in July 2013), it initially proposed that the rate of increase for 2014 to 2016 should be 5 per cent rather than the original 15 per cent.[39] Following President Putin's comments on restraining tariff increases, however, the plans for growth in all monopoly tariffs, including gas, became even more stringent. By October 2013 the strategy for industrial gas prices had been confirmed by Prime Minister Dmitrii Medvedev as a price freeze in 2014 followed by price rises linked to the previous year's rate of inflation.[40]

To conclude: the Russian authorities have realized that domestic gas prices do not need to increase materially from current levels, but, rather than allow an active market to set the price, they have decided to continue their policy of intervention to ensure that both inflation and GDP growth can benefit from lower energy prices. Figure 5.4 below compares the current low price growth outlook (assuming 5 per cent annual inflation

---

[36] 'Belousov: No longer possible to revise 2013 gas tariff increase; could slow to 5% in 2014', *Interfax*, 30 April 2013, Moscow.

[37] Pirani, S. 'Consumers as players in the Russian gas sector', Energy Comment, OIES, January, 2013, pp. 10–12.

[38] 'Belousov: No longer possible to revise 2013 gas tariff increase; could slow to 5% in 2014', *Interfax*, 30 April 2013, Moscow.

[39] 'Proposals on 2014–16 natural monopoly tariffs to be ready by August', *Interfax*, 15 May 2013, Moscow.

[40] 'Govt plans to restrict growth in natural monopolies' tariffs by previous year's inflation – Medvedev', *Interfax*, 7 October 2013.

from 2015) with the previous more aggressive plan to reach export netback parity. It highlights the dramatic shift in strategy during 2013 as well as the current irrelevance of the netback parity price target.

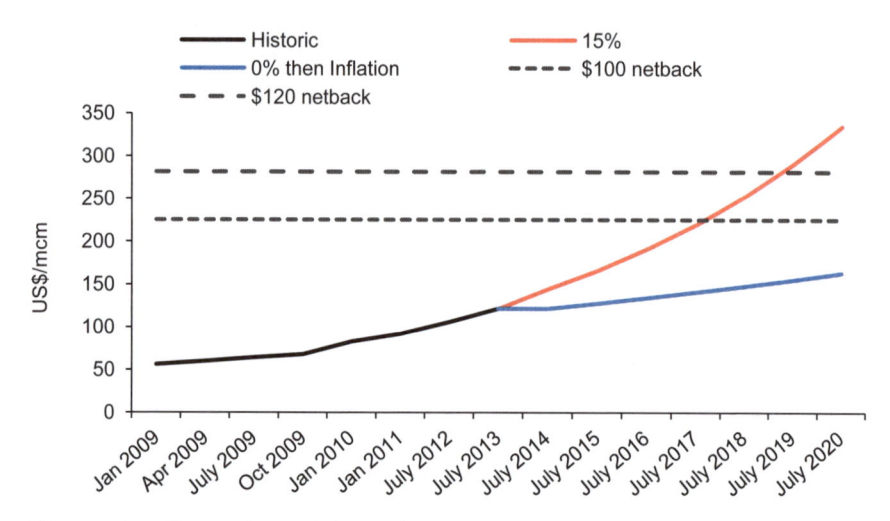

Figure 5.4: Comparison of domestic gas price growth and export netback parity levels

Source: Federal Tariff Service and author's estimates.

This new outlook for the regulated domestic gas price leaves one major anomaly in the Russian gas market: Gazprom is still forced to sell gas to domestic consumers at a higher price than is being offered by its competitors, as it cannot offer to sell gas at a price below that set by the FTS. This has two important consequences. Firstly, Gazprom is handicapped in its attempts to retain customers, as its main competitors are allowed to discount the regulated price and are therefore automatically preferred by consumers who can access their gas. The result has been that Gazprom's best customers have been cherry-picked by the Independents. This is reflected in the declining quality of Gazprom's customer base. Figure 5.5 shows that since 2007, the share of power companies in Gazprom's sales mix has declined from 37 per cent to 28 per cent, with almost exactly the opposite trend in the Municipality and Residential segment, whose share of Gazprom's sales has increased from 26 per cent to 36 per cent. The problem for Gazprom is not only that households pay lower prices but also that district heating companies in particular have a much higher proportion of non-payments. This has been reflected in a 44 per cent increase in the overdue debt from Gazprom's customers between 2011 and 2012, which now stands at RR83 billion.

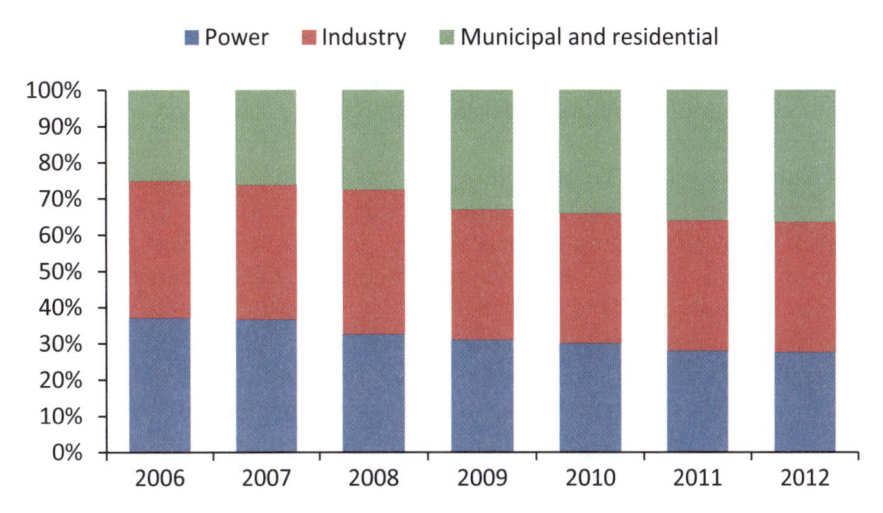

Figure 5.5: Split of Gazprom sales by major customer segment
Source: 'Gas Supply to Domestic Market', 2006–12, Gazprom presentation to pre-AGM press conference, June 2013.

The second important consequence is that Russian customers, who overall still have to buy most of their gas from Gazprom because of its continuing dominance of the gas resource base, are effectively paying a higher than 'market' price for their gas. The Independents have limited, although growing, gas supply (see Chapter 13), while Gazprom maintains a 73 per cent share of the domestic Russian supply,[41] meaning that the relatively high regulated price that it is compelled to charge is now a burden on Russian consumers, having historically been a price that offered a subsidy.

The obvious solution to this debate is the re-establishment of the Gas Exchange that was trialled in 2006–8. As discussed above, plans to achieve this are making gradual progress, with the Energy Ministry having formed a task force including Gazprom, the Russian Gas Association, St Petersburg International Mercantile Exchange (SPIMEX), and Moscow International Commodities and Currency Exchange (MICEX) to consider the creation of an exchange in St Petersburg in 2014.[42] However, it would appear that such an exchange would not be used to trade more than about 10 per cent of gas sold in the domestic Russian market if the proposed volumes

---
[41] 'Gas Supply to Domestic Market', Gazprom presentation to pre-AGM press conference, June 2013, slide 5.
[42] 'Energy Ministry sets up task force on gas trading', *Interfax*, 29 March 2013, Moscow.

(up to 17.5 Bcm from Gazprom and a figure no larger than this from the Independents) are to be believed. The idea of allowing Gazprom to offer gas at a discount to the regulated price for the first time is therefore also being considered. Deputy Energy Minister Kirill Molodtsov has stated that the Energy Ministry is considering discounts in the range of 0–30 per cent.[43] The Anti-Monopoly Service has announced its approval for the potential change, but interestingly has also stated that 'any tariff in the regulated sector must be based on market analysis … [and] should not be exclusively low'.[44] This raises the question that if Gazprom's prices can be higher or lower than the regulated price, and if Independent producers already sell gas at a discount to the regulated price, then is there a need for a regulated price at all?

We attempt to provide a pictorial answer to this somewhat complicated question in Figure 5.6, which describes a possible outlook for domestic gas prices in Russia. As described above, a 15 per cent increase in the regulated price was introduced in July 2013, and the government has now decided that there will be no increase in the price in July 2014. This means that the July 2013 average price of approximately $120/mcm will be fixed for two years.[45] Thereafter the Russian authorities have indicated that the regulated gas price will only rise in line with inflation, and we assume that this will happen for the rest of the decade, reaching a level of just over $160/mcm by 2020. However, this regulated price is increasingly likely to become a guide price for both the Independents and Gazprom – if and when the latter is given permission to offer discounts. At present Gazprom is already allowed to sell a portion of its gas at a premium to the regulated price,[46] and of course the Independents can sell at any price that consumers will bear above or below the regulated price. So if Gazprom is also permitted to price its gas below the regulated level, then, in effect, the gas price will be set in a quasi-competitive market guided by the regulated price. At the time of writing it is unclear how this process will work, and it is very likely that Gazprom will still be directed to sell a part of its gas to certain consumer segments at fixed prices. Nevertheless, from 2014, gradual moves towards a more liberal pricing environment are increasingly possible.

---

[43] 'Government to discuss Gazprom gas sales below regulated price in mid-November', *Interfax*, 28 October 2013.

[44] 'Antimonopoly service says nothing wrong with possible Gazprom discount', *Interfax*, 25 October 2013.

[45] This is an average price for industrial users in Russia. A different price is actually set for each of the country's 89 regions, and this price of $120/mcm is an approximate average that roughly equates to the price paid by consumers in the Moscow region.

[46] Any volumes classified as new purchases above the levels consumed in 2007 can be sold by Gazprom at a negotiated price above the regulated level.

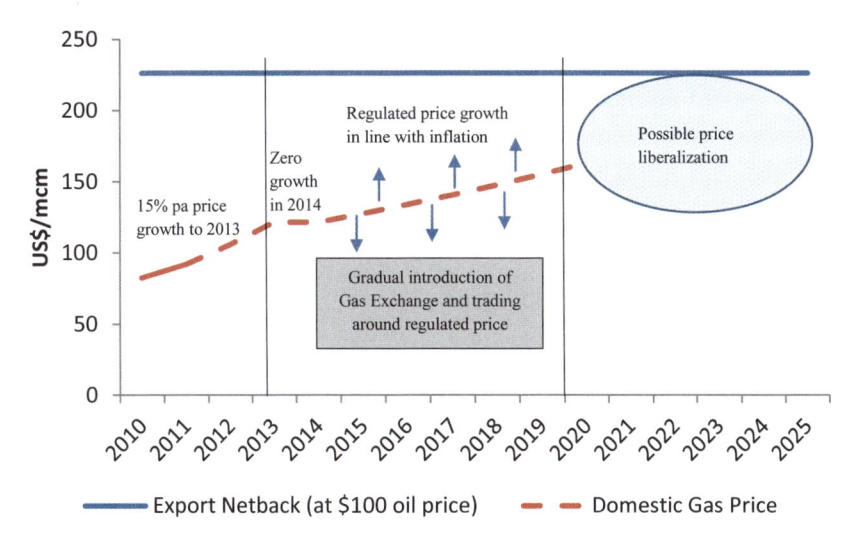

Figure 5.6: A possible outlook for Russian domestic gas prices
Source: Author's estimates, Federal Tariff Service.

Negative impact on Government Budget after changes in Mineral Extraction Tax

Discussions within the Russian government on the pace at which electricity and gas tariffs should be increased are influenced by two assumptions. On the one hand there is the view, expressed by President Putin as mentioned above, that tariff increases must be constrained in order not to damage economic recovery. This argument was recently put by researchers at the Energy Research Institute of the Russian Academy of Sciences, who found that slowing down energy price rises could add 0.8–1.3 per cent to GDP growth.[47] On the other hand, there is the problem faced by the Finance Ministry, that the slowdown in gas prices has a significant negative impact on Budget revenues.

At a time when the regulated gas price had been expected to increase towards export netback parity at a rate of 15 per cent per annum, the Russian government also decided to increase the rate of the Mineral Extraction Tax (MET), effectively a gas royalty paid per mcm of production, in order to capture up to 80 per cent of the price growth. Essentially the state had planned to use Gazprom, and to a lesser extent the Independents,

---

[47] Makarov, A.A., Mitrova, T.A. et al. *Vliianie rosta tsen na gaz i elektroenergiiu na razvitie ekonomiki Rossii* [*The influence of gas and electricity price rises on the development of the Russian economy*], Institute of Energy Research of the Russian Academy of Sciences, Moscow, 2013.

as a tax collection system aimed at extracting more revenue from gas consumers. From 2006 until 2010 MET was charged at a flat rate of RR147/mcm (approximately $5/mcm or $0.75/boe), and this was increased to RR237/mcm in 2011, with all gas producers paying the same amount of tax. From 2012, however, a longer-term plan for growing MET was introduced, with more rapid increases and a higher tax burden for the owner of the gas pipeline system (that is, Gazprom). This differential was planned to be closed over time, but in 2012 it meant that the MET rate for Gazprom jumped to RR509/mcm while that for the Independents only went up to RR251/mcm. It was anticipated that the two levels of tax would effectively equalize in 2016 or 2017 at approximately RR1200/mcm ($40/mcm or $6/boe). Given that total Russian gas production is approximately 650 Bcm/year, this increase in tax would have brought around $23 billion per annum into the Russian budget by 2017.

However, all the gas producers had been concerned about the fixed nature of the proposed tax increases, with set figures for MET being established without reference to any price changes on international or domestic markets. As a result, a lobbying process began in early 2012 and culminated in the announcement of a more flexible MET price calculation in May 2013.[48] The new tax, approved by the Duma in July 2013, has a base level of RR35/mcm but is then adjusted according to a number of key parameters, namely: the price of gas on domestic and international markets, the share of gas sold on those markets, the split of gas and liquids output from a field, reservoir depletion, depth and complexity, and finally the geographical location of fields (in order to adjust for transport costs). The overall conclusion is that the total levels of taxation will differ only marginally from the previous fixed rate of MET if the domestic gas price rises by 15 per cent per annum over the next three years, but will allow producers to pay an appropriate amount – more or less – depending on the actual price outcome.

Figure 5.7 shows a comparison between the old and new rates in 2014 for production at Gazprom's Yamburg field and Novatek's Yurkharov field, as well as for the condensate rate at Yurkharov, assuming that the gas price in 2014 has risen by 15 per cent. As can be seen, the rate for Gazprom is higher (because of its exposure to export prices), and its tax has also increased slightly more (11 per cent) than the rate for Novatek (2 per cent). However, these rates also reflect different parameters for each field, meaning that comparisons become somewhat redundant, as each asset now effectively has its own tax rate. Overall, though, the tax rates are slightly higher under the new regime on the old domestic price growth assumptions.

---

[48] 'Russian MinFin unveils gas extraction tax formula for 2014', *Interfax*, 22 May 2013, Moscow.

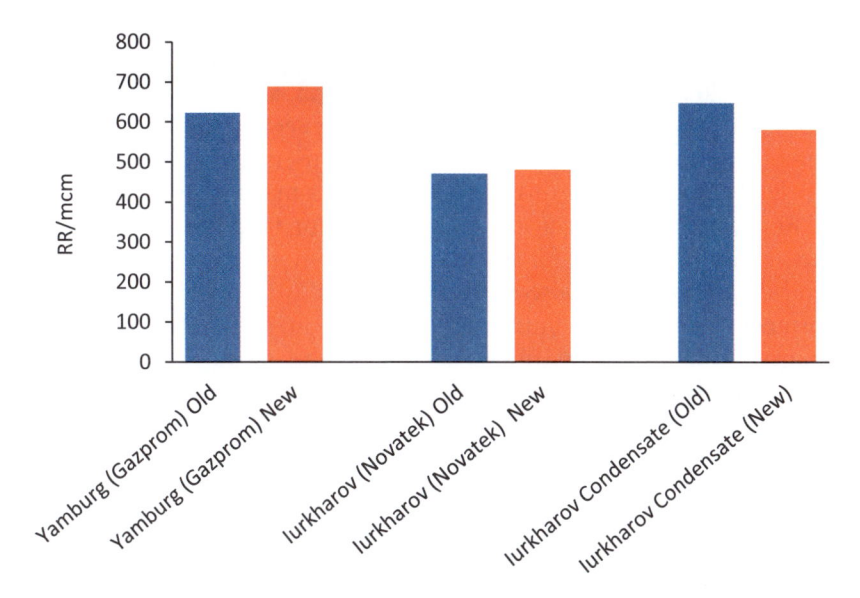

Figure 5.7: Comparison of old and new MET rates for gas production in Russia (data based on 2013 prices)
Source: Deutsche Bank Research.

The key criterion, however, is that tax revenues will remain the same if domestic prices increase by 15 per cent per annum; as discussed above, this is unlikely to be the case. As a result the government may be facing a reduction in expected budget revenues, at a time when it is also starting to have to offer more extensive tax breaks to encourage increased investment in the oil sector. Deputy Finance Minister Sergei Shatalov has estimated that the new proposed domestic gas price growth trajectory of 5 per cent per annum (rather than 15 per cent) would reduce Russia's budget revenues by approximately RR50 billion ($1.6 billion) in 2014,[49] but this differential could rise to as much as $7 billion by 2017 when the differential in prices will have widened further and production is likely to have increased.

Therefore, although the proposed changes to MET could provide a much fairer reflection of market reality for producers, and will increase the tax take for the government in comparison to 2013 levels, the benefit to the Russian budget is likely to be significantly lower than previously anticipated. This is due to the impact of lower domestic gas prices, which in turn have been brought about by the growing impact of market forces on the supply of, and demand for, gas.

---

[49] 'Slowdown in gas price growth could cost budget 50bn roubles in 2014–15', *Interfax*, 27 June 2013, Moscow.

Household tariffs

The discussion above relates entirely to the tariff for industrial users, who account for most gas sales in Russia. As can be seen in Figure 5.8, despite attempts to remove the subsidy to household users in the late 1990s and early 2000s, the price for this section of the gas market has remained at a 20–30 per cent discount to the industrial sector tariff for the past seven years. Every time a case for increasing prices more aggressively is made, political objections are raised. For example, in 2012 prices for household consumers were frozen until after the election cycle, and the current changes to industrial gas prices do not involve changes for household consumers. Indeed Gazprom noted in its latest accounts that non-payment from households rose by 47 per cent in 2012,[50] suggesting that consumers are feeling the pressure of higher prices and slower economic growth.

In response to this pressure, in 2013 the Russian government once again decided to delay increasing the gas price for household customers to a level that would remove the subsidy that is currently being funded by industrial consumers. Indeed it has even gone further by announcing that, following a price freeze in 2014, prices for household consumers may only rise at a rate equivalent to inflation less 30 per cent,[51] meaning that the discount to the industrial price will actually widen rather than close.

This strategy has a clear political motivation, and certainly contradicts the advice of research by Russian academics which suggests that the removal of the domestic gas subsidy could be a significant catalyst for stimulating economic growth. Research from the Energy Research Institute of the Russian Academy of Sciences[52] argues that removal of the household subsidy could allow a reduction of up to 20 per cent in the industrial gas price, improving the economic competitiveness of energy-intensive industries in Russia and allowing them to generate increased profits and cash flow. The subsequent trickle-down effect from these energy-intensive industries into other areas of the economy, through increased investment and intermediate consumption, could ultimately encourage the development of less energy-intensive industries and catalyse the diversification of the domestic economy that is one of the Russian government's major economic goals.

---

[50] 'Gazprom proposes solutions to rising non-payment by Russian consumers', *Interfax*, 21 March 2013, Moscow.

[51] 'Tariff freeze in 2014 could affect industry, households – Ulyukaev', *Interfax*, 9 September 2013.

[52] Makarov, A.A., Mitrova, T.A. et al. *Vliianie rosta tsen na gaz i elektroenergiiu na razvitie ekonomiki Rossii* [*The influence of gas and electricity price* ...] op. cit.

At the time of writing it appears that this long-term objective has been sacrificed in favour of the shorter-term political desire to avoid the unrest and possible instability that might be caused by any significant increase in domestic energy prices.

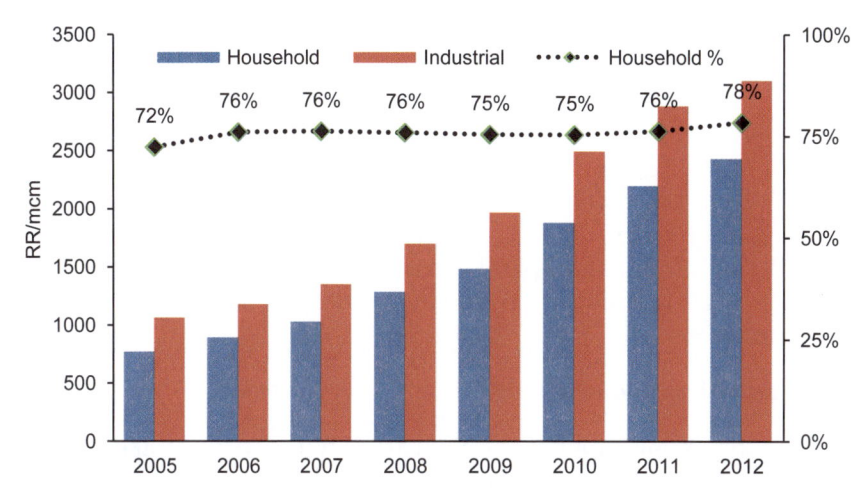

Figure 5.8: Household gas tariffs compared to industrial gas tariffs in Russia
Sources: Gazprom Bond Prospectuses in 2009 and 2012.

Future direction of regulated pricing and market reform

The Russian government has not given any indication that it might introduce a fully liberalized market by removing the regulated price altogether and, given its concerns about the impact of energy prices on inflation and economic growth, it would seem to be a reasonable assumption that this attitude will remain prevalent for the remainder of this decade. Nevertheless, if all the players in the market are allowed to buy and sell gas at above or below the regulated price, and if this is combined with the re-introduction of a Gas Exchange in 2014, then it might be possible to foresee a process of 'managed liberalization' of prices in the period 2014–20. However, further progress in pipeline regulation and the setting of transport tariffs (discussed in the next section) would also need to be an integral part of this process.

Thereafter, at some point in the 2020s, once prices have started more closely to approximate the netback parity level, it might be possible to conceive of full liberalization of domestic gas prices, and this could even occur earlier than 2020, either if economic conditions in Russia improve or if European export prices fall from their 2013 levels, bringing netback parity closer by default. However, at the time of writing, its seems most likely that the Russian government will try to maintain some control over the direction and level of domestic gas prices for the foreseeable future, while allowing increased levels of flexibility to all market participants.

Transportation tariffs and the evolution of pipeline regulation

Further reform of the domestic gas market would imply not only change in the level of regulated prices, and the possible re-establishment of the Gas Exchange, but also development of pipeline regulation. In this section we provide an overview of the transmission and storage system before discussing in turn the third-party access regime – together with proposals to change it, transmission tariffs, and the relationship of regulated prices to regulated tariffs.

Gas transmission and storage

Plans to break Gazprom into production, transportation, and trading divisions and to create a competitive gas market have been postponed to the more distant future, as noted above.[53] Gazprom still remains the dominant player in the Russian gas market, and much of this dominance comes from its status as the sole owner of the Unified Gas Supply System (UGSS).[54] Gazprom is obliged to manage the UGSS, provide for its (re)construction, functioning, and modernization, and undertake dispatch control of the UGSS and related infrastructure. The principle of indivisibility of the UGSS is firmly embedded in the Law on Gas Supply; furthermore, the Law states that technological and dispatch management of all objects connected to the UGSS can only be undertaken by the UGSS owner.[55]

---

[53] These plans were most recently reiterated in a strategy document published in January 2014 by the energy ministry, which stated that it is 'necessary to have a perspective to complete the separation of natural-monopoly activities from competitive ones in the Gazprom system, and subsequently implement the transition from regulated wholesale prices to regulation of transport tariffs'. Ministry of Energy, *Energetichskaia strategiia Rossii na period do 2035 goda (osnovnye polozheniia)*, January 2014.

[54] The 1999 Law on Gas Supply (Art.6) defines the UGSS as 'a ... complex of technologically, organisationally and economically connected and centrally managed production, transportation, storage and supply objects'. The UGSS is the main component of the Russian federal gas supply system, which also includes the regional gas supply systems, gas distribution systems, and independent organizations (Art.5). Federal Law 'On Gas Supply in the Russian Federation' (hereafter The Law on Gas Supply) No. 69-F3, 31 March 1999.

[55] The Law on Gas Supply, Articles 13 and 14.

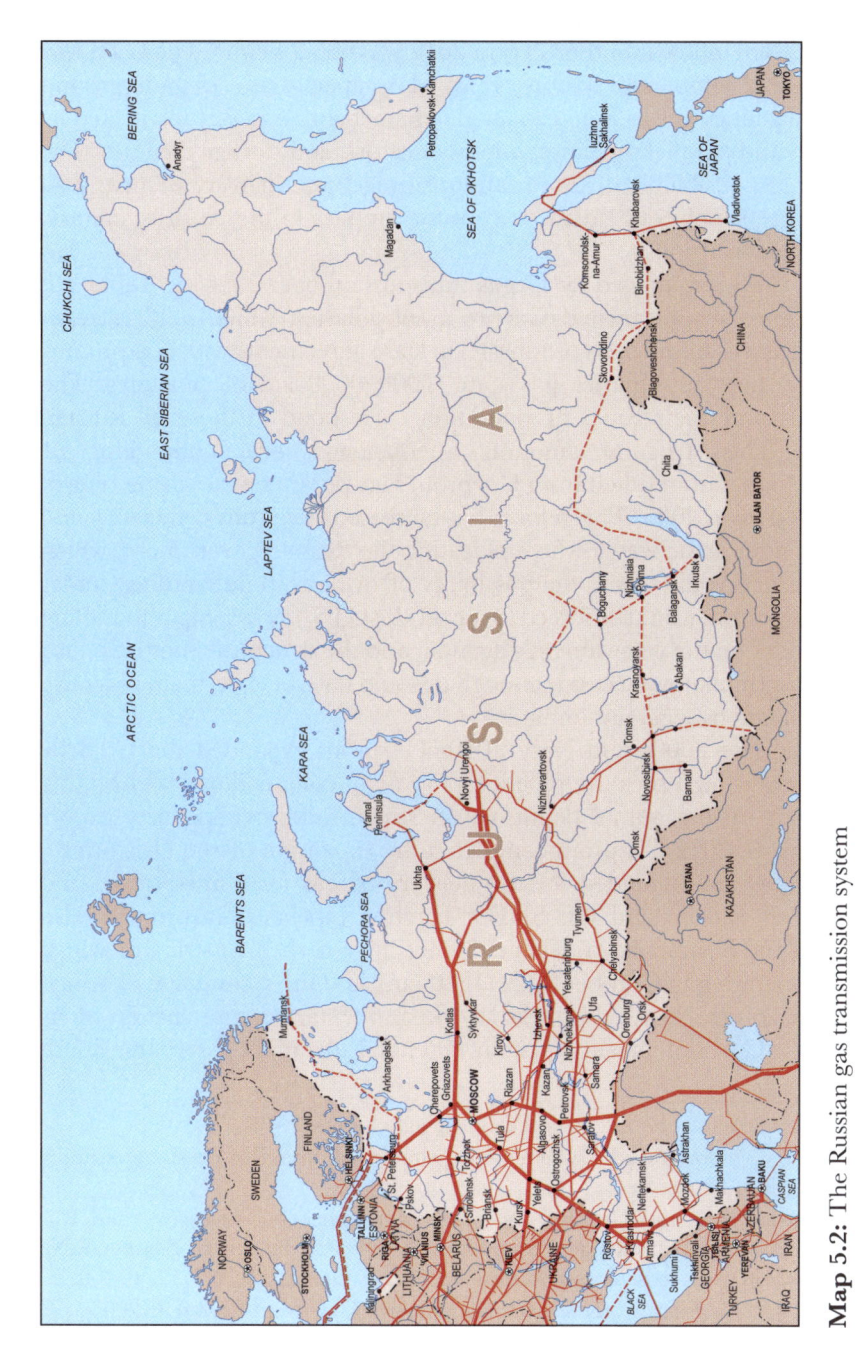

Map 5.2: The Russian gas transmission system
Source: Oxford Institute for Energy Studies.

The Gas Transmission System (GTS), a network of high-pressure pipelines with a total length of 168,300 km,[56] forms the major part of the UGSS. Total inflow into the GTS in 2012 was 666.2 Bcm.[57] The GTS also includes 222 compressor stations (43.87 MW capacity) and 25 underground storage facilities (68.16 Bcm capacity). The network is highly integrated by multiple and parallel pipelines, interconnectors, and storage facilities. The GTS is responsible for transportation, storage, and delivery of nearly all natural gas supplies in Russia (except for supplies to the Norilsk, Iakutsk, and Sakhalin regions).

The GTS is managed by 17 gas transmission operators (see Table 5.7, page 150) – wholly owned Gazprom subsidiaries, which hold relevant network assets on the basis of long-term lease agreements with Gazprom.[58] Each of these subsidiaries has some 8000–9000 km of pipelines. The boundaries of their zones of operation correspond to those of Russian regions. This structure amounts to managed legal unbundling of transmission, and resulted from Gazprom's corporate restructuring reform embarked on in 2005. This reform has increased Gazprom's efficiency and has also gone some way towards placating the Russian government, which at the time was discussing the possibility of ownership unbundling, under which Gazprom would be forced to sell off the GTS. Ownership unbundling is no longer on the government's agenda, and the structural foundation for a non-discriminatory access regime that would have been its main objective has been laid by legal unbundling.[59]

In the late 2000s and early 2010s Gazprom expanded the GTS by building new high-pressure pipelines, mostly in conjunction with bringing new fields on the Yamal peninsula on stream and creating new export corridors. In 2012 Gazprom completed the expansion of the Griazovets–Vyborg pipeline (intended for the delivery of gas for Nord Stream and north-west Russia) and the SRTO–Torzhok pipeline. Gazprom is also building the Bovanenkovo–Ukhta pipeline system, the first string of which began operation in October 2012 (1240 km), and the second line of which is under construction. Also in 2012, Gazprom started construction of its Southern Corridor pipeline system, intended to supply gas to the South

---

[56] Gazprom, Annual Report 2012. Pipeline length for end 2012.
[57] Total GTS inflow includes system inflow, withdrawals from Russian storages, and a change in GTS reserves.
[58] Seliverstov, S.S. and Gudkov, I.V. 'The Development of Electricity and Gas Networks in Russia' in Roggenkamp, M.M., Barrera-Hernandez, L., Zillman, D.N., and del Guayo, I. (eds.) *Energy Networks and the Law* (Oxford: OUP, 2012), p. 400.
[59] Stern, J.P. *The Future of Russian Gas and Gazprom* (Oxford: OIES/OUP, 2005), pp. 184, 194–5; Seliverstov and Gudkov 'The Development of Electricity and Gas Networks in Russia', op. cit., p. 402.

Stream pipeline (see Chapter 7) and to central and southern Russian regions; completion of the first string (880 km) is planned for 2015, and of the whole system (2506 km) for 2019.

Despite these recent additions, the majority of the GTS was built in the Soviet period. At the end of 2012, 74.7 per cent of pipes were more than twenty years old and only 13.2 per cent were built during the past decade (see Appendix 3).[60] This is a direct result of Gazprom's underinvestment in GTS maintenance and refurbishment, due to its financial difficulties in the 1990s: reportedly no more than 29 per cent of the required amount was made available.[61] In the 2000s, having recovered its financial position, Gazprom began to pay increased attention to the technical state of the GTS, and in March 2011 a comprehensive programme of upgrades for 2011–15 was adopted. (For details of implementation see Appendix 3.)

The existing legal framework does not include a clear set of rules for construction of incremental or new pipeline capacity. Historically, Gazprom, as the owner of the UGSS, has financed its development, while third parties compensated it through payment of regulated transmission tariffs.[62] The draft Law on Trunk Pipelines, which meant to establish *inter alia* a framework for capacity development, has remained under discussion in the Russian Duma since 1999; it was approved at the first reading but failed to progress further. Gazprom suggested that third parties should be required to contribute finance towards new/incremental construction, and their expenses compensated through transportation tariffs, although it was difficult to see how such a solution might be workable given that tariffs charged to third parties were already regulated (see below).[63] However, should new legislation be developed that would require both Gazprom and third parties to be charged non-discriminatory regulated tariffs, which would more accurately reflect their load factors, this might pave the way for a resolution of the issue of new and incremental capacity development.

Without a framework defining the source and mechanism of financing for new and incremental pipelines, it is difficult to expect Gazprom to build capacity to accommodate third parties' gas, particularly given the lack of profitability in the transmission sector at the existing level of transmission tariffs. Likewise, it is difficult to expect third parties to finance and build such capacity in the UGSS because the legal principle of indivisibility of

---

[60] Gazprom Annual Report, 2012, p. 55.
[61] Mitrova, T., Pirani S., and Stern, J.P. 'Russia, the CIS and Europe: gas trade and transit', in Pirani, S. (ed.), *Russian and CIS Gas Markets and Their Impact on Europe* (Oxford: OIES/OUP, 2009).
[62] Seliverstov and Gudkov 'The Development of Electricity and Gas Networks in Russia', op. cit., p. 412.
[63] Stern, J.P. *The Future of Russian Gas and Gazprom*, op. cit., p. 37.

the UGSS presupposes that all the networks of which it is formed, including their expansion, 'shall be owned by Gazprom, irrespective of who financed their creation or expansion', and that networks built by third parties outside the UGSS 'shall be centrally managed by Gazprom'.[64] As third parties' usage of the UGSS is set to increase, the issue of how incremental and new pipeline construction should be funded is becoming increasingly important.

Storage

Gas storage is an integral part of the UGSS and Gazprom is the sole owner of all Russian gas storage capacity. Gazprom operates 25 underground storage facilities with a capacity of 68.2 Bcm. During 2008–12, between 15.6 Bcm and 51.6 Bcm was injected into storage annually and between 18.6 and 47.6 Bcm withdrawn, depending on the severity of winter temperatures. Gazprom is also developing additional storage: 17.9 Bcm of new capacity is planned to be commissioned by 2015.[65] Gazprom's storage capacity still remains relatively small given the size of its market and extreme seasonal demand variations. This is due partly to a general problem of underinvestment in the 1990s but also, more importantly, to the fact that after the break-up of the USSR, the Russian UGSS lost storage in Ukraine and other former Soviet republics. The Ukrainian storage facilities (31 Bcm)[66] were built largely to ensure security of Soviet (and now Russian) exports to Europe. Throughout the post-Soviet period Gazprom continued to use Ukrainian storage capacity while also repeatedly attempting to acquire the Ukrainian transportation system, including its storage facilities; these attempts ultimately proved unsuccessful. Gazprom stopped using Ukrainian storage following the 'missing gas' incident in 2006;[67] its inability and unwillingness to use this storage without having an equity stake is understood to have been one of the reasons for Gazprom's difficulties in meeting nominations on some of its European long-term supply contracts during the cold spell in February 2012.[68]

Gazprom, well aware of the fact that its storage capacity is lower than it should be for optimal operation of the UGSS domestically – especially for meeting its contractual obligations on European LTSCs – has accelerated

---

[64] Seliverstov and Gudkov 'The Development of Electricity and Gas Networks in Russia', op. cit., p. 412.
[65] Gazprom's website, www.gazprom.com/about/production/transportation/underground-storage/.
[66] UkrTransGaz website, www.utg.ua/uk/activities/.
[67] For details on the 'missing gas' incident see Stern, J.P. 'The Russian-Ukrainian Gas Crisis of January 2006', Working Paper, OIES, January 2006, p. 4.
[68] Heather, P. and Henderson, J., 'Lessons from the February 2012 European gas "crisis"', Energy Comment, OIES, April 2012.

construction of new storage capacity in Russia (see above) and abroad. Aggregate active capacity used by Gazprom in non-Russian storage facilities was around 7.8 Bcm in 2012. The company is also expanding its access to storage facilities in Europe[69] and the CIS (the Abovyan storage in Armenia as well as the Osipovichskoe, Pribugskoe, and Mozyrskoe storage facilities in Belarus). Gazprom declares its strategic goal to be the increase in active capacity of its storage facilities abroad to 5 per cent of annual exports by 2030, with the priority of developing its own storage capacity.[70] In 2012 Gazprom's gas injections in non-Russian storage constituted 8 Bcm and withdrawals from them 6.2 Bcm.

In addition to its own requirements, Gazprom also provides storage services for third parties' gas for up to 10 per cent of total capacity. However, no third parties appear willing to invest in construction of new storage, for the same reasons that they are unwilling to invest in new transportation capacity (see above).

Third-party access: the existing legal framework

The Law on Gas Supply obliges the owners of all Russian gas supply systems, including the UGSS, to provide *non-discriminatory* third-party access (TPA) to available capacities in these systems to *any* party that requests it, provided that gas quality standards are met. The 1992 Presidential Decree No. 1333 also introduced the obligation on Gazprom to provide gas producers with access to the UGSS proportionally to their domestic production, provided that a single mechanism for price regulation is created.[71]

As the sole owner of the UGSS, Gazprom is obliged to manage and undertake dispatch control of the UGSS and any other objects connected to it; it thus has the exclusive right to transport third-party gas through it. The rules of TPA to the UGSS were laid down in the government's Resolution No. 858 (1997), which mandated that TPA must be provided subject to the following requirements (Art. 2.5):

- existence of transportation agreements (contracts) concluded between a third party and Gazprom (or Gazprom's transmission operators at Gazprom's request);[72]

---

[69] For details of Gazprom's European storages see Chapter 4.
[70] Gazprom, *Loan Notes* (Preliminary Base Prospectus), 12 July 2013, p. 4.
[71] Law on Gas Supply, Article 27; The 1992 Presidential Decree No 1333 'On the reformation of the state gas concern Gazprom into the Russian joint stock company Gazprom'. Notably, no single mechanism for price regulation has been established in Russia yet.
[72] The Regulation envisages the following duration of contracts: short-term (less than 1 year), medium-term (between 1 and 5 years), long-term (more than 5 years).

- availability of spare capacity in the system from the entry to exit point during the transportation period proposed by a third party,
  - 'spare capacity' is defined as technically possible capacity to accept and transport gas excluding volumes transported for Gazprom and for third parties under contracts in force at any given point in time, as well as volumes transported in line with presidential and governmental orders on compulsory supplies (Art. 1.1);[73]
- compliance of quality of the third party's gas and its technical parameters with standards and technical specifications;
- intake/offtake pipeline connections availability, with metering and quality control equipment, by the proposed date of start of transportation.

On their part, in order to obtain access to the GTS, third parties have to submit standard application forms to Gazprom which have to contain certain information, including sources of gas volumes, as well as start and end dates of supplies, intake and offtake points, documentary evidence confirming the ownership rights over gas, and evidence of the existing contractual obligation to buy or sell gas (Art. 2.6).

Although Resolution No. 858 was an important step towards establishing TPA rights, its scope and level of detail remains limited. As Gudkov notes, the Resolution does *not* regulate access to supply systems other than the UGSS; does not govern relations between the owner of the UGSS and its subsidiaries (but only with third parties); does not govern access to storage and LNG infrastructure; and does not contain anti-hoarding mechanisms or regulate connection to the UGSS.[74] It also does not set out how spare capacity is to be allocated in the event that more than one third party wants to ship its gas from one point to another.

Both Gazprom and third parties have often been critical of TPA implementation. Third parties have complained about the violation of their rights of access to the UGSS, particularly noting the lack of clarity of capacity allocation rules, the lack of cost-reflectivity of tariffs, and sub-optimality of transportation routes chosen by Gazprom for third parties' gas. Gazprom, for its part, has periodically complained that third parties have been using the network without contributing towards its development and also that they have often been 'out of balance, leaving gas

---

[73] The fact that spare capacity is defined by subtraction of capacity necessary for transportation of gas not only under Gazprom's contracts *but also under third-party's contracts*, suggests that *de jure* Gazprom does not have a priority right to access capacity.

[74] Seliverstov and Gudkov 'The Development of Electricity and Gas Networks in Russia', op. cit., pp. 401–2.

in the network which acts as a "free storage" service'.[75]

The Russian government has been a major driving force pushing towards improved implementation of Resolution No. 858 and provision of non-discriminatory TPA. As production of third-party gas has increased steadily, the government has become resolute that this gas should be marketed and sold profitably alongside Gazprom's gas; clearly, non-discriminatory TPA was a necessary condition for this objective to be met. It appears that the government has also lent its support to the Federal Antimonopoly Service (FAS), which has stated repeatedly that TPA has remained problematic and that the conflict of interest – due to Gazprom being the owner of the UGSS and the major gas producer – has led in some instances to TPA refusals.[76] The FAS launched several investigations into Gazprom's alleged discrimination against third parties:[77] some cases went in favour of third parties with Gazprom being found guilty of violating competition law (for example, in 2010, by failing to consider requests made by Gaz-Energo-Alians for access to the UGSS; and in 2012, by refusing access to the UGSS by Real-Gaz despite having been in a position to transport its gas as requested) while some went in favour of Gazprom (for example, in 2010, when Gazprom was found not guilty of violating competition law for refusing access to the UGSS by Rosneft on technical grounds). Although the fines imposed on Gazprom as a result of these cases were not substantial in financial terms, the very fact that the FAS became active in this respect is important.

The share of third-party gas transported by Gazprom through the GTS has been steadily increasing (Table 5.5) – partly as a result of government (including FAS) action; partly because more capacity became available in the UGSS due to the natural decline of Gazprom's production in the NPT region;[78] and partly due to Gazprom's willingness to accommodate the government's push for greater competition in order to avoid replacement of Resolution No. 858 by a less lenient framework (as proposed by the FAS in 2006 and again in 2012, see below). In 2012, Gazprom transported 95.8 Bcm of third parties' gas, or 14.6 per cent of GTS throughput.[79]

---

[75] Stern, J.P. *The Future of Russian Gas and Gazprom*, op. cit., p. 182.

[76] Smirnova, O. 'Anti-monopoly regulation of gas markets', presentation made at the round table 'Formation of domestic gas market: taxes, tariffs, prices, and investments', Forum 'Russia's Gas – 2012', 20 November 2012.

[77] Out of 212 antimonopoly cases in the gas sector considered in 2008–11, 28 were related to TPA issues. Ibid.

[78] Henderson, J., *Non-Gazprom Gas Producers in Russia*, op. cit., p. 222.

[79] Major third-party users of the GTS were Novatek and Itera which accounted for approximately 40.4% and 21.6% of supplies by independent suppliers, respectively. See Gazprom, *Loan Notes* (Preliminary Base Prospectus), 12 July 2013, p. 137.

Table 5.5: Gazprom's transport of its own and third-party gas, 2008–12

Year	Gas transported through the GTS (Gazprom's and third parties')	Volumes supplied to export markets (Bcm)	Volumes supplied to the domestic market by Gazprom and third parties (Bcm)	Volume of third-party gas transported (Bcm)	Russian gas production by third parties (Bcm)	Third-party gas as % of total UGSS throughput (%)*
2008	714.3	251.1	352.8	111.2	113.9	15.8
2009	598.7	195.6	335.6	66.5	120.9	11.3
2010	661.2	209.3	354.9	72.6	141.7	11.0
2011	683.2	217.7	365.6	81.5	155.8	12.0
2012	666.2	209.3	362.3	95.8	169.0	14.6

* Calculated on the basis of total GTS inflow.

Source: Gazprom Annual Report 2009, Gazprom Annual Report 2012, Gazprom Loan Notes.

Although the share of third parties' gas has remained below what it could have been, had it been allotted in proportion to production, this does not necessarily suggest discrimination on Gazprom's part. A significant share of third-parties' gas is supplied within the regions where it is being produced, and therefore does not require access to the UGSS. Overall, in the late 2000s and early 2010s, the issue of non-discriminatory TPA has become much less problematic. It is understood that most TPA requests have been granted, with only around 10 per cent of them being declined.

Since the late 2000s, as a result of both government action and market forces, increased competition – although arguably 'government-managed' – has developed between Gazprom and third-party producers in the Russian gas market, with the latter's gas becoming increasingly competitive not only in premium but also in regulated sectors (see 'Domestic gas prices and market reform' page 117). This would not have been possible without improved TPA.

Proposed changes to the legal framework

The FAS has long argued that the growing number of cases alleging discriminatory behaviour by Gazprom suggested that a more comprehensive regulatory framework than that provided by Resolution No. 858 might be necessary. The FAS has made several attempts to introduce new legislation to replace Resolution No. 858, most recently in April 2012. A new draft Resolution proposed by the FAS at that time, which laid down rules on non-discriminatory access to high-pressure pipelines,[80] remains under discussion in government. However, given that non-discriminatory TPA has recently become significantly less problematic, it is not clear whether the government will be willing to adopt the new draft.

The draft Resolution differs from Resolution No. 858 in several important respects. Most importantly, it proposes non-discriminatory access rules for Gazprom as well as for third parties; both of these would have to apply for access to spare capacity in the UGSS and other networks on equal grounds. This would create a level playing field for obtaining access proportionate to production; moreover, the draft states that a shipper does not have to be a producer. In contrast, under the existing regulatory framework Gazprom gets access to all the UGSS capacity it needs, with remaining spare capacity offered to third parties. The draft resolution suggests a 'sunset period', marked by the expiry of existing transportation contracts, after which the non-discriminatory access rules would apply. (A detailed account of the draft is given in Appendix 4.)

---

[80] Draft Resolution 'Rules On *Non-Discriminatory Access to High-Pressure Gas Pipelines in the Russian Federation*, available at the Russian Ministry of Economic Development website, www.economy.gov.ru/wps/wcm/connect/ec61c7004ad85dc2a080abaf3367c32c/pp.pdf?MOD=AJPERES&CACHEID=ec61c7004ad85dc2a080abaf3367c32c.

Transmission tariffs for third parties

Transmission tariffs charged to third parties by Gazprom for transportation of their gas through the UGSS are set by the FTS.[81] Methodology adopted in 2005 (a detailed account of which is given in Appendix 5) introduced a two-part tariff, which consisted of:

a) a fee for *usage* of the UGSS, set depending on the zones where the gas enters and exits the system (measured in currency units per mcm); and

b) a fee charged for *transportation* (measured in currency units per mcm/100km).

Since the methodology was first introduced in 2005, the tariffs have been regularly revised upwards (at a rate slightly below the rate of increase in wholesale gas prices). Here the FTS appears to have taken Gazprom's concerns into account, as the latter has long called for increase in tariffs, arguing that tariffs remained below the economically justified level of 75 Russian rubles (around $2.5)/mcm/100km.[82]

Russian tariffication methodology is not an 'entry–exit' methodology (where the shipper pays an entry charge to enter a zone and an exit charge to leave it) as defined and understood by EU regulations.[83] The Russian methodology is essentially zonal: the shipper needs to pay a fee for usage of the system which takes into account where the gas entered the system and where it left it, as well as a fee (either distance- or volume-related[84]) for transportation. The fee for usage constitutes around 70–80 per cent, and the fee for transportation around 20–30 per cent, of total revenue. Table 5.6 shows average levels of tariffs charged by Gazprom to third parties.

---

[81] Known as the Federal Energy Commission (FEC), under the responsibility of the Ministry of Economic Development and Trade prior to 2004, when it was renamed as the Federal Tariff Service (FTS) and placed under the direct responsibility of the prime minister's office, see Stern, J.P. *The Future of Russian Gas and Gazprom*, op. cit., p. 180. These regulated transmission tariffs are also charged to third parties for transportation of their gas through regional gas supply systems.

[82] Gazprom, 'Outstanding Result', *Corporate Journal Gazprom*, July–August 2011, p. 8. According to Gazprom the share of its revenues derived from transportation services rendered for third parties constituted only 2% of total sales in 2011, see Gazprom, *Loan Notes* (Final Prospectus), 17 July 2012, p. 63.

[83] With the EU consisting of a number of zones the borders of which have not yet been defined and which may or may not correspond to EU member states' borders, see Yafimava, K. 'The Third EU Package for Gas and the Gas Target Model: major contentious issues inside and outside the EU', Working Paper NG75, OIES, April 2013.

[84] See Appendix 5, Gas transportation tariff methodology, page 401.

Table 5.6. Average levels of transportation tariffs charged by Gazprom to third parties

Date of entry into force	In Russia and other Customs Union countries (ruble/mcm/100 km)	Outside Customs Union countries (ruble/mcm/100 km)
1 October 2004	19.37	$0.92
1 October 2005	23.84	$0.97
1 August 2006	26.4	29
1 March 2007	30.36	33.35
1 January 2008	36.13	40.02
1 January 2009	37.86	41.94
1 April 2009	40.44	44.79
1 July 2009	43.19	47.84
1 October 2009	45.74	50.66
1 January 2010	51.37	56.9
1 January 2011	56.15	62.19

Source: FTS website; FTS Information Bulletin No. 11 (385), 26 March 2010.

Our estimates suggest that a transportation tariff, charged as of January 2011 for transportation of gas across 2400 km (the average transportation distance for gas to a domestic Russian customer), on the basis of FTS order No 497-e/2, would be around $46–48/mcm (see Appendix 5). Given that the regulated gas price for industry in 2011 was RR3071/mcm ($104.50), it would have been possible for third parties to sell gas to customers profitably, as their cost of production is relatively low. This is a significant change compared to the situation in the mid-2000s when this would not have been possible, given the much lower regulated gas price for industry – notwithstanding the fact that the transportation charge was also lower.[85] As a result, the 'economic radius' for gas sales from the point of production – the distance within which it is profitable for third parties to sell their gas to customers – has more than doubled (to 2400 km) over the last 5–7 years. This suggests that, notwithstanding third parties' complaints about Gazprom's reluctance to transport their gas, and the aforementioned concerns about tariff methodology and selection of transportation routes, significant progress has been made towards non-discriminatory TPA during the last decade, both in respect of access and tariffs. This conclusion is also supported by the data on third parties' increasing share of gas transportation (see Table 5.5).

---

[85] Stern, J.P. *The Future of Russian Gas and Gazprom*, op. cit., p. 180.

Critique of the tariff methodology

Although the tariff methodology is very sophisticated, it nonetheless fails to address a number of aspects of third-party transportation, particularly in respect of cost reflectivity and determination of optimal routes for third-party gas.

As far as cost reflectivity is concerned, the FTS methodology's formulae for calculating both one-part and two-part tariffs have a component which takes into account the costs incurred by Gazprom (Gazprom Transgaz companies) for transportation of third parties' gas through the UGSS (so that the tariff would cover this cost as well as taxes and profit). However, the methodology does not explain how these costs are calculated. This lack of transparency has been a recurring subject of complaints by third parties, who argue that the costs used to calculate a tariff have been artificially high, leading to higher tariffs for third parties. It is impossible to know whether or not this is true, precisely because the cost base is not transparent.

Apart from the perceived lack of cost-reflectivity, third parties have also complained about Gazprom's discretion in route selection, as it is not required by law to transport third parties' gas by the shortest route. As a result, third parties could find themselves paying a higher tariff because of the potentially circuitous routes selected for them by Gazprom. Although it is a valid concern, and whereas third parties would have preferred their gas being transported by the shortest route, doing so might not necessarily support optimization of flows and minimization of the total volume of transportation in the UGSS as a whole – the objective around which the FTS tariff methodology is built. At the same time, it is impossible to conclude whether Gazprom might not be offering the shortest route to third parties because it is not an optimal solution, or because it wants to use this route for transportation of its own gas, or to make more money, or to utilize idle network capacity.

Transmission tariffs charged to Gazprom

When transporting its own gas, Gazprom charges itself a transportation tariff, the levels of which it sets internally and which are not regulated by the FTS. Table 5.7 shows the level of these internal transportation tariffs in 2005 and 2009. Interestingly, although in the 2000s third parties commonly complained that they were being forced to pay a transportation tariff far higher than Gazprom was charging itself – claiming that Gazprom charges itself between one-third and one-quarter of the regulated tariff[86] – comparison of the regulated tariffs charged to third parties with Gazprom's internal tariffs (Table 5.6 and 5.7) suggests that only in very few instances was Gazprom's internal

---

[86] Ibid.

tariff significantly lower (for example, Tiumen'-Transgaz, Perm'-Transgaz, Tat-Transgaz). In most instances the internal tariff appears to have been on a par with the regulated tariff, and in some instances even significantly higher (Kuban'-Gazprom, Tomsk-Transgaz, Astrakhan'-Gazprom).

Regulated gas prices vs regulated transmission tariffs

The question of regulated transportation tariffs needs to be considered against the background of prices for gas sold domestically. The Russian domestic gas market consists of regulated and non-regulated segments, and from 2006, the government has endeavoured to raise the regulated gas price, initially with the view of bringing it to a European netback level (as described in the section 'Domestic gas prices and market reform', above). But because of the sharp increase of oil (and hence gas) prices during the 2006–12 period, and the economic crisis of 2008–9 and subsequent financial recession, the Russian government pushed back the 2011 'deadline' for transition to European netback. Gazprom has to sell most of its gas at regulated wholesale prices, which for a long period were below the price of unregulated volumes, but in 2012 rose above those levels.

The government's consistent policy was to bring regulated domestic prices closer to the levels that would have been set by the forces of supply and demand, had those been allowed to be fully at play. This effort has only been partly successful. But the fact that under the existing legal framework only third parties' transportation tariffs are regulated –not those charged to Gazprom's subsidiaries – means that Gazprom's wholesale gas prices need to continue being regulated. This would no longer apply if non-discriminatory regulation of transportation tariffs charged both to third parties and to Gazprom were established.

The question of when this switch will be made – from regulation of prices to regulation of the transportation business for all users – remains open at the time of writing. Novatek and Rosneft have called for a more competitive marketplace where participants can have open access to customers via a transport and storage system equally regulated for all players, and have suggested that transport tariffs should start to decline rather than remain on a rising trend.[87] Meanwhile Gazprom argues that any shift to such a model must also involve all players being equally exposed to low and high margin customers, with transport tariffs gradually being adjusted to remove cross subsidies between customers.[88] Whatever the

---

[87] 'Novatek chairman calls for across-the-board competition on gas market', *Interfax*, 15 November 2013; 'Rosneft calls for reducing gas transport tariff', *Interfax*, 15 November 2013.

[88] 'Gazprom: domestic market share of independent gas producers to top 30% by 2016', *Interfax*, 19 November 2013.

Table 5.7: Transportation tariffs charged by Gazprom to its own transmission companies

Gazprom's Transmission Companies, 2005	Tariff for transportation (ruble/mcm/100 km)	Gazprom's Transmission Companies, 2009	Tariff for transportation (ruble/mcm/100 km)
Sever-Gazprom	26.54	Gazprom Transgaz Ukhta	42.68
Kuban'-Gazprom	101.5	Gazprom Transgaz Krasnodar	101.8
Surgut-Gazprom	26.55	Gazprom Transgaz Surgut	50.79
Tiumen'-Transgaz	14.6	Gazprom Transgaz Iugorsk	30.2
Len-Transgaz	35.22	Gazprom Transgaz St Petersburg	68.52
Mos-Transgaz	18.15	Gazprom Transgaz Moskva	39.27
Tomsk-Transgaz	102.3	Gazprom Transgaz Tomsk	147.13
Perm'-Transgaz	10.84	Gazprom Transgaz Tschaikovskii	19.93
Ural-Transgaz	26.42	Gazprom Transgaz Ekaterinburg	42.7
Bash-Transgaz	21.69	Gazprom Transgaz Ufa	38.78
Kavkaz-Transgaz	40.46	Gazprom Transgaz Stavropol'	71.34
Volgo-Transgaz	16.64	Gazprom Transgaz Nizhnii Novgorod	27.8
Iug-Transgaz	33.52	Gazprom Transgaz Saratov	51.49

Table 5.7: Transportation tariffs charged by Gazprom to its own transmission companies *(continued)*

Gazprom's Transmission Companies, 2005	Tariff for transportation (ruble/mcm/100 km)	Gazprom's Transmission Companies, 2009	Tariff for transportation (ruble/mcm/100 km)
Tat-Transgaz	11.93	Gazprom Transgaz Kazan'	20.69
Samara-Transgaz	17.92	Gazprom Transgaz Samara	34.88
Volgograd-Transgaz	22.45	GazpromTransgaz Volgograd	36.3
Kaspii-Gazprom	61.3	Gazprom Transgaz Makhachkala	149.88

Note. In 2005 Gazprom began reorganizing its 18 transmission companies, transforming them into 17 Gazprom Transgaz companies (hence the different names in 2005 and 2009); it is not clear whether the borders of their zones of responsibility have changed significantly in the process but the names of the new resulting companies suggest that they probably cover more or less the same pipeline geography as the old transmission companies. One noticeable change is the disappearance of Kaspii-Gazprom and Astrakhan'-Gazprom and the appearance of Gazprom Transgaz Makhachkala.

Source: for 2005 – 'Internal (wholesale) gas prices and transportation and storage tariffs for Gazprom' No 04-03-28-2005, available at www.ngvrus.ru/docs/preyskurant.pdf accessed 10 October 2013; for 2009 – adopted from the presentation by the Institute of Natural Monopolies Problems 'Costs of gas transportation through various export routes' (citing the pricelist No. 04-03-28-2009/1 'Internal (wholesale) gas prices and transportation and storage tariffs for Gazprom', available at http://ipem.ru/images/stories/Files/gaz/2_conoco.pdf, accessed 10 October 2013.

152   The Russian Gas Matrix

outcome, however, the debate for the regulator will concern how transport tariffs are managed relative to domestic gas prices; Figure 5.9 compares the historical trend of relative tariff levels since 2007.

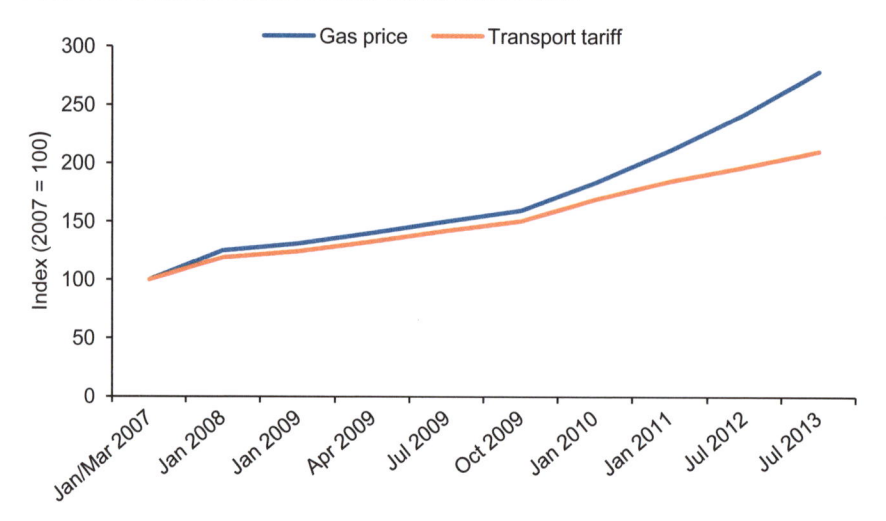

Figure 5.9: Comparison of regulated domestic gas price and gas transport tariff movements

Source: Data from Federal Tariff Service.

The starting point for Figure 5.9 is slightly confused, as the regulated gas price and transport tariffs were adjusted at different times in early 2007. But thereafter it is clear: the growth of gas prices and transport costs was very similar through 2008 and 2009, with a marginally lower growth rate for transport. However, since 2010 it is equally obvious that the regulated domestic gas price has risen much faster that the transport tariff. In 2013, for example, the gas price rose by 15 per cent and the transport tariff by 7 per cent. As a result, the wellhead netback for domestic producers has improved. Looking to the future, however, it is planned that the regulated gas price will not increase in 2014, and will rise only in line with inflation, and perhaps below, thereafter. Furthermore, many companies, including Gazprom from 2014, will be selling some gas below the regulated price, meaning that the future movement of transport tariffs will be of vital importance to their profitability. If transport tariffs continue to rise at the rate that they did in 2009–13, then margins for producers are likely to be squeezed, as the outlook for prices would appear to be for increases much lower than 7 per cent per annum. However, if transport tariffs do not continue to rise, then adequate funding may not be provided to maintain and expand the system. While this is a difficult problem for the regulator to solve, it would nevertheless appear logical that whatever tariff is decided should apply equally to all market participants who are now *de facto* competing on the basis of gas price, even though the regulated gas price continues to exist as a lever of government control in the sector.

Conclusions

The changes in the Russian market described in the three sections of this chapter may be summarized as follows:

As a result of the economic crisis of 2008–9, demand growth in the Russian domestic gas market has almost come to a halt. It slowed from an average of 2.3 per cent per annum in 2000–6 to 0.3 per cent per annum in 2007–11, and fell substantially in 2012. For the rest of this decade, academic and financial sector researchers expect growth of 1–1.5 per cent per annum at most – and this would only be the case if the economic recovery spreads beyond the natural resources sector to manufacturing and other energy-intensive economic activity. It remains possible that demand could fall further – as the example of Ukraine, where gas consumption has fallen steeply in response to a combination of high gas prices and industrial recession, shows.

Demand cuts have been caused more by the shutdown of industrial capacity than by higher gas prices. But this is not to say that rising regulated prices have not had an impact. In well-financed export-focused industrial sectors for which gas is a significant cost – chemical fertilizer production being the prime example – a demand-side response to higher prices is apparent. In industry and the power sector as a whole, the main result of higher prices has been fierce competition among gas suppliers. In this competition, Gazprom, Rosneft, and Novatek are the dominant players, with Rosneft and Novatek having had the most success in winning new customers in 2012–13 by offering price discounts.

Government policy on regulated prices has also changed as a result of the post-2008 economic situation. The aim, declared in 2006, to raise prices to the level of netback from Gazprom's oil-linked European sales prices, has effectively been abandoned. Indeed we have argued that the high level of oil prices makes that target not only unrealizable in practice but unnecessary for the development of the Russian gas market. In 2012 and 2013 non-Gazprom suppliers have been competing to sell gas at prices *below* the regulated price, suggesting that an equilibrium price has been reached. Gazprom has not been able to compete to date, as it must sell at or above the regulated price, and indeed will want a higher price to justify its new projects on Yamal. Nevertheless, in the short term it may be able to win back some market share using its cheaper Soviet-era fields, now that it has been given permission to offer prices below the regulated level.

In addition to these considerations, the government is faced with the knock-on effect of higher gas prices for electricity prices. Moreover, it regards higher gas prices as a potential brake on economic recovery. For all these reasons, the growth of regulated prices is likely to remain gradual – at or below the rate of inflation – for the rest of this decade. In that time frame, the resurrection of the Gas Exchange is possible, but may be difficult, and complete price liberalization is unlikely.

It should be emphasized that the issue of price is not limited to the level of the regulated price. The government has, so far, not addressed (i) mounting non-payment from district heating companies, a consequence of the lack of municipal services reform, and (ii) methods for addressing the cross-subsidization of household consumers. Both of these represent a mounting economic burden on Gazprom.

The reform of transport tariffs and pipeline access regulation has taken place in parallel with the reform of prices, and significant progress has been made since the late 2000s in respect of improving third-party access to the Gazprom-owned UGSS. By the time of writing, nearly all third-party requests were being granted. The introduction in 2005 of new zonal tariffication methodology also addressed some third-party concerns, although questions remain about cost-reflectivity and determination of optimal routes for third-party gas, not least because of the methodology's lack of transparency. The new regulation on non-discriminatory third-party access under discussion would address these problems, and also establish a framework for investment in new infrastructure – although, even if it is adopted, its reach will remain limited as long as the principle of the indivisibility of the UGSS remains in place. But the government's appetite for departing from this principle appears limited.

The future direction of market reform will depend on the government's decision on a number of issues: the level of the regulated price, the framework for more extensive competition, the development of pipeline regulation, and whether or not Gazprom will undergo corporate reform – in the first instance the separation-out of the transport business.

In our view, the government is unlikely to accelerate market reform as long as it sees its ability to regulate gas prices as a key lever for solving other economic problems (for example, boosting economic recovery, in particular the recovery of the manufacturing sector) and political problems (for example, avoiding social unrest caused by higher prices of gas, electricity, and municipal services). In this case, reform will continue, but at the same slow pace that has characterized it since the mid-2000s.

CHAPTER 6

SOURCES OF DEMAND IN THE RUSSIAN DOMESTIC GAS MARKET[1]

Simon Pirani

This chapter surveys the sources of demand for gas in the domestic Russian market – the power and district heating sector, which is predominant, the industrial and residential sectors, and the gas sector's own use – and discusses the factors that have shaped demand in recent years and that will shape it up to 2020. The conclusion drawn is that the 2–2.5 per cent annual average gas demand growth of the early and mid-2000s is unlikely to return, and that demand growth in the future is likely to be around 1 per cent per year or lower.

Power and district heating

More than half of the gas consumed in Russia (in 2011, 54 per cent, or 257 Bcm of 473 Bcm[2]) goes into the power and district heating sectors. The vast majority of this gas is supplied to (i) large thermal power stations; (ii) urban combined heat and power (CHP) plants, supplying electricity grids, industry, and district heating systems; and (iii) boilers that supply heat, mostly through district heating systems. Russia's CHP fleet and its district heating network were put together during the Soviet urbanization drive of the 1950s–70s and each is the world's largest by far. Its electricity grid is also among the world's largest.

A breakdown of gas supplied to the power and district heating sectors is shown in Table 6.1. (Statistics recorded by Rosstat, the national statistics agency, excerpted from Table 5.1 in Chapter 5, are shown together with those recorded by the Agency for Forecasting Energy Balances (APBE).[3]) In round numbers for 2011, 57 Bcm were supplied to wholesale generating companies (OGKs) that own large thermal power stations; 82 Bcm to territorial generating companies (TGKs) that own CHPs; and 30 Bcm to other companies, mainly in regions separate from Russia's main electricity

---

[1] With thanks to Anouk Honoré, who commented on a draft, and to Anatole Boute, discussions with whom have helped me to learn about the Russian power sector. All views expressed and mistakes made are mine.
[2] See Table 5.1 in Chapter 5.
[3] The Agency (Agentsvo po prognozirovaniu balansov v elektroenergetike or APBE) is a department of the Energy Ministry. See its website www.e-apbe.ru.

grid, that have a combination of generation assets.[4] Roughly speaking, 43 per cent of gas-for-power goes to large thermal stations, and 57 per cent to CHPs, and this is reflected in the OGKs' and TGKs' respective shares. The total volume of gas-for-power recorded by the APBE (168 Bcm in 2011), is quite a long way short of the figure recorded by Rosstat (190 Bcm in 2011), for reasons that are unclear. In addition, Rosstat records the substantial volume of gas (67 Bcm in 2011) supplied to boilers.[5]

Table 6.1: Gas for power and district heating: overview

Gas consumption (Bcm)	2006	2007	2008	2009	2010	2011
Power stations (Rosstat information)	**174.3**	**180.4**	**181.7**	**169.4**	**183.6**	**189.5**
Total to power companies (as recorded by APBE)	157.59	162.83	171.56	159.51	163.99	167.83
including: Wholesale generating companies (OGKs)	50.35	53.47	54.63	49.17	54.22	56.69
Territorial generating companies (TGKs)	83.28	83.62	84.51	78.69	81.91	81.62
Other (residual)	23.97	25.74	32.42	31.65	27.86	29.52
Boilers (Rosstat information)	**74.3**	**73.0**	**69.3**	**68.8**	**69.8**	**67.3**
Total power stations plus boilers (Rosstat information)	248.63	253.43	250.94	238.24	253.39	256.85

Source: Rosstat and APBE, Agentsvo po prognozirovaniu balansov v elektroenergetike, *Funktsionirovanie i razvitie elektroenergetiki v Rossiiskoi federatsii v 2011 godu*, [Agency for Forecasting Energy Balances, *Electricity in Russia: functioning and development, 2011*], various issues.

---

[4] The reorganization and privatization of the Russian electricity sector in 2002–8 resulted in the unified state-owned power company, United Energy Systems, being broken up. Generating assets passed to six wholesale generating companies (OGKs or Optovye generiruiushchie kompanii), that own large thermal generation stations (8.5–22 GW in each company) deemed to be of federal importance and 14 territorial generating companies (TGKs or territorial'nye generiruiushchie kompanii), that own urban CHP plants. Nuclear and hydro assets passed to two state-owned monopoly companies. Some assets of all types, in remote areas not connected to the central grid, are owned by regional generating companies. See Pirani, S. 'Elusive Potential: Gas Consumption in the CIS and the Quest for Efficiency', Working Paper NG 53, OIES, July, 2011, p. 38; RAO UES website, 'Reformirovanie elektroenergetiki', www.rao-ees.ru/ru/reforming/reason/show.cgi?content.htm.

[5] The Rosstat and APBE statistics are at variance with each other and also with the IEA's statistics. See Appendix 1, and Pirani, 'Elusive Potential', op.cit., pp. 39–44.

The evolution of gas consumption in the power sector is influenced by: the structure of the power industry and the place of gas in the fuel mix; the level of electricity demand; the results of power sector reform, including electricity price reform; and the implementation of plans to invest in new generation capacity. These issues will now be covered in turn, and then the district heating sector will be discussed.

The structure of the power industry and the fuel mix

The Russian power sector, which has been producing more than 1000 TWh of electricity per year in the 2010s, is dominated by thermal stations that burn gas and coal (including CHPs); it also has hydro and nuclear assets, electricity from which is substantially cheaper, much of which is run as base load.[6] In recent years, a little more than one-fifth of electricity has been produced by hydro stations, and a little more than one-tenth from nuclear stations. A considerable amount of generating capacity, nuclear and thermal, is under construction and planned: this is one of a number of factors that will put downward pressure on gas demand during this decade. Table 6.2 presents an outline of electricity generation and capacity.

Table 6.2: Thermal, nuclear, and hydro power

	2009	2010	2011
Electricity production (TWh)			
Thermal	652	699	714
Hydro	176	168	168
Nuclear	164	171	173
Total	**992**	**1038**	**1055**
Capacity (GW)			
Thermal	155.4	158.1	161.4
Hydro	47.3	47.4	47.5
Nuclear	23.3	24.3	24.3
Total	**226.1**	**230.0**	**233.3**

Source: Rosstat website.

---

[6] The merit order for generation capacity is set out in footnote 19 below. In 2011, capacity utilization of nuclear power stations was 81%, of hydro stations 40% and of thermal stations 51%. See Agentsvo po prognozirovaniiu balansov v elektroenergetike, *Funktsionirovanie i razvitie elektroenergetiki v Rossiiskoi federatsii v 2011 godu*, [Agency for Forecasting Energy Balances, *Electricity in Russia: functioning and development, 2011*], p. 48.

158 *The Russian Gas Matrix*

In thermal generation, gas is now completely dominant. Figure 6.1 shows that in the 1980s, as west Siberian gas production rose, gas displaced oil products from the fuel mix and overtook coal. That process continued in the 1990s, against a background of much lower electricity output. In the 2000s, gas-fired generation continued to expand, but with coal holding its own, retaining a 25–30 per cent share of electricity generation.

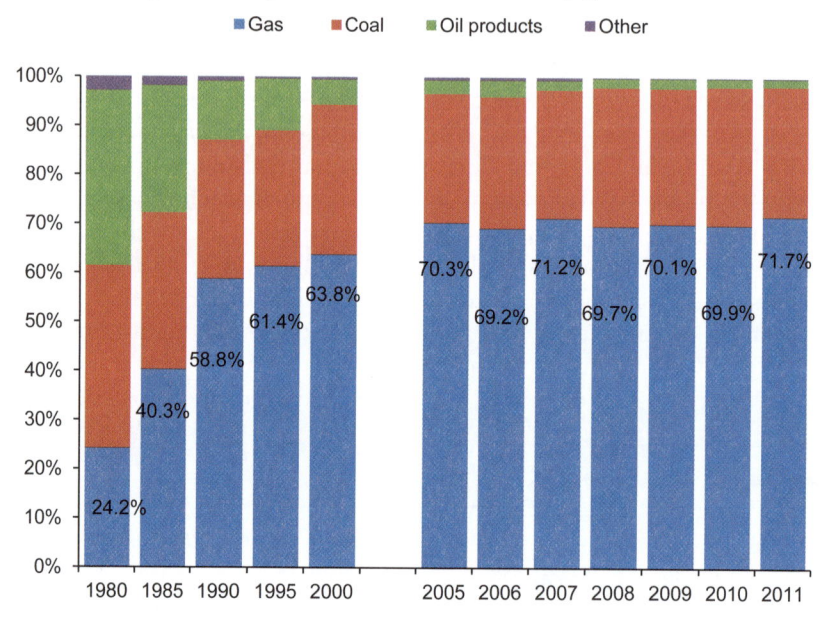

Figure 6.1: Fuel shares of thermal electricity generation
Source: APBE, *Funktsionirovanie i razvitiie elektroenergetike v RF*, various editions.

The thermal plants' fuel mix is largely determined by geography. Historically, gas was much cheaper than coal in European Russia and the Urals (which together account for about three-quarters of electricity generation[7]), and dominates thermal generation there. Coal was more readily available, and usually cheaper, in Siberia and the Far East, and it dominates generation there. Coal also has a small share of the fuel mix for thermal generation in the Urals, and in southern Russia. In central and north-western Russia its share is negligible. A breakdown of fuel shares for power generation by Federal district – showing the predominance of gas in European Russia and of coal in Siberia and the Far East – is shown in Map 6.1. (The same information is presented in Table A6.1 in Appendix 6.)

---

[7] In 2011, European Russia and the Urals accounted for 74.5% of electricity generated and 72.1% of installed generation capacity. APBE, *Funktsionirovanie i razvitie*, op. cit., pp. 37 and 160.

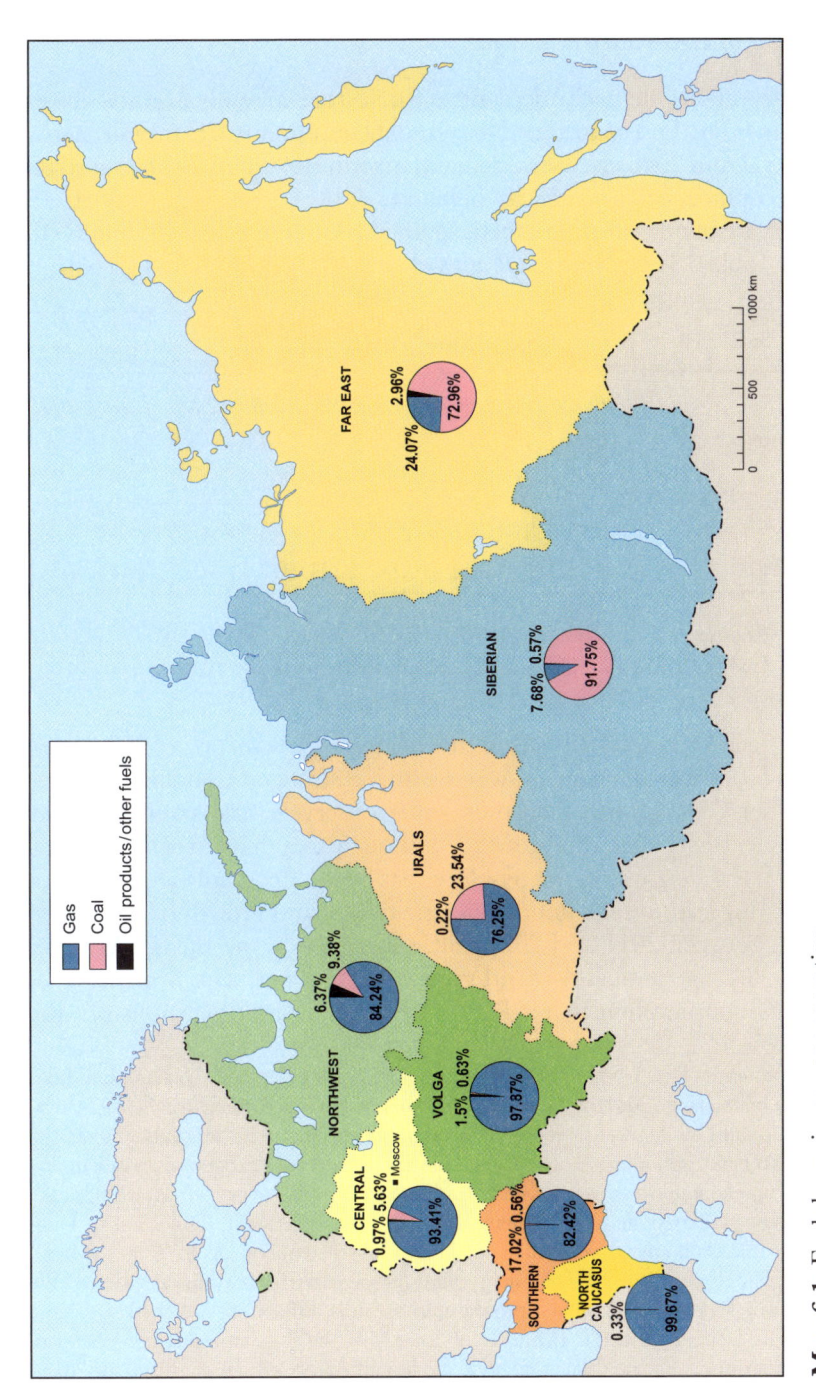

Map 6.1: Fuel shares in power generation
Source: APBE, *Funktsionirovanie i razvitie*, 2011, op. cit., p. 62.

The level of electricity demand

During the 2000s, electricity demand in Russia rose unevenly, and at a slower pace than GDP. This contributed to a gradual improvement in overall energy efficiency. From 2006 electricity demand growth slowed; in 2009, as a result of the economic crisis, electricity demand fell by 4.5 per cent. In 2010–12 electricity demand growth resumed, again at a lower level overall than GDP growth. Table 6.3 compares GDP growth and electricity demand growth.

Table 6.3: Electricity demand growth and GDP growth

	2000	2005	2006	2007	2008	2009	2010	2011	2012
GDP growth (%)	10	6.4	8.2	8.5	5.2	−7.8	4.3	4.3	3.4
Electricity consumption									
Growth (%)	3.8	1.8	4.2	2.3	2	−4.5	4.45	2	1.3
Total (TWh)	864	941	980	1003	1023	977.1	1021	1041	1053

Source: Rosstat (GDP growth); APBE, *Funktsionirovanie i razvitie 2011*, p. 146; Energy Ministry (electricity consumption growth for 2012).

The projected level of electricity demand growth up to 2020 is in dispute. Government agencies, who assume a stronger economic recovery and higher GDP growth than most other observers, also assume a faster rate of electricity demand growth – although they have recently revised their forecasts downwards. Recent forecasts by the Economic Development Ministry suggest average annual electricity demand growth of 1.95–2.08 per cent between 2011 and 2020, but, subsequently, the Energy Ministry forecasts average annual demand growth of 1.79 per cent, bringing total electricity consumption to 1151 TWh in 2019.[8] The projections of very

---

[8] The Economic Development Ministry projects growth of electricity consumption from 1041.1 billion Kwh (2011) by 19.4% to 1238.5 billion Kwh (according to its 'conservative' scenario) and by 20.9% to 1253.6 billion Kwh (according to its 'innovation' scenario) (Ministverstvo ekonomicheskogo razvitiia Rossiiskoi Federatsii, 'Prognoz dolgosrochnogo sotsial'no-ekonomicheskogo razvitiia Rossiiskoi Federatsii na period do 2030 goda', [Ministry of Economic Development of the Russian Federation, 'Projection of long-term social-economic development in the Russian Federation up to 2030'] (March 2013), March 2013, p. 156). The Energy Ministry forecast is from Ministerstvo energetiki Rossiiskoi Federatsii, 'Skhema i programma razvitiia Edinoi energeticheskoi sistemy Rossii na 2013–2019 gody', [Ministry of Energy of the Russian Federation, 'Scheme and programme for the development of the united energy system of Russia for 2013–2019'] (June 2013), pp. 2–3. There is a discrepancy between the electricity demand figures used in this document and those used elsewhere by the ministry, e.g. on its website.

slow economic recovery in Russia (see Chapter 5), suggest that electricity demand growth could be slower than this. Russian academic and financial sector researchers have taken this view, projecting for example, 1.3 per cent annual demand growth (Skolkovo Business School) or 1.4 per cent per year in the period up to 2014 (Gazprombank).[9]

Power sector reform and government policies

The reform of the power sector in the mid-2000s was a crucial turning point for Russia's economy and energy provision. The reform aimed to replace the unitary electricity provider inherited from Soviet times with (i) state holding companies to manage natural monopolies (transmission and dispatch), (ii) privatized generating companies selling electricity, generating capacity, and heat into regulated markets, and (iii) regulators for those markets. In 2006–7 most generating assets were privatized, and in 2008 the post-Soviet provider, United Energy Systems of Russia, was wound up. The wholesale electricity market, regulated by the Federal Antimonopoly Service (FAS) and Market Council, began to function.

A central consideration in the reform of the power sector was the need for investment in new infrastructure to the value of tens of billions of dollars. In the post-Soviet period, the average age of power station units has risen from 18.5 years in 1990 to 33.4 years in 2011; at the start of 2012, it was estimated that 70 per cent of the electricity transportation grid required replacement.[10] Widespread blackouts in 2005 served as a warning of the system's fragility; in 2009 the poor state of infrastructure was dramatically underlined by an accident that destroyed much of the Saiano-Shushenskaia hydro power plant in eastern Siberia and caused 75 deaths. In the run-up to privatization, ambitious plans were drawn up for investment in new generation capacity. In line with these, agreements were made with generating companies for the delivery of new capacity, under which the companies undertook legally binding investment obligations in exchange for a guaranteed rate of return (initially set at 13–14 per cent per annum) through a capacity premium, over a 15-year period. A long-term capacity market was established.[11] However, the government's overall investment

---

[9] Mitrova, T. *Vnutrennyi spros na gaz: perekhodnyi vozrast* [*Domestic demand for gas: a time of transition*], Skolkovo Business School presentation, February 2013, Porokhova, N. 'Gas and Power Markets in 2012 and New Challenges', Gazprombank presentation, March 2013.

[10] APBE, *Funktsionirovanie i razvitie elektroenergetiki v Rossiiskoi federatsii v 2011 godu*, op. cit, p.43.

[11] IEA *Russian Electricity Reform 2013 Update: Laying an Efficient and Competitive Foundation for Innovation and Modernisation*, Paris: IEA/OECD, 2013, pp. 27–9; Gore, O., Viljainen, S., Makkonen, M., and Kuleshov. D. 'Russian electricity market reform: deregulation or reregulation?', *Energy Policy*, 41 (2012), pp. 676–85 (here 679).

plans, and many of the commitments made by generating companies, were drastically scaled back as a result of the economic crisis of 2008–9.[12] A further pressure, exacerbated by the crisis and its impact on the Russian economy, was the government's hesitation in the process of implementing electricity price reform in the manner originally envisaged. It demonstrated concerns similar to those that have affected decisions on gas price increases: high prices for industry may constrain economic recovery, and high prices for the population may stimulate social unrest.

In the post-crisis period, power sector reform has changed direction. Firstly, although the goal of price liberalization has been retained, several types of price regulation remain in place. Secondly, state-owned energy companies have expanded aggressively, consolidating control over more than half of all generation assets by 2013. These companies, together with the large private ones, have begun to invest in new capacity.

Price liberalization was due to be completed in the wholesale electricity market from 1 January 2011, but price control mechanisms remain in place, and in September 2013 the government decided to extend these until at least 2016.[13] The retail electricity market is also open to competition under law, but more than half of all retail consumers purchase electricity at regulated prices through arrangements for suppliers of last resort, whose prices are regulated by regional energy commissions.[14] Finally, the regulatory authorities have been criticized for stymying the capacity market with over-prescriptive rules.[15] Against the background of these delays in electricity price liberalization, gas prices have been increasing much more rapidly than electricity sales prices (in other words, spark spreads are narrowing), and in 2012 electricity generators presented arguments to the

---

[12] Pirani, 'Elusive Potential', op. cit., pp. 50–57.

[13] The government's thinking is set out e.g. in: Ministverstvo ekonomicheskogo razvitiia i torgovlia, *Prognoz sotsial'no-ekonomicheskogo razvitiia RF na 2014 god i na planovyi period 2015 i 2016 godov*, [Ministry of Economic Development of the Russian Federation, 'Projection of social-economic development in the Russian Federation for 2014 and in the planning period 2015–2016'], pp. 125–29. The government's decision was reported in, e.g., 'Pravitel'stvo zamorozit tarify monopolii chastichno', *Vedomosti*, 20 September 2013 and Deutsche Bank, *Electric Utilities: incorporating tariff freeze into valuations*, 3 October 2013.

[14] Guaranteeing suppliers (i.e. suppliers of last resort) in 2012 supplied about 68% of the contestable customer load in European Russia and the Urals, and around 45% of the contestable customer load in Siberia. IEA, *Russian Electricity Reform 2013 Update*, pp. 58–9, referring to the Ministry of Energy.

[15] A recent report by the IEA concluded that capacity market arrangements deliver 'a form of central planning', 'incompatible with the policy goal of developing efficient, innovative and dynamic electricity markets'. IEA, *Russian Electricity Reform 2013 Update*, p. 37.

government that this was squeezing them in an unsustainable manner.[16]

At least until 2020, it seems most likely that regulation will continue to be applied to both wholesale and retail electricity prices. Ministers have argued that too rapid an increase in electricity prices could constrain Russia's painfully slow economic recovery, and could be a dangerous stimulant to inflation. On the other hand, there is an acceptance in government (i) that higher electricity prices are a precondition for further increases in regulated gas prices, and that gas prices must rise further, both to fund investment and to allow a greater tax take from gas production; and (ii) that if electricity price growth is slowed, plans for investment in new electricity capacity, which is strategically vital, will need to be amended and new sources of investment found. Other voices argue that slower price growth for both electricity and gas could be combined with the abolition of cross-subsidization of residential customers.[17]

Another aspect of the state's renewed role in the power sector is the consolidation of control by state-owned companies over most electricity generating assets. In October 2012, two of the five largest wholesale generating companies (OGK-1 and OGK-3), and one territorial generating company (TGK-11), were taken over by Inter RAO, the state energy holding company. In the same month, the Eastern Siberia energy company was taken over by state-controlled RusHydro. By the end of 2012, the four largest owners of generating capacity, all majority state owned, accounted for more than half of Russia's total capacity (134 GW out of 233 GW): Gazpromenergoholding, the electricity division of Gazprom (38 GW); RusHydro, the state-owned hydro electric company (37 GW); Inter RAO (34 GW, up from 8 GW since 2010); and Rosatom, the state-owned nuclear generator (25 GW). There has also been consolidation in the distribution

---

[16] Alfa Bank calculated average spark spreads for the year at 211 RR/MWh in 2008; 86 RR/MWh in 2009; 151 RR/MWh in 2010, 143 RR/MWh in 2011 and 47 RR/MWh in the first half of 2012. Alfa Bank *Russian Power Generators: No Catalysts, Negatives Mostly Priced In*, 15 October, 2012, p. 17. See also Gore et al. 'Russian electricity market reform', op. cit.; Boute, A. 'The Russian Electricity Market Reform: Towards the Re-regulation of the Liberalized Market', in Sioshansi, F.P. (ed.), *Evolution of Global Electricity Markets: New Paradigms, new challenges, new approaches*, London: Elsevier Academic Press, pp. 461–96, July 2013; Porokhova, N. 'Gas and Power Markets in 2012 and New Challenges', op. cit.

[17] On recent political discussions see e.g. Deutsche Bank, *Russian Gencos: spot price dynamics key to 2013 performance*, 21 February 2013. The argument on slower price growth and the abolition of cross subsidization is put by researchers at the Institute of Energy Research of the Russian Academy of Sciences in Makarov, A.A., Mitrova, T.A. et al. *Vliianie rosta tsen na gaz i elektroenergiiu na razvitie ekonomiki Rossii* [*The influence of gas and electricity price rises on the development of the Russian economy*], Institute of Energy Research of the Russian Academy of Sciences, Moscow, 2013.

sector, with the two largest grid companies being brought together in Russian Grids, another state-controlled holding.[18]

The post-reform structure of the power sector – dominated by generating companies, whether state- or privately-owned, run on competitive lines – has not only opened the way for investment and efficiency improvements, but has also transformed the way in which gas is purchased by power companies. Until the early 2000s, Gazprom had a near monopoly on sales to the power sector. In the pre-crisis period, when electricity and gas demand rose rapidly, power companies sought to purchase from non-Gazprom producers – although pipeline access to facilitate such sales was severely limited. The situation with pipeline access changed in 2009, when state-controlled OGK-1 signed a major supply contract with Novatek. A series of similar deals was concluded in 2010–11 culminating in the agreement, signed in December 2012, for Inter RAO to purchase 875 Bcm of gas over the 25-year period 2016–40. Until 2012, non-Gazprom producers were selling at a premium to regulated prices to customers who sought, for example, additional flexibility; after that date, they began to compete at price levels below the regulated prices. (See also Chapter 5, 'Domestic gas prices: catalysts for a rethink'.)

The effect of power sector reform on the fuel mix

The merit order for power generation is an important factor in determining demand for fuels. Under the regulatory system established during power sector reform, power plants are divided into base-load plants (which operate on steady load and accept whatever price for electricity is formed in the market) and peak-load plants (which are dispatched when needed), as follows. All nuclear and most hydro capacity is base load. During the heating season, CHPs operate as base-load power plants, while also meeting the demand for heat. Thermal power plants are run against the market price (in other words, they provide peak-load capacity); however they are also obliged to submit price-taking bids corresponding to their minimum technical output.[19] There is only a limited element of price competition

---

[18] Gazprom Investor Day presentation, February 2013, slide 64; Porokhova, N. 'Gas and Power Markets and New Challenges', op. cit.; company websites.

[19] The merit order is set out in government regulations as follows. First priority: 'nuclear power plants and power plants needed to ensure system reliability'. Second: 'CHP plants, technical minimum output of thermal power plants and hydro power plants'. Third: 'generators responsible for satisfying the demand of households and to whom regulated tariffs apply'. Fourth: 'generators that have committed themselves to free bilateral contracts.' Lowest: to 'remaining generators that are needed to satisfy demand, who are selected based on their market-based offers'. In addition to the constraints placed on the system by the large amount of CHP, most hydro plants

between electricity suppliers. The most significant changes to the fuel mix are expected to come about as a result of investment in new plant, which is discussed below.

In thermal generation, gas's share has grown steadily, despite government policy that it should be reduced on the grounds that excessive dependence on gas is an energy security risk. The aim of reducing gas's share in the power sector fuel balance was incorporated into Russian energy strategy from 2000; it was envisaged that, as regulated gas tariffs increased, gas would lose its price advantage over coal and cede a share of thermal generation to coal.[20] But this has not happened. Between 2006 and 2011, the regulated gas price for power companies doubled, while coal prices rose less steeply. In 2010–11 gas prices (per unit of energy produced at power stations) rose above coal prices for the first time in most parts of European Russia. Gas is likely to remain more expensive for electricity producers, but it appears that the differential is not sufficiently great to justify the costs of fuel switching. In Siberia, gas is now more than 1.5 times as expensive as coal per unit of energy produced, and in some places twice as expensive; in the Far East, the difference is minimal. In both those regions, coal remains dominant. (For detailed information on gas and coal prices for power generation, see Appendix 6, Tables A6.2 and A6.3.)

Little switching from gas to coal is now expected. Efforts to increase the share of coal-fired generation in European Russia have been hampered by problems posed by ash disposal from coal plants, other environmental issues associated with burning low-grade coal, and by the lack of rail wagons, which disrupts coal supply to power stations during the winter. The investments made since 2010 by power generators in European Russia, discussed in the next section, have focused on efficiency improvements in gas-fired generation rather than on switching to coal.

Investment in new power generation capacity

The corporate consolidation in the power sector, and the dominant position taken by state-owned companies, has laid the basis for substantial investment in new power generation capacity. New thermal power plants have been commissioned under capacity delivery agreements (CDAs) that guarantee a

---

in Russia are 'run of river' plants that operate passively and generate electricity in accordance with river flow. Gore et al, 'Russian electricity market reform', op. cit., p. 682, citing Russian government, *Power Market Regulations* 2011, decree no 1172, paras 87–9. See also Pittman. R. 'Restructuring the Russian electricity sector: recreating California?', *Energy Policy*, 35 (2007), pp. 1872–83, here p. 1875.

[20] Ministry of Energy of the Russian Federation, Ministerstvo energetiki Rossiiskoi Federatsii, 'Energeticheskaia strategiia Rossii na period do 2020 goda', (Moscow, May 2003), pp. 15–17 and 76–79; IEA *Russia Energy Survey 2002*, Paris, pp. 52–3.

specific return on investment for 15 years; new nuclear and hydro plants are being built under long-term agreements (LTAs) with the state-owned companies that operate them, where the return on investment is guaranteed over 20 years.

A substantial wave of new capacity commissioning, in 2010–11, was mainly a result of investment by private companies. After low levels of annual new capacity commissioning in the late 2000s, 3232.5 MW of new capacity was commissioned in 2010 and 4598 MW in 2011, or about 1.5 per cent and 2.1 per cent of total capacity respectively. Most of this new capacity (1661.1 MW in 2010 and 4494 MW in 2011) was combined cycle gas turbine plants (CCGT) and gas turbine plants.[21]

In 2012–13 the picture changed: the lion's share of new capacity completed, and under construction or planned for commissioning in 2013–18, was by companies that were either 100 per cent state-owned (such as RusHydro and Inter RAO) or majority state-controlled (such as OGK-2 and Mosenergo). Foreign privately-owned companies, such as E.ON Russia and Enel, played a secondary role.

The new capacity – planned or under construction with completion dates before 2020 – is about 28 GW, about 12 per cent of Russia's total. (For details, see Appendix 6.) Most of this is thermal capacity, including CCGT with efficiency levels of 55–58 per cent, compared to levels of 32–38 per cent in the plants they are replacing. The new nuclear capacity will produce a net increase in nuclear generation estimated by the Energy Ministry at about 3GW. One uncertainty is whether the government will adjust the terms of the CDAs; this issue has been under discussion, and in October 2013 a group of generating companies made a public appeal to it not to do so. Another uncertainty is the rate at which old power stations will be decommissioned; some generators are reportedly considering decommissioning large nuclear stations, in an attempt to support electricity prices.[22]

---

[21] APBE, *Funktsionirovanie i razvitiie* 2011, op. cit., pp. 35–6. Of the nine largest projects listed by APBE, two were completed by Gazprom subsidiaries (Mosenergo CHP no. 26 and TGK-1's Iuzhnaia CHP no. 22), and the rest by privately owned producers (two by E.ON Russia, two by OGK-5 Enel, and three by Fortum's Russian subsidiaries). Levels of new commissioning and decommissioning are shown in Appendix 6, Table A6.4.

[22] The Energy Ministry projects, in 2013–19, 11,268 MW of new nuclear capacity and decommissioning of 8274 MW (a net increase of 2994 MW); 18,028 MW of new thermal capacity and decommissioning of 10,712 MW (a net increase of 7316 MW); and 3476 MW of new hydro and renewable capacity. (Ministerstvo energetiki Rossiiskoi Federatsii, 'Skhema i programma razvitiia Edinoi energeticheskoi sistemy Rossii na 2013–2019 gody', pp. 31 and 39.) Analysts at Alfa Bank project 17.45 GW of new thermal capacity, built under CDAs, a net increase of 5.2 GW of nuclear capacity and 5.5 GW of new hydro capacity. (Alfa Bank, *Russian Power Generators: No Catalysts*.) For more details, see Appendix 6, Table A6.5. On generating companies' appeal, see 'Energetiki prosiat zashchity', *Vedomosti*, 7 October 2013.

Overall, it appears that the downward pressure on gas demand, resulting from new power station investment, will exceed any upward pressure. By investing in new gas-fired capacity rather than switching to coal, generators in European Russia and the Urals will maintain the very high share of the thermal fuel balance taken by gas. But the replacement of old gas-fired plant by CCGTs will result in substantial efficiency improvements, and much incremental electricity demand will be met by these improvements. And the net increase of nuclear and hydro capacity will displace all types of thermal energy to the same extent.

Conclusions on the power sector

The main factors that will shape gas-for-power demand for the rest of this decade are (i) the level of electricity demand, and (ii) the effect of investment in the construction and renovation of power stations. *The level of electricity demand* rose steadily, but more slowly than GDP, in the years prior to the economic crisis, underwent a post-crisis recovery in 2010–12, but is now rising more slowly. The Energy Ministry's most recent projection, that electricity demand will rise on average by 1.79 per cent annually, implies an economic growth rate of 3 per cent or more – which many economists believe Russia will struggle to achieve. Academic and financial sector forecasts put electricity demand growth at a lower level, for example, 1.3–1.4 per cent. *Investment in new power station capacity* will exert downward pressure on gas demand, mainly because (i) new nuclear and hydro generating capacity will displace gas; (ii) the commissioning of new gas-fired power stations, including CCGT, will result in substantial efficiency improvements; and (iii) other technical and organizational improvements resulting from power sector reforms will result in some extra efficiency gains. While gas may lose share to nuclear and hydro, it is unlikely to lose share to coal in European Russia.

A back-of-the-envelope calculation, using the Energy Ministry's assumptions, suggests that electricity demand will rise to 1172 TWh in 2020, that is about 151 TWh more than in 2011. The ministry projects 3 GW net of new nuclear capacity, which would generate something like 21 TWh, and 3.5 GW of new hydro and renewables, which would generate something like 12 TWh. The ministry also expects 18 GW of new thermal capacity. Assuming that gas's share of this new investment is about the same as its share of thermal power generation and that new gas-fired stations only replace old gas-fired stations (and not coal stations) and are 30 per cent more efficient, then the gas-fired power fleet would produce an extra 17 TWh or so. In other words, 50 TWh of the extra 151 TWh is accounted for, and the remaining 101 TWh would require up to 36 Bcm of incremental gas to produce.

However, if we assume less vigorous electricity demand growth (as most observers do), and slightly more impressive efficiency gains, it appears that incremental gas demand will be negligible. Average annual electricity demand growth of 1.3 per cent (instead of the ministry's 1.79 per cent) would imply 130 TWh of incremental electricity demand in 2020. It is completely plausible that a slower decommissioning programme in the nuclear sector could mean 6 GW of net new nuclear capacity, rather than 3 GW, and that another 10 TWh of electricity could be produced from other efficiency gains. These changes halve the expected incremental gas demand in 2020, to 18 Bcm at most, without making any more ambitious assumptions about the effect of new hydro and CCGT. Researchers in Russia have produced even lower estimates of incremental gas-to-power demand: the Skolkovo Business School put it at zero to 1.2 per cent (that is zero to 2.3 Bcm), and Gazprombank researchers, who project that new nuclear capacity in north-west Russia will displace up to 20 Bcm of gas demand, have put it at zero.[23]

Government policy, as well as the rate of economic growth, will play an essential role. In the post-reform power sector, generating companies make investment decisions under CDAs; however, firstly, the terms of the CDAs are negotiated with (and could even be amended by) the government and market authorities, and, secondly, most generation capacity is now in the hands of state-controlled companies. A secondary lever of control by government is electricity pricing regulation, although the measure that could arguably put the strongest downward pressure on demand – abolition of cross-subsidization – is unlikely to be implemented in the near future. If government policy continues to favour steady power sector modernization to underpin economic growth, as it has in recent years, gas-for-power demand growth up to 2020 is likely to remain modest; if the government implements more radical changes, which seems less likely, it could fall.

Heat production

Heat is produced in very large quantities in Russia, from (i) urban CHP plants, usually owned by TGKs; (ii) CHPs owned and used by, and often integrated into, industrial enterprises, including steel plants that recycle coke oven gas and blast furnace gas as fuel; (iii) centralized boilers, owned by district heating companies, municipal services providers, and other entities; and (iv) individual heat sources. About two-thirds of heat sold is

---

[23] Mitrova, T. Vnutrennyi spros na gaz: perekhodnyi vozrast [*Domestic demand for gas: a time of transition*], Skolkovo Business School presentation, February 2013 and Porokhova, N. 'Gas and Power Markets and New Challenges', op. cit.

produced from gas, some from coal.[24] Because of the wide array of heat producers, it is difficult to state the amount of gas consumed to produce heat: it is in the range of 160–190 Bcm/year.[25] Of this total, around 110–120 Bcm/year is for heat produced or purchased by, and sold by, district heating companies. The following discussion focuses on these volumes.

Heat is distributed by district heating companies, and delivered to households largely by municipal services providers that also supply gas, electricity, water, sewage systems, and building repair services. Heat production and distribution are characterized by extremely high losses, in large part due to ageing boilers and heat pipes. Heat consumption – most of which is in industry, residential buildings, and public buildings – is also extremely inefficient. In 2007 researchers for the World Bank estimated the potential savings in heat consumption at 144.5 million toe/year, more than 85 per cent of which it considered to be economically viable.[26]

District heating sector reform, and the potential energy savings, will take a long time to realize. But improvements in the future could reduce gas demand significantly. Change in the district heating sector could also lift an economic burden on Gazprom, which sells large volumes of gas (at regulated prices) to district heating companies. Since the 2008–9 economic crisis, there is evidence (i) that gas consumption by district heating companies is falling, but also (ii) that the level of non-payment is increasing. *Gas consumption* in centralized boilers grew more slowly than demand in any other sector in 2000–6, and fell by almost 1 per cent annually in 2007–11 (see Table 5.2 in Chapter 5). Falling population, changes in housing stock, deterioration of district heating systems, the economic recession, together with other factors, may all have contributed to the fall in consumption. The impact of rising prices, and fuel saving at municipal and household level, are also likely to be part of the picture. As for

---

[24] There are no accurate statistics of which the author is aware. The IEA's energy balances show that 65.9% of the heat produced in 2008 from CHPs and heat plants was from gas. This excludes heat that is not traded, e.g. because it is produced on site by large industrial enterprises and by autonomous sources. See Pirani, 'Elusive Potential', op. cit., p. 60.

[25] Researchers commissioned by the energy ministry estimated that '*up to* 190 Bcm/year' of gas is used in heat production ('Tsentralizovannoe teplosnabzhenie v Rossii', *Promyshlennye i otopitel'nye kotel'nye i mini-TETs* no. 4, 2012). The author, using IEA data, estimated that the two largest groups of heat producers, CHP, and boilers, in 2008 consumed 166 Bcm for heat production (Pirani, 'Elusive Potential', op. cit., p. 60). Rosstat, in its sectoral breakdown of gas consumption (see Table 5.1 in Chapter 5), includes 124–134 Bcm/year in recent years in the rows 'power stations' and 'centralized boilers', but the total of gas consumed in heat production would include considerable volumes in several other rows. See Appendix 1.

[26] World Bank *Energy Efficiency in Russia: Untapped Reserves*, (Washington: World Bank, 2008), pp. 39–46.

rising *non-payment* from district heating and municipal services companies, in 2012 Gazprom reported that it had reached 12 per cent. The most significant debtors were management companies that supply heat to residential buildings; fraudulent operators, who collect tariffs for a heating season and then disappear, had also become a significant problem.[27]

Another trend of concern is that, as district heating systems have deteriorated due to lack of investment, industrial – and above all residential – heat consumers disconnect from them and commission individual heat sources. This aggravates the problem, making district heating systems, and the CHPs and boilers that supply them, even more inefficient. Researchers for the Energy Ministry have reported that CHPs have, for the last 20 years, 'been losing their niche in the heat market' due to 'mistaken tariff policies' that disadvantaged larger systems; that by 2000 the CHPs had lost 30 per cent of their sales by volume to autonomous boilers and individual heat sources; that the industrial recovery of the 2000s had not reversed this trend; and that most CHP and large boiler complexes are now running with 20 per cent unused capacity.[28] The trend for consumers to commission autonomous heat sources *may* be one factor in both the falling consumption by boiler houses, and the increase in consumption by the residential sector, indicated in Table 5.2.

Heating sector reform

Because of the level of waste and energy inefficiency in the heat sector, it has the most *potential* for reducing gas consumption. District heating renovation projects have shown that investment in new boiler houses and heat pipelines can cut fuel consumption by 30 per cent or more, and recover investment in relatively short periods. A flagship renovation project in the Mytishi district near Moscow achieved a 31 per cent reduction in gas consumption (down to about 0.2 Bcm/year) and a 33 per cent reduction in electricity consumption, which also feeds back to gas.[29] Another example

---

[27] 'Skol'ko stoiat 6%', *Vedomosti*, 26 February 2013; 'Sostoialas' press-konferentsiia predsedatelia komiteta Gosdumy RF' 1 July 2013,www.fondgkh.ru/news/91188. html; Gazprom, Stenogramme of press conference by K. Seleznev, *Postavki gaza na vnutrennyi rynok* [*Supply of gas to the domestic market*], 28 May 2013.

[28] 'Tsentralizovannoe teplosnabzhenie v Rossii', *Promyshlennye i otopitel'nye kotel'nye i mini-TETs* no. 4, 2012.

[29] Details of the Mytishi project, see: Puzakov, V. and Polivanov, V. 'Russia', in Euroheat & Power, *District Heating and Cooling 2013 Survey*, Brussels, 2013, and World Bank *Implementation, Completion and Results Report*, IBRD-46010, 10 December 2008 on $85 million loan to Russia for a municipal heating project, 10 December 2008. The government agency for financing municipal reforms is the Cooperative Fund for Municipal Services Reform (www.fondgkh.ru).

of the scale of possible savings was provided by Gazpromenergoholding, which in 2013 acquired the Moscow United Energy Company (MOEK) (which owns most of Moscow's boiler stock and consumes an estimated 7–8 Bcm/year of gas). Gazpromenergoholding already owns Mosenergo, the city's TGK (which owns Moscow's CHPs and consumes an estimated 23 Bcm/year of gas). The company announced plans immediately to close 10 boiler houses, saving 1 Bcm/year, and in the medium term to close another 22. Their customers will be served from CHPs, once some investment in heat distribution is made.[30] On a national scale, full potential heat savings can only be realized in the context of the overall reform of municipal services provision and the efforts of local governments to renew the ageing municipal housing stock.

Technically, the upgrading of district heating systems, heat sources, and heat transportation networks is best undertaken together with the renovation or replacement of old housing stock. Institutionally, the reorganization of heat production took a significant step during power sector reform; the TGKs, established in 2006–7, manage most CHPs. But the reorganization of district heating companies and municipal services organizations is at a very early stage: the government's reform plan envisages transferring municipal infrastructure to private providers on a concession basis. The other key element is the reform of regulated heat and heat transportation tariffs, which the government envisages moving from a 'cost-plus' system to a combination of long-term contracts and RAB (regulated asset base) regulation.[31]

The outlines of heat sector reform were set out in the 2010 federal law 'on heat supply'. Further laws, which were passed in 2012–13, put in place a framework for moving from cost-plus tariff regulation to long-term tariff regulation for heat and other municipal services, and for private municipal services providers to work on a concession basis. Heat tariffs continue to be subject to 'cost plus' regulation, although they were increased twice in 2012. Three pilot projects for new forms of regulation have been started in Orel, Vladimir, and Ekaterinburg, and the Federal Tariff Service has set out a planned transition, in 2014–16, from cost-plus regulation to RAB and

---

[30] 'Nyneshniaia model' rynka v printsipe sebia izzhila', Gazpromenergoholding website http://energoholding.gazprom.ru, 24 April 2013.

[31] See also Pirani, 'Elusive Potential', op. cit., p. 66–9. Technical aspects of district heating renovation are covered in: Energeticheskii Institut im. G. Krzhizhanovskogo, *Razrabotka programmy modernizatsii elektroenergetiki Rossii na period do 2020 goda* [G. Krzhizhanovskii Energy Institute, *Development of the Programme of Electricity Modernization in Russia up to 2020*] (Moscow 2011) pp. 20–23 and 54–57.

long-term contract regulation of tariffs.[32] However, heat sector reform, like gas and power market reform, will inevitably be subject to government caution about tariff increases that affect living standards. The issue raised its head again in early 2013, with President Putin stating emphatically that the increases in tariffs for all municipal services, including heat and gas for residential consumers, were to be capped at 6 per cent per year. The tariffs fixed by the FTS in October 2013 were slightly below that level.[33]

The gradual reduction in consumption by boilers in 2007–11 indicates that the changes in the sector may be putting some downward pressure on demand. The reform of district heating and municipal services provision, which will take several more years, will eventually produce more substantial reductions in the level of consumption. This will lift an economic burden that Gazprom has carried for many years; it may eventually also result in a spread of the competition between gas suppliers that has been seen in the industrial sector to the district heating sector.

Consumption in other sectors

In addition to power and heat production, there are three other significant sources of gas demand in Russia: industry, which in recent years (2000–11) has consumed 80–85 Bcm/year; residential use (about 50 Bcm/year); and technical use and losses (about 50 Bcm/year).[34]

Industry

Gas demand in industry, like the power sector, is determined in large part by GDP growth. Since the economic crisis, industrial output has been

---

[32] Federal'nyi zakon no. 190-F3 'o teplosnabzhenii', [Federal law 190-F3 'on heat supply'] (27 July 2010); Federal'nyi zakon no. 291-F3 'o vnesenii izmenenii v otdel'nyi zakonodatel'nyie akty RF v chasti sovershenstvovaniia regulirovaniia tarifov', [Federal law 291-F3 'on the amendment of legislation in order to improve the regulation of tariffs'] (30 December 2012); Federal'nyi zakon no. 103-F3 'o vnesenii izmenenii v federal'nyi zakon 'o kontsessionnykh soglasheniiakh'', [Federal law 'on the amendment of the law 'on concession agreements'] (7 May 2013). 'Modernizatsiia kommunal'noi infrastruktury v usloviiakh tarifnykh ogranichenii', presentation by A. Chibis of the municipal sector working group of the government, July 2013; 'Tarifnoe regulirovanie v 2013 g', [Tariff regulation in 2013], presentation by S. Novikov of the RTS, July 2013; Sberbank, *Russian Gencos: what goes around comes around*, August 2013.

[33] 'Skol'ko stoiat 6%', *Vedomosti*, 26 February 2013; 'Tarify na teplo i vodosnabzhenie vyrastut na 4.2%', *Vedomosti*, 15 October 2013.

[34] See Table 5.1 in Chapter 5. There is an explanation of the sectoral breakdown used in Appendix 1.

extremely slow to recover; by 2011 gas demand in industry had returned to, but hardly exceeded, its pre-crisis level. Gas prices (now at around $150/mcm for industrial consumers) may be having some effect on demand although, as yet, there is clear evidence of this only in the chemicals sector (where gas is used as feedstock and accounts for most of the cost of production). Rising gas prices have also created an element of competition among gas producers to supply large industrial customers, just as they are competing for large power sector customers; in 2012–13 metals and chemicals producers, as well as power generators, signed long-term supply contracts with non-Gazprom producers. (See also Chapter 5, 'Domestic gas prices: catalysts for a rethink'.)

Three sub-sectors account for the bulk of industrial gas consumption in Russia: chemicals; iron and steelmaking; and cement production. Table 6.4, excerpted from Table 5.1 in Chapter 5, shows how this is reflected in the Rosstat statistics. The chemicals sector comprises the rows 'feedstock and other non-fuel uses' (almost all in chemical fertilizer plants) and 'gas processing plants'. Iron and steelmaking is included in the row 'manufacture and processing sectors', and probably accounts for about half of it. Cement manufacture is included in 'mining and extractive sector' and probably accounts for most of the gas consumption there. (That row does not include on-field consumption by the oil and gas industry.)

Table 6.4: Gas consumption in industry

	2006	2007	2008	2009	2010	2011
Mining and extractive sector	12.6	12.4	12.5	11.6	12.1	12.6
Manufacture and processing sectors	38.0	37.6	35.4	33.4	35.1	36.8
Total mining and manufacturing	**50.6**	**49.9**	**48.0**	**45.0**	**47.2**	**49.4**
Gas processing plants (ex. dry gas output)	11.0	10.7	10.9	9.8	11.1	11.4
Feedstock and other non-fuel uses	21.5	23.7	23.4	21.7	24.7	24.4
Total chemicals industry	**32.5**	**34.3**	**34.3**	**31.5**	**35.8**	**35.8**
Total domestic consumption	*456.3*	*466.7*	*462.0*	*432.6*	*462.1*	*472.8*

Source: Rosstat (excerpt from Table 5.1).

Chemical fertilizer producers, for whom natural gas purchases represent more than half their manufacturing costs, have benefited most directly from historically low gas prices and are feeling the impact of rising prices most strongly. Evrokhim, Russia's largest chemical fertilizer producer, published estimates of the costs of prilled urea (an ammonia-based fertilizer) delivered from its Nevinnomysskii Azot plant to Europe, compared to those of European and Ukrainian producers in 2010. These are shown in Table 6.5:

Table 6.5: Prilled urea to Europe: Evrokhim cost estimates, 2010

	Nevinnomysskii Azot	*European plants*	*Ukraine plants*
Gas price ($/mcm) delivered to plant	*108*	*312*	*307*
Cost per tonne of output ($)			
Gas cost	74	196	210
Other costs	63	65	65
Transportation costs	67	10	54
Import duty (EU)	26	0	26
Total	**230**	**271**	**355**

Source: Evrokhim, Annual Report and Accounts, 2010.

Table 6.5 shows that gas prices were so low in Russia that, even with higher transport costs and import duty, Russian-produced fertilizers were competitive in Europe. Ukrainian producers, however, were uncompetitive due to the high gas import price. On the other hand, gas prices in Russia are sufficiently high for Evrokhim to have been prompted (i) to invest in gas production and (ii) to decide in 2013 to invest in chemical fertilizer production in the USA, where gas prices were lower than those in Russia. Evrokhim's investment in gas production was via the purchase in 2012 of Severneft Urengoi, a Siberian producer of about 1.1 Bcm/year, roughly a quarter of Evrokhim's gas consumption. A contract with Gazprom ensures transportation of the gas to the Novomoskovskiy Azot fertilizer plant. Table 6.6 shows the cost saving reported by Evrokhim as a result of this vertical integration, including a contribution from the sale of condensate that is produced together with the gas.

Evrokhim's move upstream is perhaps significant as a pointer to the future. As regulated gas prices continue to rise, and the terms of access to the pipeline system improve, vertical integration by chemicals companies and other large gas consumers could grow, adding a further element of competition into the upstream.

Evrokhim's announcement in July 2013 that it intends to invest in fertilizer

Table 6.6: The cost of Evrokhim's own gas and its purchases

Gas costs for Novomoskovskii Azot, 2013	$/mcm	$/mmBtu
Cost of gas from Severneft Urengoi		
Gas at wellhead	46.98	1.28
MRET	24.59	0.67
Transportation	75.97	2.07
Revenue from gas condensate	*−70.46*	*−1.92*
Delivered cost to plant	77.07	2.10
Current gas cost (purchases in market)	133.95	3.65

Source: Evrokhim conference call presentation, July 2013.

production in Louisiana, USA, also indicates a possible future trend. The chemical fertilizer producers are sensitive to wide variations in international gas prices. As their raw materials (gas) costs rise in Russia, some of their international competitors have had access to gas priced at comparable levels to, or even lower than, gas in Russia: investments such as that in Louisiana are the logical competitive response.[35]

Gas processing plants that extract natural gas liquids from raw gas are, after chemical fertilizer plants, the other large group of chemicals sector gas consumers. Here, too, there has been vertical integration. Most gas processing plants are owned by Sibur, which is now controlled by the owners of Novatek. The main product of these plants is dry natural gas, most of which is sold to Gazprom; they also produce synthetic rubbers and plastics from the liquids.

For industrial consumers outside the chemicals sector, gas costs as a proportion of total costs are much lower. For iron and steel producers, who account for 15–20 Bcm/year of gas consumption, energy costs usually make up 5–10 per cent of total costs, of which 2–4 per cent is gas. To this must be added coke oven gas and blast furnace gas, which contribute about one unit of energy to the iron and steel plants for every two units contributed as natural gas.[36] In the mid-2000s, iron and steel producers made substantial investment in new plant, and consequently made energy efficiency gains.

---

[35] Evrokhim press release, 'EuroChem announces fertilizer project in Louisiana', July 2013; see also comments on new ammonia capacity in low-cost gas regions, in Evrokhim, *Annual Report and Accounts 2012*, p. 34.

[36] Gas costs as a proportion of total from publicly quoted companies' reports. See Pirani, 'Elusive Potential', pp. 76 and 128. Information on manufactured gases from the IEA, ibid., p. 105.

The third large group of industrial gas consumers is cement manufacturers, whose output rose rapidly during the 2000s due to the construction boom. They have potential for energy efficiency savings that is as yet mostly unrealized: 85 per cent of Russian capacity uses the old, energy-intensive 'wet' process. Upgrading to the newer 'dry' process has reduced energy demand by about 40 per cent in other countries. Government projections – that the cement industry could more than double its 2009 level while reducing gas consumption by 2–4 per cent – are ambitious but by no means unrealizable, although this will not happen on the timescale envisaged (by 2020).[37]

Caution should be exercised in generalizing from the cases of large gas consumers. Statistically, there is as yet no evidence of a correlation between higher energy prices and improved energy efficiency.[38] All that can be said for certain is that (i) some large companies have made investments in new capital stock (for example, steel companies during the mid-2000s) that have produced efficiency savings, even if this was not their main aim; (ii) in the chemicals sector, higher prices have significantly changed gas consumers' behaviour; and (iii) in most of Russian industry the recession has both interrupted plans for such investment, and masked any effect that it might have had.

Russian industry's recovery from the recession will increase gas demand, but this increase will be counteracted, at least to some extent, by energy efficiency improvements that the combination of higher prices and investment may produce.

Residential consumption, gasification and measures to stimulate demand

Residential consumption of gas, mostly for cooking and autonomous heat production, was the only major sector in which demand growth accelerated in 2007–11 compared to 2000–6. It accounted for 49.7 Bcm in 2011 (see Table 5.1 in Chapter 5). Gasification programmes have added to residential consumption, and this will continue during the 2010s. But there are several factors putting downward pressure on residential demand: (i) the long-term decline in total population (see 'Gas demand prospects' Chapter 5, page 113); (ii) rising prices and changing personal consumption habits; and (iii) development of infrastructure and municipal sector reform, which should improve the efficiency of consumption.

The issue of infrastructure concerns most household gas consumers, who live in multi-apartment housing blocks built in Soviet times. Gas is supplied together with electricity and heat (much of which is generated

---

[37] Ibid., pp. 77–8.
[38] Makarov, Mitrova et al., *Vliianie rosta tsen*, op. cit., pp. 7–8.

from gas), water, sewage, and building services, by municipal services supply organizations. Not only do most dwellings not have gas meters, but in multi-apartment blocks there are often no means, or only very outdated means, of adjusting the flows of heat and gas. A long-term process of municipal services reform is underway (see 'Heating sector reform' page 170). In many regions, these reforms are conducted together with housing stock renewal programmes. In the long term, these processes will lead to more efficient consumption of gas both by households and by district heating companies, but it is difficult to express the potential demand reduction in numerical terms.[39] Regulated tariffs for residential gas consumers remain at a 20–30 per cent discount to those for industrial consumers, and there is considerable political resistance to raising them. However, tariff increases will not in themselves reduce consumption; households will need to have the means to control and measure their consumption for this to happen.

The Russian government's gasification programme contributed to increases in consumption in the 2000s, and is likely to continue doing so in the 2010s. Between 2005 and 2013, the proportion of dwellings attached to gas distribution networks rose from 53.3 per cent to 64.4 per cent. The increase in urban areas was from 60 per cent to 70.1 per cent, and in rural areas from 34.8 per cent to 54 per cent.[40] There is no obvious way of determining the proportion of the increase in total consumption, or of residential consumption, accounted for by gasification. In the Far Eastern federal district (the government's gasification flagship) total consumption rose from 3.3 Bcm to 13.8 Bcm in the decade to 2011; in 2011, however, 6.5 Bcm was for upstream and technical use, 4.6 Bcm was for power and district heating, and only 0.3 Bcm was consumed by residential customers.[41] Gazprom estimates that total gas consumption in Eastern Siberia (which consumes hardly any pipeline gas) and the Far East combined will reach 22 Bcm/year by 2030, a projection significantly scaled down from earlier ones.[42]

---

[39] For information about municipal reform, see *inter alia* the site of the government's Fund for Cooperation in Housing and Municipal Services Reform, www.reformagkh.ru. It reports that 54% of dwellings are managed by private management organizations and 13% by condominiums.

[40] Gazprom, Stenogramme of press conference by K. Seleznev, 'Postavki gaza na vnutrennyi rynok' ['Supply of gas to the domestic market'], 28 May 2013.

[41] See Appendix 2, Table A.2.1. The row 'technical and other' gas includes, in the case of the Far Eastern federal district only, volumes of gas consumed on oil and gas fields, which began to be counted by Rosstat in 2007.

[42] Gazprom, Stenogramme of press conference by K. Seleznev, 28 May 2013, slide 6. The government's plan for the gas sector, drafted in 2008, envisaged gas consumption in Eastern Siberia and the Far East rising to 21 Bcm in 2015 and 31 Bcm in 2020. See *General'naia skhema razvitiia gozovoi otrasli na period do 2030 goda: proekt* (Moscow, 2008), 2/4.

The future level of gasification will depend largely on investment decisions by local government. Gasification is implemented under agreements concluded between Gazprom and the governments of Russia's 83 regions, but Gazprom – which has been investing 18–34 billion RR/year into the scheme in recent years[43] – frequently has cause to complain that local governments' investment commitments have not been fulfilled. In 2009 it revised its gasification strategy, introducing a clear requirement that schemes had not only to:

> ... meet consumer demand [and achieve] improvement of living conditions [and] the growth of [regions'] economic potential [, but also be] economically viable.[44]

Technical use and losses

Technical use of gas (in other words, as fuel for pipeline compressor stations, plus leaks from compressor stations and pipelines) is a major element of Russia's consumption, accounting for 45.8 Bcm of gas in 2011. Fuel gas accounts for more than 80 per cent of this amount; the remainder is leaks, which are particularly damaging from an environmental standpoint (because of methane's very powerful 'greenhouse' effect, more than 20 times that of carbon dioxide).[45] Political attention has recently turned to ways of reducing technical use and losses. This was epitomized in 2010 by Sergei Shmatko, then energy minister, when he argued that 'we could reduce annual consumption [by the pipeline system] by 15–20 Bcm'.[46] How rapidly such efficiency investments are made in the pipeline system is likely to depend largely on the progress of gas sector reform.

---

[43] Gazprom reported investment in gasification of RR90 billion in 2005–9, RR27 billion in 2010, RR29.07 billion in 2011 and RR33.8 billion in 2012 (from annual reports and Gazprom website).

[44] Gazprom Annual Report 2012, 64; 'Gazprom management committee approves updated concept for gasification', Gazprom press release, 1 December 2009.

[45] Research by Gazprom, Ruhrgas, Transcanada, and the Wuppertal Institute estimated that in 2004, 40.8 Bcm of gas was used as compressor station fuel, 4.8 Bcm leaked from compressor stations, and 1.4 Bcm leaked from pipelines. See IEA *Optimising Russian Natural Gas: Reform and Climate Policy* (Paris: IEA/OECD, 2006), p. 100.

[46] Ministerstvo energetiki Rossiiskoi Federatsii, Doklad ministra energetiki RF S.I. Shmatko 'O proekte general'noi skhemy razvitiia gazovoi otrasli na period do 2030', [Ministry of Energy of the Russian Federation, Report by Minister of Energy S.I. Shmatko 'on the draft general scheme for the development of the gas industry to 2030'], 12 October, 2010.

Conclusions

The steady increase in Russian gas demand (2–2.5 per cent per year) during the early and mid 2000s, which was related to steadily rising GDP and electricity demand, was interrupted by the economic crisis of 2008–9. By 2010–11 gas demand had recovered to, but not substantially exceeded, its pre-crisis level (see Table 5.1 page 110). The survey of the sources of demand in the current chapter shows that a return to robust, steady growth is unlikely in the near future. Russia's economy and infrastructure are changing in ways that could reduce demand; government policy, especially with regard to investment in the power sector, efficiency, and modernization measures, is a key factor that will influence demand; as are the expected continuation of increases in regulated gas prices, on one hand, and growing competition among suppliers on the other. Finally, the steady population growth that underpins rising demand in many countries is absent.

In the power sector, the market reform of 2006–7, and the consolidation of assets in 2011–12, has opened the way for substantial investment in new generation capacity. Competition in the electricity market has, since 2009, increasingly been matched by competition among gas suppliers for contracts with power companies and, since 2012, by such competition at a discount to regulated gas prices; these changes also stimulate efficiency measures. The construction of new nuclear and hydro capacity will displace gas from the fuel mix, and this effect will be accentuated if old nuclear capacity is decommissioned at a slower pace. Thermal generating companies have either begun construction of, or are planning, the commissioning of 18 GW of new capacity under the investment framework agreed at privatization. Provided that government policy continues to support such investment, it will lead, by 2020, to the commissioning of a substantial quantity of CCGT capacity, which will meet a considerable amount of incremental electricity demand. Other efficiency gains and fuel-saving measures, spurred by the competitive structure of the electricity and gas markets, will also make a difference. If electricity demand rises as quickly as the Energy Ministry expects (by 1.79 per cent annual average), up to 36 Bcm of incremental gas could still be needed for the power sector, but most projections suggest the requirement will be less than that, and that it could even be zero.

In the district heating sector, the potential reductions in gas consumption are proportionally greater, and demand has already started falling in the post-crisis period. But so, too, has payment discipline among district heating companies facing rising gas prices and mis-regulated heat tariffs. Much depends on the pace of reform in this sector, which is bound up with broader changes in municipal services provision, local government, and renovation of Russia's huge urban housing stock. If it is implemented under the timetable now set out by the government, it will lift a growing economic burden from Gazprom and reduce wasteful gas consumption – but significant results are not likely until the 2020s.

Residential sector consumption could also be reduced as a result of the municipal services reform, but with the same caveats about timing. In this decade, it could increase somewhat as a result of gasification, particularly in the Far East and Siberia.

Rising gas prices and competition among suppliers have not yet made a noticeable impact on levels of consumption, but among industrial customers that potential is becoming evident. Thus Russia's largest chemical fertilizer producer has both invested in upstream gas production in Siberia, and planned to establish production in the USA in response to the international impact of low gas prices there. In other sectors, noticeable efficiency improvements have been made only by big companies and usually as a result of capital stock renewal made for other commercial reasons.

Under these circumstances, Russian researchers have suggested that aggregate gas demand growth could be between zero and 1 per cent per year. Such forecasts *could* be confounded, for example, by stronger-than-forecast economic growth, especially in the manufacturing sector, or by a dramatic change in the energy balance (such as unplanned closures of nuclear power capacity or a sharp increase in the price of coal). But it is equally likely, especially if more significant efficiency gains start to be achieved, that gas consumption could fall.

CHAPTER 7

CIS GAS MARKETS AND TRANSIT

Simon Pirani and Katja Yafimava

The former Soviet countries comprise a significant, but declining, market for gas exported from Russia (including volumes contractually labelled as Central Asian, and exported through Russia). For much of the post-Soviet period Gazprom has had a de facto monopoly, and since 2006 a legal monopoly, over these sales (as it has over all exports). Prices have risen consistently in recent years, and the former Soviet countries showed some potential to become lucrative captive markets. But these markets are increasingly being influenced by those in neighbouring countries, and it now seems likely that they will become less captive. In Ukraine, the largest former Soviet market by far, Gazprom has achieved high prices, but has also suffered a sharp fall in sales volumes and has failed in its aim of establishing a downstream presence. The theme of this chapter is that while Russia will continue to be a key supplier, often *the* key supplier, to these markets, diversification of supply sources will continue and competition will intensify.

Russian and Central Asian gas is sold to Ukraine, Belarus, and Moldova in the western part of the Commonwealth of Independent States (CIS); and to Georgia and Armenia in the Caucasus.[1] (Russia also exports to Kazakhstan, both (i) gas to supply areas most easily accessed via connections from the Russian network, as swaps for exports from Kazakhstan to Russia and (ii) gas returned after processing. See Chapter 14, footnote no. 12.)

Russian exports and other gas supplies to CIS countries over the last decade are shown in Table 7.1. It shows that Russian exports reached their highest level, 89.6 Bcm, in 2006. They then fell sharply, to 60.2 Bcm in 2009, following the economic crisis. Although they appeared to be recovering up to 2011, they fell again in 2012 to 57.6 Bcm, their lowest level since 2003. At the time of writing, it appears that Russian exports to the CIS in 2013 will be no greater than in 2012; data for Ukraine suggests that they may be much lower. Table 7.1 also highlights two reasons for the decline in Russian exports: (i) the start-up in Azerbaijan of the Shah Deniz gas field in 2006, which enabled it to cover its own needs and to substitute for Russian exports to Georgia; and (ii) the steep decline of consumption in Ukraine.

---

[1] The three Baltic states that have joined the European Union are considered in Chapter 4 on Europe. Georgia initiated moves to leave the CIS in 2011 but is included here in order to give a complete picture of the situation in the Caucasus.

Table 7.1: Supplies of gas to CIS countries

		2003	2004	2005	2006	2007	2008	2009	2010	2011	2012
Ukraine	Russian imports	26.30	34.40	37.60	59.00	59.20	56.20	37.80	36.50	44.80	32.90
	Other (own production)	18.67	19.22	19.61	19.73	19.53	21.05	21.34	20.46	20.61	20.54
	Other ('reverse flow')	0	0	0	0	0	0	0	0	0	0.06
Belarus	Russian imports	16.40	18.21	19.80	20.50	20.60	21.10	17.60	21.60	23.30	19.70
	Other (own production)	0.25	0.25	0.23	0.22	0.20	0.20	0.20	0.20	0.20	0.20
Moldova	Russian imports	1.50	1.80	2.80	2.50	2.70	2.70	3.00	3.20	3.10	3.10
Azerbaijan	Russian imports	4.10	4.10	3.80	4.00	0	0	0	0	0	0
	Other (own production)	4.60	5.10	5.70	6.40	8.70	8.90	8.50	7.80	10.10	10.60
Georgia	Russian imports	1.05	1.10	1.40	1.90	1.20	0.70	0.10	0.20	0.20	0.20
	Other (Azeri imports)	0	0	0	0	0	0.80	1.00	1.50	1.50	1.53
Armenia	Russian imports	1.20	1.50	1.70	1.70	1.90	2.10	1.70	1.40	1.60	1.70
Total	Russian imports	50.55	61.11	67.10	89.60	85.60	82.80	60.20	62.90	73.00	57.63
	Other	23.52	24.57	25.54	26.35	28.43	30.95	31.04	29.96	32.41	32.90
	Total	74.07	85.68	92.64	115.95	114.03	113.75	91.24	92.86	105.41	90.52

Note. This table only shows the sources of gas for domestic consumption, and does not include, for example, Azeri exports to Turkey. Russian exports to Kazakhstan (swaps and gas returned after processing) are excluded.

Sources. Russian imports: *Gazprom in Figures* (various years); Stern, J.P. *The Future of Russian Gas and Gazprom* (Oxford: OIES/OUP, 2005), p. 69; author's estimates (Armenia and Georgia 2004). Other: *Energobiznes*/Ukrainian energy ministry; Pirani, S. (ed.) *Russian and CIS Gas Markets and their Impact on Europe* (Oxford: OIES/OUP, 2009), p. 136; Pirani, S. 'Central Asian and Caspian Gas Production and Constraints on Export', Working Paper NG 69, OIES, December 2012, p. 9; author's estimates.

There are two general points about Russian exports to CIS countries.[2] The first is that they have a political aspect, in the sense of direct government involvement in commercial issues. Throughout the post-Soviet period, Gazprom has endeavoured to raise the prices paid for these exports to levels reflective of the market – which in the western CIS was defined as netback from the oil-linked prices achieved by Gazprom at the European border. But price negotiations with some CIS importers – Ukraine, Belarus, and Armenia – are conducted alongside intergovernmental discussions covering both gas supply and gas transit, and sometimes a range of other political, economic, and even strategic and military issues. Strategic and political objectives have sometimes stymied Gazprom's commercial objective of raising export prices: for example, price rises have been delayed, or other commercial concessions made, in return for strategic concessions outside the gas sphere. In Ukraine, a substantial price concession was given in 2010, in return for the extension of the lease on the Black Sea naval base. Political factors in Belarus have kept import prices far below the European netback level, at about half the level paid by Ukraine.

The second point is that it is difficult to separate the issue of Russian gas supply to CIS markets from that of transit. Most Russian gas volumes bound for Europe are transported through Ukraine and Belarus – although their importance as transit countries is declining, mainly due to Russia's transit-diversification policy. This chapter discusses each of the CIS markets and then considers transit.[3]

Ukraine[4]

Between 2006 and 2013 Ukraine, the largest importer of Russian gas among CIS countries (and until recently the largest importer of Russian gas in the world) moved towards a market with much higher prices, lower volumes, and more competitive trading arrangements than before. Change has been rapid: in 2006, Ukraine used more than 75 Bcm of gas, most of it imported

---

[2] These arguments build on previous research by the OIES Natural Gas Research Programme. Relevant publications include Pirani, S. (ed.) *Russian and CIS Gas Markets and their Impact on Europe* (Oxford: OIES/OUP, 2009) and Yafimava, K. *The Transit Dimension of EU Energy Security: Russian Gas Transit Across Ukraine, Belarus, and Moldova* (Oxford: OIES/OUP, 2011).

[3] Simon Pirani drafted the sections on Ukraine, the Caucasus, and conclusions; Katja Yafimava those on Belarus, Moldova, and transit.

[4] Final revisions were made to this book in January 2014, before the fall of the Yanukovich government and the political and military crisis that followed. The text has not been changed to take account of these. See also footnote 1 to the Introduction, above, page 1.

at a price of $95/mcm; in 2012, after two gas disputes with Russia that led to supply interruptions, and the economic recession, consumption was more than 25 per cent lower and import prices were more than four times higher. In December 2013, as a result of negotiations at government level between Russia and Ukraine, a substantial reduction in gas import prices was agreed, as part of a political and economic cooperation package.[5] Nevertheless, we will argue here that the Russia–Ukraine gas trade is moving away from politically negotiated bilateral trading relationships, and towards relationships determined largely by market forces: the price reduction agreed was strongly influenced by, and brought Ukrainian import prices in line with, falling prices for Russian gas in Europe.

Imports

Ukraine's acute difficulties in reforming its gas sector stem from the position it occupied in the Soviet Union. In 1970 it was the largest gas-producing region, and its industry and infrastructure developed on the assumption that cheap supplies of both gas and coal would always be available. As Ukraine's gas production declined and west Siberia's rose, it seemed to be assured of continued supplies from the pipelines that crossed its territory, carrying Soviet gas to Europe. After the break-up of the Union, Ukraine became the largest of the new nations anxious to distance themselves politically from Russia, with Russia becoming determined to charge market prices for gas, as trade was internationalized. Even in the 1990s, when prices remained relatively low, Ukraine built up large debts for gas. In the 2000s, as prices rose in earnest, that problem was exacerbated. During several skirmishes in the 1990s, and the gas disputes of 2006 and 2009, Russia sought to drive prices up to a level equal to netback from the prices it achieved in western Europe. Successive Ukrainian governments accepted such price-setting in principle, but fought to slow down its implementation. Russia's ultimate sanction was to cut off gas supplies; Ukraine's ultimate

---

[5] This package of measures was agreed in talks between Russia's President Vladimir Putin and Ukraine's President Viktor Yanukovich on 17 December 2013. It included a $15 billion investment in Ukrainian treasury bonds, financed from the Russian Fund for National Well Being; a framework agreement for Russian investment in aircraft construction and other branches of Ukrainian industry; and support by both presidents for an addendum to the 2009 gas contracts, signed between Gazprom and Naftogaz Ukrainy in the presidents' presence, providing for a price reduction for gas imports in the first quarter of 2014. The details of the agreement on gas were announced after the talks by President Putin. See 'Zaiavlenie dlia pressy po okonchanii zasedaniia Rossiisko-Ukrainsko mezhgosudarstvennoi kommissii', website of the Russian president, 17 December 2013; 'Gazprom pochti vdvoe snizil tsenu gaza dlia Ukrainy', *Vedomosti*, 17 December 2013; and 'Zapomni, kak vse nachalos' ', *Zerkalo Nedeli*, 20 December 2013.

response to such cut-offs was to divert for its own use gas destined for Europe. The Russia–Ukraine gas disputes broadly followed that pattern – with the additional complication that agreements on import prices and transit arrangements were negotiated at a political level, together with other often unrelated issues between the two sides – until 2009.[6]

Within two years of the 2009 dispute, Russia's policy decision – that the cycle of disputes could only be broken by building pipelines that avoided Ukraine and Belarus – bore fruit with the commissioning of Nord Stream. (On this and other transit issues, see 'Transit' page 207.) Russia was then able to insist on the terms of the 2009 sales–purchase agreement between Gazprom and Naftogaz Ukrainy, Ukraine's state-owned national oil and gas company (which set prices not merely at the European netback level, but at a level about 10 per cent higher, and included unusually strict payment arrangements) without any serious danger of a major dispute. Ukraine was committed to this method of price setting at the very moment (early 2009) when oil prices, from which European gas prices ultimately derived, were hitting their peak.

In 2010, there was a relative easing of tension. President Viktor Yushchenko – whose pro-NATO political stance had clashed with President Putin's foreign policy, aggravating the commercial aspects of the dispute – was replaced by Viktor Yanukovich. A political agreement was then negotiated that provided for a discount of roughly $100/mcm on import prices and for a softening of some punitive aspects of the 2009 agreement in exchange for a non-gas concession (a 25-year extension on Russia's lease from Ukraine of the Black Sea naval base). Nevertheless, Russian revenue from sales to Ukraine rose again, peaking in 2011.[7] Table 7.2 shows the trends in Ukrainian gas imports during and after the gas disputes of 2006 and 2009.

Since the economic crisis of 2008–9, imports of Russian gas by Ukraine have fallen substantially. They sank to a record low in 2009, recovered in 2010–11, and fell again in both 2012 and 2013, to around two-thirds of pre-crisis levels. The high level of import prices, which even with the $100/mcm discount are comparable to those in Europe, have acted as an incentive for fuel switching and energy saving on one hand, and supply diversification measures on the other. Such measures, combined with the impact of the recession, have

---

[6] These events are analysed in detail in publications by the OIES Natural Gas Research Programme, including: Stern, J.P. 'The Russian-Ukrainian Gas Crisis of January 2006', Working Paper, OIES, January 2006, Pirani, S. 'Ukraine's Gas Sector', OIES Working Paper NG 21, June 2007, and Pirani, S. Stern, J., and Yafimava, K. 'The Russo-Ukrainian gas dispute of January 2009: a comprehensive assessment', Working Paper NG27, OIES, February 2009.

[7] Details in Pirani, S. Stern, J., and Yafimava, K. 'The April 2010 Russo-Ukrainian gas agreement and its implications for Europe', Working Paper NG 42, OIES, June, 2010.

Table 7.2: Ukraine's gas imports

	2005	2006	2007	2008	2009	2010	2011	2012	2013
Volume imported (Bcm)	55.8	57	50.59	54.6	26.95	36.47	44.8	32.94	27.9
incl: *Naftogaz Ukrainy (Russian gas)* *	55.8	57	50.59	54.6	26.95	36.47	40	24.89	12.9
Naftogaz U and traders ('reverse flow')	0	0	0	0	0	0	0	0.05	2.1
Ostchem Holding	0	0	0	0	0	0	4.8	8	12.9
Import price ($/mcm)**	44–80	95	130	179.5	232.54	257	315.5	424	414
Estimated value of imports ($ bn)	3.3	5.4	6.58	9.8	6.27	9.37	14.1	13.97	12

* 2005–9, includes central Asian volumes.

** 2005–12, prices reported by Naftogaz or derived from its reports; Ostchem Holding purchases excluded. 2013, average price derived from state statistics committee information, including all purchasers.

Source: Volumes: *Energobiznes*/Energy Ministry. Prices: press reports/ Naftogaz Ukrainy. Estimated values: author's estimates.

reduced import volumes. It could be argued that Gazprom has maintained high prices at the expense of volumes, as it has in Europe. Since 2010, the three key features of Ukrainian government policy on gas imports have been:

1. *Ukraine has consistently sought to renegotiate the 2009 contract.* It finally achieved new terms in December 2013. Russian political leaders stated repeatedly in 2010–13 that such a renegotiation was dependent on Ukraine abandoning its long-standing opposition to Gazprom taking ownership and/or control of the Ukrainian gas transportation system. But in December 2013, during the Ukrainian political crisis triggered by President Yanukovich's failure to sign an association agreement with the EU, Russia shifted its position. Russia agreed to a substantial reduction in gas import prices – from around $402/mcm in the fourth quarter of 2013 to $268.50 in the first quarter of 2014. Press reports, citing leaked copies of the agreement signed between Gazprom and Naftogaz Ukrainy, indicate that the price would henceforth be subject to renegotiation each quarter, but that all other terms of the 2009 contract were unchanged.[8]

Political commentary has focused on the fact that the gas price reduction formed part of a package of measures that included lowering trade barriers (for example, for Ukrainian steel products exports to Russia), cooperation on infrastructure projects, and Russian financial support for Ukraine ($15 billion in treasury bond purchases). For gas markets, the price level agreed is more significant. No comment has been made by either Gazprom or Naftogaz Ukrainy on the basis for the reduction. But there is no doubt that it correlates with netback from a range of European price levels. In broad terms, in late 2013 Russian gas was imported to Germany at around $450/mcm; this price, minus $100/mcm of export duty (not payable on exports to Ukraine) and $40/mcm additional transport costs, is $310/mcm. Netback from hub prices, rather than Russian import prices, was about $40/mcm lower than that, or $270/mcm.[9] We argue not that such calculations were made in the Russia–Ukraine negotiations, but that the downward pressure put on Russian import prices in Europe by lower hub prices – and the indirect effect of these dynamics on Ukraine, via 'reverse flow' (see below) – are part of the economic background against which the agreement was signed. Given the lack of information about the basis for future quarterly price setting – to say nothing of the uncertainty arising from the continuing political crisis in Ukraine and the temporary nature of some aspects of the December agreement – it is not possible at the time of writing to say anything about future price trends.

---

[8] 'O chem eshche dogovorilis' Rossiia i Ukraina', *Vedomosti*, 19 December 2013.

[9] On 11 December 2013 Platts estimated Gazprom's European border price at $12.25/MMBtu ($449.50/mcm), Gaspool Germany prices at $11.16/MMBtu ($409.57/mcm), and Dutch TTF prices at $11.20/MMBtu ($411.04/mcm).

2. *Ukraine abolished Naftogaz Ukrainy's import monopoly in 2011–12.*[10] The government has allowed Ostchem Holding, a privately owned gas trader, to import substantial quantities of gas: 4.8 Bcm in 2011, 8 Bcm in 2012, and 12.9 Bcm in the first nine months of 2013.[11] Ostchem is owned by Dmitry Firtash, former part-owner of Rosukrenergo. Another of his companies, DF Group, controls all Ukraine's major chemical fertilizer producers; they consume more than half of Ostchem's imports. The logic was not that Ostchem might access cheaper gas – customs data suggests that, if anything, Ostchem is paying a premium to Naftogaz import prices.[12] Rather, Ostchem, as a trader that is both politically well-connected and can more easily raise credit for gas purchases than Naftogaz, has gained a bridgehead in the market.[13] In 2013, two other traders, DTEK and Vetek, started to import gas from Europe. (See 'Market reform and corporate change', page 194.)

3. *The Kyiv government started consistently to consider methods of diversifying away from Russian imports.* In 2010–13 Ukraine, faced with Moscow's intransigence on price: (i) sourced imports from non-Russian sources, (ii) sought to raise Ukraine's own production, and (iii) encouraged energy saving and fuel switching to reduce gas consumption. For gas markets, the imports from non-Russian sources – and specifically from EU countries eastwards into Ukraine – may eventually prove the most significant feature of this diversification as they could lead in future to Ukraine becoming more closely integrated with European markets, even though the December 2013 Russia–Ukraine agreements may slow down or temporarily halt this process.

---

[10] Naftogaz Ukrainy's monopoly on passing imported gas through customs, introduced in March 2008, was abolished by the government's decree no. 343 of 4 April 2011; the company's monopoly over marketing imported gas was abolished by the government's decree no. 939 of 3 October 2012. 'Utechka gaza', *Kommersant'-Ukrainy*, 7 April 2011; 'U Azarova otmenili monopoliiu Naftogaza', *Obozrevatel'*, 16 October 2012.

[11] 'Ukraina i Rossiia mogut vernut' posrednika v skhemu torgovli gazom', *Vedomosti*, 13 December 2013; 'Import gaza v 2013 g. sokratilsia na 15%', UA Energy website (www.uaenergy.com.ua) 29 January 2014.

[12] In the second quarter of 2013, Naftogaz was paying \$426.10/mcm for Russian gas, while customs data for April showed that the average price of imports from Russia was \$430/mcm, suggesting that Ostchem pays a premium. In the first quarter, the premium was about \$20/mcm. Research on customs data in 2011 produced similar results. 'Tsena rossiskogo gaza dlia 'Naftogaz Ukrainy' v iiune', *Zerkalo Nedeli*, 21 June 2013; 'Ukraina v proshlom godu importirovala gaz na \$14 mlrd', *Zerkalo Nedeli*, 21 February 2012.

[13] Contractually, Ostchem buys gas that is purchased by Gazprom Germania in Central Asia, transited through Russia, and resold to Ostchem on the Ukrainian border.

Supply diversification and 'reverse flow'

Ukraine sourced gas imports from non-Russian suppliers[14] for the first time in late 2012, when small-scale deliveries began from Poland and Hungary. These imports are called 'reverse flow', because they are delivered eastwards, in the opposite direction to the large-volume flows of Russian gas. Throughout 2013, EU and Ukrainian officials made much of the potential of 'reverse flow', suggesting that up to half of the country's import requirement could soon be met in this way. Although the Russia–Ukraine agreement in December 2013 to cut import prices may have a negative effect for the 'reverse flow' business – which has suddenly been put at a competitive disadvantage – it remains a significant example of new ways in which market forces are influencing the trading of Russian gas.

The first 'reverse flow' deliveries were made under a contract between RWE Supply & Trading and Naftogaz Ukrainy: via Poland (about 4.5 million cu m/day, equivalent to 1.65 Bcm/year) and Hungary (about 5 million cu m/day (1.8 Bcm/year), with an increase to 9 million cu/m day (3.3 Bcm/year) in August 2013).[15] While Ukraine has long had a strategic commercial interest in diversifying its imports, it was the changes in European gas pricing that gave the impetus for this trade. The gap that opened up in Europe in 2009–12, between oil-linked prices (to which Gazprom's prices for Ukraine are related) and spot prices (see Chapter 3), meant that the initial 'reverse flow' deliveries were $10–40/mcm cheaper than imports from the east. Customs data showed that 'reverse flow' gas was $16/mcm cheaper at the Ukrainian border in the first quarter of 2013, and $36/mcm cheaper in the second quarter.[16]

The possibility of much larger 'reverse flow' deliveries to Ukraine from Slovakia, via one of the four strings of the Druzhba (Brotherhood) pipeline, has been publicly discussed by government, EU, and company officials. 'Reverse flow' deliveries could also be made via Romania. However, in late 2013 there appeared to be various regulatory, administrative, and technical obstacles to the expansion of 'reverse flow'. For example, Eustream, the Slovak TSO, has considered building a new interconnector to facilitate the trade, but stated in October 2013 that this project had not attracted sufficient interest. Moreover, there was strong public opposition to 'reverse flow' by Gazprom, whose commercial interests would inevitably be damaged by such expansion. Indeed, the most obvious immediate impact of

---

[14] That is, excluding arrangements made for delivery to Ukraine in the 1990s and early 2000s of gas that was nominally and contractually from Central Asia, but was delivered through Russia.

[15] Eremenko, Alla and Chubik, Andrei, 'Top-menedzhery RWE o detaliakh reversnykh postavok gaza', *Zerkalo Nedeli*, 24 June 2013; Marunich, Dmitrii, 'Dvoe v lodke, ne schitaia Evropy', *Ekspert*, 8 July 2013.

[16] 'Tsena rossiskogo gaza dlia 'Naftogaz Ukrainy' v iiune', *Zerkalo Nedeli*, 21 June 2013.

the December 2013 agreement to cut Russian import prices is that it priced suppliers of 'reverse flow' gas out of the market, and in January 2014 it was reported that 'reverse flow' has been suspended.[17]

The actual and proposed 'reverse flow' deliveries mentioned above are all physical: volumes of gas are being, or would be, metered travelling in a direction opposite to the large volumes of Russian gas going to Europe. A much greater change would be implied by 'virtual reverse flow', in other words, the re-assignment of Russian gas purchased by European buyers to delivery points in Ukraine instead of in countries further to the west. This would presumably only be possible if the legal and regulatory framework in Ukraine is transformed and its gas market effectively merged with the single European market – a reform envisaged in recent years in talks between Ukraine and the EC, in Ukraine's membership (from February 2011) of the Energy Community (the purpose of which is to enable such changes), and in Ukraine's 2010 gas market law.

Such changes could enable Ukraine to become a 'gas hub' in the future. Its enormous gas storage capacity (31 Bcm, more than half of which is a few tens of kilometres from Ukraine's western border) gives it the technical potential to do so.[18] However, progress along these lines is likely to be slow, as it involves overcoming not only legal and regulatory but also political obstacles. The legal and regulatory obstacles concern the implementation of the 2010 gas law, which has hardly made any progress in the three years since its adoption.[19] Experience shows that even in EU member states, single-market principles can take years to implement. There is little reason to believe that Ukraine, a non-EU member, could move any more rapidly. The political obstacles concern Ukraine's relationship with Russia. The two countries continue to resolve gas-related commercial issues in the context of political negotiations – most recently, in the December 2013 talks. And while it may be argued that Moscow has no business objecting either to 'reverse flow' or to Ukrainian market reforms aimed at integration with Europe, it is

---

[17] 'Ukraine hopes to diversify gas sources', *Interfax Russia & CIS Oil and Gas Weekly*, 19 June 2013; 'Interest lacking in Slovak Ukraine pipeline', *Platts European Gas Daily* 17 October 2013; 'Ukraina s nachala ianvaria ne pokupaet gaz v Evrope', *Vedomosti*, 8 January 2014.

[18] After a meeting between Ukraine's Energy Minister Eduard Stavitskii and EC Energy Commissioner Gunther Oettinger in May 2013, both sides made statements to this effect, with Oettinger implying that such a hub would be a fully functioning exchange similar to European hubs and operating under regulations that mirrored those in the EU. 'Ukraine wants test reverse flow from Slovakia', *Platts EGD* 7 May 2013.

[19] *Zakon Ukrainy Pro zasadi funktsionuvannia rinku prirodnogo gazu* [the Law of Ukraine on the Functioning of the Natural Gas Market] (law no. 2467-VI of 8 July 2010) http://zakon1.rada.gov.ua/cgi-bin/laws/main.cgi?nreg=2467-17; Tarnavskii, Viktor 'Ochevidnoe i veroiatnoe', *Energobiznes*, 30 July 2013.

clearly in Gazprom's commercial interest to frustrate the process.

In the longer term, one catalyst to change may be the expiry of the import and transit contracts between Gazprom and Naftogaz Ukrainy in 2019. This could open the way to new arrangements, although Ukraine could move more quickly than that towards integration with the European market: (i) if the differential between oil-linked and hub prices were to widen again, reviving the economic benefit of larger-volume imports from Europe; and/or (ii) if the price advantage conferred on Russian imports by the December 2013 agreement were to be lost in some other way (for example, termination of the agreement); and/or (iii) if future elections were to bring to office a political leadership more consistently focused on integration with the EU, and on distancing Ukraine from Russia.

There are two final points concerning Ukraine's efforts to diversify sources of gas. First, Ukraine has experienced extreme difficulty in opening up other new import routes apart from 'reverse flow'. In 2012, a government proposal to build a LNG regasification terminal on the Black Sea coast failed to attract any substantial international partner: the investment required, together with the uncertainty about where LNG could be sourced, proved at that stage to be prohibitive. There have also been talks with Turkmenistan about reopening direct sales, but these have produced no outcome: direct sales ceased in 2005 after Gazprom declared itself unwilling to transit the gas, and it remains unwilling.

The second point is that since 2010 the Ukrainian government has focused attention on increasing domestic gas production, the second source of supply after Russian imports. Ukraine's production has stabilized at 20–21 Bcm/year in recent years: roughly 15–16 Bcm/year from Ukrgazvydobuvannya, Naftogaz Ukrainy's onshore production subsidiary; 1 Bcm from Chernomorneftegaz, its offshore production subsidiary; 2 Bcm/year from Ukrnafta, a state-controlled oil company; and 1–2 Bcm/year from small privately owned producers. In 2011–12, the government made substantial investment in Chernomorneftegaz; the company reported $1.2 billion of capital investment (four times more than the aggregate sum for 1992–2010), it bought new drilling rigs, and projects an extra 2 Bcm/year of new production by 2015. The government also made progress in negotiations with IOCs on developing both offshore conventional resources and onshore shale and tight gas plays. A 50-year PSA was signed with Shell in January 2013 for the Iuzivska field in the Dnipr-Donetsk basin, where it is undertaking exploration together with Nadra Ukrainy. Chevron, and a consortium led by ExxonMobil, are negotiating for onshore and offshore PSAs respectively. Some exaggerated claims have been made about Ukraine's potential output: in reality, the new Chernomorneftegaz production may do no more than replace losses from the natural decline of many Ukrgazvydobuvannya fields, while successful exploration elsewhere could add 1–5 Bcm/year by 2020, with more in the subsequent decade.[20]

The Ukrainian domestic market

The high import prices paid by Ukraine since 2009, combined with the effects of the economic recession, have brought about significant changes in the domestic gas market: consumption has fallen sharply, and there has been some movement in the direction of market reform. For Gazprom, this has meant that the volumes required by one of its largest export markets have fallen sharply, as discussed above. Due to government policies and the action of domestic industrial groups, Gazprom has not made progress with its plans to establish a domestic trading company.

Table 7.3: Gas consumption in Ukraine

Bcm	2006	2007	2008	2009	2010	2011	2012
Total consumption	75.3	71.1	67.5	53.1	59	61.9	54.8
Industry	24.3	25.8	23.2	13.5	17.2	19.1	17.1
Power sector	8.6	8.4	7.5	5	7.2	7.1	6.7
District heating	12.8	10.5	10	10.1	9.5	9.4	8.8
Residential & public sector	21.3	19.2	19.6	18.9	19.7	21.6	19.2
Technical gas	8.1	7	7	5.4	5	5.1	3.8
Other	0.2	0.2	0.2	0.2	0.4	0.2	0.2
Imported gas	54.6	50.4	47.4	31.8	38.6	40.9	34.3
Own production	20.7	20.7	20.1	21.3	20.4	21	20.5

Note. The row 'imported gas' denotes the amount of gas consumed in that year that was imported, i.e. the total amount consumed minus the total produced in Ukraine. It varies from the volumes of imports stated in Table 7.2, because gas can be imported in one year, stored, and consumed in a future year.

Source: Energy Ministry/*Energobiznes*, Naftogaz Ukrainy, author's estimates.

The dramatic fall in consumption since the mid-2000s is shown in Table 7.3. The fall in 2009 is attributable less to the increase in import prices that year than to the economic recession, the impact of which was harsher in Ukraine than in almost any other country (its GDP contracted by more than 20 per cent). Consumption recovered somewhat in both 2010 and 2011, but fell sharply again in 2012 (by 11.5 per cent to 54.8 Bcm) and

[20] 'Can Ukraine finally shake its Russian energy habit?', *Gas Matters*, April 2013, pp. 26–31; Opimakh, Roman 'Ukraine's Gas Upstream Sector', *Oil and Gas Journal*, June 2013; 'Ukraine hopes to diversify gas sources', *Interfax Russia & CIS Oil & Gas Weekly*, 19 June 2013; Tarnavskii, Viktor 'Chastnyi interes', *Energobiznes*, May 2013.

in 2013 (by 8.2 per cent to 50.3 Bcm, according to preliminary statistics released by the Energy Ministry).[21]

Table 7.4 shows key prices for Ukrainian consumers, in dollars, over recent years. It shows that industrial customers have been paying import prices, plus transportation and supply charges, and a margin, whereas prices for households remain heavily discounted. District heating companies have paid regulated prices which are usually set at about 1.5–2 times the lowest household price – in other words, at a price that is far below the cost of the imported gas with which they are supplied.

Table 7.4: Gas prices in Ukraine

Annual average prices ($/mcm)	2006	2007	2008	2009	2010	2011	2012
Import price	95	130	179.5	232.54	257	315.5	424
Purchase price: industry	107.26	142.57	192.45	272.83	327.4	369.74	453.52
Purchase price: households	67.16	57.4	52.35	62	91.82	90.67	90.67

Note. There are a large number of regulated price levels, particularly for households. The prices shown here are the standard price for industry (excluding chemical plants), with VAT; and the price for households consuming less than 2500 cubic metres per year, with a meter installed.

Source: Naftogaz Ukrainy, National Commission for Energy Regulation, author's information.

Between 2006 and 2012, consumption by the industry (−29.6 per cent), power (−22.1 per cent), and district heating (−31 per cent) sectors fell far more sharply than consumption by the residential and public sector (−9.8 per cent). The reductions in industry and power sector consumption may be attributed firstly to the recession, which caused shutdowns in industry, and secondly to the impact of high import prices. The high prices stimulated (i) some fuel switching, to coal, with an increase in coal consumption about one-third as great as the reduction in gas consumption;[22] and (ii) some energy-saving investments. Officials at the European Bank for

---

[21] Table 7.3 (for 2012); Viktor Tarnavskii, 'Skol'ko gaz nuzhno Ukraine?', UA Energy website, 29 January 2014 (for 2013).

[22] In 2012, gas consumption was 15.6 mtoe down on 2006; coal consumption was 4.9 mtoe up on 2006. In 2009–12, gas consumption was on average 12.79 mtoe down on 2005–8; coal consumption was on average 0.55 mtoe up on 2005–8 (using *BP Statistical Review of World Energy, 2013*).

Reconstruction and Development, which finances and monitors these energy-saving schemes, report that most industrial enterprises in Ukraine achieve a fuel saving in the region of 30 per cent when Soviet-era manufacturing equipment is replaced with new equipment. The problem, of course, is how to fund such modernization.

The fall in consumption by district heating companies also reflects energy-saving and fuel switching, where these are possible. It is difficult, however, to project demand trends in the sector. Reforms have been postponed; general price increases were rejected but there was an *ad hoc* imposition of some price increases at the start of the 2013–14 heating season; and, as in Russia, individual households, residential apartment blocks, and sometimes whole districts are disconnecting themselves from heating systems and installing autonomous heat sources as a result of deteriorating levels of heat delivery. In 2012 the town of Uzhgorod (population 100,000) was entirely disconnected from district heating, in favour of autonomous sources.[23]

The steep fall in the consumption of technical gas (−53.1 per cent) is also significant. Most of this gas is used to fuel compressor stations on the pipeline system; the reduction of transit volumes explains part of the fall, but not all of it. There are also some schemes to reduce leakage, and, because of pricing distortions, some compressor stations are being run on electric power.[24]

Market reform and corporate change

Against this background of high import prices and falling consumption, the structure of the Ukrainian gas market has changed in three main ways: regulation is being reformed, very slowly; Naftogaz Ukrainy has been driven deeper into debt; and Gazprom's hopes of establishing a downstream presence have been frustrated, while domestic industrial groups have moved into gas trading.

Ukraine is moving the legal and regulatory framework of its gas market towards the European model, but only slowly. Crucial elements, such as the separation of Naftogaz's divisions, deregulation of the wholesale market, and the increase of prices for residential and district heating sectors above a cost-recovery level, remain distant. In July 2010 a new law on the gas market was signed by President Yanukovich. This law included: provisions for third-party access to pipeline capacity; the separation of Naftogaz's

---

[23] 'Ekonomiia po-zakarpats'ki: oblast' povnistiu vidmovilas' vid tsentral'nogo opalennia', UNIAN news agency, 22 October 2012; 'V Zakarpatskoi oblasti polnost'iu otkazalis' ot tsentralizovannogo otopleniia', *Argumenty i fakty Ukraina*, 23 October 2012.

[24] Pirani, S. 'Elusive Potential: Gas Consumption in the CIS and the Quest for Efficiency', Working Paper NG 53, OIES, July, 2011, pp. 88–90; research trips to Ukraine, February and July 2013.

production, transportation, and marketing functions; the establishment of regulatory bodies and gradual reform of price regulation; and other market reform measures.[25] The law corresponds with the requirements of the EU's second energy package, and following its adoption, in February 2011, Ukraine joined the Energy Community. However, although new access rules for gas transportation were adopted in April 2012, other aspects of market reform have yet to be implemented.

Naftogaz Ukrainy remains the predominant supplier to the Ukrainian market, accounting in 2012 for 79 per cent of supply. But since 2010 its position has been weakened. It is deeply in debt, due to a large number of inefficiencies and cross-subsidies: the IMF recently estimated that Naftogaz's negative cash balance has been running at about 20 billion hryvna ($2.45 billion) annually, and put the economic cost of inefficiencies and cross-subsidies at between two and four times that level.[26] Not only has Naftogaz's import monopoly been abolished, but the regulatory authorities have encouraged well-financed trading companies to enter the profitable sector of the market, in other words, supplying industrial and power sector customers.

Among these trading companies, only Ostchem Holding has been able to import gas through Russia. Firms affiliated with Ostchem Holding have also acquired about 70 per cent of the assets in the gas distribution sector, although reform measures that might stimulate investment are lacking. Nevertheless, other traders are becoming increasingly significant. In 2013 DTEK, which is affiliated to Ukraine's largest industrial group, SKM, and Vetek, whose owner is a politically well-connected entrepreneur, both started trading gas imported from Europe; DTEK also buys gas from production assets that it owns in Ukraine. Shell has a trading company in Ukraine that buys most of its gas from JKX, a UK-controlled production company; notably, Shell's Iuzivska PSA includes no obligation to sell gas to Naftogaz.[27]

---

[25] *Zakon Ukrainy Pro zasadi funktsionuvannia rinku prirodnogo gazu*, op. cit.; Gorska, A. 'Changes on the Ukrainian Gas Market', Centre for Eastern Studies, 21 July 2010.

[26] The IMF estimated Naftogaz's cash balance as −22.6 billion hryvna, −18.2 billion hryvna, and −20.4 billion hryvna in 2009, 2010, and 2011 respectively. For those three years, the IMF calculated Naftogaz's 'quasi-fiscal activities' (defined as the difference between actual sales, and the theoretical sales at market prices (defined as import prices plus margin for transport etc.), assuming 100% payment rates) as 42.9 billion hryvna, 56.3 billion hryvna, and 65.8 billion hryvna. IMF, *Article IV Report on Ukraine 2012*, pp. 24 and 47.

[27] 'Can Ukraine finally shake its Russian energy habit?', *Gas Matters*, April 2013, pp. 26–31; 'Firtash priobrel ocherednoi oblgaz', *Forbes Ukraina*, 19 September 2012; 'Oblgazy. (Anti)monopoliia', *Zerkalo Nedeli*, 19 October 2012; 'FGI prodolzhit privatizatsiiu oblenergo, oblgazov i TETs', *Forbes Ukraina*, 17 December 2012; 'O gazovom konflikte Firtasha i Kurchenko', *Ekonomychna Pravda*, 15 May 2012; 'Kompaniia 'regionala' budet postavliat' gaz Akhmetovu i ArcelorMittal', *Ekonomychna Pravda*, 20 June 2013; 'DTEK signed contract with PGNiG', DTEK press release, 29 October 2013.

It is significant that Gazprom has failed to establish a trading presence in the Ukrainian market. In 2008, Gazprom established a wholly-owned Ukrainian subsidiary, Gazprom sbyt Ukrainy, which, with the Ukrainian government's agreement, envisaged buying up to 7.5 Bcm of imported gas from Naftogaz Ukrainy and marketing it to industrial customers. Against a background of deteriorating Russo-Ukrainian relationships, the venture has failed to develop. It traded 1.85 Bcm of gas in 2009 and 3.24 Bcm in 2010. The firm was not mentioned in Gazprom's management reports or in its annual reports for 2011 or 2012. In 2013 it was reported that a case brought by Naftogaz against Gazprom sbyt Ukrainy for non-payment of $1.786 billion for gas delivered (the cost of about 4–4.5 Bcm at current prices) had been settled out of court.[28] Continuing friction between Russia and Ukraine over the 2009 agreement, and competition from Ukrainian business groups, have probably stymied Gazprom's progress.

Three conclusions from the changes in the Ukrainian market are:

1. Gas consumption has fallen by more than 25 per cent since 2006, under the impact of economic crisis and sharply increased import prices. Although some of this contraction is temporary, due to the recession, much is permanent: gas has been substituted by coal, or demand destroyed by improved efficiency. During 2011–13 the contraction led to a steep reduction in the volumes of Russian gas imported; the December 2013 agreement will probably reverse this trend, at least during 2014.

2. In Ukraine, the expectation that former Soviet countries could be consolidated as captive markets for Russian exports is not being fully borne out. The level of dependence on Russian gas imports remains high, but concerted efforts to reduce it have begun.

3. Economic factors, rather than political ones, have over the last 15 years moved Ukraine towards closer integration with European markets; this may have significant consequences for Russia. The differential between oil-linked and spot prices in Europe provided an impetus for 'reverse flow' deliveries and the opening up of a new source of supply for Ukraine. The influence of the market (in this case, low European hub prices) was also reflected, albeit indirectly, in the gas price reduction agreed in December 2013, even if the main driver for the agreement

---

[28] Gazprom website, annual reports and management reports; 'Ukrainskii sud: dochka Gazproma uregulirovala spor s Naftogazom Ukrainy', *ITAR-TASS*, 6 August 2013, http://itar-tass.com/ekonomika/570012; Mir TV website, 'Protivostoianiie dochki Gazprom i Naftogaz Ukrainy zakonchilos', 6 August 2013, http://mir24.tv/news/economy/7687986.

was, as most observers believed, Russia's determination to strengthen its political influence over Ukraine. In the long term, discussion about the development of an eastern European gas hub, using Ukrainian storage – which was sparked off by the 'reverse flow' trade – may be revived. This in turn raises the ultimate prospect of Russian gas being sold at Ukraine's eastern border rather than at its western border, further increasing the element of competition among purchasers. When this was discussed in 2009, during the January dispute, European buyers made it clear that they wished to leave Ukrainian transit risk with the seller, Gazprom. If in future the position changes, such an export model is conceivable. The expiry of contracts in 2019 could be a turning point.

Belarus

Import prices and import arrangements

Belarus's gas consumption was 19.4 Bcm in 2012, nearly all of which was imported from Gazprom. Domestic production is very low, at 0.22 Bcm/year (Table 7.5). The fact that gas constitutes 65 per cent of total energy consumption – with 95 per cent of power generation being gas-fired – combined with high import dependence, lack of import diversity, and little storage, makes Belarus extremely vulnerable to external supply interruptions.

Belarus's gas consumption grew continuously throughout the late 1990s and early 2000s (due to its ability to import gas at very low prices) until Gazprom made the acquisition of part of the Belarussian gas network a condition of continued low prices. The crisis of early 2004 marked the lowest ebb of the Russia–Belarus gas relationship. In January 2004, Gazprom cut gas flows to Belarus after the latter refused to conclude a new contract at an increased price and reneged on a promise to privatize the network; in February Gazprom cut gas flows across Belarus when the latter took gas from the Yamal–Europe pipeline once its contracts with Russian independent suppliers run out. A new contract, signed after the crisis, included a price increase, while Belarus also agreed to set up a gas transportation joint venture with Gazprom. However, Belarus continued to increase its consumption while delaying privatization of the network for two more years.

The January 2007 Russia–Belarus contract set the price at $100/mcm and stipulated transition to a European netback price by 2011 (with decreasing percentage discounts provided during 2007–10). Belarus also committed to sell Gazprom half of Beltransgaz, the owner of its transportation network, for $2.5 billion, with the sale to be finalized in 2010.

Table 7.5: Belarus gas consumption, imports, and prices

	2000	2001	2002	2003	2004	2005	2006	2007	2008	2009	2010	2011	2012
Consumption (Bcm)	15.9	16.1	14.7	16.7	18.3	18.8	18.8	19.7	n/d	n/d	n/d	19.8	19.4
Imports (Bcm)	17.1	17.3	17.6	18.1	19.6	20.1	20.8	20.6	21.1	17.6	21.6	20	20.3
Transit (Bcm)	n/d	24.6	27.5	33.1	35.3	40.8	44	49.5	51.4	44.6	43.2	44.3	44.3
Production (Bcm)	0.26	0.26	0.25	0.25	0.25	0.23	0.22	0.22	0.22	0.22	0.22	0.22	0.22
Import prices ($/mcm)	n/d	n/d	18	30	47	47	47	118	127	151	185	270	168

Source: Gazprom, Beltransgaz, Belarusian Ministry of Energy.

Belarus struggled to pay for gas under the January 2007 contract, even though Gazprom allowed occasional deviations from its clauses; the degree of Gazprom's benevolence decreased over time and there was a short-lived transit dispute in June 2010.[29] The new 2012–14 contract, signed in December 2011, is very favourable for Belarus, stipulating a price of $165/mcm for 2012 (just over half the price in 2011) and a $2/mcm/100 km transit tariff, while the prices for 2013–14 are Russian 'net-forward' prices (the price in the Russian Yamal-Nenets region plus transportation and storage). Belarus was able to conclude this favourable contract because: (i) in 2011 it signed the Single Economic Space (SES) gas sector agreement[30] (under which the gas price would only reach European levels by the end of 2014);[31] and (ii) it consented to sell the remaining 50 per cent stake in Beltransgaz to Gazprom (bringing the total purchase price to $5 billion).[32]

The fact that the price paid by Belarus from 2012 onwards is around a third of that paid by Ukraine and Moldova suggests that both political and economic factors are at play – with Russia attempting to create a common legal/regulatory space with Belarus which would have synchronized (and prolonged) the transition of Russia and Belarus towards prices which reflected supply and demand forces more closely. Given that the Russian government has stated (in 2013) that equal profitability in Russia will not be achieved until the end of the 2010s (see Chapter 5), it is reasonable to assume that the Russian government and Russian businesses might expect Belarus to make concessions in other sectors of its economy in return, if this transition period is to be prolonged for Belarus.

It was not until 2007 – when the new contract doubled the import price – that Belarus started to review its concept of energy security, in order to adapt to the environment of increasing prices.[33] The new concept envisaged a significant decrease of gas *share* in the energy balance, to 50 per cent, and in power generation, to 49–57 per cent, but not a decrease in consumption *volume*, estimating it at 22 Bcm in 2009.[34] In reality, consumption fell below

---

[29] For details see Yafimava, K. 'The June 2010 Russian-Belarusian gas transit dispute: a surprise that was to be expected', Working Paper NG43, OIES, 1 July, 2010.

[30] Together with Russia and Kazakhstan.

[31] Ratified by all member countries, see Yafimava, K. *The Transit Dimension of EU Energy Security*, op. cit. This concession had reflected the fact that Russia extended a grace period for its domestic consumers until 2015.

[32] Ibid.

[33] '*Ekonomia i berezhlivost*' – *glavnie factory ekonomicheskoi bezopasnosti gosudarstva*', Directive no. 3 of 14 June 2007. See Yafimava, K. 'Belarus: the domestic gas market and relations with Russia', in Pirani, S. (ed.), *Russian and CIS Gas Markets and Their Impact on Europe* (Oxford: OIES/OUP, 2009).

[34] '*Gosudarstvennaya kompleksnaya programma modernizatsii osnovnix proizvodstvennix fondov… na period do 2011*', ukaz no. 575 ot 15.11.2007 goda [Belarus].

18 Bcm in 2009, due not to successful implementation of the concept but to the economic recession. Yet Belarus's consumption – unlike that of Ukraine or Moldova – fully recovered in 2010. The less significant decrease in demand, together with the high rate of its recovery, are explained by the less severe impact of the recession on Belarus's industry, and the power sector's inability to diversify away from gas. In 2011–12 consumption remained relatively stable, although it is difficult to tell whether this is a result of the government's efforts or of a general economic slowdown. Government plays an important role, given that most of the industrial and power sectors in Belarus are state owned. The seriousness of the government's intention to prevent consumption from growing is suggested by the fact that, despite there being a favourable supply contract for 2012–14, gas consumption has not increased. This reflects a lack of certainty on the part of the government as to what prices might be after 2014.

Domestic gas sector

The Belarussian gas sector consists of two gas companies:

- *Beltransgaz* (renamed *Gazprom Transgaz Belarus* in April 2013), a 100 per cent Gazprom subsidiary responsible, *inter alia*, for high-pressure transportation and transit, storages (Osipovichskoe, Pribugskoe, and Mozyrskoe), new construction, and maintenance;

- *Beltopgaz*, a 100 per cent Belarussian state-owned company, responsible for gas distribution and domestic retail sales.[35]

Gazprom sells gas to Beltransgaz, which resells it to Beltopgaz, subsidiaries of which resell it to end users in all sectors; Beltransgaz also sells gas directly to some industrial and power sector consumers. The Economy Ministry regulates gas prices for all categories of consumers except for the residential sector, the prices for which are regulated by the government; the Economy Ministry also regulates gas transportation tariffs. The government sets the upper limit of prices at which Beltopgaz is allowed to sell gas to its final consumers, depending on a number of factors.[36] From 2010 the government began to differentiate prices (~$261/mcm) on the basis of volumes of gas consumed, with a premium of about 30 per cent on volumes consumed in

---

[35] Gazprom has occasionally expressed interest in acquiring Beltopgaz assets but no sales to this effect have taken place yet. Gazprom press conference 'Perekhod na rynochnie printsipy sotrudnichestva s respublikami byvshego SSSR', 20 June 2006. Gazprom website.

[36] Government resolution, *On gas prices*, N 122, 27 December 2012 [Belarus].

excess of government-set norms for fuel substitution;[37] from 2012 the government began to differentiate prices for most small consumers (~ $331/mcm), those consuming more than 0.5 Bcm/year (~ $294/mcm), and those consuming more than 0.5 Bcm/year as feedstock for nitrogen-based fertilizers (~$261/mcm).[38] In 2007 the government ceased to publish the prices at which Beltransgaz sells gas to Beltopgaz, and in 2012 it ceased to publish Beltransgaz's other sales prices. It appears that during 2007–11, when Gazprom's acquisition of Beltransgaz was in progress, the government was trying to squeeze Beltransgaz's margins by not increasing its markups to levels previously agreed with Gazprom (which eventually led to the June 2010 dispute) while rapidly increasing state-owned Beltopgaz's markups. In late 2011 Gazprom became the sole owner of Beltransgaz, and in 2012 the latter's markups reportedly reached $15.95/mcm.[39] This suggests that despite relatively low import prices, Beltopgaz still maintains a significant markup averaging $100/mcm.

The analysis above allows us to conclude that in the 2010s Belarusian gas consumption might not increase significantly, but will not decrease unless there is another recession. Furthermore, even if the national energy security concept is implemented as planned – and this is questionable, given its high cost ($19 billion) and the poor state of Belarus's finances – gas will still remain a vital fuel. Moreover, it will be *Russian* gas, as it is not realistic to expect alternative supplies at comparable prices in the 2010s.[40] Furthermore, it will be Gazprom gas, as long as its monopoly on pipeline exports remains intact. Overall, low price elasticity of gas demand and severely constrained ability of fuel/source diversification, together with limited affordability and implementation of energy efficiency measures, all point to the fact that in the 2010s Belarus is likely to remain a captive market for Gazprom to a significant degree. It will also remain a profitable market – at least not less profitable than the Russian domestic market – and an important transit corridor, although because of the construction of Nord Stream 1 and 2 (see 'Transit' page 207) not to a degree that would enable Belarus to use its transit status to secure significant price concessions. As the Russia–Belarus gas relationship has become increasingly asymmetric in Russia's favour, the stability of their political relationship is paramount for Belarus's gas supply security.

---

[37] In Belarus there are state-set norms for substituting imported gas by locally sourced fuels.

[38] This primarily concerned the Grodno Azot company – the major Belarusian nitrogen-based fertilizer producer.

[39] 'The price of imported gas for Belarus will be $185/mcm', *Interfax*, www.interfax.by/news/belarus/117553 accessed 9December 2013.

[40] There were discussions of Belarus participation in the Baltic LNG project, but both its economics and politics are deeply questionable.

Moldova

Import prices and import arrangements

Moldova has no indigenous natural gas production, and imports all its gas from Gazprom: 3.3 Bcm in 2011 (of which 1.15 was for Moldova and 2.15 Bcm for the breakaway region of Transdniestria) (see Table 7.6).[41] Moldova's consumption has been declining since the mid-2000s, with the exception of 2010, across all sectors of its economy. This is explained by the increase in domestic gas prices to levels fully covering the import price. In February 2011, the power sector was paying $296.6 and the industrial sector $345–394/mcm.[42] In September 2012, the average end-user tariff was reported to be around $525/mcm.[43] The payment discipline of Moldova (excluding Transdniestria), which was maintained in the 2000s, began to deteriorate after 2008: the gas debt outstanding to Gazprom increased from $88 million to $500 million between September 2011 and September 2012, although it decreased to $100 million by September 2013.[44]

By contrast, consumption in Transdniestria has been rising since the mid-2000s, while domestic prices have remained significantly below import prices. In 2012 domestic prices for industrial and power sectors increased from the range $120–150 to $257/mcm (still remaining more than $100/mcm below the import price). By the end of 2013, the widening gap between domestic and import prices resulted in Transdniestria's debt to Gazprom for gas delivered rising to more than $4 billion. Improvement in payment discipline and the likelihood of this debt being repaid are contingent on economic improvement and a settlement of the 'frozen' Transdniestria conflict, neither of which seems likely to be achieved in the short or medium term.

---

[41] In this chapter the term 'Moldova' is used to refer to the geographic region on the right bank of the river Dniestr and to the state structures of the Republic of Moldova. For detail on the 'frozen' Transdniestria conflict and its implications for Moldova–Russia gas relations see C Bruce, C. and Yafimava, K. 'Moldova's Gas Sector', in Pirani, S. (ed.), *Russian and CIS Gas Markets and Their Impact on Europe* (Oxford: OIES/OUP, 2009).

[42] See Yafimava, K. *The Transit Dimension of EU Energy Security*, op. cit. p. 275.

[43] 'Moldova Gas owes Gazprom nearly $500 mln', *Interfax Russia and CIS Oil & Gas Weekly*, 20–26 September 2012.

[44] 'Moldova may need to extend old contract with Gazprom 3 more months', 23–29 February 2012; 'Moldova Gas owes Gazprom nearly $500 mln', *Interfax Russia and CIS Oil & Gas Weekly*, 20–26 September 2012.

Table 7.6: Moldova gas consumption, imports, prices, and tariffs

	2004	2005	2006	2007	2008	2009	2010	2011	2012
Consumption (Bcm), total	n/a	2.71	2.4	2.61	2.6	3	3.2	3.25	n/a
Moldova	1.14	1.31	1.32	1.21	1.2	1.1	1.2	1.1	1.1
Transdniestria	n.d.	1.4	1.08	1.4	1.4	1.9	2	2.15	n.d.
Imports (Bcm), total	n/a	2.82	2.5	2.7	2.7	3.1	3.2	3.3	n.d.
Moldova	1.26	1.42	1.42	1.3	1.23	1.1	1.2	1.15	1.09
Transdniestria	n.d.	1.4	1.08	1.4	1.47	1.9	2	2.15	n.d.
Transit volume (Bcm)	n.d.	n.d.	n.d.	n.d.	22.9	17.7	16.8	19.9	19.6
Transit tariff ($/mcm/100 km)	2.5	2.5	2.5	2.5	2.5	2.5	2.5	3	3
Import prices ($/mcm)	80	80	135	170	232	264	250	339	394

Note: Because of the lack of data on Transdniestria's consumption during 2005–6 and 2009–12, it is assumed to be equal to its imports.

Source: ANRE website, Moldovagaz website, Yafimava, K. *The Transit Dimension of EU Energy Security*, op. cit.

Declining negotiating power and influence on import prices

The Moldovan gas transit network is part of the Balkan export corridor, carrying Russian gas to Europe, which crosses both Moldova and Transdniestria and transits around 20 Bcm/year of Russian gas to Romania, Bulgaria, Macedonia, and western Turkey. Its installed transit capacity is 44 Bcm.[45] The network is controlled and operated by Moldovagaz, of which Gazprom owns 50 per cent and holds 13.44 per cent in trust management,[46] while the Moldova state owns 35.33 per cent. Despite being a part-owner and co-operator of the network, Gazprom has been unable to enforce payment discipline in Transdniestria, but it is also unable to cut supplies to the region without interrupting supplies to Moldova and European consumers further down the pipeline. However, once the South Stream pipeline is built (see 'Transit' page 207), it will serve the same countries as the Balkan corridor does, and the transit importance of both Moldova and Transdniestria will decline sharply. Their ability to secure a better price, and accumulate further debts for gas, will also decrease, whereas their dependence on Russian gas is likely to remain high throughout the 2010s. In order to import gas from Europe, thus reducing its supply dependence on Gazprom and transit dependence on Transdniestria, Moldova intends to build a Moldova–Romania interconnector (an Ungheni–Iasi pipeline) with capacity of up to 1 Bcm and the possibility of reverse flow,[47] but construction has not yet started due to the lack of funds.[48]

Gazprom supplies gas to, and transits gas across, Moldova and Transdniestria under contracts concluded with Moldovagaz underpinned by intergovernmental agreements. The contracts stipulate volumes to be delivered to, and payments to be made by, Moldova and Transdniestria separately.[49] However, Moldovagaz's debt – whether accumulated in Moldova or Transdniestria – is listed as Moldova's debt. The 2007–11 supply contract established a formula for bringing the import price to European 'netback' by 2011, with decreasing percentage discounts to be provided during 2007–10. Upon expiry, both supply and transit contracts were extended quarterly in 2012, and in September 2013 an agreement was reached to extend the

---

[45] In 2011 transit constituted 19.8 Bcm, see 'Moldovagaz owes Gazprom $500 mn' op. cit.

[46] This 13.44% stake is owned by the Transdniestrian administration.

[47] 'Commissioner Oettinger inaugurates the works of Romania–Moldova gas pipeline', European Commission – IP/13/792, 27 August 2013, press release http://europa.eu/rapid/press-release_IP-13-792_en.htm.

[48] Available at www.vz.ru/economy/2013/9/23/651610.html, accessed 20 October 2013.

[49] *Moldova trade diagnostic study*, Center for Strategic Studies and Reforms, Chisinau-Tiraspol, 2003.

contracts until the end of 2014.[50] The failure to conclude new contracts reflects the parties' difficulties in agreeing the new terms: Moldova wishes to contract its supplies separately from Transdniestria, in order to insulate Moldovagaz from the burden of Transdniestrian debt.[51] Moldova also asked Gazprom for a 30 per cent price reduction, which Gazprom refused to consider until the issue of gas debt (mostly incurred by Transdniestria) is resolved, and assurances provided that Gazprom's investments in Moldova (its share in Moldovagaz) will be protected in the context of Moldova's obligations under the Energy Community Treaty (EnCT) to implement the Third Energy Package by 2015.[52] It was not until Moldova secured the European Commission's consent to postpone implementation of the EnCT until 2020 that agreement was reached with Gazprom on contract extensions.[53]

This analysis leads to a conclusion that whether the observed consumption trends – declining in Moldova and rising in Transdniestria – will continue to 2020 depends on a number of factors: economic (Gazprom's market access and network ownership, and Moldovan/Transdniestrian access to alternative supplies) and political (Russia–Moldova/Transdniestria relations). The most important factor remains the inability to resolve the 'frozen' Transdniestria conflict, which hampered the region's economic performance and led to accumulation of debt in excess of $4 billion (for which it cannot be held accountable, as any claims on economic entities would be unenforceable in Transdniestria, and would be transferred to Moldovagaz). Gazprom, unable to cut supplies to Transdniestria without cutting off buyers further down the pipeline, has continued to supply the region. However, this arrangement will come to an end once the South Stream pipeline is built, thus enabling Gazprom to supply Bulgaria, Macedonia, and western Turkey directly (see 'Transit' page 207). Transdniestria's debt accumulation would no longer be tolerated and the region would either have to raise its domestic tariffs and

---

[50] 'Moldova to extend existing gas contract with Gazprom', *Interfax Russia and CIS Oil & Gas Weekly*, 5–11 September 2013; also see 'Moldova did not want to be like Romania and extended its old contract with Gazprom' (in Russian), http://oilgasfield.ru/news/85565.html.

[51] 'Supply and transit contracts with Moldova extended to cover the first quarter of 2012' (in Russian), Gazprom press release, 29 December 2011, available at: www.gazprom.ru/press/news/2011/december/article126953; 'Moldova ready to extend gas supply contract, in no rush to sign a new agreement', *Interfax Russia and CIS Oil & Gas Weekly*, 25–31October 2012.

[52] The Third Package requires that no vertically integrated company (such as Gazprom) can control gas transportation assets in the EU.

[53] Reportedly, the Russian energy minister, Novak, suggested that no price reductions will be considered until Moldova denounces its membership of the EnCT, see 'Moldova seeking 30% discount on Russian gas – minister', *Interfax Russia and CIS Oil & Gas Weekly*, 6–12 September 2012.

pay for gas imports in full, or reduce consumption dramatically. Moldova would remain the only country receiving all its gas via Transdniestria, and hence would be well advised to build the interconnector with Romania.

Importers in the Caucasus

Russian gas exports to the Caucasus have fallen sharply in recent years (see Table 7.1), due mainly to the start-up of Azerbaijan's Shah Deniz field in 2006. This enabled Azerbaijan to substitute its own gas for Russian imports and to supply most of Georgia's requirements. Azerbaijan has imported no Russian gas since 2007. In Georgia, the deterioration in political relations with Russia hastened efforts to substitute Azeri volumes for Russian ones. Only Armenia remains dependent on Russia for all its imports.

In *Armenia* Gazprom has, since 2006, controlled, and subsequently increased its stake to 80 per cent of, the national gas company Armrosgazprom. Gazprom's exports, of between 1.4 Bcm/year and 2.1 Bcm/year in recent years, are all purchased by Armrosgazprom. Import prices rose in April 2010 from $170/mcm to $180/mcm. In 2012, import prices were reported at $220/mcm, and in 2013 they were due to rise to $270/mcm. During the year, negotiations between Armenia and the Customs Union (Russia, Belarus, and Kazakhstan) on accession intensified; Armen Movsisian, Armenia's energy minister, stated that accession would result in a 30 per cent reduction in import prices, as Russian export duty would no longer be charged.[54]

In *Georgia*, Russian imports fell to 0.15–0.2 Bcm/year from 2008 – in other words, they were limited to the volumes traditionally exported to Georgia in lieu of fees for the transit of Russian gas to Armenia. Since then, Azerbaijan has accounted for 90 per cent of Georgian imports. Most of the Azeri volumes form part of a barter arrangement (gas for oil and gas transit services). Some additional volumes are bought at bilaterally negotiated prices; a press report based on customs information indicated that these prices were in the region of $240–270/mcm in 2012–13.[55]

---

[54] Armrosgaz website (2010 prices); 'Armeniia vedet peregovory s Gazpromom', www.oilpipelines.ru, 21 May 2013; 'Rossiia predlagaet podniat' dlia Armenii tsenu na gaz', www.oilcapital.ru, 17 October 2012; 'Armenia svernula na sever', *Vedomosti*, 4 September 2013; 'Gas Prices Reduced as Armenia Joins Customs Union', *Asbarez Armenian News*, 8 October 2013.

[55] Henderson, J., Pirani S., and Yafimava, K. (2012). 'CIS Gas Pricing: Towards European Netback?' in Stern, J.P. (ed.), *The Pricing of Internationally Traded Gas* (Oxford: OIES/OUP, 2012), pp. 178–223; 'V ianvare-fevrale 2013 goda eksport gaza iz Azerbaijana sostavil 19.5% ob"ema za ves' 2012 god', *Business Caucasus*, 7 May 2013 businesscaucasus.com/node/2257.

Given the political factors, it is difficult to speak of direct competition between Russia and Azeri exports – but there is indirect competition, which appears to hold prices fairly close to each other. While Armenia and Georgia have sourced their imports from Russia and Azerbaijan respectively, in line with political considerations, the prices they pay are similar.

Transit

As long as the western CIS countries were the only corridor for Russian gas exports to Europe, and were therefore crucial for Gazprom's ability to deliver on its European long-term supply contracts (LTSCs), they had significant negotiating power vis-à-vis Gazprom, by virtue of being able to threaten to reduce transit flows in the event of not getting their preferred price and volume terms for their own gas imports. As a result, the transit dimension remained one of the weakest links of European gas security in the 2000s, characterized by a series of transit disputes – with Ukraine (January 2006, March 2008, and January 2009), Belarus (February 2004, January 2007, August 2007, and June 2010) and Moldova (January 2006). These culminated in the January 2009 Ukrainian crisis, which resulted in the shutdown of 80 per cent of the transport capacity for Russian gas exports to Europe for two weeks.[56]

During the January 2009 crisis, Russian gas supplies to Ukraine were cut off on 1 January, and to Europe on 7 January, and they did not restart until 20 January.[57] For two weeks, no gas flowed to Europe via Ukraine. This had a range of consequences for different European states, with southeast Europe bearing the major brunt, including a humanitarian emergency. Ukraine, although left without supplies for three weeks, had reversed its pipeline network so that it was able to move the gas accumulated in storage facilities in western Ukraine to consumers in eastern Ukraine. Overall, Ukrainian consumers did not suffer shortfalls, but Ukraine's reputation as a reliable transit state was destroyed. For its part, Gazprom also suffered huge reputational, as well as financial, damage because of loss of sales on its LTSCs.

The European response to the crisis was one of non-intervention. The European Commission called it a 'purely commercial' conflict that should be resolved bilaterally, as there was no unanimous political backing from the EU member states for it to participate in dispute resolution. A monitoring

---

[56] For an overview and analysis of all Russia–western CIS transit disputes in the 2000s see Yafimava, K. *The Transit Dimension of EU Energy Security*, op. cit.

[57] For detailed analysis of the January 2009 Ukraine gas crisis see Pirani, S. Stern, J., and Yafimava, K. 'The Russo-Ukrainian gas dispute of January 2009' op. cit.

mission was set up by the Commission (for which European utility companies provided most of the monitors). This did not lead to a resumption of flows, and the crisis made it evident that the Commission had little political leverage with either Ukraine or Russia and was unable or unwilling to provide the finance to resolve the crisis. It also showed that the concept of solidarity between EU Member States, as embodied by the 2008 Second Strategic Energy Review, failed to work and demonstrated the Commission's inability to assist in a gas (energy) security crisis.

Yet the Commission had important instruments – within the Energy Charter Treaty (ECT) and within the EU–Russia and EU–Ukraine Energy Dialogues – to request a greater transparency of contracts, and transparency in relation to flows at metering stations on the Russia–Ukraine and Ukraine–EU borders, *before* the crisis. This knowledge would have been helpful during the crisis, when parties disputed the very existence of a transit contract. It might also have avoided a situation where there was no independent verification of volumes coming into, and out of, Ukraine. This would have made it clear whether or not transit commitments had been honoured, an issue that could have been raised in evidence in arbitration proceedings. Also, the Commission could have initiated the establishment, within the ECT framework, of a transparent volume allocation system in Ukraine, to ensure that volumes nominated for transit could be traced through the system. (The design of the Ukrainian network does not allow for the physical separation of gas in transit from gas destined for domestic consumers.)

The existence of the ECT, and the fact that it was signed and ratified by Ukraine, did not prevent the latter from violating its transit provisions on non-interruption and non-reduction of transit flows. The ECT dispute prevention and dispute resolution mechanisms remained unused, despite the fact that the ECT secretariat issued statements before and during the crisis, recalling ECT principles on uninterrupted transit and reiterating that the secretariat 'stands ready to support the work of an independent conciliator' in the case of a transit dispute 'should the parties call for it'. The inability and/or unwillingness of the ECT parties to use the ECT was one of the main reasons why, in late 2009, the Russian government terminated its provisional application of the ECT, and announced its intention not to become a contracting party. This ended hopes of establishing a common governance structure covering all parties involved in the Europe–Russia gas trade.

The January 2009 transit crisis was an unprecedented gas security event, which elevated the concept of transit to the top of energy security thinking, and demonstrated the ultimate failure of all existing instruments – both bilateral and multilateral – to ensure European transit security.[58]

---

[58] Yafimava, K. *The Transit Dimension of EU Energy Security*, op.cit., p. 1.

After the crisis, Gazprom, which adopted its transit diversification policy in the mid-2000s following its smaller-scale disputes with Belarus (in 2004) and Ukraine (in 2006), accelerated construction of the transit-avoidance pipelines, Nord Stream and South Stream (see Chapter 3).

A greater number of transit-avoidance pipelines would translate into increased power for Gazprom to arbitrage between all transportation routes, potentially making western CIS pipelines the routes of last resort. In turn, this would spell a dramatic change in Russia–CIS negotiating dynamics, with a decline of the ability of western CIS countries to interfere with transit, as well as in the impact of such interference, while these countries would still remain Gazprom's largely captive market (at least) during the 2010s. The size of these markets might become smaller – this is particularly the case for Ukraine (see 'Ukraine' page 183) – but pricing will remain the major source of conflict between Gazprom and western CIS countries, and the latter's ability to get better price terms by using their transit power will decline. Furthermore, having accomplished only very limited (fuel and route) diversification, and having witnessed their transit power being reduced, they stand to lose significantly more than they might gain, should they choose to interfere with transit flows. This makes any future interference with transit unlikely, but not impossible.

The new transit-avoidance gas transportation geography, which would cost Gazprom (and its European partners) dear in terms of investment, and would cost western CIS countries dear in terms of foregone transit revenues and negotiating power, is a direct consequence of the two decades-old failure of Russia and western CIS countries to find a lasting solution to the transit problem, of which the January 2009 crisis was only the last straw.

Although no new transit disputes have taken place in Ukraine since the January 2009 crisis – largely because of the supply contract's very strict terms, which include the activation of a pre-payment clause in case of a single missed payment – Gazprom's relations with Ukraine have remained difficult. Ever since the January 2009 contracts were concluded, Ukraine has tried to revise them, in particular to reduce the price and the volume and to relax the pre-payment clause. In April 2010, Ukraine secured a price discount of about 30 per cent (as export duty relief) in exchange for prolonging the Russian Black Sea naval base lease[59] – choosing this political arrangement rather than a commercial one (a Naftogaz–Gazprom joint venture to own and manage the Ukrainian gas transportation network).

The fact that the price discount was codified through both commercial (contractual amendment) *and* political (intergovernmental agreement) arrangements suggests that any worsening of political relations between

---

[59] Pirani, S. Stern, J., and Yafimava, K. 'The April 2010 Russo-Ukrainian gas agreement' op. cit.

the two countries threatens to unravel the Black Sea fleet agreement. Given the strong upward dynamics of oil prices (to which the gas price is linked in the 2009 contract formula) in the 2010s, even with the discount secured under the Black Sea fleet deal Ukraine's import price has remained very high – $315.5/mcm in 2011 and $424/mcm in 2012 (see Table 7.4). Ukraine continued its attempts to revise the price downwards, achieving a result with the reduction of the import price to $268.50/mcm in December 2013; this reduction was codified as an addendum to the January 2009 supply contract.[60]

The negotiations on forming a joint venture to own and manage the Ukrainian transportation network have been going on since the early 2000s but have never come to fruition – mostly because all Ukraine's political forces have viewed ownership over the network as a symbol of national independence, not to be parted with.[61] Over time, Gazprom's enthusiasm in relation to the joint venture has waned, whereas its determination to build South Stream has increased, with the first string scheduled to become operational by late 2015. Gazprom has also been interested in acquiring a stake in the Ukrainian storage facilities (which it stopped using following the 'missing gas' incident in 2006) because the February 2012 cold spell (when Gazprom was unable to meet some nominations on its European LTSCs, partly because it could not use the Ukrainian storage) demonstrated its need for such storage.[62]

However, Ukraine has refused all Gazprom's offers of joint ownership of the network, and has stated that it would start arbitration proceedings against Gazprom in order to revise the price stipulated by the 2009 supply contract; it also announced that it would liquidate Naftogaz (which is a party to the contract) claiming that this would make the contract invalid.[63] None of this took place, and Ukraine's rhetoric softened in late 2012, with its prime minister admitting that Naftogaz would lose its case should it refer the contract to arbitration. Importantly, although Ukraine has always paid in full for gas it imported under the January 2009 contract,[64] in 2012 it neither took nor paid for its shortfall under contractual take-or-pay (TOP)

---

[60] Eremenko, A. et al, 'Zapomni kak vse nachinalos', *Zerkalo Nedeli*, 20 December 2013, http://gazeta.zn.ua/internal/zapomni-kak-vse-nachinalos-_.html

[61] For details on the history of negotiations over the gas transport consortium, see Yafimava, K. *The Transit Dimension of EU Energy Security*, op. cit., pp. 139–215.

[62] Heather, P. and Henderson, J., 'Lessons from the February 2012 European gas "crisis"', Energy Comment, OIES, April, 2012.

[63] M. Selivanova, 'Scrapping Naftogaz to avoid gas wars with Russia', *RIA Novosti*, 2 September 2011, http://en.ria.ru/analysis/20110902/166386364.html.

[64] However, Ukraine has missed its payment deadline for August 2013 imports, see 'Dobrota zakanchivaetsia', *Vzgliad*, available at http://vz.ru/economy/2013/10/29/657214.html.

obligation of some 42 Bcm, importing from Gazprom only 27 Bcm. Ukraine refused to pay when Gazprom presented it with a $7 billion bill and Gazprom has not (yet) moved to enforce this through arbitration.[65]

Gazprom's relations with Belarus have been marred by several disputes, although they were of much smaller magnitude than those with Ukraine, and the February 2004 and the June 2010 disputes had a very limited impact on European consumers. Gazprom's freedom of manoeuvre, in terms of its ability to narrow the gap between the prices paid by Belarus and the prices paid by Ukraine and Moldova (and by European consumers), has always been limited by the political considerations of the Russian government, which viewed Belarus as a (sometimes difficult) political ally, given the latter's lack of EU and NATO aspirations (in sharp contrast with Ukraine and Moldova).

Nonetheless, as of the early 2000s, Gazprom began to apply its strategy of 'Europeanization' of western CIS prices to Belarus as well. The January 2007 contract, which envisaged a phased transition towards the European netback price by 2011, is confirmation that this strategy was partly successful. At the start of 2011, by which time Belarus's sale of 50 per cent of Beltransgaz to Gazprom was completed, Belarus was paying $270/mcm for its gas imports, five times the price it paid in 2006, although the contract was only selectively enforced and its parties' adherence to it was influenced by broader political and commercial considerations. The fact that Belarus was 'allowed' to deviate from the contract and accumulate debt in 2009, but not in 2010 (as witnessed by the June 2010 dispute),[66] indicates a decreasing degree of Russian government tolerance towards Belarus. This may have resulted from Belarus not recognizing Abkhazia and South Ossetia, granting asylum to the ousted Kyrgyz president, and boycotting the Customs Union. In late 2011, partly helped by the June 2010 dispute, Belarus gained the consent of the Russian government to achieve equal profitability of gas sales simultaneously with Russia by 2015. The 2012–14 contract, underpinned by the SES gas sector agreement, stipulated a price of $165/mcm in 2012 (three times lower than the price paid by Ukraine) and a formula-based price in 2013–14 (a Russian domestic price with the cost of transportation and storage added); Belarus also sold its remaining 50 per cent stake in Beltransgaz to Gazprom for $2.5bn. The post-2014 future is unclear.

Belarus will remain an important transit territory for Russian gas exports to Europe: both the Yamal–Europe (33 Bcm capacity[67]) and Beltransgaz

---

[65] 'Energy Community cannot help Kyiv in dispute with Gazprom over unpurchased gas', *Interfax Russia & CIS Oil and Gas Weekly*, 28 February–6 March 2013.
[66] Yafimava, K. 'The June 2010 Russian-Belarusian gas transit dispute', op. cit.
[67] Used exclusively for transit of gas (to Poland, Germany, the Netherlands, and Belgium).

(Northern Lights) (51 Bcm[68]) transportation networks run across its territory. However, given that both of these networks are owned by Gazprom – which built the former, and finalized acquisition of the latter in 2011 – Belarus's transit *power*, in other words the ability to interfere with transit in order to secure lower gas prices, has declined. It declined even further with Gazprom's completion of the Nord Stream 1 and 2 pipelines across the Baltic Sea which provided the company with the means to reroute gas flows away from Belarus (and Ukraine); a feasibility study has been conducted into construction of Nord Stream 3 and 4. That Gazprom views Belarus as a more secure transit corridor following its acquisition of Beltransgaz, is suggested by Gazprom's considering the building of a Yamal-2 pipeline (15 Bcm) across Belarus (parallel to the Yamal–Europe pipeline) which would reach the Belarus–Poland border and then go south to Slovakia to connect with the existing network in Hungary, thus further diversifying away from Ukraine (see Chapter 3).[69] Although the Belarussian state retains physical control over both the Northern Lights and Yamal–Europe pipelines, as they are located on Belarussian territory, the Belarus transit corridor has become significantly more secure with Gazprom becoming the single owner and operator of the pipelines.

Gazprom's relationship with Moldova could be considered the least problematic of all three western CIS countries, as there has been no transit interference there since 1999 – mostly because Gazprom was able to become a joint owner and operator of the Moldovan network – were it not for the fact that Gazprom has been unable to impose payment discipline on Transdniestria. The territory owes Gazprom nearly $4.5 billion, and, although there is little prospect of this being paid, Gazprom has been unable to cut supplies to Transdniestria without cutting off countries further down the pipeline. Gazprom has refused to consider a price reduction for Moldova until the issue of gas debt is resolved, and assurances are provided that its stake in Moldovagaz is protected in the context of Moldova's obligations under the Energy Community Treaty (see 'Moldova' page 202).[70]

---

[68] This is its projected capacity, whereas operational capacity is estimated at 48 Bcm. Used both for domestic supplies and for transit (to Poland, Lithuania, and the Russian city of Kaliningrad).

[69] Gazprom website, 'about subsidiaries', Gazprom Transgaz Belarus, (www.gazprom.com/about/subsidiaries/list-items/gazprom-transgaz-belarus); 'Gazprom and EuRoPol GAZ to cooperate under Yamal–Europe-2 gas pipeline project', Gazprom website. However, in October 2013, Gazprom's deputy CEO Golubev stated that Gazprom is in 'no hurry' to build either Nord Stream 3 and 4 or Yamal–2 until there is 'the qualitative recovery of demand in Europe', see 'Gazprom awaiting recovery in demand before building new pipelines in Northern Europe', *Interfax Russia and CIS Oil & Gas Weekly*, 31 October–6 November 2013.

[70] The Third Package requires that no vertically integrated company (such as Gazprom) can control gas transportation assets in the EU.

Importantly, should South Stream be constructed, both Moldova and Transdniestria will find themselves in an extremely vulnerable position. Russian gas previously delivered to Bulgaria, Macedonia, and western Turkey via Moldova and Transdniestria could be redirected via South Stream. Moldova and Romania would then remain the only countries which would continue to receive Russian gas via Transdniestria (in Moldova's case, all of its supplies), and hence be vulnerable to the latter's transit interference. It is conceivable that, once Gazprom no longer has to rely on the Balkan corridor for transit, it might stop supplies to Transdniestria in order not to incur additional debt. Under these circumstances, if Gazprom continued to supply gas to Moldova, there would be a risk that Transdniestria might divert this gas for its own purposes. In this case, Gazprom might choose to stop supplies to Moldova as well, once the existing contracts expire, should the amount of Transdniestrian non-payment outweigh the amount of Moldovan payment, as there would be no commercial rationale for continuing supply – although there might be political reasons. As far as supplies to Romania are concerned, it would be relatively easy to construct a pipeline link to connect South Stream with the Romanian network.

As argued elsewhere by the author, a general pattern has emerged which characterizes Russia–western CIS gas relations during the 1998–2013 period: a breakdown in political relations ('space of places') caused a breakdown in contractual relations ('contractual' space) which in turn led to a breakdown in gas flows ('space of flows').[71] Although attempts have been made to create an overarching 'legal/regulatory' space whereby all parties – European, Russian, and western CIS – involved in gas trade would subscribe to the same set of rules (in the form of the ECT) which would strengthen transit security, these attempts have failed. An attempt to create a similar 'legal/regulatory' space covering European and (some of) western CIS parties (in the form of the EnCT) is ongoing, but its potential success will inevitably be limited since Russia, as one of the main suppliers of gas to both western CIS countries and Europe, would be absent from it.

Ultimately, Gazprom chose transit avoidance as the solution to the transit security problem. Yet, this is not a complete solution, as Gazprom's transit-avoidance pipelines are likely to face significant regulatory obstacles in Europe, emanating from the Third Package[72] (see Chapter 4.) Also, worryingly for western CIS countries, the Russian transit-avoidance strategy will intensify the economic, and hence political, disconnect between them and Europe. Europe's transit security will no longer be

---

[71] Yafimava, K. *The Transit Dimension of EU Energy Security*, op. cit., p. 319.

[72] Yafimava, K. 'The EU Third Package for Gas and the Gas Target Model: major contentious issues inside and outside the EU', Working Paper NG 75, OIES, April 2013.

a function of western CIS countries' supply security, and Europe will be less dependent and hence less interested in the success (or failure) of western CIS gas sector reforms. At the same time, western CIS countries will remain significantly (although to varying degrees) dependent on Russian gas, at least during the 2010s, and therefore the stability of their political relations with Russia will be paramount for their supply security. Given that political considerations have always been present in the relations of Russia/Gazprom with all three western CIS countries – if only because western CIS countries were often unwilling or unable to make commercial concessions and preferred making political ones – it may be concluded that an increased degree of western CIS economic dependence on Russia is likely to lead further towards a decrease in their political freedom.

Conclusions

Russian gas exports to CIS countries rose to a peak of 89.6 Bcm in 2006 and in the years since the economic crisis have fallen to two-thirds of that level. Some of the reduction is long-term or permanent, and includes the displacement of Russian exports to Azerbaijan and Georgia by Azeri-produced volumes. A major factor is the fall in gas consumption in Ukraine, caused by the conjunction of high Russian export prices and a severe economic recession. Russia's policy of maintaining high export prices is clearly one of the background reasons for the decline in the gas trade; another is the long-term loosening of trade relations between former Soviet states – a general trend maintained over the two post-Soviet decades.[73]

The reduction in gas consumption in Ukraine – by a quarter or more since 2006 – is remarkable. More than 20 Bcm/year of demand has been destroyed. The high prices charged for Russian imports in 2009–13, combined with the effects of the economic crisis, led to the shutdown of industrial capacity and a sharp fall in industrial consumption, but also to substantial fuel-switching (to coal) and to some efficiency savings.

Ukraine's serious difficulties in the post-crisis period also triggered, for the first time, political efforts to increase domestic gas production and to diversify sources of gas supply. On production, agreements have been signed and some investments made, but the potential is for an impact on the gas balance in the 2020s, rather than in this decade. Supply diversification efforts, on the other hand, led to the start-up in 2012 of 'reverse flow' deliveries via Poland and Hungary, which, despite their modest scale, raised

[73] See, for example, Mitrova, T., Pirani S., and Stern, J. 'Russia, the CIS and Europe: gas trade and transit', in Pirani, S. (ed.), *Russian and CIS Gas Markets and Their Impact on Europe* (Oxford: OIES/OUP, 2009), especially pp. 397–400.

the prospect of Ukraine integrating more closely with European markets over the long term. The December 2013 Russia–Ukraine agreements, which provided for a steep cut in the price of Russian gas imports, as part of an economic and trade cooperation package, were motivated by Russia's broader political aim of strengthening its political influence over Ukraine – but the gas price reduction sharply undercut 'reverse flow' suppliers, and brought import prices close to the level of netback from the European market, illustrating the way in which market forces at work in Europe, albeit indirectly, are influencing the trade of Russian gas.

The reduction in demand for Russian gas observed in Ukraine could go further. To the extent that prices remain high, consumers in Belarus and Armenia, for example, could make efforts to switch fuels. Sourcing alternative gas supplies requires substantial investment, and time – although 'reverse flow' to Moldova, for example, is feasible. Nevertheless, the CIS importers will continue to need substantial volumes of Russian gas for the foreseeable future.

Russia's policy of transit diversification has fundamentally changed its relationship with Ukraine and Belarus, the largest CIS importers. During the gas disputes with Ukraine (2006 and 2009) and Belarus (2007), those countries were prepared, if faced with supply cut-offs triggered by failure to pay bills and/or pricing issues, to divert to their own consumers Russian volumes bound for Europe. The completion of the Nord Stream pipeline, and the likelihood that at least some of the South Stream line will be built, largely removes this lever from the transit countries' hands. While the main aim of transit diversification was to forestall further supply disruptions, it also strengthens Gazprom's hand in negotiations with the transit countries. It makes the prospect of disputes involving supply disruptions less likely, although if political and economic circumstances were sufficiently negative they can not be ruled out.

Another possibility that may influence the western CIS markets is that of a prolonged period of relatively low gas prices in Europe. This could weaken Russia's bargaining position, at least in as much as it could intensify efforts in Ukraine and Moldova to access imports from the west.

Whichever direction prices move in, transit diversification will surely further diminish Russia's gas trade with the western CIS countries.

CHAPTER 8

ASIA: A POTENTIAL NEW OUTLET FOR RUSSIAN PIPELINE GAS AND LNG

James Henderson

Russia has huge eastern gas resources, which include giant fields in the Siberian and Far East Federal Districts as well as prospective licences offshore in the Arctic seas (see Table 8.1). These would seem to be ideally placed to meet the rapid growth of gas demand in north-east Asia – particularly in China where gas consumption has more than doubled in the past five years[1] and is set, potentially, to triple again by 2030.[2] However, despite more than 20 years of negotiations, the only current Russian gas exports to the east are in the form of LNG from the Sakhalin-2 project, while the prospects for pipeline exports remain mired in negotiations over price. However, the growing importance of the Asian gas market to Russia and to Gazprom has been highlighted by the continued problems surrounding gas exports to Europe and to the former Soviet Union (Chapters 3, 4, and 7), the rapid changes taking place in the domestic gas market in Russia (Chapter 5), and the imminent emergence of significant new global gas competition. In essence, the Russian government and Gazprom may only have a short window of opportunity to get their strategy on volumes, price, and flexibility right or the chance to sell significant gas into China and Asia may be lost for a decade or more. It is clear that some members of the Russian political elite, including Igor Sechin, CEO of Rosneft, and Leonid Mikhelson, CEO of Novatek,[3] understand this problem and, as a result, Gazprom is not only facing competition from global gas suppliers but also from its own domestic competitors for the right to sell gas into the world's fastest growing energy economy. The resolution of the issues that underpin the question of how and whether Russia can finally grasp the clear opportunity on its eastern borders may ultimately also provide an insight into the future structure of the Russian gas industry as a whole.

---

[1] Data from *BP Statistical Review of World Energy 2013* (London: BP, 2013).
[2] US EIA *International Energy Outlook 2013* (Washington: US Energy Intelligence Agency, 2013), p. 103.
[3] 'Novatek eyes big role in Russian LNG exports to China as Sechin seeks market liberalisation', *Oil and Gas Eurasia*, 7 March 2013, sourced from www.oilandgaseurasia.com/en/news/ on 8 November 2013.

Table 8.1: Russia's eastern gas reserves and resources

	ABC1 Reserves (Bcm)	C2 Reserves (Bcm)	Resources (Bcm)	Total (Bcm)
Siberian Federal District	2,600	3,600	31,700	37,900
Far East Federal District Onshore	1,400	1,100	12,000	14,500
Sakhalin	900	300	5,400	6,600
East Siberian and Laptev Seas			5,600	5,600
Chukchi and Bering Seas			2,700	2,700
Total	**4,900**	**5,000**	**57,400**	**67,300**

Source: Gazprom Presentation 'Gazprom in Eastern Russia', June 2013.

A brief history of Russia–China gas negotiations

The first Memorandum of Understanding between Russia and China on the issue of gas exports was signed in 1994 between CNPC, the Chinese state oil and gas company, and the Russian Energy Ministry.[4] It identified the possibility of an eastern gas export option from the Kovykta field in East Siberia, although this immediately highlighted a problem that dogged negotiations for the subsequent decade: Kovykta was not owned by Gazprom, the only company allowed to export gas from Russia. Nevertheless, the initial agreement was underpinned by a subsequent visit by Russia's Prime Minister Viktor Chernomyrdin to Beijing in June 1997, where a general agreement that a future gas contract could deliver 25 Bcm/year over 30 years was reached.[5] In the following year, the issue of Gazprom's lack of eastern gas assets was addressed by broadening the negotiations to include a western pipeline route from West Siberia through the Altai region to Xinjiang in western China, and in 1999 a feasibility study covering both the eastern and western routes was initiated; this involved Russia, China, and also South Korea, which had been identified as another potential buyer of Russian gas.[6]

---

[4] Henderson, J.A. 'The Pricing Debate over Russian Gas Exports to China', Working Paper NG 56, OIES, October, 2011.

[5] Paik, K.-W. 'Northeast Asian Countries' Oil and Gas Relations with Russia', presented at Clingendael Energy Conference 'The Geopolitics of Energy in Eurasia: Russia as an Energy Lynch Pin', 22–23 January Clingendael International Energy Programme/Clingendael Asia Studies (The Hague: Clingendael Institute, 2008), p. 18.

[6] Paik, K.W., 'Pipeline gas introduction to the Korean peninsula', Chatham House Report, January 2005, p. 11.

However, the confusion over the likely source of Russian gas supplies was encapsulated in the parallel negotiations being carried out by Gazprom, with its preference for a western route, and Sidanco (which became part of TNK-BP in 2003) which owned the Kovykta field and was promoting the eastern route. The negotiations around Kovykta were always likely to be undermined by the lack of Gazprom involvement (and, by inference, a lack of Russian government backing) and this ultimately proved to be the case. In 2004 Gazprom refused to support the development of the field and challenged the validity of the licence.[7] As a result gas sales discussions ceased and Gazprom ultimately agreed to buy TNK-BP out of the field in 2007.[8]

As the Kovykta partners faded from the gas export dialogue, Gazprom continued to pursue its own agenda. In 2002 it had signed a Memorandum of Understanding to become a partner in the West–East gas pipeline in China that would have provided an ideal link with its proposed western Altai export route from Russia.[9] Indeed discussions between Gazprom and CNPC matured to the point where a Strategic Co-operation Agreement was signed by the two companies in October 2004, paving the way for a series of regular committee meetings to continue the negotiations.[10] These culminated in a visit by President Putin to Beijing in March 2006, at which a protocol on gas exports was signed. It envisaged a scheme including both pipeline routes and a total planned volume of 68 Bcm/year, 38 Bcm of which would go through the eastern pipe and 30 Bcm through the west, with a planned start date of 2011.[11]

However, by December of the same year the first major disagreement between the two parties had emerged, with price being the key area of dispute.[12] Essentially it became clear that one of Gazprom's main reasons for promoting a western pipeline route from West Siberia was to allow it to link its exports east to China with those west to Europe, creating what it hoped would be intense price competition. Furthermore Gazprom also made it clear that it expected to make the same margin on its sales to China

---

[7] 'TNK-BP gas field development faces suspension', *Financial Times*, 20 September 2006, London.

[8] 'Gazprom pays $770m for TNK-BP gas field', *Financial Times*, 1 March 2011, Moscow.

[9] Data from Gazprom website at http://gazprom.com/about/history/chronicle/2002/, accessed on 1 April 2011.

[10] Paik 'Pipeline gas introduction to the Korean peninsula', op. cit., p. 19.

[11] 'First gas supplies to China may begin in 2011 – Miller', *Interfax*, 21 March 2006, Moscow.

[12] 'Russia, China have significant differences on price for Russian gas – Gazprom official', *Interfax*, 21 December 2006, Moscow.

as it was making on sales to Europe, despite the fact that the transport distance to Shanghai is more than 3000 km greater. Not surprisingly, the Chinese negotiators wanted to take the extra distance inside China into account in the pricing arrangements, leading to a $50/mcm mismatch in price expectations.[13] A potential solution appeared in 2007 with Gazprom's purchase of the Kovykta field in Irkutsk,[14] which suggested that Gazprom might be persuaded to focus on the eastern export route. This possibility was further underlined by the publication of Russia's 'Eastern Gas Programme' in the same year, which emphasized the importance of developing the country's gas resources towards the Pacific coast.[15] The economic development of Russia's eastern regions was made a government priority, and the development of gas supply for both domestic and export markets – focusing on local gasification, the development of LNG facilities on Sakhalin Island, and the construction of both the East and West export pipelines – was seen as a vital component of the planned growth.

Unfortunately the economic crisis of 2008 not only delayed Gazprom's purchase of Kovykta but it also caused a hiatus in the entire development plan for Russia's gas industry in the East, as funds were in short supply and management of cash flow became a priority. Nevertheless, by October 2009 stability had been restored and talks had re-opened, with a framework agreement being signed that again defined the principle of a two pipeline export plan and outlined the commercial and technical parameters to be used, including an agreement that prices would be linked to the Japan Crude Cocktail (JCC) oil price benchmark.[16] However, despite the confidence of Gazprom CEO Aleksei Miller that 'pricing is the only issue [to be resolved] here',[17] the differences between the two sides had widened due to the development by China of a more diversified import strategy. In 2006, while Russia and Gazprom had been arguing about price, CNPC had started the process of dramatically improving its bargaining position by negotiating significant gas import deals with Central Asia, in particular Turkmenistan (see Chapter 14), and also by developing links with Burma via pipeline to the south, and with multiple LNG suppliers via the

---

[13] Henderson 'The Pricing Debate over Russian Gas Exports to China', op. cit., p. 6.
[14] 'Gazprom pays $770mm for TNK-BP gas field', *Financial Times*, March 2011, Moscow.
[15] 'Gazprom in Eastern Russia, Entry into Asia Pacific Markets', Gazprom Press Conference, 17 June 2009, Moscow.
[16] JCC is the Japanese benchmark oil price against which LNG contracts in Asia tend to be linked in long-term supply contracts.
[17] 'Gazprom, CNPC sign frame agreement on gas supply', *Interfax*, 12 October 2009, Moscow.

construction of regasification facilities on its eastern seaboard.[18] As a result, although another 'agreement' with Russia was signed in September 2010 that once more specified volumes, dates, take-or-pay levels, the period of increasing deliveries, and the level of guaranteed payments, with plans to sign a final export agreement, which would include price details, in mid-2011,[19] China was now in a much stronger position to reject Russia's proposed western export route and to demand a more competitive gas price.

An important milestone was reached in March 2011 when Gazprom finally completed the purchase of the Kovykta field from TNK-BP and was therefore in a better position to agree to China's request for an eastern rather than a western export route. This also allowed Gazprom to shift the pricing negotiations away from a comparison with Central Asian gas in the west, towards higher priced LNG in the east.[20] However, although this resolved the pipeline issue, with China effectively winning the debate, it still left the negotiating parties as much as $100/mcm apart on price when they met for discussions in July 2011 and again in October of the same year, with Gazprom still apparently insisting on netback parity with prices in Europe.[21] As a result, despite more positive rhetoric from then Russian Prime Minister Putin that 'we are nearing the final stages as to the delivery of raw gas to the Chinese market',[22] by the end of 2011 Russia was no closer to sealing a deal for pipeline sales to China and the rest of north-east Asia. In contrast, Turkmenistan had signed an initial agreement, subsequently confirmed in 2013, to boost its gas exports to China to an ultimate peak of 65 Bcm/year.[23]

The foundation for the Russia–China negotiating positions

The long-term inability of Russia and China to reach an agreement over gas sales clearly reflects the entrenched bargaining positions of both sides as they have attempted, over more than a decade, to achieve the best deal for themselves. It appears that Russia views China as a huge and growing market which is desperate for more gas, and therefore wants to charge

---

[18] US EIA, China Report, April 2013, www.eia.gov/countries/cab.cfm?fips=CH, accessed on 9 September 2013.

[19] 'Gazprom, CNPC sign expanded terms of gas deliveries to China', *Interfax*, 27 September 2011, Moscow.

[20] 'Gazprom buys Kovykta operator's assets', *RIA Novosti*, 1 March 2013, Moscow.

[21] 'Russia insists on netback parity for gas prices with China, Europe', *Interfax*, 24 October 2011, Moscow.

[22] 'Russia and China near final stage in gas talks', *Interfax*, 10 October 2011, Moscow.

[23] 'Turkmenistan to boost gas supplies to China by 25 Bcm/year', *Interfax*, 24 October 2011, Moscow.

a premium price for supply from an obvious source of imports from East Siberia. On the other hand, China sees Russia's East Siberian gas as a stranded asset with China as its only realistic market, and therefore wants to use what it regards as a strong bargaining position to drive down the price of potential imports. It is worth considering the key planks of both parties' negotiating positions in more detail, before moving on to discuss the state of the Russia–China dialogue as of the time of writing (January 2014).

One important characteristic of the Russia–China dialogue is that the Chinese politicians and senior company executives very rarely comment on the negotiations, leaving it to their Russian counterparts to make optimistic comments about the timing of a deal. This reticence, combined with the clear difficulty over the issue of price noted above, would appear to suggest that, although China may see the logic of importing gas from Russia, its negotiating strategy is based on a willingness to take a long-term view of developments in the global gas market, rather than rushing to commit to a long-term, geopolitically sensitive pipeline deal with Russia at the wrong price.

One likely reason for China's apparent reluctance to rush into a deal is its own, somewhat complex, domestic gas price reform. Most gas prices in China remain regulated in a rather intricate fashion through the value chain, with different administrations having the right to set prices at various stages of production and sale.[24] Essentially, relatively low wellhead prices have historically been set for producers by the central government, to which have been added fixed transport charges, to establish a minimum city-gate price in various regions. The local administrations have been able to negotiate for gas at around this minimum price and then charge a margin for sale on to industrial and residential consumers. Larger power generation and industrial customers have also been able to negotiate individual deals with producers, but the net result has generally been a relatively low gas sales price. The Chinese authorities have tended to re-set wellhead prices on a relatively *ad hoc* basis, with the timing often influenced by political considerations. However, the shift in China's energy economy towards becoming a net importer of gas in 2009–10 has catalysed an initiative to increase prices in order to reduce losses being incurred by the Chinese state companies – which were being forced to purchase high-cost imports and re-sell the gas at lower domestic prices. A 25 per cent price increase in 2010 started to address this issue,[25] but no further changes were made in the subsequent two years, meaning that in 2012 PetroChina incurred a Y42 billion (US$6.7 billion) loss on the difference between the

---

[24] Higashi, N. *Natural Gas in China: Market Evolution and Strategy* (Paris: IEA, 2009).

[25] 'China's gas price hike a small step towards reform', *Reuters*, 1 June 2010, Beijing.

cost of its gas import purchases and the price at which it could sell to customers in the domestic market.[26]

These continuing losses forced the Chinese authorities to act again in 2013, when the average regulated wellhead price was increased by a further 15 per cent to reach Y1.95 per cubic metre, which after transport costs means that the final gas price in many coastal regions has now reached $10–12/MMBtu (US$360–430/mcm).[27] Although this was still below the average spot import price for the first half of 2013 (which stood at approximately $14.50/MMBtu ($520/mcm)[28]) it has provided further evidence that the Chinese authorities do want to bring the domestic gas price up towards international levels. Furthermore, the Chinese government has decided that regulated prices will now apply only to the volumes of gas that were purchased in 2012 ('base gas'), and that any additional 'new' volumes will be priced at market levels, with the intention that both prices will converge in 2015.[29] This market level will be based on the tariffs established by a trial gas pricing project in the Guangdong and Guangxi regions that started in December 2011, in which the gas price has been set relative to a basket of international fuel oil and LPG prices in order to provide a further benchmark connected to the global energy market.[30] This has essentially meant that these two regions have been paying an oil-linked gas price, with a slight discount, for the past two years. In a move that confirms the strategy laid out in the 12th Five Year Plan, released in December 2012, the Chinese government has now decided that all prices should transition towards this price formation mechanism over the next few years.[31] For the next two years at least it appears that this will happen via government decree as well as market mechanisms, and significant increases in the regulated price in 2014 and 2015 are anticipated; it could rise to Y2.80 per cubic metre, implying an average price on the eastern seaboard of $13.5–15/MMBtu, or close to the LNG spot price.[32] At this point the Chinese authorities may be able to create a firm basis for all negotiations

---

[26] 'PetroChina 2012 profits down, production up' in *China Daily*, 22 March 2013, sourced from www.chinadaily.com.cn/business/2013-03/22/content_16333814.htm on 1 August2013.

[27] 'China, India gas price reforms open the door to more LNG imports', *Reuters*, 3 July 2013, Beijing/New Delhi.

[28] Data from Energy Intelligence Group at www.energyintel.com/ accessed on 11 November 2013.

[29] O'Sullivan, S. *China Gas: The Price Is Becoming Right* (London: Trusted Sources, 2013).

[30] 'Has China finally picked up the pace of natural gas reform?', Norton Rose, April 2012,.

[31] 'China may see more pricing reforms after government hikes gas prices by 11%', *Platts*, 4 December 2012, Singapore.

[32] Price hike spurs China gas plays, lifts reform hopes', *Reuters*, 2 July 2013, 'Singapore.

with gas importers, essentially citing a domestic price on the east coast for use in netback calculations to any delivery point for gas into China. However, until this price level has been achieved, Chinese negotiators may remain reluctant to sign any long-term import deals that would commit them to oil-linked pricing for significant volumes, and this appears to have been one of the reasons for the delay in reaching agreement with Russia.

Meanwhile, Gazprom negotiators apparently have regarded the rapid growth in current and forecast Chinese gas demand, which is forecast by the IEA to rise from 132 Bcm in 2011 to 307 Bcm in 2020 and 370 Bcm in 2030,[33] as making an overwhelming case for gas imports from Russia if China is to meet its economic and environmental goals. However, China has strengthened its bargaining position by developing a variety of gas supply options in order to avoid becoming over-reliant on one source. Over the past five years it has developed a multi-directional gas supply strategy that involves not only imports from multiple sources but also increasing levels of Chinese company investment in international upstream gas assets. Indeed so successful has this strategy been that supplies from Russia would no longer appear to be a short-term necessity, but rather as needing to fight for their place in the Chinese gas supply and demand balance, as shown by Figure 8.1.

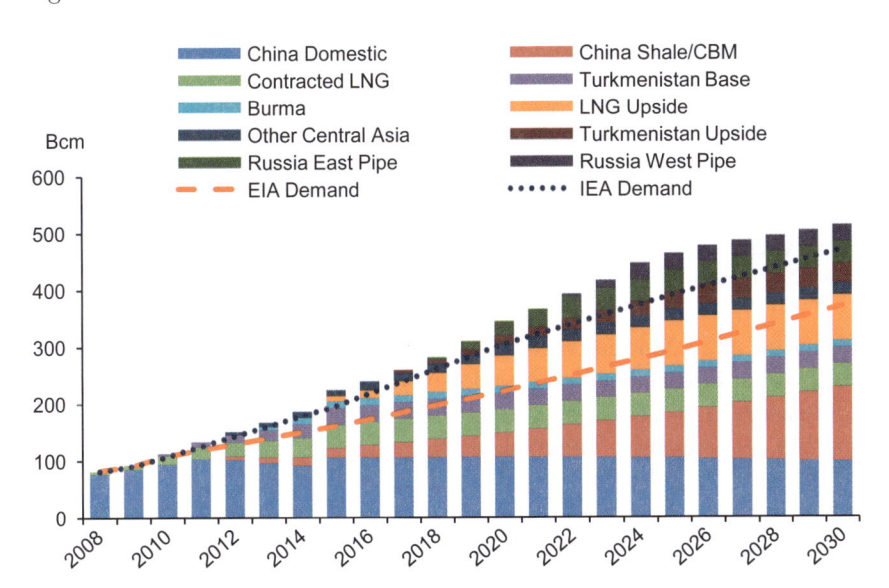

Figure 8.1: An estimate of the Chinese gas supply and demand balance to 2030
Source: EIA International Energy Outlook 2013, IEA World Energy Outlook 2013, CNPC, author's estimates

---

[33] IEA World Energy Outlook 2013, p.103

Figure 8.1 shows two rather contrasting demand forecasts from the US Energy Information Administration (EIA) (2013) and the IEA (2013) together with actual and estimated figures for the various sources of gas supply which could meet that demand. Currently, the main source of supply is conventional domestic gas production, which is expected to remain relatively flat throughout the forecast period, while domestic unconventional output is expected to contribute an increasing share, although perhaps not growing as rapidly as had been hoped. The Chinese authorities had originally set a target for shale gas production to reach 6.5 Bcm by 2015[34] and 50–80 Bcm by 2020, but both these goals now seem unlikely to be met given the technical, financial, and logistical challenges that need to be faced in such a relatively short timescale.[35] As a result, the supply forecast above sees Chinese unconventional production largely being dominated by coal bed methane (CBM) to 2020, with shale gas output only reaching 14 Bcm by the end of the decade but then jumping to 76 Bcm by 2030. However, the development of all of China's unconventional resources is at a very early and uncertain stage, and so the total of 125 Bcm seen for combined shale and CBM output by 2030 could be a large over- or underestimate that creates a significant level of uncertainty. What we do know, however, is that one catalyst for the recent price rises in China has been the need to encourage the development of this resource,[36] and as a result it seems that CNPC negotiators will remain keen to avoid any excessively high-cost commitment to long-term supplies that might undermine the country's hopes for increasing indigenous supply.

Irrespective of the country's domestic production potential, however, its import requirement will continue to grow. Under the supply forecasts used in Figure 8.1, which envisage total conventional and unconventional production of 147 Bcm by 2020 and 227 Bcm by 2030, China's gas import needs will expand from 14 Bcm in 2010 to 140 Bcm in 2030 under the EIA demand forecast and to 240 Bcm by 2030 under the IEA forecast.

The Chinese authorities have taken steps to meet this demand via a number of pipeline and LNG sources. Imports from Turkmenistan began in December 2009 via the 1833 km Central Asia–China pipeline that crosses Uzbekistan and Kazakhstan to join the West–East pipeline on the border of China's western province of Xinjiang.[37] The total capacity of the pipeline corridor will reach

---

[34] 'China set to miss targets for shale gas field development', *Financial Times*, 2 July 2013, London.

[35] Gao, F. 'Will there be a shale gas revolution in China by 2020?', Working Paper NG 61, OIES, April, 2012.

[36] 'China vigorously promoting shale gas exploration, development', *Oil and Gas Journal*, 3 May 2012, Shengdu.

[37] Henderson, J.A. 'The Pricing Debate over Russian Gas Exports to China', Working Paper NG 56, OIES, October, 2011.

55 Bcm/year by 2014 and 80 Bcm/year by 2020. Two 15 Bcm/year lines have been built and a third 25 Bcm/year line is under construction. Under agreements signed between China and Turkmenistan in September 2013, a fourth 25 Bcm pipeline – going via Uzbekistan, Tajikistan, and Kyrgyzstan to Kashgar rather than via Kazakhstan to Khorgos – will be added, with deliveries starting in 2016 and full capacity utilization in 2020. Turkmenistan's exports to China are now contracted to reach 65 Bcm by 2020, and the main source of gas will be the newly developed Galkynysh field, which came on stream in mid-2013.[38] In addition, by 2020 there could be supplies of 10 Bcm/year from Uzbekistan, plus a similar volume from Kazakhstan, if both those countries achieve their production and export goals, although Simon Pirani[39] highlights a number of constraints that could prevent this. Nevertheless, imports of Central Asian gas to China are set to rise sharply over the next five to ten years, with the Chinese authorities favouring purchases from a region in which they have significant influence and which has allowed them to take equity interests in the gas fields that are providing them with a strategically vital source of energy.

While Central Asian gas provides gas supply from the west, imports from the south are being supplied via a new pipeline from Burma, where China again has both political and commercial interests. The 12 Bcm/year line was opened in July 2013[40] and, despite construction work being disrupted by the armed conflict between the Burmese government and separatists in Kachin province, is expected to provide a secure source of relatively low-cost imports into southern China from gas fields in the Bay of Bengal. The gas line has been built in tandem with a 440 kbpd oil pipeline, highlighting the strategic nature of the investment for China and ensuring its priority as an energy import route. Meanwhile on the east coast China has adopted a gas import policy based on LNG, with current contracts suggesting that supply will reach at least 40 Bcm by 2016 and probably more than this when spot cargoes are included.[41] Indeed the list of current operational and planned LNG regasification terminals scheduled to be built on China's east coast suggests that LNG imports could reach as much as 100 Bcm/year by the end of the current decade.[42]

---

[38] 'Turkmenistan has high hopes as giant gas field nears output', *Reuters*, 22 May 2013, Ashgabat; 'China asserts clout in Central Asia with huge Turkmen gas project', *Reuters*, 4 September 2013; 'China ups Turkmen supplies', *Platts International Gas report*, 9 September 2013, p. 25.

[39] Pirani, S. 'Central Asian and Caspian Gas Production and Constraints on Export', Working Paper NG 69, OIES, December 2012.

[40] CNPC Press Release, 31 July 2013, 'Myanmar–China gas pipeline starts delivering gas to China'.

[41] Kushkina, K. *Golden Age of Gas In China* (Moscow: The US Russia Foundation, 2012).

[42] US EIA, China Report, op. cit.

Furthermore, Chinese companies have begun to participate in the LNG projects that will supply the country with gas in future. In the USA, where the prospect of significant LNG exports has increased as four liquefaction projects have now received non-FTA export licences,[43] Sinopec has taken a 33 per cent stake in shale assets owned by Devon Energy, and CNOOC has acquired assets from Chesapeake in the Eagle Ford shale and also in Colorado and Wyoming, while in Canada CNPC has an interest in Shell's Groundbirch licences in British Columbia.[44] In Australia – the first country to begin exporting LNG to China – CNOOC increased its interest in BG's Queensland Curtis LNG project to 50 per cent in early 2013,[45] while CNPC has established a presence through its ownership of a 20 per cent interest in the Fisherman's Landing scheme, and Sinopec has acquired 15 per cent of the APLNG project in Queensland. Perhaps most interestingly, though, CNPC has also purchased a 20 per cent interest in Area 4 of the emerging gas play in Mozambique offshore east Africa,[46] demonstrating its desire to encourage the development of as diverse a range of gas supply as possible.

Within this compass of China's gas supply, shown in Figure 8.2, Russia would seem to sit neatly as the northern option, and this fact has probably been the source of Russian confidence that a deal will ultimately be concluded. However, the supply and demand balance in Figure 8.1 suggests that this is not a guaranteed outcome, and the potential for an abundance of gas supply may actually be causing the Chinese to delay the negotiations. Russia could ultimately have a significant role to play if China's indigenous supply (especially unconventional supply) disappoints, if Uzbekistan and Kazakhstan fail to provide any export gas, if new LNG projects hoping to offer supplies are very high cost and therefore uncompetitive on price, and if demand reaches the high end of expectations. However, if one or all of these outcomes fails to materialize, then there would appear to be a clear risk that an aggressive Russian marketing strategy could leave it excluded from Asia's fastest growing gas market.

---

[43] 'US picks up steam on natural gas export projects', *Reuters*, 11 September 2013.

[44] Henderson, J.A. 'The Potential Impact of North American LNG Exports', Working Paper NG 68, OIES, October, 2012.

[45] 'BG Group signs $1.93bn LNG supply, equity deal with CNOOC', *Reuters*, 6 May 2013, Sydney.

[46] 'CNPC to buy stake in ENI Mozambique assets for $4.2 billion', *Bloomberg*, 14 March 2013, London.

Asia: a Potential New Outlet 227

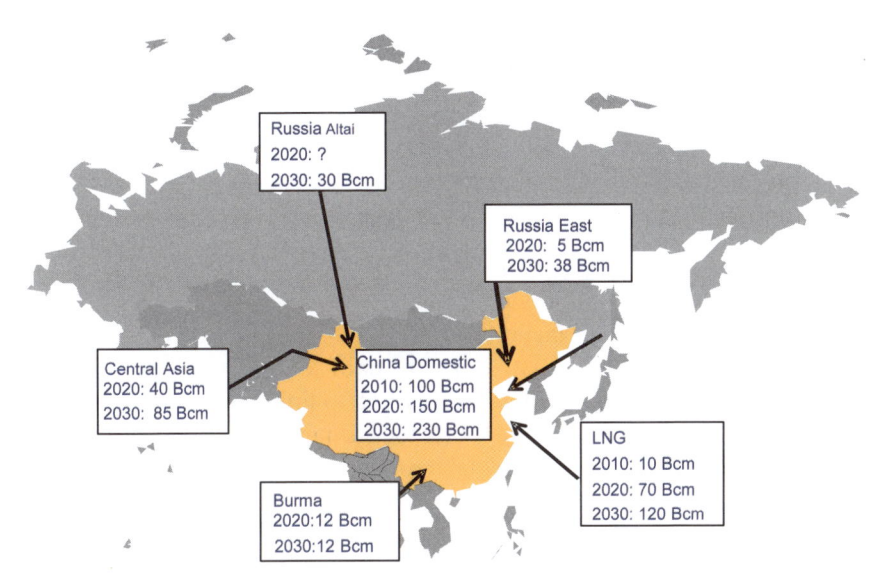

Figure 8.2: Possible supply volumes to China as part of its multi-vector gas import policy
Source: Author's estimates based on IEA, EIA, and CNPC forecasts.

The situation in 2012–14: increased urgency but LNG the main focus

The continuing delay in negotiations, combined with China's strategy of establishing a diversity of import options other than those from Russia, has created increasing frustration in the Russian government. This finally manifested itself in a speech by the newly re-elected President Putin in April 2012, when he emphasized the need for Russian energy companies to meet the challenges of the new global energy economy.[47] In particular, he focused on the impact of North American shale gas, and underlined that this was not only having an impact on the domestic US economy but was also starting to influence gas markets in Asia and Europe. He specifically mentioned competition in new markets such as Japan, India, and China. These comments, combined with the approach of the APEC summit hosted by Russia in Vladivostok in September 2012, catalysed an increased level of dialogue on the eastern gas issue, with Gazprom CEO Aleksei Miller meeting Jiang Jiemin, the head of CNPC, in April followed by a Heads of State summit when Putin visited Beijing in June.[48] Further meetings

---

[47] 'Russia has to meet challenge of shale revolution – Putin', *Interfax*, 11 April 2012, Moscow.
[48] 'Gazprom and CNPC discuss oil and gas supplies to China' Gazprom press release, 27 April 2012.

took place in July and September between senior company representatives, but throughout the entire period positive statements from the Russian side were not matched by agreement from the Chinese side.[49]

Essentially it appeared that, although both sides had agreed that the oil price could be used for indexation purposes, disagreement continued over the base starting price and the exact relationship between the gas and oil price (the 'slope', in LNG market terminology[50]). In order to strengthen its bargaining position, Gazprom decided to demonstrate to China that it had other sales options by announcing that LNG would become its priority, with an expansion of the facilities on Sakhalin and the development of a new facility at Vladivostok cited as the main focus.[51] Gazprom thereby hoped both to create demand for its gas and to link the price negotiations to high Asian LNG prices, but both tactics failed to alter the Chinese position, which became increasingly robust as China explored its own diversification strategy of alternative domestic and imported gas supply.

Eventually, in October 2012 both Russia's Energy Minister Aleksandr Novak and President Putin were forced to acknowledge these trends, with Novak announcing that a revised Eastern Gas Strategy would take into account the potential of China's shale gas resources and its new import options via pipeline and LNG.[52] Meanwhile Putin stated that:

> A brutal fight is unfolding among gas exporters for short-term and long-term contracts [and insisted that Russian companies, and especially Gazprom] are simply obligated to take these trends into consideration, to clearly imagine how the situation will develop not just in the next two to three years, but throughout the upcoming decade.[53]

Within a week, the president was in discussion with Gazprom's CEO Miller about the company's new east-facing development plans, and was urging specific action that would accelerate both the development of domestic infrastructure and the export facilities that could provide Russia

---

[49] 'Gazprom to hold latest round of negotiations in July for gas supply to China', *Interfax*, 25 June 2013, St Petersburg.

[50] LNG contracts include a base price and a slope by which the gas price increases relative to the oil price. A 17% slope is approximately equivalent to a 1:1 relationship between the oil and gas price, and most contracts are negotiated with a slop in the range 12–16%.

[51] 'LNG priority in supplies to Japan, eastern China', *Interfax*, 18 June 2012, Moscow.

[52] 'Putin tells Energy Ministry to adjust General Gas Scheme to 2030 and Eastern Gas Programme', *Interfax*, 23 October 2012, Moscow.

[53] 'Putin orders Gazprom to prepare key principles of gas export strategy', *Interfax*, 24 October 2012, Moscow.

with flexibility of sales destinations and the potential to increase export volumes dramatically over the next decade.[54]

Russia's eastern development options

The Putin–Miller discussion focused on much of the infrastructure which has been at the heart of the Eastern Gas Programme that was formally approved by the Russian government in September 2007,[55] but with some new priorities to reflect the changing market dynamics. The initial plan highlighted four production centres that would provide the basis for gas development in East Siberia and the Russian Far East (see Map 8.1): the Sakhalin Centre, based on the offshore projects around the island with a particular focus on the Sakhalin-1, 2, and 3 projects; the Irkutsk Centre, with the main asset being Gazprom's Chaiandinskoe (Chaianda) field; the Irkutsk Centre, based around the Kovykta field now owned by Gazprom; and the Krasnoiarsk Centre, where Gazprom has some small fields such as Sobinskoe that could link into the existing UGSS trunk pipeline system in the west of the country.[56] The total resources of these four regions have been estimated by Gazprom at 66 Tcm, of which approximately 5 Tcm is categorized as proved ABC1 reserves,[57] and the estimated production potential is estimated to be in excess of 120 Bcm/year.[58]

Although a significant proportion of this gas is targeted for export markets in Asia, as will be discussed later, gasification of the domestic market in Russia is also a vital component of the Eastern Gas Programme. As highlighted by Jonathan Stern and Michael Bradshaw,[59] this gasification programme is a response to increasing concern in the Russian government about the depopulation and economic degradation of a region of the country that borders on the fastest growing economy in Asia, namely China. Although from a western perspective the threat of a passive or active 'invasion' of eastern Russia by China seems unlikely in the short term, it is perceived as a very real geopolitical threat in Moscow. As a result, the industrial regeneration of Russia's Far East is seen as a core political

---

[54] 'Gazprom gets political push to start Chayanda project', *Interfax*, 30 October 2012, Moscow.
[55] Russian Minister of Energy and Industry Viktor Khristenko signed approval order number 340 to formally create the Eastern Gas Programme.
[56] Paik, K.-W. *Sino-Russian Oil and Gas Co-operation* (Oxford: OIES/OUP, 2012).
[57] Gazprom presentation ahead of 2013 AGM, 13 June 2013, 'Gazprom in Eastern Russia, Entry into Asia-Pacific Markets', Moscow.
[58] Paik *Sino-Russian Oil and Gas Co-operation*, op. cit, p. 88.
[59] Stern, J., and Bradshaw, M.K. 'Russian and Central Asian Gas Supply for Asia', in J. Stern (ed.), *Natural Gas in Asia* (Oxford: OIES/OUP, 2008), pp. 220–78.

230 *The Russian Gas Matrix*

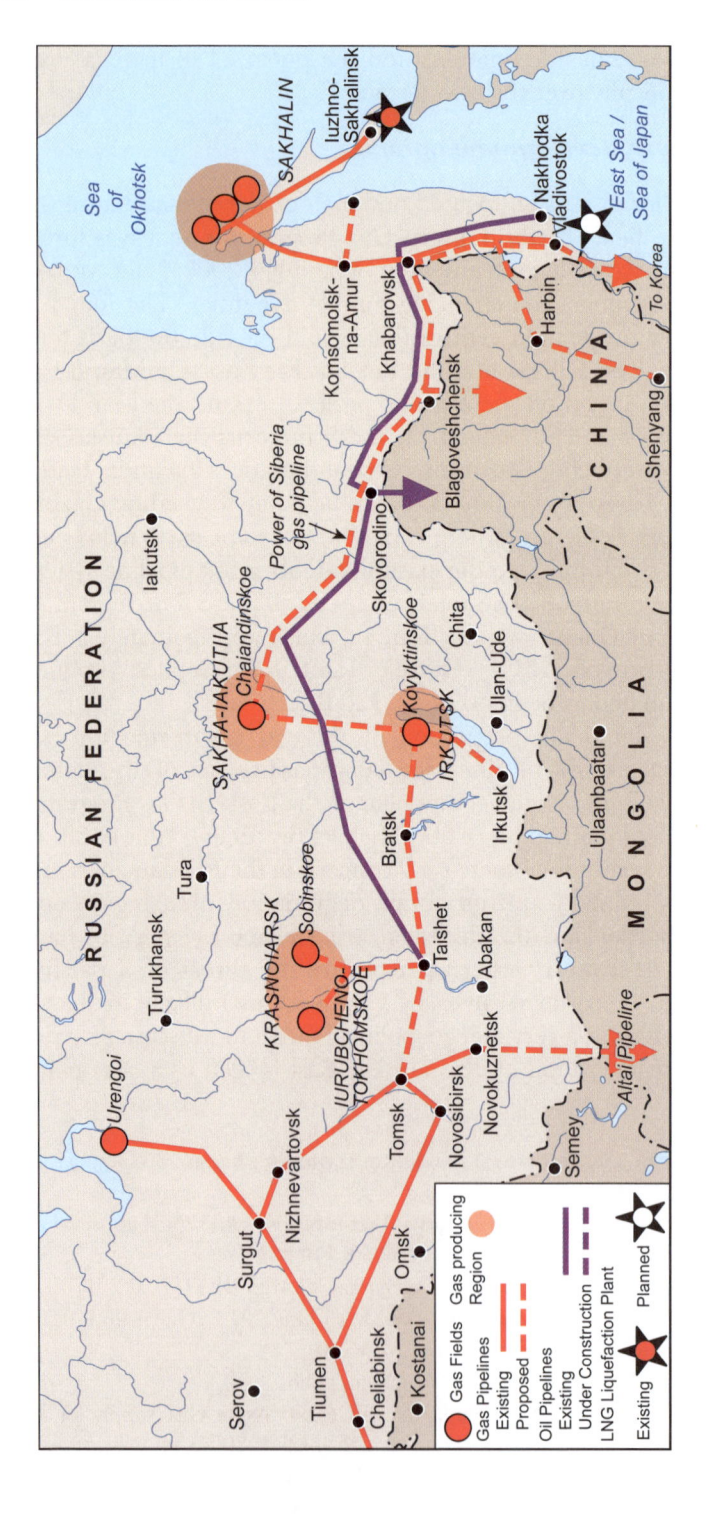

Map 8.1: Russia's Eastern Gas Programme development plan
Source: Gazprom.

objective.[60] The oil sector has played its role in the form of new pipeline infrastructure – the East Siberia-Pacific Ocean (ESPO) pipeline to Vladivostok and China – and field developments to encourage increased export sales and the development of domestic refineries. The development of the gas industry in Russia's eastern regions must therefore be viewed through a domestic political lens as well as from a commercial and geopolitical export perspective.

With this important proviso, which complicates many of the investment decisions that Russia and Gazprom are now facing in the east, it is nevertheless the case that increased exports to Asia are a fundamental plank of the current Russian Energy Strategy, published in 2009. This document foresees more than 140 Bcm of gas production in the Russian East by 2030, and exports to Asia increasing from zero in 2008 to 75 Bcm by 2030, accounting for more than a quarter of all non-CIS exports at that date (Figure 8.3).[61] These exports will not only provide diversification from Russia's reliance on a somewhat stagnant European market, but will also provide vital extra revenues for the Russian budget through the payment of export taxes on pipeline sales to non-CIS markets. With this in mind President Putin has appeared, from October 2012 onwards, to be very keen to encourage Gazprom to accelerate the development of the Eastern Gas Programme.

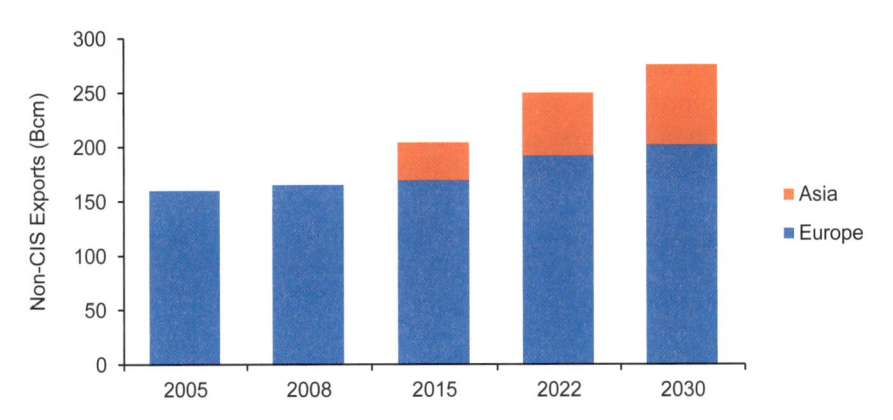

Figure 8.3: Russian gas export forecast from the Energy Strategy to 2030
Source: Russian Energy Strategy to 2030.

---

[60] Paik *Sino-Russian Oil and Gas Co-operation*, op. cit, pp. 1–7.
[61] Russian Energy Strategy to 2030, published 13 November 2009, Government Approval No. 1715.

The 'Power of Siberia' pipeline as a foundation for Eastern expansion

Gazprom has significant gas reserves in East Siberia, and the development of its two major assets, Chaianda and Kovykta, will be discussed in Chapter 12 as part of its overall production strategy. However, as far as the development of Russian gas in the East is concerned, the construction of transport infrastructure is the key issue. As noted above, the development of domestic gas infrastructure and greater demand from Russian customers is one priority, but any project will need to be underpinned by export sales. Gazprom, under the instruction of President Putin, has been keen to develop an export model similar to that used by the ESPO, namely to sell hydrocarbons via a direct link to China but also to provide market flexibility by gaining access to the Pacific coast and the markets of Asia and even South America. The ESPO runs from Taishet in Krasnoiarsk to the port of Nakhodka near Vladivostok, with a spur to China at Skovorodino, providing a route for oil from East and West Siberia to multiple Asian markets.[62] A similar plan is envisaged for gas, with the recently named 'Power of Siberia' pipeline set to run initially from the Chaianda field in Sakha via Blagoveshchensk on the Chinese border to Vladivostok (see Map 8.1). This 3200 kilometre pipeline with a 1420 millimetre diameter would have an ultimate capacity of 61 Bcm/year, with a planned start-up date of late 2017 and an initial cost estimate of RR770 billion (c.US$24 billion).[63] The final investment decision for the project was announced in October 2012 shortly after President Putin's meeting with Gazprom CEO Miller, in an apparent move to demonstrate that Russia planned to proceed with its eastern project with or without an agreement to sell gas to China,[64] but this attempt to improve Russia's bargaining position clearly required an alternative sales outlet to be available.

In order to provide a credible balance to potential pipeline sales to China this alternative needed to access the existing LNG market in Asia, and as a result Gazprom fixed upon the construction of a liquefaction facility at the end of the Power of Siberia pipeline at Vladivostok. The Gazprom Board of Directors has approved a 10 million tonne two-train facility, that at the time of writing is being assessed in tandem with a Japanese consortium (the Japan Far East Gas Consortium) and is planned to be online in 2019, with the option of a third 5 million tonne train to be

---

[62] Data from Transneft website at
http://eng.transneft.ru/projects/118/10020/, accessed on 11 November 2013.
[63] 'Gazprom gets political push to start Chayanda project', *Interfax*, 30 October 2012, Moscow.
[64] See www.gazprom.com/about/production/projects/pipelines/ykv/, sourced on 31 July 2013.

added at a later date.[65] However, even as the announcement of investment approval for the project was made, questions were being asked about the logic of bringing gas 3200 kilometres from East Siberia to the Pacific Coast when an alternative source of supply is available through an existing pipeline from Sakhalin.[66] Indeed the entire debate about the future of Russia's eastern gas strategy is rather summed up by the confusion over the Vladivostok LNG plant, which appears to rely on a pipeline which only makes commercial sense if an export deal with China is also concluded.

The underlying logic of the need for a gas sales agreement with China to justify the development of Gazprom's East Siberian assets is underscored by the planned capacity of the Power of Siberia pipeline at 61 Bcm/year – that is, 52 Bcm of gas exports (38 Bcm of gas exports to China, and 14 Bcm to the first two trains of Vladivostok LNG), plus 9 Bcm for the domestic market. It would therefore seem that Vladivostok LNG should only be developed as part of an all-or-nothing strategy in East Siberia as, although Gazprom has occasionally declared its readiness to build a pipeline from East Siberia to Vladivostok just to supply the LNG plant, the commercial logic of spending so much money on such a long pipeline to a 10–15 mt liquefaction facility would seem to be incontestably flawed. Indeed Gazprom has effectively admitted as much by confirming that 'Power of Siberia' will not be built unless a deal with China is signed.[67] However, another layer of complexity and confusion has been introduced by the possibility of supplying Vladivostok LNG from reserves offshore Sakhalin Island, although the debate over this idea has again highlighted the conflict between political and commercial objectives in Russian gas strategy.

Gazprom's Sakhalin question – expansion, co-operation, or diversion?

Sakhalin Island is already home to Russia and Gazprom's one existing gas export project to Asia, the Sakhalin-2 LNG facility,[68] which reached peak output of around 10.5 mm tonnes per annum in 2011. Sakhalin Energy, which operates the licence and controls the marketing of the gas, has arranged sales with the major importers in north-east Asia, including companies in Japan, Korea, Taiwan, and China itself, establishing a price benchmark of sorts for Russian gas in the region. But the main question for

---

[65] 'Gazprom approves Vladivostok LNG investment', *Oil and Gas Journal*, 25 February 2013.
[66] 'Gazprom looks east to restore fortunes as US shale gas booms', *Financial Times*, 6 June 2013.
[67] 'Gazprom reports slow progress on China deal', *Interfax Energy*, 4 September 2013.
[68] Gazprom website at www.gazprom.com/about/production/projects/deposits/sakhalin2/ accessed on 11 November 2013.

the future is whether Sakhalin gas is set to play a greater role in Russian exports, and if so from what location. This issue has become particularly relevant since the completion of the Sakhalin–Khabarovsk–Vladivostok (SKV) pipeline in 2011, the initial purpose of which was to supply Vladivostok and the intervening domestic markets with up to 5 Bcm/year of Sakhalin gas. The source of this gas is Gazprom's Sakhalin-3 project (see Chapter 12), and the potential for new discoveries on the licence to supply up to 30 Bcm/year of gas has provided some flexibility for Gazprom as it considers its Eastern LNG strategy. It has the option to develop the SKV pipeline to its full design capacity of 30 Bcm/year and supply gas to the Vladivostok LNG facility – which may need more than 20 Bcm/year if it ultimately reaches its full 15 mtpa capacity – as well as increasing sales into the domestic market. Alternatively, Gazprom could decide to use at least part of the increased output to provide gas for an expanded Sakhalin-2 liquefaction facility which Shell, the project operator, has been promoting for some time.[69] Gazprom had announced that it would decide whether to expand Sakhalin-2 by June 2013, but its failure to do this has raised questions about whether Vladivostok LNG is now the favoured option.[70] However, the question of whether Vladivostok LNG will receive its gas from Sakhalin or from East Siberia also appears to be undecided, with conflicting announcements being made on a regular basis. During 2013, senior Gazprom managers referred first to one and then to the other as intended sources of gas for Vladivostok LNG.[71] Then Energy Minister Novak suggested that Sakhalin-3 gas would be used to feed Sakhalin projects and not Vladivostok, and implied that the choice between the two was essentially binary, as there was not enough gas to supply both from Sakhalin. He also appeared to concede that Vladivostok's gas would come from East Siberia, as the economics of the Chaianda field development and the Power of Siberia pipeline rely upon LNG exports to the Pacific – only for Aleksandr Medvedev, head of Gazprom Export, to implicitly contradict that view.[72] (The gas resources and infrastructure on Sakhalin Island are shown in Map 8.2.)

---

[69] 'Shell upbeat over its Russian future despite setbacks', *Nefte Compass*, 7 February 2013, Moscow.

[70] 'Chances fade for Sakhalin 2 expansion', *Nefte Compass*, 18 July 2013, Moscow.

[71] 'Gazprom guns for Vladivostok LNG', *Nefte Compass*, 28 February 2013, Moscow; 'Gazprom keen on Kirinsky', *Interfax*, 19 April 2013, Moscow; 'Gazprom LNG Vladivostok company to be registered soon', *Interfax*, 9 July 2013, Moscow.

[72] 'Kirinsky gas to feed Sakhalin, not Vladivostok', *Interfax*, 19 July 2013, Moscow; 'Changes in global energy markets and perspectives on the development of the oil and gas sector in Sakhalin', Aleksandr Medvedev, Sakhalin Oil and Gas Conference, 27–29 September 2013.

Map 8.2: Licences and fields at Sakhalin Island

Source: Henderson, J., *Non-Gazprom Gas Producers in Russia* (Oxford: OIES, 2010), p.191.

The Sakhalin question is further confused by the historic negotiations between Gazprom and the Sakhalin-1 partners (ExxonMobil and Rosneft) over the use of that project's extensive gas potential, estimated at 350 Bcm of proved and probable reserves.[73] The partners have consistently been keen to export their gas, and signed a prospective agreement with China as long ago as 2006,[74] but Gazprom has refused to support a deal that would have broken its export monopoly. It has made offers to purchase the gas at low domestic prices, but these have been rejected. The most logical conclusion for both parties would be to use Sakhalin-1 gas in an expanded Sakhalin-2 liquefaction project, and the fact that this solution has not, and most likely will not, be agreed upon is due to the animosity between players that is a significant aspect of the continuing conflict over Russia's eastern gas strategy.

Uncertain demand and new competition are creating difficult commercial decisions

This apparent confusion between Gazprom's LNG options and their alternative sources of supply may seem to reflect a lack of strategic thinking at Russia's largest gas company. But in reality it is the result of huge uncertainties, both in the negotiations with China and other potential buyers of Russian gas in Asia, and also within the Russian gas sector itself, where competition between rival suppliers to the Asian market is emerging. This uncertainty has combined with the continuing political drive to increase investment in East Siberia and the Russian Far East, and the ongoing need to increase export revenues to bolster the Russian budget, to produce heated debate within the Russian political elite, but as yet no concrete decisions have been reached.

As far as negotiations with China are concerned, 2013 and early 2014 have seen the continuation of the trend of optimistic forecasts of an imminent deal from the Russian side, but with no agreement being signed. Following President Putin's intervention into the discussion of Gazprom's eastern export strategy in late 2012, he also took a leading role in the first negotiations with China in 2013, as in March he welcomed the new Chinese President Xi Jinping to the Kremlin in his first overseas trip since his appointment.[75] Once again substantive discussions on a gas deal appeared to take place, with volumes, terms of deliveries, and gas quality all being

---

[73] Data from Rosneft website at www.rosneft.com/Upstream/ProductionAndDevelopment/russia_far_east/sakhalin-1/ accessed on 2 August 2013.

[74] 'Exxon agrees to sell gas from Sakhalin 1 to China', *Wall Street Journal*, 23 October 2006, Washington.

[75] 'Russian and Chinese presidents to meet on Friday', *Interfax*, 22 March 2013, Moscow.

re-confirmed, while the possibility of a pre-payment for gas deliveries was also discussed in an attempt by the Chinese to provide some financial assistance for field development and pipeline construction.[76] However, price remained a sticking point, with the Russian side still reluctant to concede on any price formation mechanism that might affect their relationships with customers in Europe. As stated by Gazprom CFO Andrey Kruglov:

> Gazprom is not willing to deliver gas to foreign markets at a price less profitable than in other markets.[77]

However, another interpretation of Gazprom's firm stance on oil-linked prices for sales to Asia and a requirement of netback equivalence with sales to Europe is that it does not want to create a precedent in Asia which European customers might use to renegotiate prices further. As a result, a Memorandum of Understanding was signed outlining a potential 30-year gas supply contract for delivery of up to 38 Bcm/year via an eastern pipeline route from 2018, but final confirmation was postponed.[78]

During 2013, the two sides publicly traded arguments on price. Gazprom CEO Miller said that he hoped that an agreement would be reached by the end of the year, using a price formation mechanism independent of US spot prices;[79] a senior CNPC manager said, on the contrary, that a price formation mechanism that includes a link to US gas prices was desirable; President Putin then reiterated Russia's firm stance on oil-linked pricing during a speech at the Gas Exporting Countries Forum,[80] and Gazprom suggested that it would rather make no agreement with China – and abandon the Power of Siberia project – than do an unfavourable deal.[81]

In September 2013, when the two sides met again at the G20 meeting in St Petersburg, another 'agreement' was signed, again confirming all the technical details of a deal.[82] However, in October further encouragement that a final agreement could be reached was received from two sources.

---

[76] 'Gazprom, CNPC discussing advance payment for gas supplies via eastern route', *Interfax*, 22 March 2013, Moscow.

[77] Ibid.

[78] 'Gazprom, CNPC sign memorandum on eastern route gas pipeline supplies to China', *Interfax*, 22 March 2013, Moscow.

[79] 'Gazprom to ink contract with China independent of US spot market prices', *Interfax*, 19 June 2013, St Petersburg.

[80] 'Hub-based pricing a big challenge for exporters – Putin', *Interfax*, 1 July 2013, Moscow.

[81] 'Better not to sell gas to China under contract in unfavourable situation – Gazprom', *Interfax*, 27 June 2013, Moscow.

[82] 'Gazprom, CNPC ink main terms for natural gas delivery to China', *Interfax*, 5 September 2013, St Petersburg.

Firstly, the Chinese government conceded publicly for the first time that its shale gas resources might not meet the timetable set in previous five year plans – implying that imports would be needed to fill the gap and providing clear encouragement to Russia.[83] Secondly, Gazprom CEO Miller and CNPC Chairman Zhou Jipin met in October and, according to Gazprom at least, agreed that the price for gas exports from Russia should be based on a formula involving oil products.[84] However, although the negotiations seemed to have some positive momentum, there was still no certainty over a final agreement,[85] and indeed Miller subsequently extended the deadline from end-2013 to the Chinese New Year (the end of January 2014).[86] Most recently, at a meeting in Beijing in January 2014, the prospects for a final signing were pushed back to May 2014, when President Putin is scheduled to be in Beijing himself,[87] with the hope that senior political intervention may provide the final impetus required for a deal. However, as of January 2014 it remains unclear whether the key commercial parameter of the gas price will be agreed in time.

Given this continuing uncertainty, the commercial logic of Gazprom's statement (that it will not build a pipeline until agreement has been reached) would appear to be irrefutable. The economics of the development of the Chaianda, and eventually of the Kovykta, gas fields, plus the construction of the Power of Siberia pipeline and a liquefaction plant at Vladivostok, can only be optimized with a combined piped gas and LNG sales strategy. Figure 8.4 shows the author's calculations for the breakeven gas price for the various schemes involving Russia's East Siberian gas fields and various export routes, and as can be clearly seen the breakeven price of LNG at Vladivostok, in the absence of a gas sales agreement with China, would be about $14.50/MMBtu before any sea transport or regasification costs are taken into account. This rather high price pertains even if one assumes that both the Power of Siberia pipeline costs are reduced by a third to account for a lower capacity requirement, and that East Siberian field developments are also downgraded accordingly. In contrast, the average breakeven price for the combined project (assuming a 10 per cent discount rate) could be as low as $10/MMBtu, although this would probably be reflected in a somewhat higher LNG price at Vladivostok to account for the additional

---

[83] O'Sullivan, S., and Henderson, J. 'UPDATE China Gas – End of the shale dream, but not of the growth plan', Trusted Sources, 8 October 2013.

[84] 'Gazprom can begin talks on western route for gas delivery to China, having agreed on eastern – Miller', *Interfax*, 22 October 2013.

[85] 'Thorny price issue remains in Russia–China gas deal', *Energy Wire*, 6 September 2013.

[86] 'Miller – China gas contract may be signed before Feb', *Interfax*, 17 December 2013.

[87] 'Gazprom, CNPC earmark Putin's May visit to China for contract signing, *Interfax*, 22 January 2014.

transport and liquefaction costs. The breakeven price for pipeline sales to China alone would be dependent on the exact timing of volume requirements and the need for the Kovykta field to be developed in addition to Chaianda, but we estimate that it would need a price of approximately $11/MMBtu. This implies that it would be possible as a stand-alone project, but only at a price much higher than Chinese negotiators currently appear prepared to sanction.

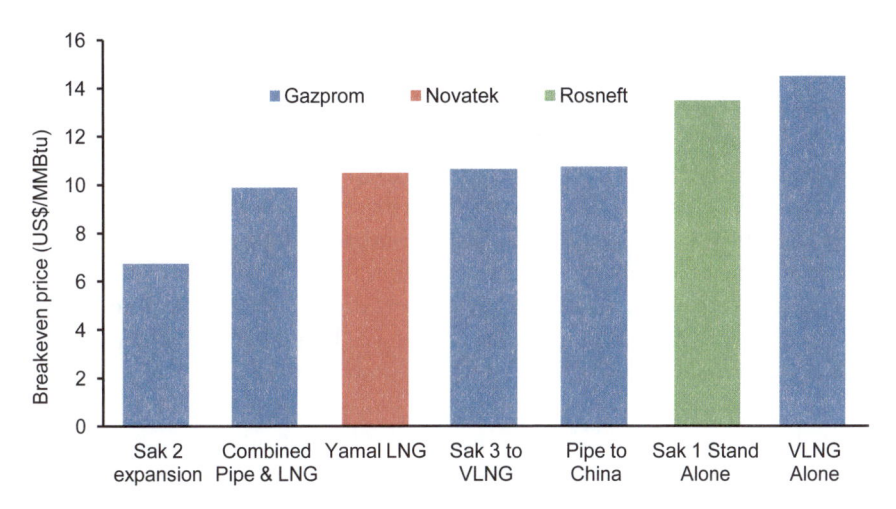

Figure 8.4: Estimates of breakeven gas prices for Russian gas to Asia
Source: Author's estimates.

LNG into Asia – another opportunity for Russia, but which companies will exploit it?

Although Gazprom appears to have multiple options for gas exports to China, the continuing lack of progress on a pipeline deal means that Russia could still find itself excluded from a significant role in the Chinese gas market. However, even if a deal for pipeline gas is not forthcoming in the short term, Russia can still compete for new LNG contracts or spot sales that are included in the 'LNG Upside' bar in Figure 8.1 above. Indeed, a few spot LNG cargoes from the Sakhalin-2 project have already reached the Chinese market, and Gazprom's planned Vladivostok LNG terminal clearly has China as one of its main targets. Beyond China, the Japanese and Korean markets are already receiving the bulk of Sakhalin-2's LNG output, and Gazprom has been in negotiation with both countries to expand supplies as it develops new resources. India has also become an attractive target for Russian gas, with Gazprom's trading company Gazprom Global LNG signing a 20-year deal to supply GAIL with 2.5 mtpa in 2012,[88] while spot cargoes have also been sold to countries as diverse as Thailand and Kuwait.

However, the growing importance of LNG as a potential driver of Russian gas exports to Asia raises the intriguing possibility that, despite its strategic role in negotiations with China and its existing sales from the Sakhalin 2 project, Gazprom may not ultimately be the driving force of Russia's expansion into the Asian gas market. Two other Russian companies, Novatek and Rosneft, are planning to develop major LNG projects in Russia with a focus on sales to the Asian market (see Map 8.3), and both companies appear to be receiving support in their plans from the federal government (see Chapter 13 for more detail). Indeed Rosneft CEO Igor Sechin (a close ally of President Putin) who acts as Secretary of the Presidential Energy Commission, has argued that Gazprom's LNG export monopoly should be lifted specifically so that his company can export gas to Asia. Furthermore, he added that as Asia is not yet a major market for Gazprom, such a move would not harm it, with the clear implication that other Russian companies should be allowed to compete for customers there.[89] He further underlined this point by confirming that the proposed LNG liquefaction plant at Sakhalin-1 would specifically target Japan and South Korea.[90]

The Russian government appears to be adopting a similar approach, with the debate around the potential liberalization of LNG exports no longer focused on whether it should happen (as the legislation has now been approved by the government[91]) but rather on how any future sales could be ring-fenced to Asia rather than Europe. Despite the difficulty in actually achieving such a result in practice, the Russian Energy Ministry has proposed the formation of a commission to co-ordinate LNG exports, to avoid competition between different suppliers,[92] with Minister Novak stating that:

> ... for now there is no need for independent producers to ship more gas to Europe as Gazprom ... has a strong position there

but implying that any Russian company could export LNG to other markets such as Asia.[93] Novatek and Rosneft have done their best to avoid

---

[88] 'Gazprom, India's GAIL agree 20 year LNG sales deal', *Reuters*, 1 October 2012, Dubai.
[89] 'Sechin requests liberalisation of LNG exports for Shelf, doesn't think Gazprom will suffer', *Interfax*, 13 February 2013, Moscow.
[90] 'Rosneft expects to be producing LNG on Sakhalin no later than 2018', *Interfax*, 11 April 2013, Moscow.
[91] 'Russian govt. approves bill on LNG export liberalization', *Interfax*, 30 October 2013.
[92] 'Energy Ministry could co-ordinate gas supplies, LNG exports when market liberalised', *Interfax*, 19 April 2013, Moscow.
[93] 'Govt. could resolve disputes arising from liberalisation of LNG exports', *Interfax*, 9 July 2013, Moscow.

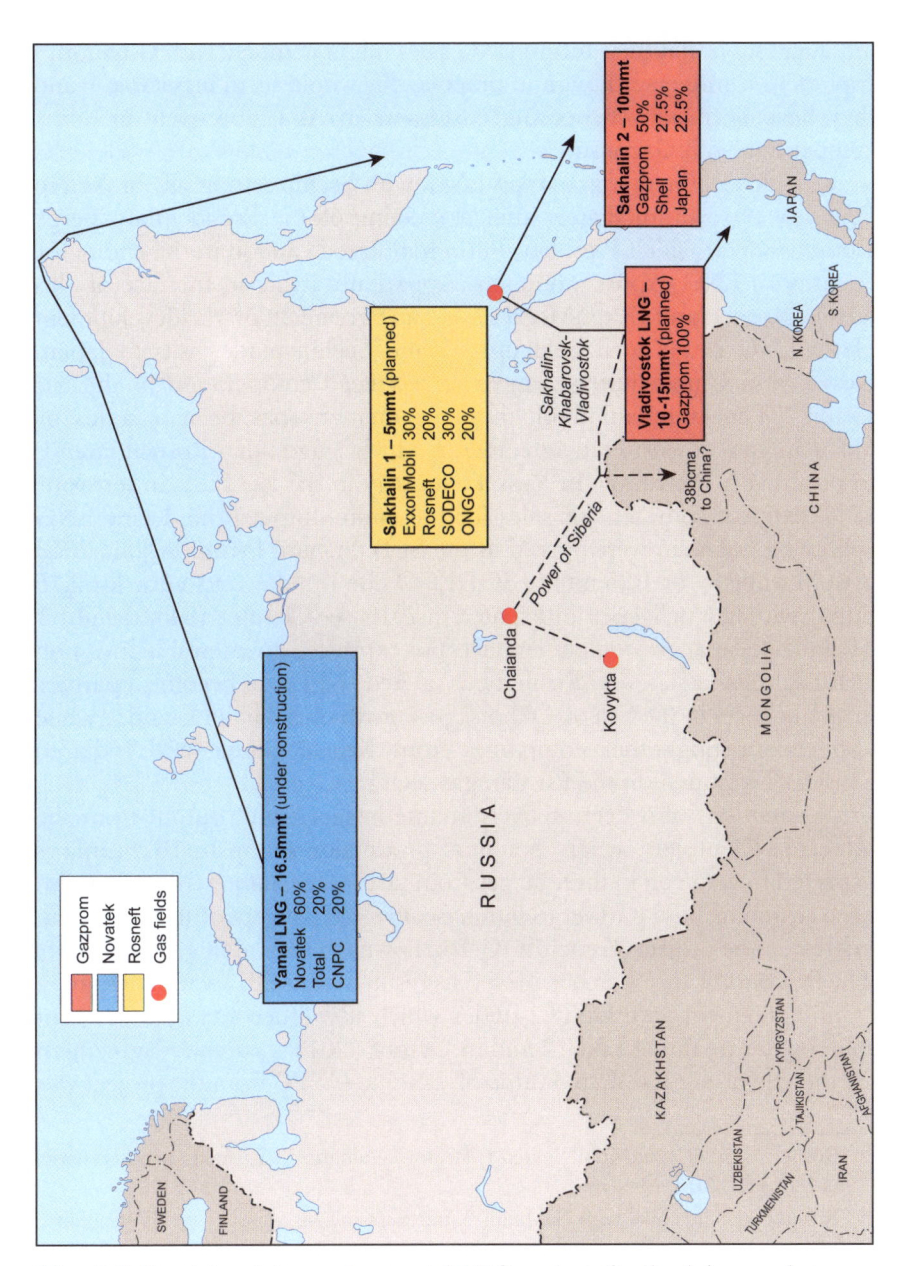

Map 8.3: Russia's existing and potential LNG projects for the Asian market
Source: Author.

any suggestion that their future LNG sales might compete with Gazprom's exports to Europe – although in practice this would seem inevitable – and they have been actively seeking customers in Asia, effectively in direct competition with Gazprom.

Rosneft was the first to market LNG, with Sechin setting off for Asia in February 2013, immediately after a meeting of the Presidential Energy Commission at which President Putin had urged caution in the ending of Gazprom's LNG export monopoly. Apparently flying in the face of this advice, Rosneft and ExxonMobil signed an agreement on the development of an LNG complex at Sakhalin-1[94] and Sechin made visits to Japan, Korea, and China during, which he discussed co-operation in the gas sector.[95] The culmination of these negotiations has been a series of upstream and downstream agreements, which suggest that Rosneft intends to play a very active role in Russia's strategy in the Far East. In terms of LNG exports, contracts for sales for the entire output from a new LNG facility on Sakhalin were agreed at the St Petersburg Forum in June 2013 with Marubeni (for 1.25 mtpa), SODECO (for 1 mtpa), and Vitol (for 2.75 mtpa), with first deliveries anticipated in 2019 (see Chapter 13 for details).[96] Meanwhile Japanese companies have also expressed an interest in investing in Russian gas assets, and Rosneft has invited INPEX to become a partner in two licences in the Sea of Okhotsk just north of Sakhalin Island,[97] while also encouraging other companies from Korea, China, and India to consider investment in the Russian gas sector.

Novatek has also been actively marketing its future output from the Yamal LNG project, where potential production of up to 16.5 mtpa is expected to come on in three stages from 2017 (see Chapter 13). It has also been searching for a partner to purchase up to 30 per cent of its 80 per cent interest in the project (French IOC Total owns the remaining 20 per cent). The first part of this search achieved its result in June 2013 with the signing of an agreement with CNPC, under which it will become a 20 per cent shareholder in Yamal LNG,[98] and in October 2013 a purchase agreement for the sale of not less than 3 mtpa of LNG to CNPC from the project was

---

[94] 'Rosneft and ExxonMobil extend strategic alliance', Rosneft press release, 13 February 2013, Moscow.

[95] 'Rosneft to hold talks with Sakhalin 1 partners on gas supplies for LNG plant', *Interfax*, 11 April 2013, Sakhalin.

[96] 'Rosneft urges Energy Ministry to ensure export liberalisation for three LNG contracts', *Interfax*, 21 June 2013, Moscow.

[97] 'Rosneft, INPEX agree to work together on Sea of Okhotsk shelf', *Interfax*, 29 May 2013, Moscow.

[98] 'Novatek brings CNPC into Yamal LNG, signs long term supply contract with China', *Interfax*, 21 June 2013, St Petersburg.

confirmed with the signing of a Heads of Agreement.[99] The overall deal was completed in January 2014, once the LNG export monopoly had been lifted. It marks a crucial moment for the Yamal project, as it makes possible significant financial support from Chinese banks. However, it also potentially reiterates to the Russian administration the fact that Gazprom may not be the preferred Russian partner for Asian gas importers, given the pointed contrast between CNPC's long, and to date fruitless, negotiations with Gazprom and its new deal with Novatek. Furthermore, according to Russian Energy Minister Novak, Novatek has apparently also secured further agreements for sales of up to 70 per cent of the Yamal volumes, including deals with India and various trading companies.[100]

Conclusions

Russia's relationship with the Asian gas market was initially catalysed by Sidanco, in partnership with BP, who were keen to develop the Kovykta field in Irkutsk in the early 2000s and open a pipeline route to China. Gazprom's export monopoly meant that this goal became unrealizable without the state gas company's involvement, and for the past decade Gazprom has sought to reach an agreement that can link Russia's huge East Siberian resource base with the growing demand for gas in China and north-east Asia. However, despite numerous memoranda of understanding and strategic agreements, no contractual commitments have been made, mainly because the two key players, Russia and China, each view themselves as having the stronger bargaining position.

Gazprom sees rapidly growing Chinese gas demand driven by economic growth and environmental concerns and therefore assumes that Russian gas must be a necessity to satisfy consumers and complete China's 'compass of diversification' with a northern import option. It therefore feels justified in demanding a high oil-linked price for its gas, which would also allow it to satisfy its other strategic imperative, which is to avoid undermining its export strategy in Europe. CNPC, on the other hand, clearly views East Siberian gas as a stranded resource with no other realistic market than China, and therefore is not inclined to pay a premium price for it. This inclination is reinforced by two other strategic issues. Firstly, China is in the middle of a process of reforming its domestic gas pricing structure and is therefore reluctant to take on expensive long-term import contracts that could exacerbate the losses that CNPC and others are already making on existing imports. As a result, CNPC is demanding some link to market prices in any future pipeline-based contract with Russia, and although this link is being

---

[99] 'Conclusion of heads of agreement on LNG supply with CNPC', Novatek Press Release, 22 October 2013.

[100] 'Novatek tentatively contracts half of LNG – minister', *Interfax*, 2 July 2013, Moscow.

characterized by the potential adoption of US Henry Hub prices as a benchmark, in the future it could equally become a link to a domestic gas hub on the east coast of China, where pipeline imports, LNG deliveries, and indigenous supply all converge.

Secondly, over the past five to ten years China has developed a multi-vector supply strategy that has seen it bring gas from Central Asia in the west, Burma in the south, and LNG on its east coast, to combine with indigenous conventional and unconventional production to meet its growing demand. Furthermore, Chinese companies have begun actively to invest overseas to ensure that the country has an interest in many of the new projects that could supply imported LNG, further strengthening its security of supply. This strategy has been so successful that, under some demand scenarios, there may be no place for high-priced Russian pipeline gas until well into the next decade or beyond, and as a result the Chinese negotiators presumably feel little urgency to close a deal – unless the price is competitive and other concessions, such as equity participation in upstream assets, are made. The one big uncertainty in this picture is Chinese domestic supply, with the development of unconventional resources being critical to the overall picture, and although recent announcements suggest that the expectations for shale gas in particular may be downgraded, CNPC still appears to be prepared to wait to see if development plans are successful, rather than commit to high-cost imports at this stage.

However, although the outcome of negotiations for pipeline exports into China, and potentially on to South Korea, remains unclear at the time of writing, Russia clearly still has the opportunity to sell gas into China and the rest of Asia in the form of LNG. Gazprom has seen this as an opportunity to demonstrate to its Chinese counterparts that it does have alternative markets for its eastern gas, and has promoted the development of the Vladivostok LNG terminal as a means of monetizing this resource. However, the 3200 kilometre distance from the gas fields of East Siberia to the liquefaction plant clearly undermines the economic rationale for this scheme, unless a deal with China is signed that can underpin the construction of the 'Power of Siberia' pipeline. This has led to a rather confused Gazprom LNG strategy in the east. The source of gas for Vladivostok LNG has switched from East Siberia to the new discoveries in the Sakhalin-3 licence and back again, while the potential expansion of the Sakhalin-2 project adds complexity to the debate.

Finding a resolution is a complex issue for Gazprom as it combines not only economics but also the geopolitics of relations with Asia and the domestic political need to encourage growth via new infrastructure in Russia's eastern regions. However, the lack of certainty means that

Gazprom has so far failed to secure a significant place in the Asian gas market. By contrast, Novatek and Rosneft are starting to take on the role of marketers and sellers of Russian gas in the region. Both companies have prospective new LNG schemes and have been told by government that, even though Gazprom's monopoly of LNG exports has been lifted, they should avoid competing with Russian exports to Europe if at all possible. This has left Asia as their prime target, and both companies have signed deals with buyers from the major importing countries, including China.

An interesting, but as yet unanswerable, question is whether the initial success of Novatek and Rosneft can be a precursor to either company, but more particularly Rosneft, being given a greater role in Russia's eastern gas strategy if Gazprom continues to procrastinate. Rosneft has already set an example of how pipeline deals in the region can be done, through its successful exploitation of the ESPO line, and it has a burgeoning relationship with China as sales of crude and oil products continue to increase. At this stage Gazprom, whose strategy is of course largely controlled by President Putin, would not appear to be under any threat of losing its dominant role, and indeed the president has been keen to point out that the company is close to signing deals with Japan and China. However, at the time of writing in January 2014 the timetable for a final agreement with China has been pushed back once more, this time to May 2014, raising further doubts about Gazprom's ultimate ability to get a deal done. While it seems most probable that 2014 will see a final conclusion to the Gazprom–CNPC negotiations, if Gazprom does fail to secure an export deal, which is becoming increasingly important both to the company's strategy and to future Russian budget revenues, then it is perhaps not inconceivable that a change of strategy in favour of increased non-Gazprom involvement in Russia's Asian strategy may rise up the political agenda.

CHAPTER 9

SUMMARY: THE INFLUENCE OF MARKETS ON RUSSIA'S GAS SUPPLY STRATEGY

Simon Pirani

This brief chapter summarizes, and points to common themes in, Chapters 3–8, which dealt with the European, Russian, CIS, and Asian markets into which Russian gas is sold; and highlights ways in which changes in these markets influence Russian gas supply. This provides a link for the reader to Part 3 (Chapters 10–15), in which the sources of gas supplied to the Russian gas balance (Gazprom's production, the non-Gazprom producers, and Central Asian and Caspian imports) are analysed.

In many ways it is easier to see the vast differences between the European, Russian, CIS, and Asian markets for Russian gas than to see the similarities. Throughout the history of the Soviet and Russian gas industry, up to the present, a starting-point for analysis was the completely different economic functions of Russian gas in its two main markets, European and domestic. The principal function of the exports to Europe was to support the state budget and the development of new gas resources by maximizing revenues; Gazprom's export strategy in 2009–12, of resisting concessions on price at the expense of volume, reflects this traditional role. Although the Russian state has used its strong position in gas markets, particularly in eastern Europe and former Soviet countries, to strengthen its hand politically, revenue-raising has been and remains the most important function of gas exports (Chapters 4 and 7). In contrast, the economic function of gas in the domestic market throughout the post-Soviet period has been, above all, as a cheap input for industry (with a view to enhancing its development) and (directly and via cheap electricity and heat) as a social benefit to the population.

Price levels are another gauge of the deep differences between the Russian and European markets. While the Russian price reforms, begun in the mid 2000s, have gone a considerable way to closing the gap (Chapter 5), the differences in the prices paid by consumers remain striking. Russian power sector and industrial customers paying the highest domestic prices, however far they are from the Siberian gas fields, still pay less than half the prices charged to the best-placed European gas consumers. Russian domestic prices now compare unfavourably with those in the USA (Chapter 1) – but that is due to the exceptional American market circumstances brought about by the 'shale gas revolution'.

Notwithstanding the differences, the argument made in this book is that *common trends* in *all* the markets into which Russian gas is delivered are framing the evolution of supply: 'markets are driving change', as the title says. Four such trends run through Chapters 3–8: changes in demand arising from the 2008–9 economic recession; the influence on demand, in volume terms, of governments' energy policies; changes in market regulation and the political framework; and changes in price formation and price level. The extent to which they are common to the different markets may be summarized as follows:

Changes in demand arising from the recession have been crucial. In Europe, the sharp fall in demand, combined with supply-side developments (the 'shale gas revolution' and availability of extra LNG), drove down gas prices formed on a market basis. This provoked a crisis between gas buyers and Gazprom over oil-linked pricing in long-term sales contracts. There followed the first serious reverses in the constant expansion of Russian gas sales to Europe since they began, and an uncertain outlook for the rest of this decade (Chapter 3). Falling demand due to the recession was also a central factor in Ukraine, the largest CIS gas importer (Chapter 7), and in the Russian domestic market, where steady pre-crisis demand growth of 2–2.5 per cent per year gave way to a 0.3 per cent annual average in the period since then (Chapters 5 and 6). It is Asian demand that continues to expand rapidly, and the future of Russia's gas sector will be determined partly by the outcome of its long-running negotiations on sales to China: this could still turn out to be a huge missed opportunity, or a factor of regeneration, or a combination of the two (Chapter 8). In the meantime, the reductions in European and CIS demand have already led to the adoption of a more gradual timetable for the Yamal peninsula development, Russia's largest new supply-side venture since the opening up of west Siberia in the 1970s, and they also underlie the new, competitive relationship between Gazprom and the non-Gazprom producers.

The influence on demand, in volume terms, of governments' energy policies is more complex. In Europe, government support for renewables has helped to subdue gas demand, and increase uncertainty, raising the possibility that gas demand is in 'terminal decline' (Chapter 4). In Ukraine, the government's policy of diversifying away from Russian gas imports has begun to make a noticeable impact, reducing those imports substantially in 2009–13 and helping to force a price concession in the agreement of December 2013 (Chapter 7). In Russia, the situation is different: policies aimed at reducing gas's share of the primary energy balance have failed, and while new nuclear power generation capacity will challenge gas in some areas, gas will continue to dominate thermal power generation, district heating, and energy consumption by industry (Chapter 6). On the other hand, energy market reform and efficiency policies are among the factors that will help keep demand growth low. Government policy is important in China: the drive to shift power generation from coal to gas, as well as economic growth per se, remains a magnet for Russian gas imports, despite the failure of the negotiations so far.

Changes in market regulation and the political framework are very different in the European, Russian, and CIS markets. In Europe, successive legislation aimed at market liberalization has posed major challenges to Gazprom's commercial strategy (Chapter 4); in Russia, regulatory reform of both the electricity sector and energy (especially gas) markets, has been a less direct but no less powerful catalyst of change (Chapter 5); in the CIS, the influence of the importing countries' regulatory reform has been less significant and less straightforward, but the transformation of the political framework for gas transit – and specifically, the Russian government's support for transit diversification – has suppressed some of the economic distortions built into the system left over from Soviet times and strengthened market forces (Chapter 7).

Changes in price formation and price level, which have themselves been influenced by the economic recession, are arguably the most important common theme. In Europe, the shift away from oil-linked price formation towards gas-to-gas pricing on 'hubs' has resulted in substantial changes to Gazprom's commercial relationships (Chapter 3). In the CIS, most cross-border import prices have moved away from bilaterally negotiated levels towards price linkage with the European market. In Russia – at least for power generation, heat, and industry – the heavy discounting of consumer prices is being phased out. Prices were to have been linked to those in Europe, and although that aim has now been superseded, prices have risen to above cost levels and are increasingly reflective of supply and demand (Chapter 5). Despite the considerable differences between these transitions, there are two underlying themes. First, the prices at which most Russian gas is sold – whether export prices or domestic consumer prices – did not directly reflect market fundamentals at the beginning of the 2000s, and by the 2010s 'that failure to reflect fundamentals has become important and, in many countries, unsustainable'. Second, as the Ukrainian case shows and the Russian case is beginning to show, 'it will become increasingly difficult for countries which participate to any significant extent in gas trade [...] to divorce their domestic prices from international levels', as recent research by the OIES Natural Gas Research Programme concluded.[1] These trends – towards prices that reflect market fundamentals, and away from domestic prices that are held at levels unreflective of international prices – are driving change in the supply side.

---

[1] Stern, J.P. 'Conclusions', in Stern, J.P. (ed.) *The Pricing of Internationally Traded Gas* (Oxford: OIES/OUP, 2012), here pp. 487–8.

Internationally, while the large regional gas markets (Europe, the former Soviet Union, Asia, the Americas) remain separate, they are starting to be linked by the growth of the LNG trade. This trend will have manifold indirect impacts on the supply strategy of Russia, one of the world's largest producers – for example, via the knock-on effects of the US 'shale revolution' on the European market, and a possible opposite effect in the future (Chapter 4). If and when Russian exports to Asia expand substantially, the impact of gas market globalization will become more direct. There is no direct linkage between the prices of Russia's current exports to Asia – LNG from Sakhalin – and prices in the European and CIS markets, and in this decade any linkage will remain negligible or weak. But the situation could change dramatically during the 2020s (Chapter 8). By then, gas will be flowing to China in substantial volumes from Central Asia, possibly from all three producing countries; Russia may have started up LNG exports via several other routes; and decisions may have been taken about a pipeline route to China.

The way in which Russia's supply strategy is shaped by these common factors – lower demand in the west, growing demand in the east; government energy policies; changes in market regulation and political frameworks; and changes in price formation and price levels, and linkage between regional markets – is the subject of Part 3.

Part 3
Russia's Gas Supply Strategy

CHAPTER 10

SOURCES OF RUSSIAN GAS SUPPLY

James Henderson

This is the first of six chapters that discuss the sources of supply for the Russian gas balance, and the impact upon these of the changes in demand described in Part 2.

Any mention of Russian gas supply immediately leads to an analysis of its dominant player, Gazprom, which has effectively controlled the industry during the entire post-Soviet era. The company was created in 1989 by the then Minister of Gas Viktor Chernomyrdin, and became a joint stock company in 1992, before its partial privatization in 1993,[1] but its links to the Russian government have always remained strong. Indeed many have argued that it has continued to act as a quasi-Ministry of Gas for the past 20 years, struggling to balance the need to satisfy the political objectives of its masters and the commercial goals of its shareholders. Nevertheless, Gazprom's control over the vast majority of Russia's Soviet era gas assets has historically given it a huge competitive advantage over any potential rivals, due to the fact that it inherited fields at the peak of their productive capacity where almost all the capital expenditure was already a sunk cost. Indeed, this fact was the key reason why the company was able to so successfully provide the foundation upon which Russia's collapsing economy was able to lean in the chaotic period of the 1990s. Gazprom did not have to spend much money to maintain a level of gas production that could sustain the needs of the Russian population (even in winter) and also of Russian industry, essentially by providing gas to both at heavily subsidized prices.

Gazprom's control over the Russian gas sector was based not just on its dominance over supply, however, but also on its ownership of the trunk pipeline system and, as a result, its monopoly of gas exports to both CIS and, more importantly, non-CIS countries.[2] Under Russian law it is obliged to offer proportional access to third parties for the transport of gas domestically, but in effect the lack of transparency concerning capacity

---

[1] Stern, J.P. *The Future of Russian Gas and Gazprom* (Oxford: OIES/OUP, 2005), p.170.
[2] Gazprom's monopoly over pipeline exports was enshrined in law in 2006 (Stern in Pirani, S. (ed.), *Russian and CIS Gas Markets and their Impact on Europe* (Oxford: OIES/OUP, 2009), pp. 79–80), but prior to that date it had a de facto monopoly thanks to its ownership and control of the trunk pipeline system in Russia.

utilization in the system has often been used as an excuse to avoid moving any gas that Gazprom has regarded as a threat to its business. This effective control over gas movements in Russia has been combined with its sole access to the much higher prices available for gas exports to Europe (albeit somewhat offset by its requirement to sell low-priced gas in Russia). Gazprom has therefore dominated the Russian gas sector historically, through its control of the country's gas resources, productive capacity, control of infrastructure, and access to high-value revenues.

However, even as early as 2005, when Jonathan Stern published *The Future of Russian Gas and Gazprom*,[3] he noted that the future of Russian gas supply would become much more a question of balance between a number of different sources of supply than one of dominance by a single player, with 'Independent' gas production and imports from Central Asia playing increasingly important roles. As has been described in the chapters above, the Independent producers have indeed increased their significance as sellers of gas into the Russian domestic market, and as Figure 10.1 shows, Central Asian imports have also had a role of fluctuating significance.

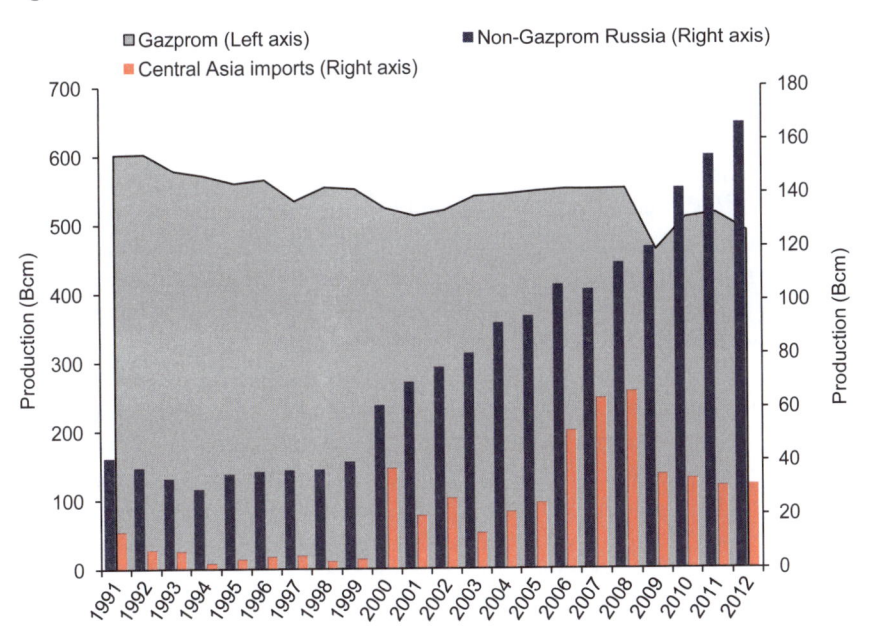

Figure 10.1: The main elements of Russian gas supply since 1991
Source: *Interfax*, Energy Intelligence Group, Russian Ministry of Energy.

---

[3] Stern, J.P. *The Future of Russian Gas and Gazprom*, op. cit., pp. 203–4.

Throughout the 1990s Gazprom accounted for 92–95 per cent of all supply into the Russian gas balance, with the remainder being made up of a minimal amount of imports from Central Asia (3–5 Bcm per annum) and associated gas from the oil producers, which at around 40 Bcm/year was the total of non-Gazprom output in Russia. Clearly with this share of overall production, combined with its political backing and ownership of the pipeline system, Gazprom dominated decision-making in relation to alternative supplies. But, as can be seen from Figure 10.1, its needs did begin to change towards the end of the decade as the gradual decline in its own production left it needing to turn to third parties for extra supply to meet growing demand from its domestic and export markets. In a sense, this period around the turn of the millennium was the first example of markets starting to drive Gazprom strategy, although without impinging on Gazprom's overall control of the Russian gas sector.

Two key events coincided to cause this change in the supply balance, namely the onset of significant decline at Gazprom's major Soviet fields[4] (the Big Three: Medvezh'e, Urengoi, and Yamburg) and the start of the rebound in European and Russian gas demand after the 1998–9 economic crisis. With demand growth in Europe being especially strong, thanks to the rapid expansion of gas-fired power across the continent,[5] Gazprom was very keen to encourage producers to provide gas for the Russian domestic market in order to free up more of its own gas for export, thus avoiding the need to develop high-cost fields on the Yamal peninsula for as long as possible. Although, as will be described in Chapter 11, Gazprom did start production in 2001 at one new super-giant field, Zapoliarnoe, its overall strategy was to offset the underlying decline in its existing fields with a series of low-cost alternatives – satellite fields close to its own infrastructure, cheap Central Asian imports, and third-party domestic gas that it could pass through to Russian consumers. In the meantime, Gazprom was able to recover its financial stability, which had suffered significantly in the 1990s due to non-payments from domestic customers and lower European revenues resulting from the fall in oil prices after the economic crisis.

As a result, both third-party production and imports from Central Asia jumped higher in 2000, with the former continuing this trend throughout the subsequent 12 years, while the performance of the latter has been much more volatile. The situation with Central Asian imports is somewhat complicated because in the early years of this period Gazprom had, in effect, delegated this business to a series of trading companies, which

---

[4] Landes, A. *Gazprom: The Dawning of a New Valuation Era*, Renaissance Capital, Moscow, 2002, p.59.

[5] Honoré, A. *European Natural Gas Demand, Supply and Pricing: Cycles, Seasons and the Impact of LNG Price Arbitrage* (Oxford: OIES/OUP, 2010), p.27.

included Itera and Eural Trans Gas. As a result, until 2005, the increase in imports from Central Asia reflects the demand for gas from the FSU countries to which Itera and Eural Trans Gas were exporting it. Gazprom reclaimed control of the import–export trade in 2005 following a dispute over gas sales from Turkmenistan. From then, Gazprom would buy gas to level out imbalances in its own supply portfolio, and also to manage the political leverage that Russia could exert through being the only purchaser of gas from the region, and therefore a major provider of revenue. However, as we discuss in Chapter 14, the balance started to shift in 2006 for two key reasons. Firstly, Russia's (and Gazprom's) need for gas became more urgent as demand continued to rise and Gazprom's strategy of delaying investment in new fields started to cause concern over its ability to meet consumer needs. Not only did imports of Central Asian gas increase but so did the price paid, until it effectively reached export netback parity. By this time Gazprom had rather lost control of the situation, and this was emphasized during and after the 2008–9 economic crisis when it could no longer afford to pay high prices for imports and took desperate measures to halt the flow, including refusing to re-start imports from Turkmenistan after a pipeline explosion in 2009.[6] In addition, the second reason for the shift in balance between Gazprom and the Central Asian gas suppliers had also become apparent: Turkmenistan had already signed its first gas export deal with China and the Turkmen–China pipeline was under construction. Following the events of 2009 this east-facing strategy was accelerated and all the Central Asian countries turned east to find an alternative, and securer, source of demand in Asia, with the result that Russia and Gazprom have essentially been relieved of much of their influence in the region.

As far as third-party producers in Russia are concerned, they were initially encouraged either to sell extra gas to Gazprom, or to sell gas to consumers at premium prices above Gazprom's low regulated price. Novatek led the 'Independents', as they are officially called in Russian legislation, in co-operating with Gazprom in this way, always providing gas as directed rather than offering any form of competition to the state-controlled sector leader. As a result, even though the decline in Gazprom's own production was halted in the mid-2000s, the growth in Independent production accelerated, as Gazprom increasingly turned to this apparently subordinate alternative source of supply.

However, just as a combination of events – Gazprom's decision to pay netback prices for imported gas, followed by the impact of the 2008–9 economic crisis – changed Gazprom's relationship with Central Asian producers, so another – President Putin's call in 2006 for domestic prices to

---

[6] 'Turkmens say Gazprom broke deal', *The Moscow Times*, 13 April 2009.

also reach export netback parity[7] – catalysed a change in strategy from Independent producers. Although Independent gas production had been increasing since 2000, this had largely been at the behest of Gazprom as it sought to defer its spending on high-cost new fields. However, once it became clear that domestic prices would be on a continuing and rapid upward slope towards a target of export netback parity, the prospects for sales into the Russian domestic gas market looked even brighter, and sparked the non-Gazprom sector to adopt a more proactive strategy. As a result, although Gazprom itself had called for the change in pricing strategy, to generate increased revenues for the development of the high-cost fields on the Yamal peninsula, rising prices also encouraged another natural market reaction, namely increased output from alternative producers. Unfortunately this emergence of a domestic incentive to increase output also coincided with the sudden halt, and then decline, of demand for Gazprom's gas on export markets following the 2008–9 financial crisis and the surge in LNG imports to Europe as a result of the US shale revolution. As a result, the seeds for a significant change in the balance of power in the Russian gas sector had been sown. Demand had fallen just as investment in supply was increasing, with the inevitable consequence that, over time, too much Russian gas was made available to the market.

The following chapters discuss the consequences of this sequence of events, first addressing the impact on the sector's major player, Gazprom. As Figure 10.1 shows, Gazprom's production in 2012 is more than 110 Bcm lower than the level of 1992. This decline has not been linear, and has reflected many factors including shifts in demand patterns across Gazprom's markets and the switching of the company's production away from Soviet-era fields towards a combination of satellite fields and large new developments. Fundamentally lower Gazprom output is not necessarily a sign of the company's failure, of course, and indeed we will argue that its strategy of managing a portfolio of equity and third-party production has been very logical. However, in Chapter 11 will also discuss how Gazprom, having finally decided that it needs to underpin its own future gas production by investing in the Bovanenkovo field on the Yamal peninsula, now finds itself tied into a high-cost development strategy at a time when the market for Russian gas is becoming much more competitive. Faced with the possibility of becoming a marginal swing producer towards western markets, Gazprom has sought out alternative routes to growth, and Chapter 12 analyses how the Asian countries could offer significant potential to sell the company's large East Siberian and Far East reserves into a rapidly growing market. However, the chapter also discusses how the domestic political, vested interest, geopolitical, and management issues that have

[7] Henderson, J. 'Domestic Gas Prices in Russia', op. cit., p.13.

undermined Gazprom's business in its traditional markets could also affect its ability to adopt a commercial strategy in the East – potentially having a negative impact on the strong financial base that the company has built up over the past few years thanks to high commodity prices.

Chapter 13 then discusses the role of the Independents, highlighting how Novatek and Rosneft have come to the fore in recent years and are effectively creating a new triumvirate in the Russian gas sector. Their impact in the domestic market has been discussed in Chapter 5, but on the supply side both companies have plans to double production over the rest of the current decade and, perhaps more importantly, are keen to sell this increased output in the export as well as the domestic market. Novatek's growth model is especially reliant on its Yamal LNG project, with its sales to Asia and Europe, and Rosneft also sees the Asian gas market as a driver of company growth, where its ability to succeed in a region where it already has strong relationships through its large oil exports could provide a significant indicator concerning the future of the Russian gas sector as a whole.

In Chapter 14 we review the declining role of Central Asian imports as part of the Russian gas balance, as the influence of China in the region increases. However, despite the relatively low volumes of gas arriving in Russia, we highlight two important features of Russia's relationship with the region. Firstly, we note that Russian companies are becoming involved in exports of gas from the region to China, providing more potential competition to Gazprom. Secondly, and perhaps more importantly, we question whether Gazprom's continuing purchase of gas from Central Asia at high prices may be an attempt not only to retain some political influence for Russia, but also to affect the export price to China, from which it could benefit if and when it agrees its own export deal with the world's fastest-growing gas market.

Chapter 15 then provides a conclusion to the supply section, highlighting in particular how the reaction of consumers across all the markets for Russian gas, combined with the willingness of various producers to react in an increasingly competitive fashion, is causing fundamental changes in the shape of the Russian gas supply portfolio.

CHAPTER 11

GAZPROM'S WEST-FACING SUPPLY STRATEGY TO 2020

Jonathan Stern and James Henderson

Introduction

Throughout its history Gazprom's supply strategy has been dominated by the need to provide gas to markets either in the west of Russia or in foreign countries west of Russia. The concept of east-facing production only emerged in the very recent past, when the Sakhalin-2 LNG plant went into operation in 2009, but the reality remains that for the remainder of the 2010s the vast majority of Gazprom's output will be transported westwards from its core production area in the Nadym Pur Taz (NPT) region of West Siberia. Indeed this chapter will argue that Gazprom's strategy has essentially been fixed for the next few years by its decision to develop the huge gas resources on the Yamal peninsula, which will, by 2030, provide more than half of its gas output. However, it is important to understand that Gazprom's future shape has been largely determined by decisions taken both inside and outside the company over the past twenty years, and also by market forces beyond Gazprom's control. Before we consider the state of the company's current portfolio and its future trajectory, we should briefly review how the company developed during its early years and what forces have driven its strategy to date.

In the aftermath of the collapse of the Soviet Union, as the Russian economy went into sharp recession and oil production fell by almost half, Gazprom and the gas industry provided a relative haven of stability, supplying more than half of the country's energy needs from the legacy assets it received from its Soviet inheritance. Although gas production did fall by 11 per cent from its 1990 high of 643 Bcm to a 1997 low of 570 Bcm, this was a minor blip compared to the 42 per cent fall in oil output, and was just reflecting the reduced demand from a contracting economy. Furthermore, during the 1990s, Gazprom was producing around 95 per cent of all Russia's gas, with the remainder being provided by associated gas from oil companies (see Chapter 13), and the foundation of Gazprom's stable output was provided by three giant fields that had been developed in the Soviet era, namely the 'super-giants' Medvezh'e, Urengoi, and Yamburg. These three had produced almost all of Gazprom's gas during the latter part of the Soviet era, and even by 2000 still comprised 75 per cent of the company's output. However, the increasing maturity of the fields (Medvezh'e had come onstream in 1972, Urengoi in 1978, and Yamburg in 1986)[1] combined with a lack of investment during the 1990s – as Gazprom struggled with non-payment issues and the need to provide finance to the Russian state – meant that by the early to mid-2000s their production was declining at around 18–25 Bcm/year.[2] Gazprom attempted to address this issue by developing small satellite fields around its three super-giants in order to exploit the infrastructure that already existed in West Siberia. But by 2001 it had become clear that this strategy was not working, as the company's output hit a post-Soviet low of 512 Bcm.

Gazprom's initial response was to undertake its first major development project of the post-Soviet era, bringing a fourth super-giant field, Zapoliarnoe, on stream in 2001, and combining this with the continued development of satellite fields. This combination allowed Gazprom to create total additional capacity of 218 Bcm/year, which enabled its overall production to recover and stabilize at a plateau of 550–560 Bcm/year.[3] Zapoliarnoe itself had an initial production capacity of 100 Bcm, but this was increased to 115 Bcm in 2007, and further to 130 Bcm in 2013, following development of its Valenginian reserves, making it the largest single producing field in Russia.[4] Map 11.1 shows the three original super-giant fields in Gazprom's portfolio as well as the newer Zapoliarnoe, as well as the other major gas fields in Western Siberia which together still account for more than 90 per cent of Russian gas production.[5] Indeed even as late as 2012, all West Siberian production came from a relatively small group of fields in the Nadym Pur Taz (NPT) region, on and around the Taz peninsula between the Ob and Taz Bays. The map also shows the huge pipeline network constructed during the Soviet era to transport the gas from these

---

[1] Stern, J.P. 'The Russian Natural Gas "Bubble": Consequences for European Gas Markets', RIIA Chatham House, 1995, p. 7.

[2] Stern, J.P. *The Future of Russian Gas and Gazprom* (Oxford: OIES/OUP, 2005), Tables 1.3 and 1.5, 6 and 9.

[3] Presentation by Alexander Ananenkov at the Gazprom Press Conference 2009 production. For details of the Gazprom production associations and reserves at the fields which they control, as well as how production at the different associations developed from 2009–11, see Appendix 7, Tables A7.1 and A7.2.

[4] 'Gazprom brings Zapolyarnoye field up to targeted 130 Bcm/y'. *Interfax*, 10–16 January 2013, p. 53. For an explanation of the different geological horizons of Russian gas fields, see Stern, J.P. *The Future of Russian Gas and Gazprom*, op.cit., Appendix 1.2, pp. 63–5

[5] In 2012 Western Siberia accounted for 92.6% of Gazprom production. The other 7.4% of production came from North Caucasus (2.6%), Volga/Urals (3.6%), North West 0.5%, and Central Siberia (0.7%). OAO Gazprom, *Annual Report 2012*, p. 45.

fields to markets in western Russia and abroad. Two major corridors (northern and central) bring gas out of West Siberia via an enormous high-pressure network of large diameter (1440 mm or 56 inch) pipelines. This additional inheritance from the Soviet era has been both a boon and a burden for Gazprom, providing it with control over the gas sector in Russia but also leaving it with a huge capital expenditure obligation. (Transportation is discussed in Chapter 5, and its development also referred to in Chapter 12.)

However, despite the success of Zapoliarnoe and Gazprom's satellite field strategy, it was recognized that the dominant role of West Siberia in the Russian gas supply portfolio would inevitably go into decline at some point. Indeed that time was seen as having arrived, as the last major Gazprom field in West Siberia, Iuzhno-Russkoe, was developed in the mid-2000s. Even Gazprom acknowledged that the additional development of new regions would be required to prevent production from declining relatively quickly.[6] However, at that stage the company did not appear to have a specific answer to the problem, and although its proven reserve and A+B+C1 resource base was huge (Table 11.1), the company's development plan remained unclear and in 2005 was characterized by one of the authors as the lack of a 'future supply roadmap'.[7]

This view was endorsed more generally by authors who argued that, although Gazprom had more than adequate reserves, it had not paid enough attention to developing them. Instead, at the behest of the Russian government, it has invested hugely in Russian oil and electricity sectors, and also in ambitious overseas projects in African, Middle Eastern, and Latin American countries. The general contention was that Gazprom had no specific plans, and had made no specific investments, which would allow the company to replace gas production from declining fields with new production from the Yamal peninsula and offshore fields.[8] With the Russian economy, and hence energy demand, recovering after the 1998–99 economic crisis, and with gas consumption in Europe continuing to grow rapidly due to increased demand from the power sector, this caused serious concern both inside and outside Russia. Starting from the mid-2000s, increasing alarm was expressed by a number of commentators, Russian and non-Russian, that Gazprom had not invested sufficiently in future production in order to guarantee that it could meet its market

---

[6] This was confirmed by Gazprom's own data, Stern, J.P. 'Future Gas Production in Russia: is the concern about lack of investment justified?', Working Paper NG 35, OIES, 1 October 2009, Chart 1, p.3.

[7] Stern, J.P. *The Future of Russian Gas and* Gazprom, op. cit., p.202.

[8] For a review of production prospects in the late 1990s and early 2000s see Stern ibid., pp. 5–19.

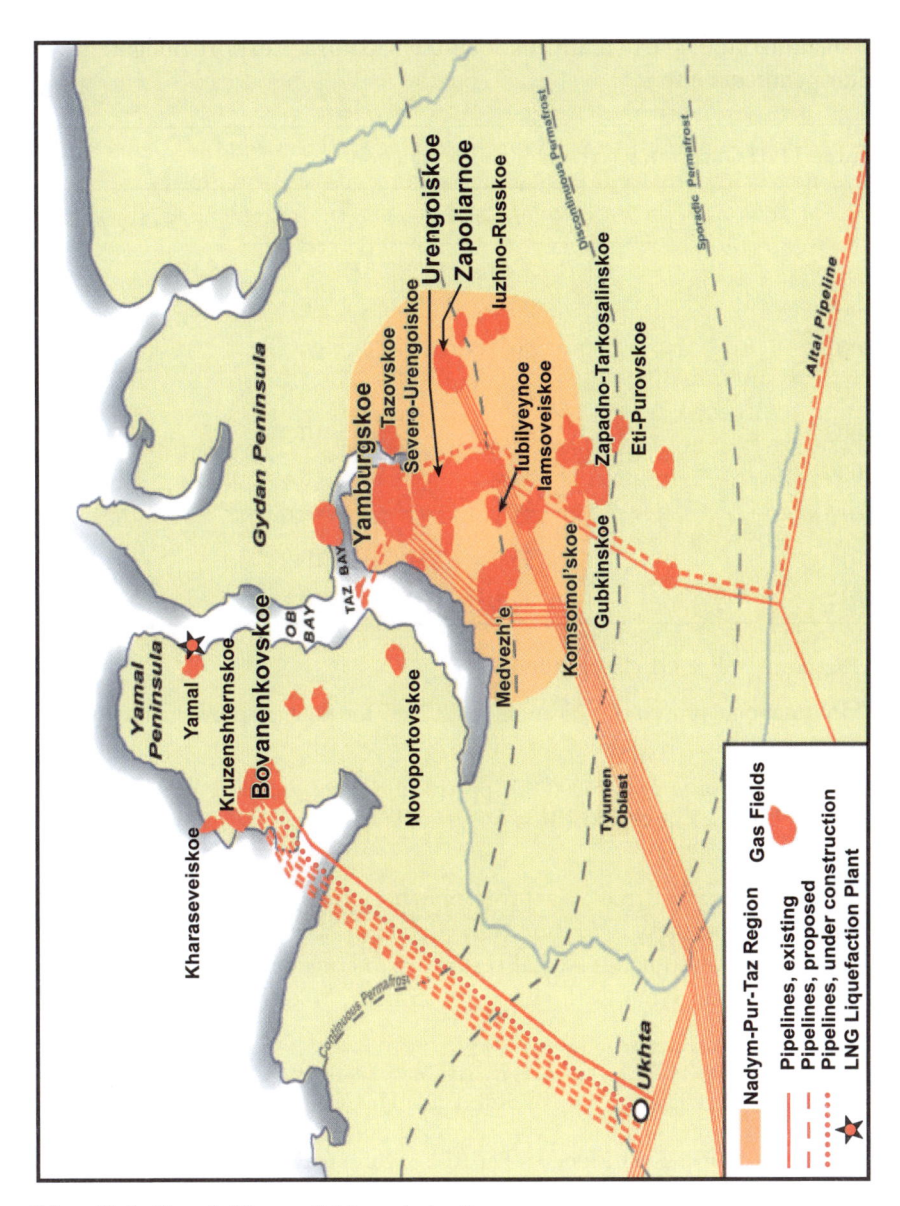

Map 11.1: Yamal–Nenets fields and pipelines
Source: Oxford Institute for Energy Studies.

obligations and, particularly, its long-term contract export obligations to European customers.[9]

Table 11.1: Gazprom gas reserves and production

	Reserves (Tcm)		Production (Bcm)
	A + B + C1*	Proven**	
2005	29.1	20.7	555.0
2006	29.9	20.7	556.0
2007	29.8	20.8	548.6
2008	33.1	18.2	549.7
2009	33.6	18.6	461.5
2010	33.1	19.0	508.6
2011	35.0	19.2	513.2
2012	35.1	19.1	487.0

* Russian A + B + C1 classification.

** International proven classification, for 2005–7 the data are proven and probable reserves.

Sources: *Gazprom in Figures 2003–7*, pp. 18 and 32; *Gazprom in Figures 2008–12* Factbook, pp.9 and 18.

In fact the criticism of Gazprom may have been somewhat unfair, as in reality the company was not only debating a number of alternative development options but was also starting to consider the implications of changing market dynamics for its supply strategy.[10] Company presentations[11]

---

[9] See among others: IEA *Optimising Russian Natural Gas: reform and climate policy* (Paris: IEA/OECD, 2006), pp. 33–4; Riley, A. 'The Coming Russian Gas Deficit: consequences and solutions', Centre for European Policy Studies, Policy Brief 116, Brussels, October 2006; Milov, V. 'The Risk of Russian Energy Supply Disruptions', *Energy Policy*, 2006; Dienes, L. Natural gas in the context of Russia's energy system', *Demokratizatsiya*, Fall 2007, pp. 408–29; Simmons, D. and Murray, I. 'Russian Gas: will there be enough investment?' *Russian Analytical Digest*, 27 July 2007, pp. 2–5; Helm, D. *The Russian dimension and Europe's external energy policy*, September, 2007; Noel, P. 'Beyond Dependence: how to deal with Russian gas', European Council on Foreign Relations, Policy Brief, November 2008.

[10] Locatelli, C. 'EU Gas Liberalization as a Driver of Gazprom's Strategy', Paris: IFRI/Russia-NIS Centre, February, 2008, pp.15–22.

[11] For example Gazprom Investor Day 2009, slide 13.

and press conferences[12] from 2006–9 show that Gazprom was considering not only the development of the Yamal peninsula, and in particular the Bovanenkovo field, but was also discussing development of the Shtokman field in the Barents Sea and the exploitation of deeper wet gas and condensate reserves beneath its existing fields in West Siberia. Furthermore, the ultimate decisions to proceed with Yamal and to delay the Shtokman development were taken at a time when European demand was continuing to grow (before the 2008–9 financial crash), but also as the impact of US shale gas was starting to hint at a reduced need for LNG imports into North America. As a result, although Gazprom's indecision in the mid-2000s about its supply strategy did appear to put at risk its ability to meet demand, with hindsight the company's reluctance to rush into major capital expenditure commitments seems entirely justified.

Table 11.2: Gazprom production projections, 2010–20 (Bcm)

	2010	2011	2012	2013	2014	2015	2020
Gazprom production (actual)	508.6	513.2	487	487*			
Gazprom projections (with date of projection)							
2009		507	510	523		615	620
2010			529	542	566		620
2011			519	521	549	570	650
2012				529	541	548	650
2013 (Feb.)					496	518	518	540
2013 (Dec.)						491	482	

* Preliminary.

Sources: Gazprom Annual Reports, Press Conferences, Managers' Briefings, and Investor Day Presentations. Projection figures have been rounded.

Unfortunately for Gazprom, however, pressure to ensure that production would not go into short-term decline but would instead grow rapidly to 2015 led to a decision to invest in Yamal development being taken in the period 2006–7, just prior to the economic crisis in Europe that led to the falling gas demand described in Chapters 3 and 4. As a result, Gazprom's

---

[12] For example, pre-AGM press conference held by Alexander Ananenkov on 14 June 2007.

production did go into decline, from a high of 556 Bcm in 2006 to a low of 487 Bcm in 2012, but for reasons that were entirely different from those predicted by the pessimistic commentators in the mid-2000s. As Table 11.1 shows, Gazprom's production plateaued and then declined, at first slowly and then sharply, with the arrival of the global recession, recovering briefly in 2010–11, and then declining sharply again in 2012, despite the fact that Gazprom's top management in charge of production claimed that the company, thanks to its investments in new fields, had the capacity to produce 600 Bcm.[13] Furthermore Table 11.2 shows that since 2010, the company's short-term production projections have tended to be overstated in relation to actual results, in the case of 2012 by 35–55 Bcm, due to a combination of reduced demand in all of its markets, and availability of alternative supply. Gazprom had done its best to respond to this change in circumstances, focusing on improving performance at existing fields – for example, production at Zapoliarnoe was increased to 130 Bcm, while use of enhanced recovery at other older fields also encouraged output to hold up better than had been predicted.[14] However, the change in strategy came too late to prevent Gazprom from investing in the development of huge new projects – although it managed to delay the start of Bovanenkovo by one year from 2011 to 2012 – and by the early 2010s, this meant that it had a significant surplus of production capacity.

This chapter considers these supply developments in more detail, in particular focusing on the period from the mid-2000s, and the company's likely production trajectory and options up to and beyond 2020, inevitably focusing on 'west-facing supply' to European Russia, the FSU, and European markets as this will continue to dominate the company's and Russia's gas sector.[15] It will consider the imminent shift in Gazprom's production base away from its traditional West Siberian heartland towards the more remote and expensive Yamal peninsula; address the issue of the company's initial failure to enter the Atlantic Basin LNG market and its current plans to try again; discuss the continued lack of a wet gas strategy, despite the fact that it has served companies such as Novatek so well (see Chapter 13); and finally contemplate a potential shift in the company's strategy towards the growing gas markets in Asia. This latter topic provides the link to Chapter 12, where we will discuss Gazprom's longer term plans: the importance of political influence over its strategy, its diversification into the oil and power sectors, and the chances of significant reform at the company.

---

[13] Gazprom Press Conference presentation 'Mineral and Raw Material Base Development', May 2013.

[14] See Appendix 7 for information on production from individual associations.

[15] Chapter 8 deals with supplies to Asian markets, including LNG exports.

The Yamal peninsula: foundation of Gazprom's west-facing production

From the early 2000s, speculation about future Gazprom production rested on a judgement of how quickly (and successfully) the new super-giant fields on the Yamal peninsula, starting with Bovanenkovo and followed by the smaller Ob–Taz Bay fields (Map 11.2), could be brought into production. Up to the mid-2000s, there was debate within Gazprom as to whether it would be more beneficial to move immediately to these very large, but very high cost, new fields, or rather to develop the smaller fields located around, and deeper horizons of, the existing NPT producing fields.[16] This debate was effectively ended in October 2006 when Gazprom's management committee approved the investment rationale for the development of Bovanenkovo and gave the instruction to commence work on the project. The formal launch of the officially titled 'Yamal Megaproject' did not take place until December 2008, but by then construction of key infrastructure to access the peninsula had begun and the first part of the pipeline linking the new fields to the UGSS had been laid.[17] This event was generally regarded as a victory for advocates of Gazprom's traditional upstream strategy, led by Gazprom Deputy CEO Alexander Ananenkov, with a focus on big long-term projects rather than the more innovative smaller field strategy that was apparently advocated by other members of the Gazprom management team.[18]

Gazprom divides the Yamal peninsula fields (Map 11.2) into three groups:

- The Bovanenkovskoe zone, where production is projected to reach 220 Bcm/year, includes the following fields: Bovanenkovo, Kharaseveiskoe and Kruzenshternovskoe (for which Gazprom holds the licences);

- The Tambeiskoe zone, where production is projected to reach 65 Bcm/year, includes the following fields: North-Tambeiskoe, West-Tambeiskoe, Tasiiskoe, Maliginskoe (for which Gazprom holds the licences); South Tambeiskoe (which will provide the gas for Novatek's Yamal LNG project, see below) and Siadorskoe.

---

[16] The advocates of these views on the Gazprom Board were Vlada Rusakova who favoured the smaller fields (and in 2013 retired from Gazprom to become Rosneft's Director of Gas Marketing), and Alexander Ananenkov who favoured and oversaw the first phase of Yamal production, and was then retired from the company for governance reasons (see Chapter 12). Many of the smaller, deeper fields with liquid content were subsequently developed by other producers (see Chapter 13).

[17] Gazprom press release, 3 December 2008, 'Gazprom launches Yamal megaproject'.

[18] 'There is no alternative to Yamal Gazprom's Alexander Ananenkov says', *LNG World News*, 20 July 2011, sourced on 11 November 2013 from www.lngworldnews.com/russia-there-is-no-alternative-to-yamal.

266 *The Russian Gas Matrix*

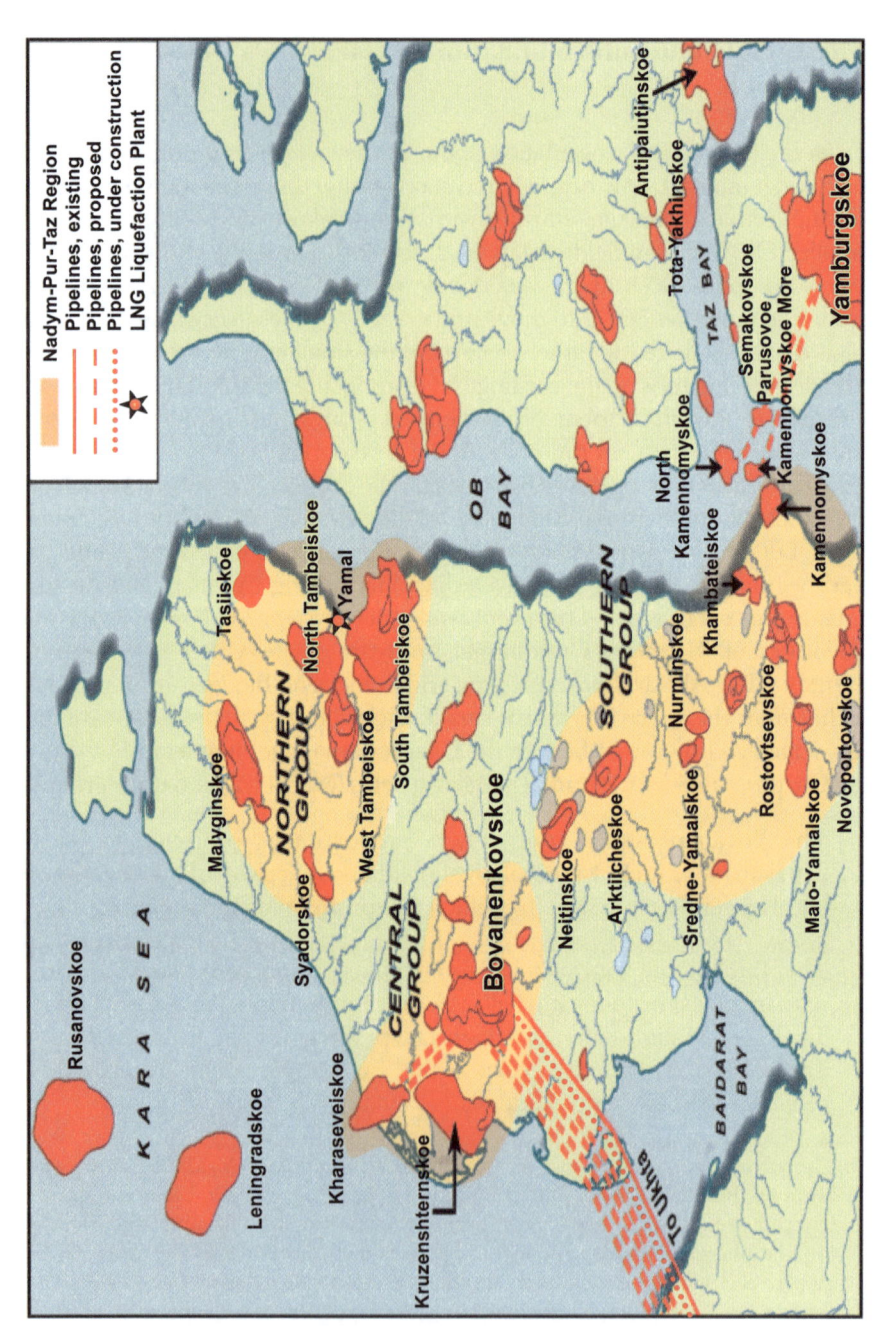

Map 11.2: Yamal peninsula and Ob–Taz Bay fields
Source: Oxford Institute for Energy Studies.

- The southern zone, where production is projected to reach 30 Bcm/year, includes the following nine fields: Novoportovskoe (for which Gazprom holds the licence), Nurminskoe, Malo-Yamal'skoe, Rostovtsevskoe, Arktiicheskoe, Sredne-Yamalskoe, Khambateiskoe, Neitinskoe, Kamennomysskoe (onshore).

The reserves of the major Yamal peninsula gas fields (Bovanenkovo 4.9 Tcm; Kharaseveiskoe, Kruzenshternovskoe, and South Tambeiskoe a total of 3.3 Tcm) are certainly enormous, but even at this magnitude they are still significantly smaller than the Soviet-era super-giants of Urengoi (10 Tcm) and Yamburg (8 Tcm) and therefore will not reach the plateau production levels, or longevity at plateau, of those fields.[19] In addition, their location presents two significant challenges not previously faced by Russian (or any other large-scale hydrocarbon) development: the much harsher permafrost conditions encountered on the peninsula, which is significantly further north than the NPT fields; and the transportation challenge of creating a new pipeline corridor running 1100km from the field (through these permafrost conditions) to the existing trunk pipeline system.

Furthermore, even before development work could be started, a major transport system had to be constructed to deliver pipe and equipment to the peninsula. This was achieved by the construction of the 525km Ob–Bovanenkovo railway,[20] completed in 2012, which needed to be sufficiently robust to withstand the instability of ice conditions in Baidarat Bay, as well as the combination of shallow water and soft seabed conditions that made pipe-laying very awkward.[21] As a result of these logistical and geographical issues, the development costs of the new fields on Yamal are inevitably higher than similar sized fields in the relatively benign NPT region.

Despite these issues, in 2011 Gazprom senior managers foresaw total production from the Yamal peninsula reaching 125 Bcm by 2020 and almost 300 Bcm by 2030, providing Gazprom with the means not only to

---

[19] Urengoi production peaked at over 300 Bcm and Yamburg at nearly 180 Bcm in the mid-1980s and 1990s respectively; in the 2000s they were both still producing more than 140 Bcm/year. By contrast the original plans for Bovanenkovo saw it peaking at 140 Bcm for only two years and then declining slowly over the following decade. Stern, J.P. *The Future of Russian Gas and Gazprom*, op.cit., Tables 1.4, 1.5 and 1.7 7–14.

[20] The scale of the logistical challenge and the harshness of the conditions can be grasped by looking at the Gazprom account and photos of the operation of the railway, *Railroad significance for the Yamal Megaproject implementation*, www.gazprom.com/about/production/projects/mega-yamal/railway/.

[21] The Yamal brochure states that: 'despite an insignificant depth there are frequent spells of stormy weather, complex sea floor sediments and frost penetration reaching the bottom in the winter. These conditions [only] permit working activities in the Bay for [a few] months a year'.

replace its depleting fields in West Siberia but also to grow overall production towards the targets of 650–750 Bcm that had been set at that stage.[22] Production is expected to come on stream from the three zones in the order in which they are listed above, with Bovanenkovo and its surrounding fields providing the initial surge in production through to the mid-2020s, before the Tambeiskoe and southern zones provide additional gas in the period up to 2030 and beyond. Furthermore, this total does not include the potential offshore developments in the Ob and Taz Bays – for example the North Kamennomysskoe and Kamennomysskoe More fields – shown in Map 11.2. Plateau production from these fields has been estimated to reach around 30 Bcm/year, with initial expectations being for production to start in 2020 and 2023. Gas will be carried by a conventional transportation system to the trunk pipeline corridor via the Yamburg field.[23]

The full detail of the infrastructure requirements was announced at Gazprom's 2009 Annual General Meeting, by Alexander Ananenkov, deputy CEO in charge of production.[24] The first phase would create two 1420mm pipelines from the Bovanenkovo field across Baidarat Bay to Ukhta (connecting to the Northern Lights network – see Map 11.2). These pipelines will operate at a much higher pressure than the rest of the UGSS (120 atmospheres compared with 75 for the rest of the high-pressure network) and carry up to 60 Bcm/year of gas each, allowing 115 Bcm/year (the full capacity of the Bovanenkovo field) to be produced through just two pipelines.[25] Production from Bovanenkovo officially started in October 2012, reaching 4.9 Bcm that year, and will expand as compressor stations are added to the pipeline infrastructure. Eleven compressors will be added to the first pipeline in 2013, completing the initial phase of the development,[26] with one more line then needed to allow Bovanenkovo to reach full output. A further four lines will then be required in the second and third Yamal peninsula phases, in order to increase production to the ultimate target of 250–300 Bcm, combined with the ultimate development of all three field groups.

---

[22] Gazprom Investor Day, February 2011, 'All you need is gas', slide 11.

[23] Gazprom Loan Notes 2013, p.128; 'Gazprom ups Q1 IFRS earnings 5.3% to 380.7 bln rubles as forecast', *Interfax*, 29 August–4 September, 2013, p. 39.

[24] The project was subsequently set out in much greater detail on the company's website in two documents: the *Yamal Megaproject brochure*, www.gazprom.com/f/posts/25/697739/book_my_eng_1.pdf (this brochure was written in 2007and thus anticipates the work ahead rather than reviewing what has been achieved); and the *Bovanenkovskoe field brochure*, www.gazprom.com/about/production/projects/deposits/bm/ The information in these paragraphs is based on those brochures which were accessed on 2 August 2013 (the date may be important because the Yamal brochure in particular will be rewritten).

[25] Requiring high resistance, specially coated K65 (X80) grade steel pipe.

[26] 'Gazprom launches long-awaited Bovanenkovskoe field', *Interfax*, 18–25 October 2012, pp. 54–6; Gazprom Loan Notes 2013, pp. 140–1.

As a result, by 2012, Gazprom had established the gateway to the Yamal peninsula fields and therefore access to potential production of 300 Bcm/year. However, with uncertainty about future demand growth in Gazprom's three major markets – domestic (Chapter 6), former Soviet Union (Chapter 7), and European (Chapters 3 and 4) – and the prospect of substantially increased production from Independent producers (Chapter 13), it remains unclear when this potential will be fully realized. Gazprom's recently revised supply outlook to 2030 certainly suggests that the company itself is now questioning the criteria on which future supply strategy should be based.

Gazprom's supply outlook to 2020 and beyond

In 2011 and 2012 Gazprom consistently painted a picture of overall company production increasing to a level of 650 Bcm in 2020 and 750 Bcm by 2030,[27] in line with the expectations for total Russian production forecast in the Russian Energy Strategy to 2030, published in 2009.[28] According to Investor Day presentations, in 2020 over 600 Bcm of this output was expected from onshore Russia, including new Yamal output, with a further 30 Bcm from the Shtokman field in the Barents Sea, plus a small portion from East Siberia that will be sold into the Asian market.[29] However, in February 2013 Gazprom announced a significant downgrade of its forecasts, as shown in Figure 11.1, with total output only expected to reach 540 Bcm in 2020 and 625–700 Bcm in 2030, approximately 100 Bcm below the previous forecast in each case.

In effect, Gazprom has acknowledged lower future demand for its gas, and incorporated 1–3 year delays in all of its major field developments. The plateau production level of 115 Bcm/year at Bovanenkovo was not, in early 2014, expected to be reached until 2021, compared with an expectation a year ago of 2018. Indeed Figure 11.1 shows production from the Yamal peninsula of 90–100 Bcm in 2020.[30] Delays in the peak output

---

[27] Gazprom Presentation to Investors, February 2011, 'All you need is gas', slide 11; Gazprom Presentation to Investors, February 2012, 'Size does matter', slide 9.

[28] Energy Strategy of Russia to 2030, Government Decree No. 1715, 9 November 2009.

[29] In comparison, the IEA (New Policies) Scenario in 2012 for total Russian production was 704 Bcm for 2020 and 808 Bcm for 2030. *World Energy Outlook 2012* (Paris: IEA/OECD, 2012), Table 4.5, p.138.

[30] 'Gas Market Integration for a sustainable global growth', Aleksei Miller, Speech at the World Gas Conference, June 2012. Delays of 1–3 years at other Yamal peninsula and Ob–Taz Bay fields include Kharaseveiskoe, Kruzenshternovskoe, Pestsovoy, Severo Kamennomysskoe and Kamennomysskoe More. OAO Gazprom Annual Report 2012, p.51; 'Gazprom puts off production plateau for projects totalling 200 Bcm', *Interfax*, 30 May–5 June 2013, p. 35.

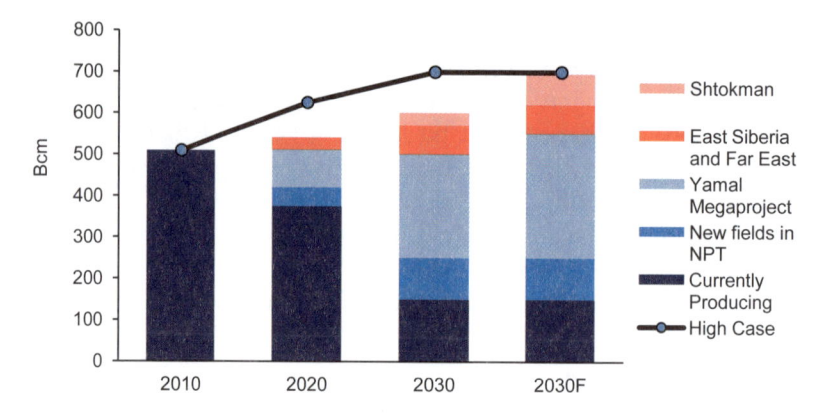

Figure 11.1: Gazprom projections of gas production to 2030
Source: Gazprom Investor Day Presentation 2013.

of other Gazprom fields are detailed in Table 11.3, further indicating that Gazprom believes that it will need to act as the swing producer for Russian gas supply in the wake of slowing demand growth and the availability of alternative (lower cost) Independent gas supply. This marks a dramatic change from the strategy that prevailed until 2009, when it was assumed that, should the market become oversupplied, Gazprom's gas would take priority and that third-party suppliers would be the losers.

Table 11.3: Changes in estimates of peak production dates for Gazprom fields

Field name	Old peak	New peak
Bovanenkovskoe	2018	2021
Kharaseveiskoe	2022	2023
Severo-Kamennomysskoe	2022	2024
Kamennomysskoe More	2025	2027
Iuzhno-Kirinskoe	2020	2023
Kirinskoe	2015	2018
Pestsovoe	2019	2020

Source: *Interfax*.

These projections imply that Gazprom's production may not recover to the levels of the mid-2000s (see Table 11.1) before 2020. Prior to that date (and in the low case prior to 2030) very little increase in production is anticipated for western markets – such as European Russia, the FSU, and Europe – presumably because Gazprom has realized that previous estimates of demand for its gas now look over-optimistic. Indeed the only reason for a significant change in the production estimate from Gazprom's current West Siberian and Yamal fields would be if an agreement were to be reached between Gazprom and CNPC to go ahead with the Altai pipeline project, which could open up the possibility of moving substantial quantities of West Siberian gas to eastern markets. As discussed in Chapter 8, at the time of writing this was not under active discussion. But it does highlight a key strategic point that will be taken up in Chapter 12, namely that eastern gas sales are likely to become increasingly important as a source of growth for Gazprom.

An additional conclusion from Figure 11.1 is that for the remainder of the 2010s, Gazprom's supply profile will be met by the Bovanenkovo field plus a little over 40 Bcm from new fields in the NPT region. As a result, Gazprom's production tasks for delivery of gas westward to 2020 seem likely to be relatively minor, and given that the major investments for the opening up of the Yamal peninsula fields have already been taken, the company's main job will be the gradual phasing in of Bovanenkovo gas over the remainder of the 2010s. Indeed the company's main task in both the short- and medium-term would seem to be to ensure that it does not accelerate output too rapidly and cause an oversupply in the market – rather than to develop new fields – and to this effect it has already announced downgrades in short-term production targets. Bovanenkovo will now be producing 41 Bcm in 2014 and 43 Bcm in 2015 compared to previous estimates of 68 Bcm and 90 Bcm.[31]

During the 2020s, production from currently producing fields (excluding Bovanenkovo) is expected to have declined to 140 Bcm, and although new West Siberian NPT production of more than 100 Bcm is planned to be developed, total NPT production will have declined to 240 Bcm in 2030, from more than 430 Bcm a decade earlier. Meanwhile over the same period, Yamal peninsula production will be expanded, under current plans, to 250–300 Bcm, but in either the low or high case scenarios this will not lead to any dramatic increase in overall production above the levels seen in 2010 or 2020. Essentially, Yamal gas will act as a buffer to replace declining NPT production, with any anticipated growth seen as coming from the Shtokman field or eastern fields. As the future of the Shtokman project remains very unclear, however, having been effectively 'consigned to future generations' by Gazprom (see below),[32] one apparent conclusion is that Gazprom's west-facing supply strategy has effectively been determined, and there is limited flexibility to change it.[33]

---

[31] 'Gazpromu nevygodno mnogo dobyvat'', *Vedomosti*, 17 October 2013.
[32] 'A cap on Gazprom's ambitions', *Financial Times*, 5 June 2013.
[33] This may not be the case for Eastern Siberia/Far East projects where decisions and timing may depend on contracts being signed for exports of pipeline gas and LNG (see Chapters 6 and 7).

Another very important issue is that 70 per cent of Gazprom's new production will be relatively high cost, with the suspicion that it may even not be profitable, a point strongly made in a Sberbank analysis of Gazprom's reserve portfolio:[34]

> Out of this 35 Tcm [of Gazprom reserves], 53% is in nonproducing greenfield properties that require gigantic infrastructure capex, while another 11% are in high sulphur high opex deposits in Orenburg and Astrakhan, which together produce less than 30 Bcm and have been in a state of decline since 1987. This leaves the remaining 36% of Gazprom's total reserves, or about 12.7 Tcm, in the camp of wet and dry gas in the cheap brownfield region (NPT). Stripping out wet gas, the dry Cenomanian deposits on which Gazprom was mostly riding since 1972 have less than 20 years life left with an average 64% depletion rate.
>
> To ... offset the natural decline on the dry brownfield, Gazprom needs to either get a lot wetter, or become much greener ... Going greenfield is value-destroying, as investors have learned by the sad lesson of Yamal ... Hence the only value-accretive move for Gazprom is to dive deeper into the wet world ... wet is the only type of gas for the company that is profitable and growing ... while other gas is either shrinking or value-destroying.

These judgements make two very important points: the significance of liquids for the profitability of gas developments, as exemplified by non-Gazprom producers (see Chapter 13), and the importance of not paying too much attention to the size of reserves without giving sufficient consideration to the cost of developing that resource and delivering it to markets.[35] However, because it has made the bulk of the investment in the first phase of Yamal peninsula development, Gazprom has essentially locked itself into a supply development path, certainly up to 2020 and potentially up to 2030, from which it will be very difficult to deviate. While the 40 Bcm of production from new NPT fields required by 2020 will be significantly lower cost than Yamal peninsula production, this is only around 30 per cent of new capacity. In addition, once the second Bovanenkovo–Ukhta line is completed, there will be a strong temptation to increase production – given the size of the investment which has been

---

[34] A+B+C1 reserves are roughly equivalent to proven plus probable reserves in the western classification. Sberbank *Russian Oil and Gas – the unconventional issue*, October 2012, pp. 13–14.

[35] A good example of this is the Astrakhan field close to the Caspian Sea with a resource base of 3–5 Tcm but production of 12 Bcm/year (of which only 6 Bcm is sales gas) due to environmental problems. 'Gazprom: environmental factors constrain production increase at Astrakhanskoye field', *Interfax*, 9–15 May 2013, p. 34.

made to establish the Yamal fields and pipeline corridor[36] – rather than develop new NPT fields (even at overall lower unit cost).

In terms of wetter gas, Gazprom had identified the likely shift in its portfolio away from dry gas as early as 2010, when it produced analysis showing that the share of dry methane gas would fall from 73 per cent in 2008 to only 37 per cent by 2030, to be replaced by gas richer in the more liquid components such as ethane, propane, and butane.[37] Furthermore, it is already producing 15 Bcm/year from the deeper (Valenginian at around 3000 metres) horizons at the Zapoliarnoe field. The Achimov horizons of the Urengoi field (at depths of 3600–3800 metres) went into production in 2008 in a joint venture with Wintershall, and in 2012 the German company increased its acreage in these deposits through an asset swap with Gazprom in exchange for equity in Wingas.[38] As a result, Urengoi Achimov is expected to produce 8.7 Bcm/year in the 2010s and more than 36 Bcm/year in the 2020s, but Gazprom's reluctance to extend its wet gas strategy is demonstrated by the fact that the Achimov deposits at Yamburg are still waiting to be appraised.[39] This caution with regard to wet gas reflects not only the company's history as a dry gas producer from traditional Cenomanian fields but also the fact that, as already highlighted, it will have more than enough gas from Yamal to meet its needs for the next decade at least. If Gazprom is already reining in output from its huge new investment in Bovanenkovo, then why would it invest more in an alternative source of supply? As a result, despite the superior economics of wet gas production, as exemplified in the economics of Novatek's business model (see Chapter 13), Gazprom's current strategic direction would suggest that it is unlikely that it will make up more than 10–20 per cent of the company's total output by 2030.

This will place the company at a strategic disadvantage, because the wetter gas produced by the Independents has a lower delivered breakeven cost to domestic and (theoretically) export markets, thanks to the boost that field economics receive from the 'liquids credit'. A full cycle discounted cash flow analysis of the Bovanenkovo field, for example, would suggest that the delivered price of gas to Moscow is approximately $140/mcm, and to the German border this would rise to around $265/mcm after

---

[36] This includes not only the direct investment in fields and pipelines but also in the Ob–Bovanenkovo railway which is believed to have cost $4.5bn.
[37] Gazprom presentation, February 2010, 'Gaining Momentum', slide 12.
[38] Gazprom Annual Report 2008, p.5; 'Gazprom, BASF sign asset swap agreement', *Interfax*, 8–14 November 2012, pp. 55–6.
[39] Gazprom Annual Report 2012, p.51; 'Gazprom wants to appraise gas content in Achimov', *Interfax*, 8–14 November 2012, p. 54; Gazprom Loan Notes 2013, p. 121.

transport costs and export tax.[40] These figures are of course somewhat speculative, but they have been confirmed in conversation with Gazprom as being close to the company's own estimates. By contrast, the cost of gas delivered from an average wet gas field in Nadym Pur Taz could be as much as one-third lower than this, thanks to the sale of liquids, with the actual outcome being, of course, dependent on the level of the oil and condensate prices.[41] Furthermore the very profitable economics of wet gas production are underlined by further studies by analysts at Sberbank, who suggest that the payback time from a well on Novatek's Iurkharovskoe field is less than 100 days.[42]

Aside from conventional gas, the country's huge unconventional gas resource base (estimated by the IEA at 35Tcm) has raised interest about the possibility of unconventional production.[43] However, the outlook for production from these resources appears to be relatively modest, as although Gazprom is, for example, developing coal bed methane (CBM) in the Kuzbass Basin, the expectation is that it will eventually only produce 4 Bcm/year.[44] Substantial shale and tight gas production over the next two decades is also unlikely due to the size, availability, and lower cost of the company's extensive conventional gas reserves.[45] Indeed, the only situation which might affect this judgement would be if very sizeable and low-cost shale reserves were discovered in the west of the country, adjacent to existing pipelines.

Interestingly, however, although unconventional gas does not seem to have any significant immediate future in Russia, the same may not true of shale oil, which Gazprom Neft and Shell have begun to develop at

---

[40] Discounted cash flow analysis carried out by James Henderson, a co-author of this chapter, and checked with Gazprom IR department in Moscow. Also cross-checked with analysis from Wood Mackenzie Consultants.

[41] Discounted cash flow analysis carried out by James Henderson and cross-checked with Wood Mackenzie Consultants.

[42] Sberbank Research, 'Russian Oil and Gas', op.cit.

[43] *World Energy Outlook 2013* (Paris: IEA/OECD, 2013), Figure 3.5, p. 116 gives the unconventional resource base at around 32 Tcm at the end of 2012: roughly 50% tight gas, 25% CBM, 25% shale gas. By unconventional gas is meant principally CBM and shale gas; in Russian gas literature there is a tendency to refer to any gas with associated liquids as 'unconventional', see Sberbank 'Russian Oil and Gas', op. cit.

[44] Gazprom Annual Report 2012, p.47.

[45] *World Energy Outlook 2012*, op. cit., Figure 4.5, p.142, estimates unconventional gas production in Russia around 50 Bcm in 2035 accounting for about 7% of total gas production in that year, of which about two-thirds is tight gas and the rest CBM, no shale gas production is foreseen. 'Russia to begin developing shale gas only in 30 years – Fedun', *Interfax*, 24–30 March, p. 34.

Verkhne-Salym in the Khanti Mansi region.[46] In addition, the Bazhenov tight oil formation has intrigued petroleum developers since the Soviet era, and with the application of modern technology could prove highly prospective.[47] Gazprom's involvement, albeit indirect, in this potential new source of liquids supply may provide a pointer to its overall liquids strategy. As discussed above, the company appears to be reluctant to accelerate wet gas production, but it would not appear to have rejected the idea of diversification into liquids altogether, seeing this rather as being achieved through its 98 per cent-owned subsidiary Gazprom Neft. Although this is not directly relevant to the company's west-facing gas development, it is a theme to which we will return in Chapter 12 as we discuss Gazprom's overall strategic direction and influence in the Russian energy sector.

Gazprom's LNG strategy: undermined by western markets

As early as the 1970s, Soviet leaders contemplated the possibility of sending LNG to both the US and Japanese markets. Although these plans were undermined by the arrival of the Reagan administration and the deregulation of US gas markets, the concept that Russia needs to be a player in the LNG market if it is to have a full role in the global energy economy has remained.[48] In the post-Soviet era, Gazprom naturally inherited Russia's LNG strategy. The possibility of an LNG export plant at Murmansk was raised as early as 1995,[49] but the real momentum came in the 2000s as initial plans for development of the Shtokman field were developed.

Gazprom first officially acknowledged that it needed to take a more global view of the gas market in its annual report for 2003, when it discussed its plans for an LNG scheme to serve the North American market, where indigenous production was in decline and LNG imports were expected to rise significantly over the following decade.[50] Reference was also made to the growing Asian market, but Gazprom's primary interest was the USA, which, prior to the shale gas revolution, was building large numbers of LNG regasification facilities in anticipation of a rapid increase in gas

---

[46] Gazprom Loan Notes 2013, p. 105. Henderson, J. 'Tight Oil Developments in Russia', Working Paper, WPM 52, OIES, October 2013.
[47] Gordon, D. and Sautin, Y. 'Opportunities and challenges confronting Russian oil', Carnegie Endowment for International Peace, 2013.
[48] For history of LNG projects up to the mid-2000s, see Stern, J.P. *The Future of Russian Gas and Gazprom*, op. cit., pp. 162–5. Of the LNG projects mentioned in that source only the terminal based at the Kharaseveiskoe field has been abandoned.
[49] Stern, J.P. 'The Russian Natural Gas "Bubble"', op. cit., p. 27.
[50] Gazprom Annual Report 2003, pp. 4–6.

imports. By 2004 this initial mention had turned into a strategy to supply remote markets (but primarily the USA) based on the development of Shtokman, and Memoranda of Understanding were signed with a number of US and European companies to develop the field and to advance Gazprom's position in the LNG market. An initial agreement was also made with PetroCanada to construct an LNG plant on the Baltic Sea near St. Petersburg, which would have provided an outlet for gas piped from West Siberia to access global markets in liquefied form.[51]

Gazprom had clearly understood that LNG could provide it with the opportunity to diversify its gas sales portfolio beyond its traditional pipeline customers, and in 2005 the company sought to exploit one of the other main advantages of having an LNG sales option, namely the flexibility to trade gas according to market demand and prices. With this in mind, its London-based subsidiary Gazprom Marketing and Trading (GMT) undertook its first spot LNG trades with an initial sale of one cargo into the Cove Point terminal in the USA and a swap deal with Gaz de France that allowed Gazprom to exchange gas piped to Europe for another LNG cargo, which was also delivered to the USA.[52] Since 2005 GMT continued to trade actively on the spot market (since 2008 via its wholly owned subsidiary Gazprom Global LNG), and has also signed a number of longer-term contracts including a deal with KOGAS to supply eight cargoes during 2013 and 2014[53] and another, signed in October 2012, to provide Indian company GAIL with 2.5 million tonnes (mt) of LNG per annum for 20 years.[54] Figure 11.2 shows the level of Gazprom's spot LNG trading since 2005, with sales peaking at 2.3 mt in 2011 and the diversity of buyers expanding to include China, Japan, and Taiwan as major customers in 2012.

Having established its trading subsidiary and made its first foray into the LNG market, Gazprom expanded its vision for the company in the global gas market and in 2006 announced that it would be developing an 'aggregator model' that would see it combine sales of equity LNG from company-owned projects with traded cargoes secured in the open market or via swap deals.[55] It announced plans to become a major player in the LNG market and by 2009 this ambition had crystallized into a government target, announced in the Energy Strategy to 2030, that by the end of the plan period Russia (and by inference Gazprom) should be exporting 15 per cent

---

[51] Gazprom Annual Report 2004, pp. 26–27.
[52] Gazprom Annual Report 2005, p. 59.
[53] 'GMT Singapore and KOGAS sign mid-term LNG agreement', Press release from GMT, 9 October 2012.
[54] 'GMT Singapore and GAIL sign 20-year LNG supply deal', Press release from GMT, 1 October 2012.
[55] Gazprom Annual Report 2006, p.57.

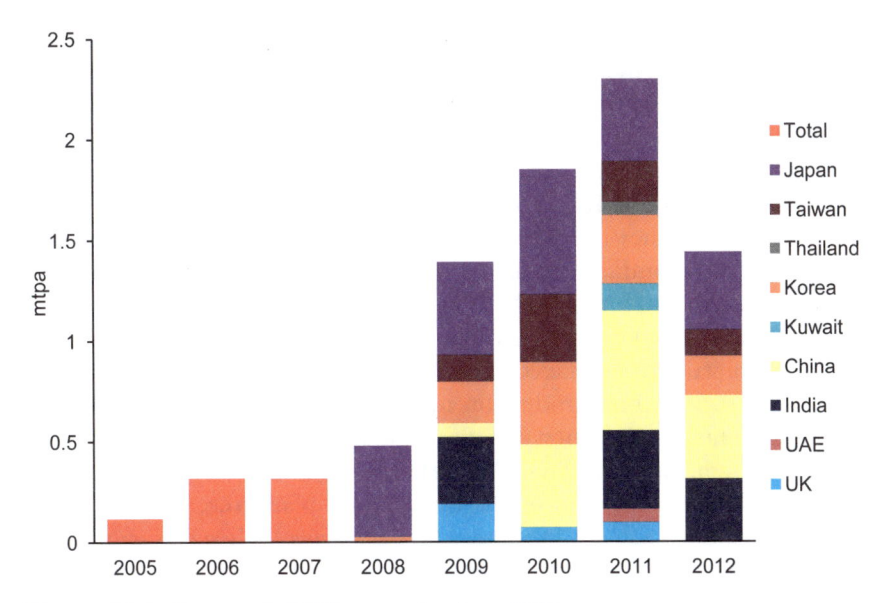

Figure 11.2: Gazprom's LNG spot sales since 2005
Source: Gazprom Data Book 2012, Gazprom Annual Reports 2005–7.

of its gas in the form of LNG, implying sales of 53–59 Bcm/year (39–43 mtpa). However, in order to achieve this goal and to establish itself as a genuine player in the market, Gazprom obviously had to develop some LNG projects of its own, and as far as the Atlantic Basin in the west was concerned, the Shtokman field had always appeared to be the primary goal. However, the technical and geographical difficulties involved in developing the field meant that it was always likely to be a long-term project.

Shtokman – preserved for future generations?

In the first half of the 2000s it appeared that North America, and in particular the USA, would become the major new consumer of LNG imports, as its indigenous production went into decline; a large number of regasification terminals were built, and a number of new projects in the Atlantic Basin were prepared to meet this expected demand. One of those projects was Gazprom's Shtokman field in the Barents Sea, 550km offshore Murmansk and to the east of the Norway–Russia maritime border.[56] Over the past 25 years, this field has been associated both with LNG exports and with pipeline gas supplies to north-west Russia and Europe via what became the Nord Stream pipeline (see Chapter 3). In Gazprom's 2005 Annual

---

[56] A substantial amount of information about the field and the project can be found on the Shtokman AG site http://shtokman.ru/en/.

Report, this 3.9 Tcm field was identified as the priority project for Russia's LNG strategy, with the potential for 70 Bcm/year of production to be split between LNG and pipeline sales. The first stage of the project was expected to produce 22.5 Bcm/year, enough to create 15 mtpa of LNG sales from a liquefaction plant located on the mainland at Murmansk. By 2007 two foreign partners, Total and Statoil, had been chosen to join Gazprom in the development of the field, working via a joint operating company that allowed them to contribute to the technical and financial issues surrounding a very complex and remote development without actually owning a share in the licence, which remained with Gazprom alone.[57]

The project plan envisaged a final investment decision being taken in March 2010, with first production in 2013/14, but discussions over the development plan, combined with the dramatic change in the profile of the US gas market, caused delays. Debates over the balance between pipeline gas and LNG, the location of the liquefaction plant, the possibility of a remote field development by subsea tie-back, and the issues surrounding the maintenance of a platform and crew in iceberg-strewn northern waters 650 kilometres from the shore, all combined to increase the estimated cost of the project from an initial $15 billion in 2005 to more than three times that figure by 2012.[58] At the same time, the gas price in its target market had fallen from a high of $9/MMBtu to around $3/MMBtu.[59] As a result, in August 2012 Gazprom and its partners announced that they had ended their cooperation at the field because all the parties believed that the project costs were too high to make the field economic to develop.[60] Essentially, the framework agreement under which the three companies operated jointly came to an end without a final decision to proceed having been made, and it was decided not to roll over the agreement. Statoil took the decision to write off the $336 million that it had invested in the project and to walk away from its partnership with Gazprom.[61] Total did not make such a definitive statement of intent, and has continued to discuss alternative approaches to the development of the field, with a plan to produce a new road map for the project being announced in March 2013.[62]

---

[57] 'Gazprom, Total and StatoilHydro create Shtokman Company', Statoil press release, 21 February 2008, Oslo.

[58] 'Gazprom to think technology for new Shtokman partnership' 2b1st Consulting, 15 August2012, sourced from www.2b1stconsulting.com/ on 15 October 2013.

[59] Data from US EIA, www.eia.gov/dnav/ng/ng_pri_sum_dcu_nus_m.htm, accessed on 5 December 2013.

[60] 'Gazprom freezes Arctic project', *Financial Times*, 29 August2012.

[61] 'Statoil writes of $336mm Shtokman investment', *Reuters*, 7 August2012,.

[62] 'Gazprom, Total discuss implementation of Shtokman project', *LNG World News*, 29 March 2013.

However, Gazprom itself has provided two conflicting outlooks for the field since then which have again demonstrated the company's apparent confusion over how to react to the changing global gas market. Firstly the company's CFO Andrey Kruglov stated that the South Kirinsky field in the Sakhalin-3 licence had become Gazprom's main development priority and that Shtokman would be left 'for future generations' to develop. As he rightly explained:

> Gazprom has been trying to develop the project for the past ten years. They planned to liquefy part of the extracted gas and send that to the US … [but] US shale gas has definitely undermined Shtokman that was oriented on the US market.[63]

However, in the same month (June 2013) Gazprom also announced that it was considering reviving the Shtokman project, with the options under discussion including a dramatic reduction in the LNG component and a re-focus on pipeline sales to Europe via the Nord Stream pipeline.[64]

As a result, prospects for the Shtokman field remain unclear. The market for LNG in the Atlantic Basin in 2013 suggested that such a high-cost and complex project would not be economic for at least a decade, and probably not for a great deal longer unless it could be aimed at non-North American markets. Therefore, the most likely outcome is that no gas will come from Shtokman until well into the 2020s, with Gazprom continuing to debate the prospects for this strategic field over the next few years. The possibility of pipeline sales to Europe exists but, as noted above, Gazprom has more than enough gas supply to serve markets over the next decade without developing an expensive new offshore field such as Shtokman.

Other proposed LNG projects

Another casualty of the emergence of US shale gas, with its implications for global LNG trade and gas prices in Europe, was Gazprom's plan in 2007 to build an LNG terminal on the Baltic Sea. The concept involved constructing a stand-alone liquefaction facility to receive gas via the main trunk pipeline from West Siberia in order to provide an alternative outlet for Russia's core gas fields to global markets. A 5mt plant was planned and a shortlist of potential international investors was drawn up which consisted of PetroCanada, BP, ENI, Iberdrola, and Mitsui, with PetroCanada being particularly enthusiastic and offering Gazprom a share in the Gros Cacouna regasification terminal in Quebec in return for a 25 per cent stake in the

---

[63] 'Shtokman project to be delayed', *Moscow Times*, 4 June 2013.

[64] 'Gazprom keeps optimising Shtokman field development project', Gazprom press release, 18 June 2013.

Baltic scheme.[65] However, in February 2008 the Gazprom Board decided that the project would be 'economically inadvisable' due to falling prices in the USA (the primary target for the LNG), and the economic crisis in Europe which was leading to a stagnation in gas demand.[66]

In 2013, however, Gazprom reconsidered its position and announced, at the St Petersburg International Economic Forum, that it had signed an agreement with the Leningrad region to build an LNG terminal on the Baltic coast, most probably at Ust Luga, but this is unconfirmed so far. Gazprom CEO Aleksei Miller confirmed that a 10 mt plant is planned, with two sites currently under technical and commercial appraisal ahead of an investment decision being taken in 2014.[67] The current schedule then envisages a construction phase lasting until 2018, when first LNG would be produced. Few other details about the project are available, and it remains to be seen whether Gazprom's announcement at the Forum was a genuine strategic move back into Atlantic Basin LNG, or a knee-jerk reaction to the increasing competition it faces from Rosneft and, in this case, Novatek which is likely to be selling LNG into Europe from its Yamal project. However, the announcement in late 2013 that Gazprom will build a gas pipeline to an industrial zone near Ust Luga, suggests that infrastructure is being put in place to enable the project to go ahead.[68] The gas will presumably be sourced from fields in West Siberia, as with the previous incarnation of the scheme, and Gazprom has suggested that customers will be sought not only in Europe but also in the emerging gas markets of South America.

The smallest proposed liquefaction scheme is the Pechora LNG plant, which is run by an independent company, Pechora LNG LLC, owned by the Alltech Group.[69] Russia's increasing focus on LNG as a strategic resource has brought the project to the attention of the Energy Committee of the State Duma; the authorities in the Nenets region, where the facilities would be located, support the scheme.[70] Gazprom may also become involved in the project as a potential response to emerging competition from its domestic rivals.[71]

---

[65] Gazprom shortlists 4 firms for $3.7bn Baltic LNG', *Reuters*, 19 April 2007 '.
[66] 'Gazprom drops Baltic LNG plan', *Reuters*, 7 February 2008.
[67] 'Gazprom aims to launch LNG plant on Baltic in 2018', *Reuters*, 26 June 2013.
[68] 'Gazprom to build gas pipeline to Ust-Luga – Leningrad region governor', *Interfax*, 26 September–2 October 2013, p. 36. The proposed pipeline has a capacity of 7 Bcm/year which would be insufficient for an LNG terminal but could presumably be expanded.
[69] More information about the project can be found at the website of Pechora LNG. www.pechoralng.com/index-eng.html.
[70] 'Port to be built as part of Pechora LNG project at Indiga in Nenets region – governor', *Interfax*, 23 March 2013.
[71] 'Gazprom to Pechora LNG', *Barents Observer*, 10 February 2012.

The original concept for Pechora LNG, conceived in 2011, was the development of two fields in the Nenets Autonomous District of northwest Russia – Kumzhinskoe and Korovinskoe – the first of which is located in the Pechora river delta and the second in Pechora Bay. The combined proven and probable reserves of the fields total 111 Bcm of gas plus 30mmbbls of condensate, and the production potential is estimated to be 4 Bcm/year of gas plus 4.5kbpd of liquids.[72] The current plan for the monetization of the gas reserves would involve the construction of a 2.6 mt liquefaction plant at a cost of approximately $4 billion, with first output planned for the end of 2018.[73] Because the project was small and its investors relatively unknown (Alltech was previously involved in West Siberian Resources, a small E&P company), its future was regarded as relatively uncertain until 2013, when it appeared to take on new momentum.[74] Gazprom has been in talks to take on a more active role, and one of its design institutes, Giprospetsgaz, has been involved in the creation of twelve development options for the facility. Furthermore, Russian Energy Minister Novak mentioned Pechora LNG as a 'real, viable project' during a discussion of future gas export plans, and plans for construction of a port at Indiga to service the scheme have been discussed between the regional authorities and Gazprom.[75]

If Gazprom does become formally involved in Pechora LNG, then it is possible that the project could be expanded to include the liquefaction of gas from the company's fields on the Yamal peninsula, via a new spur pipeline from the main gas trunk pipeline at Ukhta. If this were to be the case, then plans to expand the liquefaction facilities from 2.6 mt to 8 mt at a cost of an extra $8 billion could be implemented.[76] However, no specific details of either Gazprom's involvement or the timing of a final investment decision have yet been announced and so, in late 2013, the Pechora LNG project remained realistic, but somewhat speculative.

Conclusions on west-facing LNG projects

Gazprom outlined its LNG strategy in its 2006 Annual Report: it aimed at becoming a seller of LNG direct from Russian projects and a purchaser

---

[72] Data sourced from www.pechoralng.com/project-pechora-lng/reserve-base.html, accessed 5 December 2013.

[73] Data sourced from www.pechoralng.com/project-pechora-lng/lng-plant.html, accessed on 5 December 2013.

[74] 'Pechora LNG considering foreign investors for project', *Interfax*, 25 January 2013.

[75] 'Russian Energy Ministry says new LNG projects possible', *Interfax*, 27 September 2013.

[76] 'Pechora LNG gets 'positive' Gazprom review after Shtokman delays', *Bloomberg*, 25 September 2012.

and trader of LNG in the global market.[77] At that time, Gazprom Marketing & Trading had been established in London and was already carrying out spot sales into the Atlantic Basin market, and the prospects for the development of LNG projects at Shtokman and on the Baltic Sea coast near St Petersburg seemed relatively optimistic. However, the change in market circumstances that has been a constant theme throughout this book has undermined Gazprom's strategy, with the USA set to become an exporter rather than importer of LNG, and with energy demand in other Atlantic Basin markets having been hit by the post-2008 economic downturn. As a result the Shtokman project, which has also suffered from technological and cost challenges, appears to have been postponed indefinitely, while only in 2013 did the Baltic LNG project begin to show signs of being re-kindled, after having been postponed in 2008.

Therefore it would seem that Gazprom's LNG strategy reflects the same issues as its overall production strategy: project delays in the west, in the face of reduced demand for Russian gas. Once again, eastern markets would seem to offer the best hope of expansion, and this theme will be discussed in more detail in Chapter 12. It is clear that Gazprom itself had identified this potential outcome in the mid-2000s – when it demanded the right to enter the Sakhalin-2 project with a 51 per cent interest,[78] and in its recent pronouncements that the fields on its Sakhalin-3 licence offer greater short-term upside than the Shtokman project, despite the fact that the former are much more recent discoveries. These factors both indicate the extent to which the company's LNG focus has switched from west to east.[79]

However, another challenge to Gazprom's LNG plans has emerged in the form of competition from domestic rivals, with Rosneft and Novatek both promoting projects that have been encouraged by the removal, in December 2013, of Gazprom's monopoly on LNG exports. In particular, Novatek's Yamal LNG project offers competition to Gazprom in western as well as eastern markets (discussed in Chapter 13), and the re-kindling of the Baltic LNG concept discussed above could be interpreted as a Gazprom response to this threat. It remains unclear in 2014, however, whether Gazprom will re-assert its position as a potential seller of LNG into the Atlantic Basin, or if Russia's LNG strategy will instead be led by an Independent producer.

---

[77] Gazprom Annual Report 2006, p.57.
[78] 'Shell bows to Kremlin pressure on Sakhalin project', *New York Times*, 11 December 2006.
[79] 'Gazprom eyes Kirinsky as Shtokman replacement', *Interfax Energy*, 3 June 2013.

Conclusions

Supply strategy in the 2010s

Following a period of reliance on its Soviet legacy fields in the 1990s, by the mid-2000s Gazprom's supply strategy was that the original super-giant fields would, after a transitional period (during which time a number of smaller fields would fill the gap), be replaced by new super-giant fields on the Yamal peninsula. However, this investment decision unfortunately coincided with a dramatic change in the shape of Gazprom's main export markets due to the shale revolution in the USA and the economic crisis in Europe, with the result that by 2012 Gazprom's production had declined from the 556 Bcm of gas seen in 2006 to only 487 Bcm, despite its management having claimed that it could have produced 600 Bcm. Even allowing for some exaggeration, this means that the company had a surplus of more than 100 Bcm of gas supply and had clearly over-invested in production capacity, failing to anticipate – in common with all other observers (Russian and non-Russian) – reduced market requirements and increased availability of competing gas supplies. As a result Gazprom's strategy for west-facing gas supply – in other words, gas to be delivered to European Russia, western CIS countries, and European markets – in the 2010s seems relatively clear. In the face of demand in the company's major markets being unlikely to increase greatly, and because supply from other Russian producers is forecast to grow substantially, Gazprom production may well not be required to exceed the levels of the mid- to late-2000s (500–550 Bcm/year) and may even fall below that range (as it did in 2012 and 2013).

Gazprom's supply strategy appears to be further 'set in stone' because in the late 2000s and early 2010s the company made a massive investment in the first phase of Yamal peninsula production ('the Yamal Megaproject'), as a result of which the Bovanenkovo field started production at the end of 2012. This will have an ultimate initial production capacity of 115 Bcm/year, rising to a possible 140 Bcm as deeper reservoirs are developed. However, by 2013, the company foresaw only 90 Bcm of Yamal peninsula gas, together with additional production capacity from new fields in the NPT region of only 40 Bcm/year, being required by 2020, despite a decline of well over 100 Bcm of output at existing NPT fields. Given the amount of money invested in Yamal infrastructure and field development (which is estimated at up to $100 billion[80]) this potential excess of capacity is clearly not optimal for generating maximum returns, but it does provide Gazprom with an element of flexibility to respond to different industry outcomes.

---

[80] 'Gazprom's Empire at the End of the Earth', *Bloomberg Business Week*, 9 February 2012.

For example, should projections of market demand prove too pessimistic, or production from other Russian producers too optimistic, requiring Gazprom to produce more gas, the company could accelerate Bovanenkovo production and reach first phase capacity (115 Bcm/year) more rapidly. But a more likely outcome is a slower phasing in of new production over the remainder of the decade to 2020, with Gazprom's own production forecasts for 2014–16 providing clear evidence that this is already occurring.

Supply options beyond 2020

Gazprom's plan for the 2020s, as of 2013, is that the second and third phases of Yamal peninsula development will increase production to 250–300 Bcm/year, while 100 Bcm/year of production from new NPT fields will be also needed. These projections imply an aggregate total of 530–560 Bcm/year of gas from Gazprom's west-facing fields (mainly in West Siberia, including Yamal and NPT). This range is well above the production level of 2012–13, and towards the upper level of Gazprom's output in the 2005–8 period and its forecast output for 2020 (Tables 11.1 and 11.2). As a result, an important conclusion is that 'west-facing' pipeline supply requirements may already have peaked, with output basically remaining flat for the next 20 years. The one major caveat is that the huge resources available in the Yamal region will provide Gazprom with the continuing flexibility to accelerate development activity and increase production, if market circumstances change. The company's high-case production scenario demonstrates this optionality and also suggests that it is based not only on Yamal reserves but also on the Shtokman field.

However, even if the base case outlook for production volumes in the west is relatively flat, the same is not true for costs, as any of the forecast supply configurations would leave Gazprom's portfolio dominated by Yamal peninsula production, the delivered cost of which is likely to be at least 50 per cent higher than its current production base in West Siberia.[81] Although in the 2020s Gazprom would have the option to delay or abandon additional Yamal peninsula field development, focusing instead on wet gas in NPT or accelerating development of Ob/Taz Bay fields such as Severo-Kamennomysskoe and Kamennomysskoe More, the reality is that having made the initial investment to open up the peninsula, it will be difficult for the company to pursue any supply strategy other than maximizing production from those fields.

With hindsight in 2013, therefore, it seems to have been a major mistake for Gazprom to move so quickly to develop Yamal peninsula gas. Given

---

[81] In 2013 we estimate breakeven cost of Yamal gas delivered to German border at $7.50/MMBtu compared to approximately $4–5/MMBtu for existing Gazprom production in Nadym Pur Taz.

what we now know about demand and alternative supply, from a cost perspective, it would have been far better for Gazprom to wait and either purchase additional gas from other producers, or develop its own smaller fields, or deeper horizons at existing fields with a higher liquid content. However, it is all too easy to make a harsh judgement on the decision to progress Yamal peninsula production. Less than a decade ago all observers – Russian and non-Russian – were forcefully expressing alarm about what they perceived to be Gazprom's future supply crisis in relation to (what were perceived as) ever-increasing demands in all of its markets – domestic and foreign. The fact that the 2013 situation looks completely different should not obscure the fact that, at the time, no doubts were expressed about the strategic or financial wisdom of opening up the Yamal peninsula for future generations of Gazprom's customers. Nevertheless, given the decisions that have already been made about Gazprom's west-facing strategy and the prospective demand for its gas, questions about the company's growth profile and its competitive position in the global gas market remain very relevant. If production for western markets appears to be relatively flat and the necessary investments have already been committed, then the obvious answer would be a diversification east, and as we will discuss in the next chapter it would seem that this is where Gazprom's future may now be decided.

CHAPTER 12

THE DYNAMICS OF GAZPROM'S FUTURE STRATEGY

James Henderson and Jonathan Stern

This chapter offers an analysis of how Gazprom has responded, and may respond in future, to the commercial challenges it faces as gas markets evolve and to the political issues arising from its status as a government-controlled company. In Chapter 11 we showed that Gazprom's west-facing strategy is effectively settled – the company having, to an extent, suffered from the coincidence of its decision to develop the Yamal peninsula fields with a significant shift in demand for Russian, and in particular Gazprom's, gas in the Atlantic Basin, Europe, the FSU, and in Russia itself. Gazprom will essentially use the Yamal Megaproject as a swing producer to adjust its own output in the face of lower-than-anticipated demand in export markets, increased competition from Independent producers in the domestic market, and the possibility of increased global supply competition in the second half of the 2010s. Given the analysis of expected demand growth and price formation across Gazprom's markets, outlined in Chapters 3 to 7, and the increased supply competition the company faces – at a time when its own production strategy is set to become increasingly reliant on higher-cost gas – the outlook would appear to be rather bleak. Furthermore, Gazprom faces another major dilemma, described in Chapter 1, about how to balance the commercial necessity to respond to market pressures with the political necessity of supporting the aims of the Russian government and interests associated with it.

This chapter will use two lenses to view the key issues that Gazprom faces – its financial position and its current attempts to diversify its production base towards the east. We will highlight the continuing importance of export sales to Europe and the FSU, but will emphasize that the gradual decline in their share of revenues points to a need to find an alternative market for gas export sales – most likely in the east. We will also note the increasing importance of non-gas sales (from liquids and power) in Gazprom's revenue stream, but will question whether these changes, which appear to be politically as well as commercially motivated, may distract the company from its core gas sector objectives. An analysis of Gazprom's capital expenditure trends will demonstrate a shift away from upstream investment towards the transport sector, again seemingly driven by political objectives and vested interests, as well as by commercial rationale, but this will also be seen to highlight the plans for eastern expansion. We adopt a broad definition of vested interests, ranging from influential businessmen owning companies that can benefit from Gazprom's spending to the political elite. They can gain advantage through control over the distribution of Gazprom's cash-flows to lower-ranking individuals, who take advantage of rent-seeking opportunities along the entire gas value chain in which Gazprom operates.[1] The breadth and depth of Gazprom's business across the Russian economy offers multiple money-making opportunities for these vested interests, who as a result are not incentivized to encourage any reform of the sector and are in fact motivated to maintain the status quo to their own benefit for as long as possible.[2]

Despite the presence and influence of these vested interests, however, Gazprom's balance sheet offers evidence of a robust financial position at present, but the company's underlying cost trends and the risk of taxes being increased, together with the requirement for high future capital expenditure, point to a potentially more difficult future. A question is raised about whether investments will be prioritized on a commercial or political basis. From the perspective of Gazprom's eastern supply strategy, we ultimately pose the question of whether the company, in its current form, can satisfy all the requirements placed upon it by a diverse mixture of interested parties. If it cannot, then one possible consequence could be significant reform of the company.

A perspective on Gazprom's financial situation

Despite its problems in both the domestic and export markets, particularly in the period after 2008, Gazprom has managed to maintain a fairly robust financial position. Total gas revenues have maintained a rising trend (Figure 12.1) even as overall production and export volumes were declining (see Chapters 3 and 11) – this is mainly thanks to the steady increase in prices across all markets. However, there has been a noticeable shift in the split of gas revenues due to changing price and volume dynamics. Russian

---

[1] For example the Russian newspaper *Vedomosti* ('Podriadchiki Gazproma prodolzhaiut bogatet", 28 August 2013) drew attention to the holdings of three men with significant influence among the Russian political elite: Gennadii Timchenko, who now owns 50% of Stroitransgaz, a major contractor for Gazprom; and Ziiad Manasir and Arkady Rotenberg, who both own companies (Stroigazkonsulting and Stroigazmontazh respectively) that between them are reported by *Vedomosti* to account for half of Gazprom's investment budget.

[2] Grant, C. 'Stifling progress in China and Russia', Centre for European Reform, 2012·

288 *The Russian Gas Matrix*

domestic gas prices increased by 15 per cent per annum in the period 2009–13 (see Chapter 5), providing a significant boost to the company's domestic revenues despite an overall 7.5 per cent decline in sales by volume.[3] This pattern has been equally marked in the FSU market, where Gazprom has imposed significant price increases in some markets, notably Ukraine (see Chapter 7). Meanwhile in Europe, although oil-linked contract prices rose during the 2008–12 period, Gazprom both lost volumes and was forced to adjust its prices down (including offering rebates) in contract re-negotiations (see Chapter 3). As a result, the share of European sales in gas revenues has fallen from 60 per cent in 2008 to 53 per cent in 2012, while the share of FSU sales rose from 17 per cent to 19 per cent, and the share of domestic sales from 23 per cent to 29 per cent.[4] Although European sales recovered in 2013, as volumes and prices rebounded, this historic trend highlights one of the reasons for Gazprom's search for new export markets in the east.

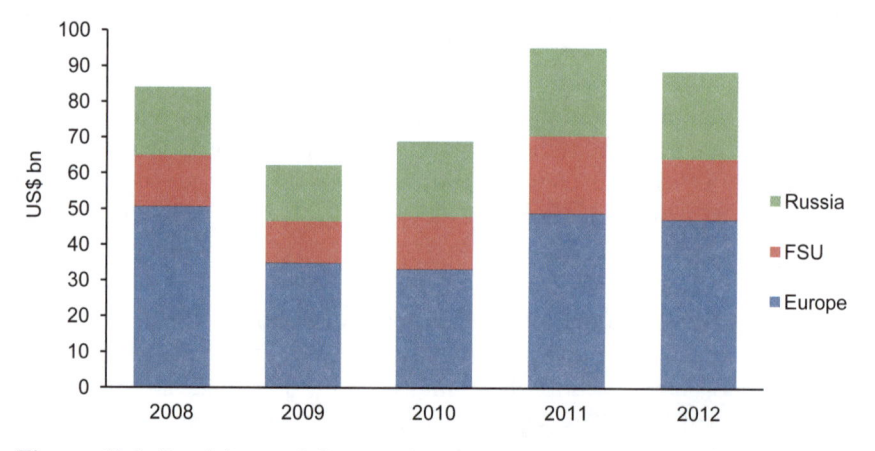

Figure 12.1: Breakdown of Gazprom's gas revenues, 2008–12 (US$ bn)
Source: Graph created from data extracted from Gazprom Management Reports: 2007, p.2; 2008, pp. 94–5; 2009, pp. 26–7; 2010, p.30; 2011, p.32; 2012, p.35.

However, an increasingly important element of Gazprom's revenues is in non-gas sales, particularly power and liquids (Figure 12.2), with the result that in 2012 gas sales only accounted for just over half of the company's total overall revenues. The biggest non-gas contribution came from the sale of crude oil, condensate, and refined products, which provided a larger

---

[3] Gazprom Group sales of equity gas in Russia declined from 297 Bcm in 2007 to 275 Bcm in 2012 according to the company's Databooks for 2011 and 2012.

[4] Data calculated from figures in Gazprom's IFRS reports for the years 2007 to 2012.

share of revenues (31 per cent) than gas exports to Europe, for the first time in the company's history. Power and heat sales accounted for 7 per cent of revenues (up from only 2 per cent in 2007), with other sales (including transportation) accounting for the remaining 6 per cent. As a result, it is clear that Gazprom is becoming a much more diversified company, with non-gas revenues and sales to non-European markets playing an increasing role. Indeed, the importance of liquids is emphasized by its contribution of almost one-third of revenues, despite accounting for only 10 per cent of overall production in 2012,[5] and its significance could increase further as the company's oil subsidiary Gazprom Neft is aiming to double output from 50 to 100 mmtoe by 2020. This goal, when combined with much slower expected growth in gas production and sales, could substantially increase the share of liquids in total revenues.[6]

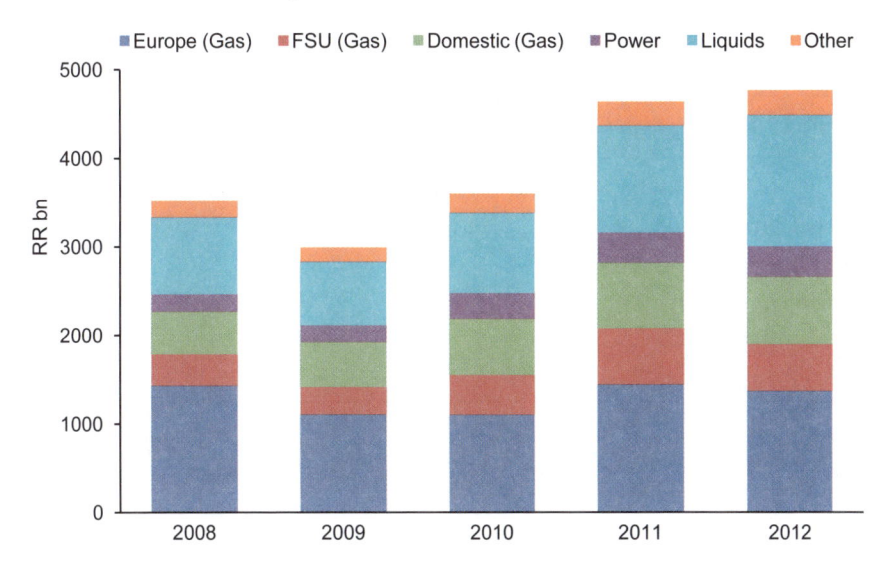

Figure 12.2: Breakdown of Gazprom's total overall revenues, 2008–12 (RR bn)
Source: Gazprom IFRS Accounts, 2008–12.

While the make-up of Gazprom's revenues has been relatively volatile, its margins have been quite stable, with an earnings before interest, taxation, depreciation, and amortization (EBITDA) margin in the range 35–42 per cent and a net (profit) margin averaging 26 per cent in the period 2007–12. However, despite the fact that the company made almost $40 billion in profit in 2012 the financial figures did reveal some declining

---

[5] Production data from Gazprom's Management Report for 2012, p.7.
[6] Gazprom Loan Notes 2013, p.102.

trends, with revenues falling by 3 per cent in dollar terms and net income down by 15 per cent (again in US dollars) compared to 2011, while the company's margins fell to their lowest level since the 2008–9 economic crisis.[7] Table 12.1 highlights two key reasons for this outcome, namely rising costs of production ('Prime Cost' in Table 12.1), which have increased by 60 per cent over four years, and a sharp increase in mineral extraction tax. This raises two key issues about Gazprom's future: firstly, whether it can control its costs by increasing efficiency and reducing the amount of 'incidental losses' that are widely believed to result from corrupt practices;[8] secondly, whether the Russian government will increasingly turn to Gazprom for extra revenues from rising taxes and dividend payments, undermining the company's ability to operate on a commercial basis. The answers to both these points are, to some extent, linked, because if Gazprom does not improve its cost control and profitability, then it will struggle to meet the government's financial requirements.

Table 12.1: Increase in Gazprom's production and transmission costs

	2009	2010	2011	2012	2013 prelim
Cost of production*:					
Prime cost	381.5	385.5	427	536.1	613.8
Mineral extraction tax	147	147	237	509	602
Total production cost	529	532.5	664	1045.1	1215.8
Cost of transmission**	38.5	40.2	42.5	48.7	53.6

* rubles/mcm

** rubles/mcm/100 km

Source: Gazprom Financial and Economic Press Conference, 2013, Slide 5.

---

[7] Profit margins calculated by author from profit and loss account in Gazprom IFRS financial statements for the years 2007 to 2012 by dividing net profit by total revenues to generate a net profit margin.

[8] Aslund, A. 'Gazprom's demise could topple Putin', Peterson Institute for International Economics, 9 June 2013, claims that investment analysts in Moscow believe that have identified $30–40 billion per annum of what they term 'value destruction' and what he defines as waste and corruption. Other examples of similar accusations can be found in reports such as 'Gazprom shackled by corrupt practices', *Reuters*, 6 January 2011, and 'Russia's wounded giant', *The Economist*, 23 March 2013, which claims that 70% of Gazprom's capital expenditure budget is siphoned off in corrupt practices.

The question of cost efficiency is particularly relevant when Gazprom is compared with its emerging domestic gas competitors, with Novatek providing the most immediate example. As President Putin has recognized, Russian gas companies are now operating in a much more competitive world, in which they will have to fight to prosper, with cost competitiveness being one key criterion for success.[9] Gazprom has often been criticized as being less efficient than its peers, not least by Rosneft CEO Igor Sechin, who has in the past called for Gazprom to 'increase its effectiveness'.[10] This is emphasized by a comparison of margins generated in the Russian gas business. Throughout the six years to 2012 Gazprom has had significantly lower margins than its closest domestic competitor Novatek (Figure 12.3). Although it can be argued that their business models are rather different, in fact the bulk of both companies' revenues are made up of oil and gas sales. Indeed in 2012 Gazprom's performance should have been the better of the two, as it enjoyed access to higher prices in export markets, while also having a higher average selling price in the domestic market than Novatek. That its EBITDA margin declined to 33 per cent while Novatek's comparable figure was 45 per cent offers some reflection of the competitiveness of both companies.[11] This clear difference in performance has not gone unnoticed by either Gazprom's competitors or domestic politicians, with even President Putin demonstrating a willingness to encourage competition in both the domestic and export markets in order to improve Russia's overall position.[12]

The combined issues of costs, political objectives, and vested interests is also evident in Gazprom's capital expenditure. A consideration of both historic data and future expectations can provide some interesting indicators of Gazprom's strategy. Table 12.2 gives a breakdown of the company's spending on capital projects since 2006, during which period the figure has more than tripled in ruble terms. Although the overall amount declined slightly from 2011 to 2012 in dollar terms, the most recent total still amounted to almost US$49 billion (although only $43 billion of this was reported in the company's IFRS accounts due to tax offsets).[13] Furthermore the breakdown given in Table 12.2 is interesting; a significant trend has

---

[9] 'Putin orders Gazprom to prepare key principles of gas export policy', *Interfax*, 24 October 2012.
[10] 'Sechin says Gazprom must raise its game', *Reuters*, 20 June 2010.
[11] Calculation based on data in Gazprom IFRS financial statements for 2012, p.5 (Consolidated Statement of Comprehensive Income) and Novatek IFRS financial statements for 2012, p.5 (Consolidated Statement of Income).
[12] 'Putin calls to phase out Gazprom monopoly on LNG exports', *Bloomberg*, 13 February 2013.
[13] Gazprom Management Discussion and Analysis of IFRS accounts for 2012, p.44.

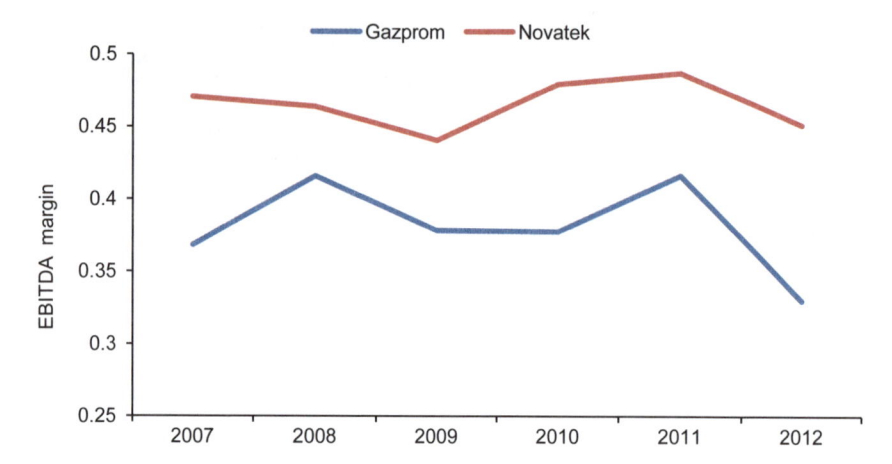

Figure 12.3: Comparison of Gazprom and Novatek EBITDA margins, 2007–12
Source: Author's calculations from Gazprom and Novatek IFRS Reports 2007–12.

emerged over the past decade, with expenditure on production declining from around one-third in the late 2000s to less than one-fifth by 2012. Over the same period, gas transportation infrastructure – both domestic and export – has come to dominate Gazprom's corporate agenda, accounting for 46 per cent of total capital expenditure in 2012.

Table 12.2: Gazprom's total capital expenditures (RR bn)

	2006	2007	2008	2009	2010	2011	2012
Transport	242	198	280	289	503	905	706
Gas production	131	186	275	253	282	288	286
Refining	21	39	94	89	92	140	171
Crude oil production	56	66	55	59	92	97	139
Power generation/sales			34	36	54	85	70
Distribution	23	39	43	30	45	51	61
Storage			9	11	21	25	23
All others	13	49	32	23	22	36	79
Total	**486**	**577**	**801**	**811**	**1111**	**1628**	**1536**

Sources: Management Report OAO Gazprom 2012, p.44; 2011, p.41; 2010, p.35. Management's Discussion and Analysis of Financial Condition and Results of Operations 2007, p.8; Gazprom Investor Day Presentation 2013.

Given the increasing remoteness of Gazprom's new projects, which are located outside the core producing areas, spending on new infrastructure is likely to remain a key priority, with many pipeline and LNG export projects being proposed for the rest of the 2010s and into the 2020s.[14] However, although this remains an important theme – and one that, as we will discuss later, has drawn much criticism from investors due to the perceived political and vested interest connotations attached to much of the company's spending on pipelines in particular – it is not the only important trend to be gleaned from Gazprom's capital expenditure forecast. Although 2012–13 investment figures show that spending on Yamal production and transportation has peaked (at least until the next phase of development),[15] the company's most recent overall estimate envisages average capital expenditure to 2020 of RUR 700–900 bn per annum ($25–30bn) in its gas business alone,[16] with a further $7–8bn per annum to be spent by Gazprom Neft.[17] Additional funds are likely to be used in the power generation and 'others' business sectors, and as a result a number of commentators estimate total company expenditure in the range $40–45bn for the next five years.[18]

Table 12.3 shows Gazprom's breakdown of its gas sector investments for the period 2013–30 by project and by sector, and immediately highlights the fact that the share of spending on transportation will remain high. With projects such as South Stream, Power of Siberia, and the expansion of both Yamal–Europe and Nord Stream all being actively considered, this is perhaps no surprise. But it does again raise the question of the viability of these projects and whether they are being driven by a political or commercial rationale.

The importance of Yamal is evident, as 28 per cent of all spending between 2013 and 2030 will be undertaken to develop the fields and related infrastructure on the peninsula, emphasizing that, as described in Chapter 11, Gazprom's west-facing supply strategy is tied to the future of this 'Megaproject'. However, the emergence of a new upstream area is also clear, with East Siberia and the Far East attracting 19 per cent of all capital expenditure over the next two decades. The growth of Gazprom's overall

---

[14] Gazprom presentation to investors, February 2013, 'Gas for the Future', slides 10–12.
[15] Gazprom Financial and Economic Policy Press Conference, 27 June 2013, www.gazprom.com/press/conference/2013/economic-policy/ Slide 11 shows that production and transportation investment connected with Bovanenkovo production halved during that period. There are no major new investments in production and the big increase in transportation investments are related to South Stream.
[16] Gazprom presentation to investors, 8 February 2013, slide 12.
[17] Ibid., slide 59.
[18] For example 'Gazprom – Cheap for Reasons', *UBS*, 9 October 2013, p.30; 'What's Right With Gazprom', Sberbank, July 2013, p.52.

Table 12.3: Gazprom's investments in the gas sector 2013–30 (%)

	2013–20 by major project	2013–30 by major project	2013–30 by business segment
Shtokman project	1	7	
Yamal Megaproject	25	28	
Eastern Siberia and the Far East	22	19	
Transportation	14	10	
Reconstruction of transportation	8	8	
Existing production	14	11	
Other (exploration, new production and drilling)	16	16	
Transport and storage			47
Production			31
Exploration			8
Processing			8
Other			6

Source: Gazprom Investor Day Presentation 2013.

production is closely related to its expansion into this new region, and the development of the Eastern Gas Programme (see Chapter 8) also has significant domestic political and foreign policy implications, as Russia seeks to expand its global energy influence towards Asia, and in particular China. The key question for Gazprom is whether it will ultimately be able to establish the same dominance in Russia's emerging eastern gas business as it has historically enjoyed in the west, or whether the combination of commercial and political pressures that it faces will again inhibit its progress or even undermine its position completely.

A final point in relation to Gazprom's financial position is that it starts the next period of its evolution from a fairly robust base. Even though the company has increased its annual spending three-fold over the past six years, it has nevertheless managed to finance its capital expenditure from its operating cash flows rather than through increasing its debt burden. Only in the crisis year of 2009 did the company's free cash flow[19] become

---

[19] Free cashflow is defined as post tax cashflow plus change in working capital minus capital expenditure, calculated by the Author as (EBIT*(1–tax rate))+ depreciation + change in working capital – capital expenditure, where EBIT is Earnings before interest and tax.

negative, although between 2008 and 2012, it fell – from $18 billion to $2 billion – as spending increased faster than profits (Figure 12.4). Nevertheless Gazprom's balance sheet has remained robust, with total debt since 2007 in the range of $45–60 bn ($48 billion in 2012) and net debt of $35 billion, leading to a net debt:equity ratio of only 12 per cent and a net debt to EBITDA ratio of 0.69, both well within the acceptable parameters for a company of Gazprom's size.[20] Indeed it would be fair to say that Gazprom could double its current total debt burden and still remain within the limits regarded as acceptable to most international energy companies.[21]

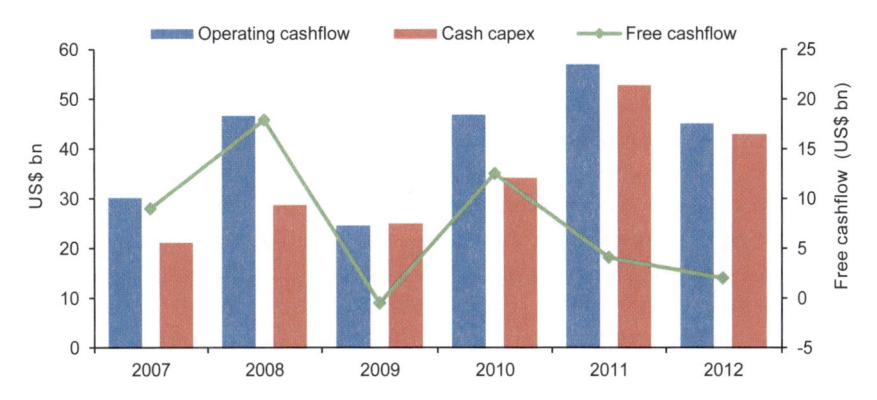

Figure 12.4: Gazprom's free cash flow, 2007–12
Source: Calculated by Author from Gazprom IFRS reports 2007–12.

However, even this strong current financial position does not necessarily bode well for Gazprom. As will be discussed below, the Russian government will increasingly have to turn to the gas sector for extra budget revenues, and is likely to do this through two main sources – increased taxes[22] and increased dividend payments. While Gazprom will not be alone in having to pay higher dividends, as the Russian administration is keen to force all state-owned enterprises to pay out a higher proportion of their profits,[23] it will nevertheless once again raise the recurrent theme of the company's overall strategy.

Is Gazprom's main priority to be run along commercial lines to generate higher profits and therefore dividends? Alternatively, should it focus

---

[20] Derived from Gazprom balance sheets in IFRS Statements for the years 2007–12: 2007, p.3; 2008, p.3; 2009, p.3; 2010, p.3; 2011, p.3; 2012, p.3.
[21] A net debt / EBITDA ratio of less than 2 is generally regarded as more than adequate.
[22] 'Putin says natural gas producers may face higher tax burden', Business Week, 17 April 2012.
[23] 'Russia tightening state dividend rules before asset sales', Bloomberg, 25 June 2013.

specifically on its political objectives of supporting the domestic economy: providing higher taxes to the Russian budget, being a lever for central government in regions across Russia, and acting as a tool of foreign policy? Or is its main goal to provide funds for vested interests through its capital expenditure programmes and hugely cash generative operations? If the answer is that it should do all of these things, then is it possible to achieve this in a much more competitive domestic and global gas market? Clearly this is a very complex problem for both the Russian government and Gazprom's management. One clear example of where all these issues are coming together in the short to medium term, is over the company's decisions on how to invest in and exploit the opportunities in the Asian gas market.

Gazprom's supply plans for Eastern Siberia and the Far East

In one sense, Gazprom's decision as to whether or not to invest in production from East Siberia and the Far East of Russia would appear to be relatively simple. As described in Chapter 8 there is clearly a potential market for Russian gas in Asia, both as piped exports to China and as LNG to the other major importing countries of the region. The Russian Energy Strategy to 2030 outlines a plan for Russia's overall energy production and export mix to be diversified towards the east, because the opportunity for growth in the mature western markets has diminished by comparison. (See Chapter 8, 'Russia's eastern development options'.) The purely commercial question for Gazprom, therefore, is: does it have sufficient gas assets to provide a competitive long-term source of supply to these markets, and the ability to bring them onstream in a timely fashion? Here we argue that the answer to the first part of the question appears to be an unqualified yes, while the answer to the second part must be tempered by a number of more subjective external factors.

The foundation for Russian gas exports to Asia is likely to be the construction of the 'Power of Siberia' pipeline that will run from East Siberia to the Pacific Coast, from where the gas would be sent as LNG to markets such as Japan and Korea, while a pipeline spur to China would also carry specific exports to that market (see Chapter 8, and specifically Map 8.1). Although the economics of a pipeline to China can work independently, the ideal solution would be a combined upstream and infrastructure project across the Far East of Russia that would combine the benefits of piped gas and LNG exports from the region. In terms of the upstream supply for this scheme, throughout the 2000s it appeared that the Kovykta field in Irkutsk would be the main source of gas exports to China, but progress was undermined by the fact that the field was owned by

TNK-BP rather than Gazprom – while the latter had exclusive export rights.[24] Although Gazprom now owns Kovykta[25] it appears that it has now decided that the primary supply will come from Chaiandinskoe (Chaianda), a field that has been a Gazprom asset for longer and which is located closer to the likely entry point into China at Blagoveshchensk.

Chaianda has a total reserve base of 1.3 Tcm of gas, and the plan to develop it was established at a joint meeting with the Sakha regional government in December 2012, where it was announced that more than 400 wells would be drilled and extensive processing and storage built, allowing peak production to reach 25 Bcm/year of gas plus 30kbpd of oil.[26] First gas output is envisaged in 2017, with the total field development cost estimated at RR430 billion (c.US$13.5 billion), but both of these estimates are complicated by the fact that Chaianda also contains a large volume of helium, which as a strategic asset under Russia law must be utilized and not flared. The plans to extract, process, and store or sell the helium are very complex and unresolved technical issues remain, which already appear to be causing delays.

However, the Chaianda field only has the capacity to meet around 40 per cent of the supply necessary to fill the 'Power of Siberia' pipeline and as a result the Kovykta field remains a key part of Gazprom's eastern supply portfolio, despite the fact that it is no longer the first priority. Kovykta is located in the Irkutsk region 800 kilometres south of Chaianda. It is actually larger than Chaianda, with its 1.5 Tcm of reserves having the capacity to produce up to 35 Bcm/year of gas at peak production.[27] Although no specific cost estimates have been produced by Gazprom, the previous field owners (TNK-BP) estimated capital expenditure at approximately $7 billion as long ago as 2005. More recent estimates are in the region $11–15 billion.[28] In addition a new pipeline connecting the field to the Power of Siberia trunk line would also need to be constructed, at an approximate cost of $5 billion. The timing of first production from the field will depend upon market conditions, and on sales agreements being signed, but should be technically feasible by the end of the current decade, although once again the issue of helium processing and storage will need to

---

[24] See Henderson, J.A. 'The Pricing Debate over Russian Gas Exports to China', Working Paper NG 56, OIES, October, 2011, pp. 6–10.
[25] 'Gazprom pays $770mm for TNK-BP gas field', *Financial Times*, 1 March 2011.
[26] 'Gazprom presents Eastern Russia's biggest Chayanda project in Yakutia', Gazprom Press Release, 11 December 2012, Yakutsk.
[27] 'Gazprom studies Kovykta resources in Chayanda project', *Interfax*, 26 October 2011, Moscow.
[28] Estimates from Wood Mackenzie Consultants, UBS, Bank of America Merrill Lynch, Deutsche Bank.

be resolved before gas output can commence. However, the latest announcements from Gazprom suggest that the phasing of its eastern projects will more likely see first gas from Kovykta in 2024, once Chaianda output has reached its peak.[29]

Gazprom also owns some other smaller assets in East Siberia that could be used to supply the Asian export market, but these are more likely to be used to meet the anticipated growth in domestic demand in eastern Russia as the government's plans to gasify the region are implemented alongside the export projects. The Chikanskoe field,[30] located close to Kovykta, has estimated reserves of just under 100 Bcm and could produce up to 3.5 Bcm/year of peak production into the local market if required.[31] Further to the west, exploration work is continuing at more than 10 existing and prospective fields in Krasnoiarsk region, and these will almost certainly be directed at the domestic market; the company also owns some small fields on the Kamchatka peninsula.[32] However, although all these assets could and may be developed on a stand-alone basis for local consumers, in reality the commercial and political success of Russia and Gazprom's Eastern Strategy will rely on the success of major export projects.

The second source of gas to supply these projects is Sakhalin Island, where Gazprom's existing Sakhalin-2 project is already producing and exporting 10.5mt of LNG per annum.[33] However, the key debate for the future of the company's eastern expansion concerns how the newly discovered reserves on the Sakhalin-3 licence, where Gazprom currently has a 100 per cent interest, will be exploited. The first Sakhalin-3 field to come online is Kirinskoe, which is Russia's first subsea development and produced its initial gas output in October 2013.[34] With about 160 Bcm of gas reserves and 5.5 Bcm/year of peak production potential, the field is too small to support an export project on its own but has, instead, been designed to fill the initial capacity of the SKV pipeline, which links Sakhalin Island with Vladivostok.[35] The rather small volume of gas flowing through

---

[29] 'Gazprom planning to launch Kovykta in 2024', *Interfax*, 20 September 2013.

[30] See 'Irkutsk gas production center', Gazprom website. www.gazprom.com/about/production/projects/deposits/gas-production-center/, accessed on 31 July 2013.

[31] Gazprom Bond Prospectus, June 2012, p.114.

[32] For details see 'Resources of Kamchatka', Gazprom website. www.gazprom.com/about/production/projects/deposits/kamchatka, accessed on 18 November 2013.

[33] For data on Sakhalin-2 see 'Sakhalin II', Gazprom website www.gazprom.com/about/production/projects/deposits/sakhalin2/, accessed on 15 November 2013.

[34] 'Gazprom launches production at offshore Kirinskoye field', *Interfax*, 24 October 2013.

[35] See 'Sakhalin III', Gazprom website. www.gazprom.com/about/production/projects/deposits/sakhalin3//, accessed on 31 July 2013 (for reserves); Gazprom Bond Prospectus, June 2012, pp.115–16 (for peak production).

the pipe does make the $16 billion cost of the 1,800 kilometre pipeline seem rather extravagant,[36] but this politically inspired domestic project could also catalyse further exports of Sakhalin gas via a southern route. Indeed, the announcement in March 2013 that exploration and appraisal work at the South Kirinskoe field had discovered 560 Bcm of additional reserves on the licence, with the potential for the total ultimate resources to reach 1.4 Tcm, has changed the situation dramatically.[37] Estimated production from Sakhalin-3 now has the potential to reach up to 30 Bcm/year, giving Gazprom the option to develop the SKV pipeline to its full design capacity and provide an alternative source of gas supply to the Vladivostok LNG facility (which may need over 20 Bcm/year if it ultimately reaches its full 15 mt capacity) as well as increasing sales into the domestic market.

However, a potentially more commercial solution – which Shell, the project operator, has been promoting for some time – would be for Gazprom to use at least part of the South Kirinskoe output to provide gas for a third train at its Sakhalin-2 liquefaction facility.[38] The economics of expanding an existing LNG plant on a relatively short timeframe appear more attractive than the construction of a new greenfield plant almost 2000 km from its gas supply.[39] Gazprom initially announced that it would make a decision on whether to expand the Sakhalin-2 project by June 2013, but its failure to do this has raised questions about whether the project will now go ahead, or whether Vladivostok LNG is the favoured option.[40] Conversely, whether Vladivostok LNG will receive its gas from Sakhalin or from East Siberia also appears to remain undecided, with conflicting announcements being made on a regular basis. As recently as February 2013 it appeared that the gas to supply the new liquefaction facilities would come from East Siberia,[41] but then in April 2013, Gazprom surprised a group of potential Japanese buyers by announcing that, instead, feed-gas for the plant would be coming from the newly upgraded South Kirinskoe field.[42] This was

---

[36] See 'Russian Gazprom reveals cost of its gas pipeline investment', OSW website. www.osw.waw.pl/en/publikacje/eastweek/2011-05-18/russian-gazprom-reveals-costs-its-gas-pipeline-investments, accessed on 31 July 2013.

[37] 'Gazprom plans to start production from South Kirinskoye in 2018', *Interfax*, 15 August2013; 'Gazprom boosts Sakhalin reserves', *Nefte Compass*, 28 March 2013.

[38] 'Shell upbeat over its Russian future despite setbacks', *Nefte Compass*, 7 February 2013, Moscow.

[39] For an economic comparison of Sakhalin-2 expansion versus Vladivostok LNG from Sakhalin see Figure 8.4 in Chapter 8.

[40] 'Chances fade for Sakhalin 2 expansion', *Nefte Compass*, 18 July 2013, Moscow.

[41] 'Gazprom guns for Vladivostok LNG', *Nefte Compass*, 28 February 2013, Moscow.

[42] 'Gazprom keen on Kirinsky', *Interfax*, 19 April 2013, Moscow.

apparently confirmed in July 2013 by Gazprom Deputy CEO Vitaly Markelov, who stated that the gas would come from the Sakhalin-3 project, at least initially.[43] Within two weeks, however, Energy Minister Novak was suggesting that Sakhalin-3 gas would feed Sakhalin projects, and not Vladivostok – implying that the choice between the two was essentially binary, as there was not enough gas to supply both. Indeed, he appeared to concede that Vladivostok's gas would come from East Siberia, as the economics of the Chaianda field development and the Power of Siberia pipeline rely upon LNG exports to the Pacific.[44]

Ultimately the decisions on Gazprom's eastern supply options will be driven as much by domestic politics and competition between various state organizations, as by commercial rationale. In terms of the potential growth in Gazprom's production profile over the next twenty years, it would seem that East Siberia and Far East Russia can provide the boost that is unlikely to be found in the company's already established western supply outlook. As can be seen in Figure 12.5, although eastern gas may have a rather limited impact in the years up to 2020, Gazprom could be producing as much as 100 Bcm/year from its eastern assets by 2030. This outcome will, of course, depend on markets being found for the gas and, as discussed in Chapter 8, this is by no means guaranteed in the short term. As a result, by 2020 Gazprom may only be producing gas from Sakhalin-2 and the Kirinskoe field at Sakhalin-3, with all other East Siberian gas postponed into the next decade. Nevertheless, the potential for its eastern strategy to provide significant production growth for Gazprom and for Russia is clear. The question remains, however, as to whether Gazprom will be the company to exploit this opportunity.

Gazprom: structure, ownership, political control and the rise of competitors

One of the most important issues in Gazprom's project-related decision-making is the fact that the company is required to take instructions from the government, and specifically from the president. This situation stems not only from the company's current importance to the Russian economy but also to its history over the past 50 years, as the company has essentially emerged as a corporatized continuation of the Soviet Ministry of Gas.[45]

---

[43] 'Gazprom LNG Vladivostok company to be registered soon', *Interfax*, 9 July 2013, Moscow.

[44] 'Kirinsky gas to feed Sakhalin, not Vladivostok', *Interfax*, 19 July 2013, Moscow.

[45] Stern, J.P. 'The Russian Natural Gas "Bubble": Consequences for European Gas Markets', RIIA Chatham House, 1995, p.4.

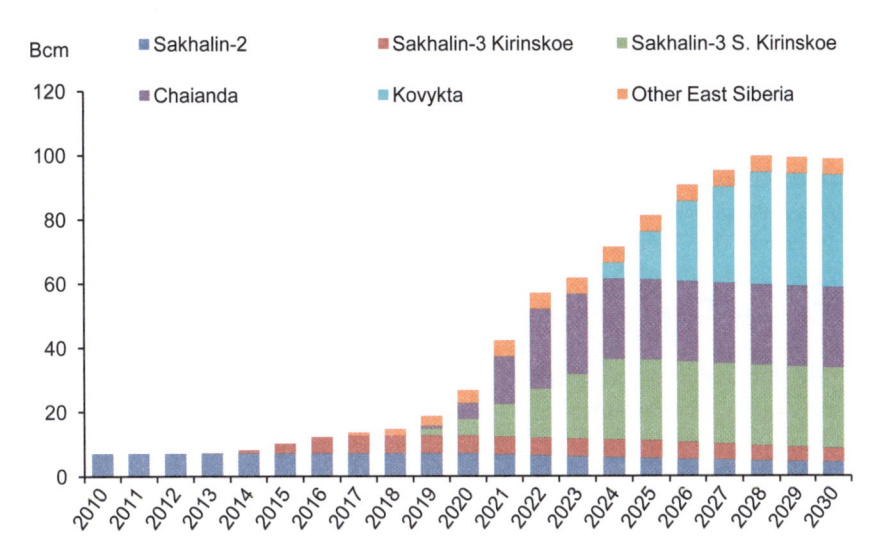

Figure 12.5: Gazprom's potential eastern gas production
Source: Author's estimates based on Gazprom forecasts.

One of the great achievements of its Soviet legacy is that, despite Russia's major fields having been discovered in extremely inhospitable terrain, thousands of kilometres from domestic and export markets, they were nevertheless developed successfully and linked to customers at home and abroad via an enormous trunk pipeline system.[46] As a result, Gazprom is an extraordinary industrial concern. It is the largest gas-producing company in the world, and it also controls and owns a vast transport network with 168,000 km of high-pressure pipelines, 222 compressor stations, 64 Bcm of working gas storage, and 654,000km of low-pressure distribution networks.[47] Furthermore, it supplies a very large proportion of the Russian domestic market in addition to having the world's largest pipeline exports. Indeed, the company is unique in the gas world, not just for the scale of these operations, but for the fact that it participates in the entire gas chain. This places it in the position of being an upstream producer, transportation company, utility company, and exporter, with all of these functions carried out on a huge scale. And from a political perspective, its other great achievement has been that it managed to use the legacy of the Soviet gas industry to maintain its fields and pipelines during the economic collapse post-1991, enabling it to sustain the country through the crisis of the 1990s

---

[46] Only Canada (and more recently China) has a similar geographical configuration, but on a much smaller scale.
[47] 'Mineral and Raw Material Base Development', Gazprom Press Conference presentation, Slide 8, 2013.

almost single-handedly, when production elsewhere in the energy balance was in freefall. Gazprom therefore acquired a reputation for being able to deliver large projects prioritized by the leadership for political reasons. However, the efficiency with which these projects are delivered, and their cost – in comparison with international experience – are questionable, and the legacy of prioritizing political necessity over commercial effectiveness still haunts the company today.

Nevertheless, as Russia made the transition from the Yeltsin era of the 1990s to the Putin era of the 2000s, the level of state control over Gazprom rose. In the 1990s Gazprom's CEO Rem Viakhirev was a hugely influential figure in Russian politics, as he effectively provided funds on demand to the government budget, and provided the energy needed to keep industry working and people warm in winter.[48] Importantly Viakhirev also held the balance of power at Gazprom, as Yeltsin had granted him the voting rights to the company's 11 per cent of treasury shares.[49] As soon as Putin became President in 2000 the re-assertion of state control over Gazprom became a priority. Viakhirev was replaced by Aleksei Miller, a friend and former colleague of Putin from St Petersburg.[50] Miller's first job was to end, and then reverse, the asset stripping at the company that had seen a number of key fields sold off at below market prices; Putin was also keen to ensure that the Russian state fully owned more than 50 per cent of the company's shares.

To that end, and also with a view to boosting Gazprom's market value, as well as ensuring government control, it was announced that Gazprom would sell its treasury shares to 100 per cent state-owned company Rosneftegaz while at the same time ending the 'ring-fence' which had prevented foreign shareholders owning any domestic shares in the company,[51] limiting them to ADRs traded in London.[52] As a result 10.74 per cent of Gazprom's shares were sold to Rosneftegaz, giving the state an effective controlling interest: 50.002 per cent.[53] However, because 0.889 per cent of this interest was held by Rosgazifikatsiia, a company that was controlled but only 75 per cent owned by the state, the government's actual

---

[48] 'Rem Vyakhirev: Gas Oligarch' *Financial Times*, 15 February 2013.
[49] Colton, T.J. and Holmes, S. *The State After Communism: Governance in the New Russia* (London: Rowan & Littlefield, 2006), p.232.
[50] Goldman, M. *Oilopoly: Putin, Power and the Rise of the New Russia* (Oxford: OneWorld, 2008), p.104.
[51] 'Big bang ends Gazprom ring fence', *Upstream*, 23 September 2005.
[52] ADRs are American Depository Receipts, which act as a vehicle to trade non-US stocks on international markets. Until 2005 non-Russians were only allowed to own Gazprom ADRs rather than Gazprom domestic shares, with the results that the ADRs traded at a significant premium to the local shares.
[53] For shareholdings at the end of 2011 see 'Shares', Gazprom website. www.gazprom.com/investors/stock/.

equity interest was in fact 49.78 per cent. In order completely to reinforce its position, in July 2013 it was reported that Rosneftegaz had purchased a further 0.23 per cent of Gazprom shares, bringing its holding up to 11 per cent which, with the addition of the Russian Federal Property Agency's 38.373 per cent and the 75% of Rosgazifikatsiia's 0.889 per cent owned by the Russian government brought the state's equity share to above 50 per cent.[54]

Given the importance of the energy sector in the Russian economy (Chapter 1), this desire for government involvement and control is hardly surprising, but gas infrastructure projects have a particularly important place in the government's domestic and international priorities. Gazprom is required to meet the gas needs of the population, with specific gasification targets for regions and the country as a whole (Chapter 5), and it is also required to undertake export projects that meet political as well as company objectives. In the west, the construction of the Blue Stream and Nord Stream pipelines (by-passing transit states such as Belarus and Ukraine) was initiated under government instruction, and the further plan to build the 63 Bcm South Stream pipeline to complete the transit-avoidance plan has a similar motivation (see Chapter 4).

The involvement of the state in the development of Russia's eastern assets is also very clear; from the adoption of the Eastern Gas Programme as a government programme in 2007 to President Putin's order to Aleksei Miller in 2012 to re-focus on the east and build the Power of Siberia pipeline.[55] Furthermore the Sakhalin–Khabarovsk–Vladivostok pipeline is another clear example of politics overriding commercial priorities, with Gazprom being forced to spend $16 billion on a pipeline with an initial capacity of only 6 Bcm/year, in order to satisfy domestic demand in the Primorsky region and to supply Vladivostok with sufficient gas to meet the power needs of the APEC conference held there in September 2012.

It would therefore appear that just as the instruction to build the South Stream pipeline (a multi-billion dollar project that is unlikely to carry substantial new Russian gas to Europe but whose main goal appears to be that of removing Ukrainian political and economic control over Russian gas exports to Europe – see Chapter 4) is putting Gazprom's western strategy under financial strain, so government involvement in Gazprom's eastern plans also seems to be putting the company at a competitive disadvantage. The confusion over whether Gazprom should focus on political or on commercial priorities is exemplified by the indecision over the future of the Power of Siberia project where, in spite of Putin's instruction, in 2013 Gazprom on the one hand claimed that the pipeline

---

[54] 'Gazprom confirms state has increased stake in gas giant to controlling', *Interfax*, 11–17 July 2013, p. 31.
[55] 'Gazprom unveils $38 billion project to conquer Asia', *Reuters*, 29 October 2012.

will not be built until a pipeline sales contract with China had been signed, while on the other still seeing East Siberia as a long-term source of gas for the Vladivostok LNG plant.[56] This uncertainty over Russia's strategy, which is compounded by the fact that the Power of Siberia pipeline also has a domestic goal of encouraging industrial development and regional gasification,[57] is not only causing confusion among potential buyers of Russian gas in Asia, but is also allowing Gazprom's domestic and international competitors to gain a strategic advantage.

In terms of Gazprom's position within the Russian political environment, the most relevant issue in terms of competition is the rise of Rosneft – which appears to be openly challenging for a lead role in the Asian market.[58] The company's CEO, Igor Sechin, has argued that third-party sales of Russian gas in the east would not offer competition to Gazprom because it does not have a position in the market yet.[59] The government seems to have accepted this, and in 2013 ended Gazprom's export monopoly on LNG exports (see Chapter 13). As a result, both Rosneft and Novatek are progressing with LNG projects (see Chapters 8, 11, and 13), and are competing to sell gas into the same Asian markets as Gazprom. Indeed, Novatek has taken CNPC (Gazprom's potential counterpart in China) as a 20 per cent partner in the Yamal LNG project, and Rosneft has contracted to sell gas to Japanese utilities who could also be customers for LNG from Gazprom's Vladivostok project. (See Chapter 8.) The trends that have been so apparent in the Russian domestic market, with Gazprom losing market share to rivals who are not encumbered with political obligations, may also be repeating themselves in the Asian gas market.

However, it seems unlikely that Putin would want to see Gazprom undermined too significantly, as his political strategy tends to revolve around creating a balance of power between different interest groups.[60] Furthermore, his support for Gazprom is clear not only from debates over the company's position in the European export market, where he has naturally taken its side against the perceived threat of EU legislation, but also in the fact that he appears to take personal control of many of the company's strategic decisions, with clear examples concerning the South Stream pipeline project

---

[56] 'Gazprom reports slow progress on China deal', *Interfax*, 4 September 2013; 'Vladivostok LNG: Promoting Development of East Siberian Gas', Wood Mackenzie Consultants, 2013.

[57] See 'Power of Siberia: Yakutia and Irkutsk gas production centers', Gazprom website. at www.gazprom.com/about/production/projects/pipelines/ykv/, sourced on 15 November 2013.

[58] 'Rosneft on the lookout for Gazprom's misfortunes', *Interfax*, 7 June 2013.

[59] 'Rosneft says Exxon's LNG project intact', *The Moscow Times*, 26 September 2013.

[60] 'Gazprom and Rosneft: Putin's Balancing Act', *Natural Gas Europe*, 5 June 2013.

and expansion into the Asian gas market.[61] As a result, it appears very unlikely that his preferred route would be to see Gazprom undermined by domestic competition. However, we argue that in return for this support, Gazprom is expected to perform well on a number of levels.

Firstly, it needs to provide more revenues for the Russian budget. At present, the gas sector only provides around 5 per cent of budget revenues on an average annual basis, compared to around 45 per cent from the oil sector (see Chapter 1). However, the threat of a decline in oil production from the core West Siberian area has meant that the Russian government has had to offer tax incentives to encourage the development of new greenfield regions.[62] As fields in these regions come on stream, paying lower taxes, the gas sector will need to provide an increased revenue stream to compensate. It can do this through paying higher taxes domestically via increases in mineral extraction (MET) royalty tax, which are being introduced,[63] (see Chapter 1) and by increasing export taxes through higher export sales, especially to Europe and, in the future, Asia. Given the reduced outlook for European gas demand and therefore limited prospects for expansion of Russian exports to Europe (Chapter 3), substantial future export revenue growth is most likely to come from Asia, if Gazprom can ultimately enter the market successfully. This will require significant investment on which Gazprom must make a reasonable return, because the second way in which it will increase its contribution to the budget is via higher dividend payments, half of which will go to its largest shareholder, the Russian government. Legislation is already being introduced to force state-owned companies to pay out 25 per cent of their internationally audited (IFRS) profits, as opposed to their Russian accounting (RAS) profits, and this payout level may increase to 35 per cent by 2016.[64] However, for this to make a real impact, Gazprom will need to be a profitable company. With questions being asked about the viability of Gazprom's new politically-driven projects in a more competitive global gas market (see Chapter 1), it is unclear whether it will be able to balance all its objectives successfully.

The current situation in Asia may provide some clues as to the potential outcome. If Gazprom can successfully agree a gas export deal with China and re-assert its position as the leading Russian company exporting both piped gas and LNG to Asia, then this is likely to indicate that its position in the Russian gas sector is under no immediate short-term threat. However,

---

[61] 'Kremlin shields Gazprom from EU probe', *Financial Times*, 11 September 2012.

[62] Henderson, J. 'Tight Oil Developments in Russia', Working Paper WPM 52, OIES, 8 October 2013, pp.12–14.

[63] 'Russia Gas: Tax shift to help producers from Gazprom to Novatek', *Bloomberg*, 22 May 2013.

[64] 'Gazprom, Transneft to shift to IFRS-based dividends in 2016 – draft budget', *Interfax*, 2 October 2013.

if it continues to delay the signing of a deal and allows its domestic rivals, Rosneft and Novatek, to become the prime movers in Russia's entry into the Asian energy market – where Rosneft already has a strong foothold via its oil exports to China[65] – then perhaps we may see the beginning of a more significant re-shaping of the Russian gas sector.

Reform of Gazprom

The reform of Gazprom has been a major agenda item for all Russian governments since the late 1990s, and various options to break up Gazprom into its constituent parts have been debated and eventually rejected.[66] However, this does not mean that no reforms have taken place, but rather that they were more along the lines of the continental European model, rather than the more radical Anglo-North American example that many Russian and foreign critics were seeking.[67] The reform of Gazprom to date can be seen in two stages: the first starting when Aleksei Miller took over from Rem Viakhirev as president in 2001; this period focused on improving governance, regulatory procedures, financial accounting, and transparency.[68] Stage 2, which Miller announced had begun in 2004, focused on corporate restructuring, with the aim of forcing the company's wholly owned subsidiary companies to focus on their main function, while divesting non-core activities. As a result, by 2012 the company's structure had been significantly simplified, with eight wholly-owned production (Dobycha) companies; seventeen wholly-owned transmission (Transgaz) companies; a sales subsidiary (Mezhregiongaz); a holding company for distribution company assets (Gazprom Gazoraspredelenie[69]), one processing company (Gazprom Pererabotka); two servicing and repair subsidiaries (Podremont Orenburg and Urengoi), and one storage subsidiary (Gazprom PKhG).[70] Thus a considerable amount of management and legal unbundling has been achieved, but the key questions

---

[65] 'Rosneft to deliver 30mt of oil to China in 2014', Interfax, 29 October 2013.

[66] For these debates see Stern, J.P. *The Future of Russian Gas and Gazprom* (Oxford: OIES/OUP, 2005), pp. 184–7.

[67] This refers to the period up to the mid-2000s. By the 2010s, continental European utilities had themselves been forced to adopt many structural elements of the Anglo-North American model.

[68] For details see Stern, J.P. *The Future of Russian Gas and Gazprom*, op. cit., pp. 188–92.

[69] A holding companies with 7 branches, 155 subidiaries, 20 affiliated distribution companies, and Gazprom Transgaz Kazan (in Tatarstan). Gazprom Annual Report 2012, p.63.

[70] OAO Gazprom, IFRS Consolidated Financial Statements – 31 December 2012, pp. 48–9.

of efficiency and alleged corruption, which have also been at the heart of the Gazprom unbundling debate, remain unresolved.

Governance and alleged corruption remain key issues

Gazprom governance has been problematic and much-criticized since the company's creation. In the early years, the criticism focused on the 'family business' nature of the Viakhirev management, which had few financial controls.[71] After Miller became CEO in 2001 the previous (mainly Soviet-era) management were replaced by individuals (mainly) from St Petersburg with existing links to the political and corporate elite. However, despite the adoption of new governance principles by the company, allegations and complaints about poor governance and corruption continued.[72] For example, in 2008 former Deputy Prime Minister (responsible for energy) Boris Nemtsov and former Deputy Energy Minister Vladimir Milov went public with assertions that Gazprom's subsidiaries had been sold off below cost to entities affiliated with the current President of Russia.[73] Furthermore, governance issues continue to surface in public from time to time – for example the sudden resignation of Deputy CEO and head of production Aleksander Ananenkov in December 2011, allegedly due to holding equity in companies which supplied Gazprom with equipment and services.[74]

Equally important, though, are the instances where President Putin himself has been alleged to have a personal relationship with individuals who subsequently became prominent in domestic gas and export businesses – the most often cited being Gennadii Timchenko CEO of Gunvor, and Matthias Warnig, Managing Director of Nord Stream.[75] For example Stroitransgaz, which is now owned by Timchenko,[76] has close links with Gazprom (which retains a small equity interest),[77] and has been involved in

---

[71] Viakhirev himself, in an interview shortly before his death, hit back with a scathing account of the motivations of those in and around government in the 1990s. 'Ekskliuzivnoe interviu Rema Viakhireva: Putin kogda uslishal, chto ia ukhozhu, tak obradoval'sia', *Russian Forbes Magazine*, 11 September 2012.

[72] OAO Gazprom Annual Report 2012, pp. 108–9.

[73] Nemtsov B. and Milov, V. *The Nemtsov White Paper, Part II, Putin and Gazprom*, 2008; for a less inflammatory account see Pirani, S., *Change in Putin's Russia: Power, Money and People* (London, Pluto Press, 2010), pp. 95–6.

[74] 'Ananenkov leaves Gazprom, Markelov from Tomsk to take his place', *Interfax*, 29 December 2011–11 January 2012, pp. 33–4.

[75] 'Thirteen: Vladimir Putin's Friends, Colleagues, Neighbours and Acquaintances', *Novaia Gazeta*, January 2011.

[76] 'Timchenko buys Stroitransgaz', *Nefte Compass*, 4 June 2009.

[77] 'Companies with Gazprom's participation and other affiliated entities', Gazprom website. www.gazprom.com/about/subsidiaries/list-items/.

a number of major infrastructure projects. One in particular, the Sakhalin–Khabarovsk–Vladivostok pipeline, was regarded by a number of commentators as excessively expensive and a source of significant wasted overspend to the benefit of contractors.[78] While it is impossible to say how Gazprom's management may or may not be linked to specific deals, the presence of influential figures from the Russian elite in the gas industry does point to the influence of the president across the sector as a whole, suggesting that he is intimately involved in establishing the balance of power.[79] Putin's influence is, of course, also evident at Gazprom in his continued support for company CEO Miller, with whom he worked in St Petersburg. Miller has been at the helm of Gazprom for more than a decade and looks to have no immediate challenger, despite having looked periodically shaky due to corporate and personal health problems.[80] Putin clearly regards him as a trusted lieutenant – a man with no personal political ambitions who carries out orders competently – and might have difficulty finding a similar candidate for such an important position. However, this means that Putin himself must take some specific responsibility for the continuing allegations of corruption and poor governance. But his most recent intervention on the subject, when asked whether there were corrupt elements within Gazprom – 'there probably are, but police should catch them and throw them in prison'[81] – hardly seems a satisfactory response to what is perceived as a widespread problem both within and outside Russia.[82]

From this it may be concluded that the status quo in the gas sector, with Gazprom remaining the dominant player, but with well-connected third parties able to take advantage of specific opportunities, is the preferred

---

[78] 'Is Gazprom losing its market' *Rus.business news*, 21 January 2013, described the SKV pipeline as 'insanely expensive' and reported that the Federal Anti-Monopoly Service had alleged that Gazprom manipulated the contract bidding process to exclude multiple suppliers.

[79] Pirani, S., *Change in Putin's Russia*, op. cit., pp. 95–6.

[80] 'Gazprom's demise could topple Putin' *Bloomberg*, 9 June 2012,.

[81] 'Corrupt Gazprom officials, if there are any should be caught, imprisoned – Putin', *Interfax*, 26 September–3 October, 2012, pp. 48–9. There have been particular problems and accusations in relation to Gazprom's purchase of large diameter pipe. 'Stepashin expects Gazprom check to produce "interesting results"', *Interfax*, 9–13 May 2013, pp. 31–2.

[82] For specific allegations linking Putin's and First Deputy Prime Minister Shuvalov's financial interests to Gazprom see: Belton, C. 'A realm fit for a tsar'; and Belton, C. 'Investments raise question of Kremlin cronyism', *Financial Times*, 1 December 2011 and 28 March 2012. For a more general and somewhat extreme view of Gazprom's corruption problems see Aslund, A. 'Why Gazprom resembles a crime syndicate', Peterson Institute for International Economics, 2012, and Aslund, A. 'Gazprom's demise could topple Putin', op. cit.

option for the political elite. Some individuals and their companies, specifically Igor Sechin and Rosneft in particular, and even some leading politicians, have been pressing the case for an increased role for the gas Independents. It is likely, however, that any successful argument for change will not be based on an anti-corruption drive (despite occasional attempts to clamp down on corruption in Gazprom, such as those inspired by President Putin's statements in 2012[83]) or an ideological desire to improve Gazprom, but rather on the direct need for the Russian economy to have a more competitive and proactive gas sector. If Gazprom can start to deliver growth in exports (especially to Asia), competitive gas to the domestic market, and increased revenues to the Russian budget then it will almost certainly survive in its current form for the foreseeable future. If, however, it appears that Russia is losing its place in the global gas market and is not providing the Russian economy with adequate revenues and a competitively priced gas supply, then it is possible that the commercial argument for change may overcome the political desire for maintenance of the current system. In any case, change is likely to be slow, as the appetite for 'big bang' style reform has been dampened by memories of the privatization process in the 1990s. But the outcome of the debate about Russia's role in the Asian market may provide clues to the overall direction of the Russian gas sector.

Oil and Power – successful diversification or major distraction?

Notwithstanding the debate about reform, and even possible break-up, of Gazprom outlined above, the company has in fact spent the past decade accumulating oil and power assets as part of the Russian government's drive to re-assert its control over the energy sector. To an extent, the acquisitions have also mirrored the company's own ambition to expand both upstream and downstream, with a prime example being the purchase of Sibneft, now Gazprom Neft, in 2005.[84] Gazprom has been a producer of small amounts of condensate since its inception,[85] but its desire to become a more rounded energy company was first seen in the proposal for it to merge with Rosneft in 2004.[86] In the event, the merger met significant opposition from Rosneft's then CEO, Sergei Bogdanchikov, and it ultimately had to be abandoned because of possible international litigation in relation to Rosneft's acquisition of assets from the Russian oil company Yukos (which had been accused of tax evasion, made bankrupt, and its

---

[83] 'Putin slams Gazprom on corruption', Gazeta.ru, 2 October 2012, at http://en.gazeta.ru/news/2012/10/02/a_4796441.shtml.
[84] 'Gazprom buys Sibneft stake for $13.1bn', *Financial Times*, 28 September 2005.
[85] Gazprom produced c.280 kbpd of condensate in 2012.
[86] 'Gazprom in state merger with Rosneft', *Financial Times*, 14 September 2004.

CEO imprisoned).[87] However, in 2005, shortly after the failure of the merger, Gazprom acquired the oil company Sibneft, and subsequently created Gazprom Neft, which by the early 2010s was producing nearly 60mtoe per year.[88]

Gazprom Neft has become a significant player in the Russian oil sector, producing more than 1 mmbpd of oil and condensate in 2012 and accounting for just under 10 per cent of the country's overall liquids output. Interestingly, Gazprom Neft is not only an oil producer but also seems to be becoming an important part of Gazprom's wet gas strategy, having formed a 50:50 joint venture with Novatek to invest in gas condensate assets.[89] The first of these assets is Severenergia which came on stream in 2012.[90] It remains to be seen whether Gazprom Neft is encouraged to invest in other similar assets. In any case, it has effectively become Gazprom's 'liquids diversification', and, as mentioned above, makes a major contribution to the 30 per cent of Gazprom's 2012 revenues generated from liquids. This comes not only from its upstream production but also from its downstream interests in five refineries, as well as retail businesses in Russia, the CIS, and Europe. Gazprom Neft has significant growth plans across all of its businesses, aiming to produce 100mmtoe of oil and gas, refine 70mt of oil, and sell 40mt of oil products by 2020.[91]

Furthermore, Gazprom Neft, in which Gazprom owns a 98 per cent stake, also allows the parent company to maintain some balance in the debate over its position relative to Rosneft. One example of this can be found in Asia, where Gazprom Neft is seeking dramatically to increase its oil exports from around 1mt in 2012 to 8mt by 2015, in order to supply a refinery in Vietnam in which it has invested.[92] In addition, owning an oil subsidiary may also allow Gazprom to compete more effectively in offshore regions, in particular the Arctic, by transferring some of its more difficult assets, such the Prirazlomnoe and Novoportovskoe oilfields, onto a management team more focused on their development.[93] This could start

---

[87] For a complete account of the merger, its failure, and aftermath see Gustafson, T. *Wheel of Fortune: the battle for oil and power in Russia* (Harvard University Press, 2012), pp. 336–46.

[88] For more details of the company's history and business see 'About us', Gazprom website. www.gazprom-neft.com/company/.

[89] 'Novatek, Gazpromneft set up Yamal JV', *ICIS Heren*, 28 July 2010.

[90] 'Novatek, Gazpromneft JV acquires 51% of Severenergia for $1.8bn', *Interfax*, 30 November 2010.

[91] Gazprom Investor Day Presentation, February 2013, slide 45.

[92] 'GazpromNeft could ship 6mt through ESPO in 2 years – Transneft', *Interfax*, 11 November 2013.

[93] 'Gazpromneft goes offshore Arctic', *Barents Observer*, 28 February 2013.

to deflect some of the attention away from the inefficient development of these assets to date, but could also beg the question as to whether it might not be preferable for Gazprom Neft simply to be an independent, state-owned company in its own right, removing all distraction from Gazprom's management. At present this does not appear to be an option, although it may become a possibility if more systematic reform of the Russian gas sector becomes a reality.

Gazprom has also developed significant positions in the Russian power and petrochemical sectors, again combining its role as a vehicle of political control with its commercial ambition to become a more vertically integrated company. It has become a particularly large player in the power generation sector through its subsidiary Gazpromenergoholding, which is the largest electricity producer in Russia, with 38 GW of installed generating capacity (17 per cent of the national total).[94] This has allowed Gazprom to diversify its gas price risk, given the close link between the two industries in Russia, although one might question how successful this strategy has been given that power and heat only contributed 7 per cent of Gazprom's total revenues in 2012. However, from a political perspective, following the privatization of the Russian electricity industry in 2007, it has allowed the Russian government to re-assert greater control over the sector, where three of the top five generating companies are now state-controlled (see Chapter 6). As a result, one might again question whether political objectives have overruled commercial logic.

Conclusions

We concluded from Chapter 11 that Gazprom's strategy in the west is relatively fixed, thanks to the company's decision to invest heavily in the Yamal Megaproject, and that fields on the Yamal peninsula, being relatively high-cost compared to Russian alternatives, will effectively be used as swing production depending on levels of demand in Russia's main domestic and export markets. This may, of course, work in Gazprom's favour, if alternative sources of supply do not materialize as expected, or if demand rebounds faster than currently anticipated, but it does leave the company somewhat at the mercy of its lower-cost competitors and of external market forces.

These potential risks are not reflected in the company's current financial situation, which is fairly robust. Although profits declined slightly in 2012, Gazprom nevertheless generated almost $40 billion of net income, and its balance sheet has a relatively low level of net debt. However, rising costs

---
[94] Gazprom Investor Day Presentation, February 2013, slides 62–71.

and an increasing tax burden are more worrying, especially as there is no sign of these trends receding in the near future, while the company's capital expenditure burden has also been growing rapidly and focusing more on transport infrastructure than exploration and production assets. This reflects not only the fact that investment in the Yamal development has now peaked, but also that Gazprom's plans to invest more in pipelines such as South Stream add little commercial value. A key question, therefore, is whether Gazprom can maintain the balance between financial profitability and political necessity, especially as it seems likely that it will be required to pay an increasing level of dividends over the next few years.

Such dilemmas are being played out in Gazprom's eastern strategy, where the company's domestic and geopolitical goals as a state-controlled entity, and its need to compete with both domestic and international peers in the oil and gas sector, are potentially in conflict. Russia's Energy Strategy to 2030 paints a clear picture of increased exports to Asia, and this objective fits well with Gazprom's own need to find a secure source of volume and revenue growth, as its western markets become increasingly competitive. It would also provide an important source of new tax revenues, as Russia faces the possibility of declining oil revenues over the next decade. The outcome of Gazprom's negotiations with CNPC are thus of much more than mere corporate relevance.

However, the complexity of the export discussions, combined with the added need to improve gasification in the Russian Far East and support industrial growth there through the construction of new infrastructure, appears to be causing confusion over which assets Gazprom is prioritizing. If a deal on pipeline exports with China is concluded, then the construction of the Power of Siberia pipeline can underpin both the Vladivostok LNG plant and resolve the issue of where Sakhalin-3 gas would best be used, as it would then be free to supply an expanded Sakhalin-2 LNG plant. However, if no deal is concluded by the end of 2014, then indecision could continue, raising concern about Gazprom's ability to efficiently provide a significant supply of gas for a new market. Any extended indecision could not only affect the willingness of Asian buyers to make contractual commitments, but could also undermine Gazprom's overall status – especially as two of its major competitors in the domestic market, Rosneft and Novatek, appear to be progressing their own projects rapidly.

Given President Putin's acknowledged influence over Gazprom, one might have expected that the Russian government would have a definitive opinion on this debate, but at the time of writing this does not appear to be the case. While continuing to encourage Gazprom to complete a deal with China, the government has also responded to the promptings of Novatek and particularly Rosneft (and its influential CEO Igor Sechin) by removing the monopoly on LNG exports to allow competition between Russian companies in the Asian market. There is no sign, as yet, that this is meant to be seen as a precursor to Gazprom's overall position being undermined, as it appears that the Russian political elite, and especially Putin himself, are keen to see a balanced status quo maintained, with Gazprom at its heart. However, the balancing act that Gazprom is being asked to perform – between behaving as a competitive commercial entity on the one hand and a state-owned utility on the other – seems to be becoming much harder in the face of the rising challenges it faces. In addition, the distractions of a growing oil business and the largest power business in Russia could also undermine the ability of Gazprom's management to meet all its commercial and political objectives, although they do provide some significant diversity of revenues. One possible conclusion, therefore, is that if it becomes clear that Gazprom cannot do all that is asked of it, then a restructuring of the company and the gas sector may become a political, as well as a financial, necessity. Of course reform and a 'break-up' have been discussed many times before and have been rejected equally frequently, and there is certainly no sign that it is the preferred option. However, developments in the east of Russia could provide an important signal as to whether an alternative, more radical, option is being considered.

CHAPTER 13

NON-GAZPROM RUSSIAN PRODUCERS: FINALLY BECOMING TRULY INDEPENDENT?

James Henderson

A brief history of third-party gas supply in Russia

Until the mid-2000s, Gazprom dominated the Russian gas sector to such an extent that the concept of 'independent' third-party suppliers was a misnomer. Any producer of gas in Russia, or supplier of imports into Russia, relied on Gazprom to allocate room in the annual Gas Balance, to provide space in the pipeline, and usually to buy the gas itself. Russia's state-owned gas company enjoyed an effective monopoly as distributor of gas to the domestic and export markets. Indeed, Gazprom's position was underpinned by its position as the sole exporter of Russian gas to non-CIS countries as this was, and still is, where the bulk of its revenues are generated. However, in this chapter we will argue that Gazprom's dominant role in the Russian gas sector is being challenged by the growing volume of third-party gas produced and sold in an increasingly competitive market place, where customers have a real choice as to whether to buy gas from Gazprom or from suppliers who can genuinely start to be called 'Independent'.

This shift in the Russian gas market has been a very gradual one, but it is nevertheless informative to trace briefly the changes that have taken place in the post-Soviet era. In the immediate aftermath of the collapse of the Soviet Union in 1991, the collapse in GDP led to a decline in energy demand and to a general lack of investment in infrastructure, as revenues and cash flows fell sharply. In the oil sector, this led to a collapse in oil production,[1] but in the gas sector vigorous efforts were made by government to prevent such an outcome, because both economy and population were highly reliant on gas (accounting for half of total primary energy supply[2]) which effectively stopped people from freezing in the depths of winter. Gazprom was kept as a single entity, rather than being broken up as the oil sector was, and became the foundation of the Russian energy economy. It provided funds to the government on request, sustained significant non-payments from residential and industrial customers, and invested in

---
[1] From 11mmbpd in 1990 to 6mmbpd in 1997 according to data from the Ministry of Energy.
[2] IEA *Russia Energy Survey 2005* (IEA, Paris, 2005).

infrastructure, but most importantly, it kept gas flowing to the domestic market in spite of the losses it was making there.[3] In return, it was granted a monopoly on sales to export markets, from which it could generate the profits from high oil-linked prices to subsidize its domestic market losses.

From a supply perspective Gazprom was able to rely on its Soviet heritage: its three huge low-cost gas fields (Medvezh'e, Urengoi, and Yamburg). The gradual decline in output from these fields was matched by the falling demand in the Russian domestic market that resulted from the slump of the early 1990s. As a result, the supply–demand balance for Russian gas was maintained without much need for increased upstream investment throughout the 1990s. During this period Gazprom's 'Big 3' fields accounted on average for 75 per cent of Russian gas supply, with another 15–20 per cent supplied from other Gazprom fields, and the remainder from associated gas produced by Russian oil companies, with a varying amount of imports from Central Asia.[4] Gazprom was completely dominant.

This picture started to change after the economic crisis of 1998–99. The devaluation of the ruble and a rising oil price created the environment for a recovery in Russian GDP and a consequent increase in energy, and especially gas, demand. Although this was beneficial for Gazprom's finances, as domestic prices started to rise and non-payment declined, it did start to put pressure on gas supply: demand and supply from Gazprom's main fields were now moving in opposite directions. Although Gazprom did bring some new fields onstream, in particular Zapoliarnoe, it also began to become increasingly reliant on sources of third-party gas. As discussed in Chapter 11, this initially took the form of increased imports from Central Asia, but gradually also incorporated purchases of third-party domestic supply, including more associated gas from the oil companies, but also some new dry gas from independent producers. Novatek, which has become Russia's second largest gas supplier, produced its first gas in 2003 and by the time of its initial public offering (IPO) in 2005 had more than doubled its initial production to 25 Bcm/year.[5]

Although Gazprom developed a strategy of replacing declining production from its core fields with new output from smaller satellites, it was also facing the reality that it would soon need to develop the high-cost supergiant fields in its portfolio, in particular those on the Yamal peninsula. However, it was keen to delay the large expenditure requirements for these fields for as long as possible. Following the global economic crisis in 2008–9, gas demand in Europe fell sharply and it became unclear whether there would be sufficient demand to justify the huge capital expenditure required

---

[3] Landes, A. 'Gazprom – Ready for Lift Off', Renaissance Capital, Moscow, 2001
[4] Stern, J. *The Future of Russian Gas and Gazprom* (Oxford: OIES/OUP, 2005).
[5] *IPO Prospectus*, Moscow: Novatek, 2005.

to develop these assets. Gazprom therefore increasingly relied on third-party gas, and the market share of Independent producers in Russia increased from 10 per cent in 2000 to more than 20 per cent by 2009.[6] At the same time, Gazprom's dependence on domestic third-party supply increased further, because imports from Central Asia suddenly became much more expensive, after Turkmenistan, Uzbekistan, and Kazakhstan achieved prices based on a European export netback in their negotiations with the Russian government and Gazprom.[7] The impact on Gazprom was to deprive it of a cheap source of alternative supply and to increase its reliance on third-party domestic producers to fill gaps in the supply–demand balance.

In the late 2000s, this reliance was very much seen as cooperation between third-party suppliers and Gazprom, with the former keen to emphasize that they were helping to meet domestic demand in order to free up more Gazprom gas for export sales.[8] However, three key factors have now changed this dynamic. The first is that demand for Gazprom's gas in export markets has been in decline since 2008. Sales to Europe fell from 168 Bcm in 2008 to 139 Bcm in 2012, while exports to FSU countries declined even more dramatically from 96 Bcm to 66 Bcm in the same period.[9] A natural assumption might have been that this extra gas would have been sold by Gazprom into the domestic market, crowding out third-party producers who had previously been reliant on Gazprom for access to customers and for pipeline capacity to move their gas. However, the impact of the second and third key factors, namely rising domestic gas prices and increased third party access to the trunk pipeline system, have changed the dynamics of the domestic market completely, encouraging the Independents to compete, rather than co-operate, with Gazprom.

In Chapter 5 we have described how regulated domestic gas prices have been rising consistently since 2000, and that although President Putin's 'netback parity' target is no longer seen as realistic, it has nevertheless acted as a catalyst to push the domestic gas price towards a level reflecting supply and demand. The most important piece of evidence for this assertion is that in 2012, for the first time ever, Independents began to offer gas at prices that were discounted from the regulated price. As Gazprom has historically been mandated to sell its gas at the regulated price, albeit with a few options to offer minor flexibility, this effectively put it at a competitive disadvantage for the first time in its history. Figure 13.1 shows the average

---

[6] Data from various editions of *Nefte Compass*, published by Energy Intelligence Group, and from company sources.
[7] See Chapter 11 for a fuller discussion of this issue.
[8] See 'Novatek equity stake sold to Gazprom' Novatek Press Release, 29 September 2006.
[9] Gazprom Databook 2012.

price at which Novatek has been selling gas to consumers versus Gazprom's average price since 2004, and demonstrates clearly that in 2012 it offered an average 2 per cent discount, although many of its new contracts are believed to offer 5–10 per cent discounts to the regulated gas price.[10]

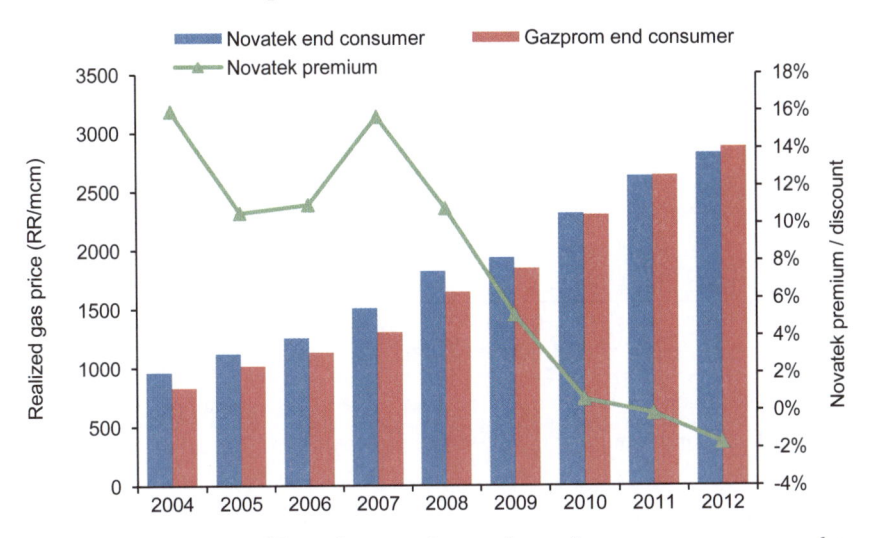

Figure 13.1: Average Novatek gas price to domestic consumers compared to average Gazprom price

Source: Novatek and Gazprom Management Reports, 2004–12.

The significance of this change for the independent gas sector in Russia is that it has effectively opened the whole domestic market for competition between suppliers. Historically, Gazprom has always been the first logical source of gas for consumers, not only because of its huge production but also because of its enforced low regulated price. Indeed Gazprom would have been forced to supply the entire market at regulated prices if the Russian government had not decided to fix the volumes of regulated gas sales at the levels supplied in 2007.[11] As a result, third-party producers have only ever 'competed' in the premium gas segment, where consumers were prepared to pay higher prices for extra gas. However, from 2012 this situation has been turned on its head. Consumers would now prefer to buy independent gas at a discount to the regulated price, rather than Gazprom's now higher-priced regulated gas. Of course this is not possible for much of the time because of Gazprom's huge share of the supply portfolio in Russia,

---

[10] 'Henderson, J. Evolution in the Russian Gas Market: The Competition for Customers', Working Paper NG 73, OIES, January, 2013, p.27.

[11] Russian Govt. Directive No.333 on Improvement of State Gas Price Regulation, dated 28 May 2007.

but the shift in momentum is clear. Independent suppliers now produce 27 per cent of Russia's gas and satisfy almost 40 per cent of domestic demand.[12]

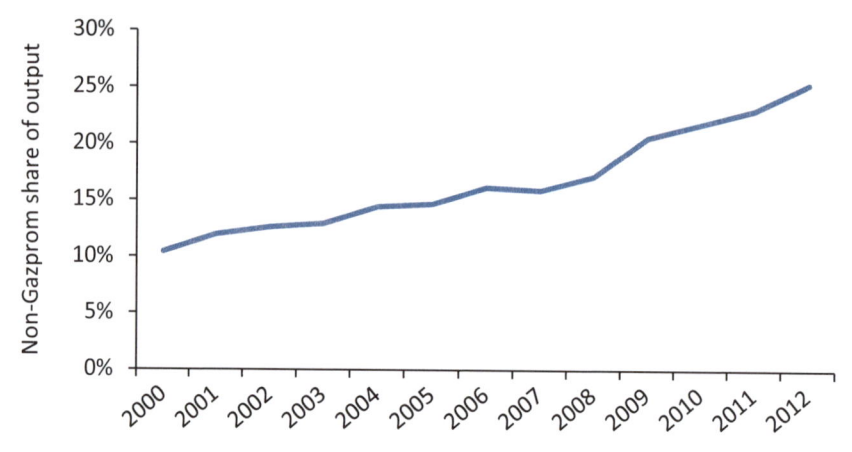

Figure 13.2: Share of Independent producers in Russian gas supply
Source: Company reports and Ministry of Energy Data.

This increase in Independent producer market share could not have happened without the third key factor listed above, namely the increasing amount of trunk pipeline capacity made available to third-party suppliers. Under the Gas Supply Law, adopted as early as 1999,[13] Gazprom had been mandated to provide third-party access to its trunk pipeline system, but while its own output from its core fields in West Siberia had remained high it was always able to use the excuse of lack of available capacity to undermine attempts by independent producers to move their gas. However, a combination of declining output from Gazprom's major fields and increasing government insistence that TPA rules should be applied correctly (which included a demand from President Putin himself[14]) have seen Independents now able not only to gain access to the pipeline system more freely, but also to sign long-term contracts with consumers. This latter change has been vital because it has allowed producing companies to secure long-term sales to underpin new projects, which had previously been impossible as Gazprom had been reluctant to guarantee pipeline access for more than 1–3 years. Contracts with durations of up to 15 years can now be signed with consumers, ensuring that investments in new fields can

---

[12] See Chapter 5 on the Russian gas market for explanation of market share statistics.
[13] Burgansky, A. *Stand and Deliver: Oil and Gas Yearbook 2010* (Moscow: Renaissance Capital, 2010), p.180.
[14] 'Gazprom considers easing access to pipelines after Putin rebuke', *Bloomberg Business Week*, 25 September 2008.

be made with more commercial security. (See also Chapter 5.)

One final catalyst for rising independent production has been the fact that the economics of the fields owned by non-Gazprom gas companies tend to have features that allow them to be very competitive: they may be close to existing pipeline infrastructure; they can also be close to consumers where distribution networks have been developed in regions outside West Siberia; the gas may be associated with oil production and therefore have priority by benefiting from the government demand that gas flaring be reduced; and most importantly, the gas is often combined with condensate production. This latter point is most important, as the production of liquids in combination with gas helps to underpin the economics of the overall field, allowing the supplier to price gas very competitively, especially when the revenues from liquid sales are related to the current high level of oil prices. This production of 'wet gas' has provided a particular competitive advantage for Independent producers, as Gazprom has tended to focus on its more traditional dry gas output. Novatek, in particular, has demonstrated the high margins that can be generated from combined gas and condensate sales.[15]

As a result of all these factors, Gazprom has found that third-party production – which it encouraged in the mid-2000s as it attempted to delay its own capital expenditure plans while keeping supply and demand in balance – has now become a competitor rather than a source of cooperation. The size of Gazprom's production base of course ensures that it will be a major player for many years to come, and its status as a state company continues to give it advantages, but the strategies of third-party producers in Russia are becoming truly independent of the former quasi-monopoly supplier. Below we will discuss in more detail how the ranks of the 'Independents' have now swelled to include not only pure gas companies such as Novatek, but also oil companies that are keen to diversify their production and have noted the high margins that can now be made in the gas sector. Furthermore, the opportunity to access export markets for gas is also emerging (see also Chapter 8). This further increases the incentive to compete with Gazprom, and as a result it seems likely that the growth that the independent gas sector has experienced in the past 12 years will continue for the remainder of the 2010s at least.

Novatek – the largest independent producer continues to expand

Although Novatek was not the first serious independent gas producing company in Russia (Itera, formed in 1992, claims that honour)[16] it nevertheless

---

[15] According to the company's 2012 financial statements 10% of Novatek's production was in the form of liquids but it generated 32% of total oil and gas revenues.

[16] Henderson, J. *Non-Gazprom Gas Producers in Russia* (Oxford: OIES, 2010).

has been the most active player to exploit the emerging market niche for third-party producers over the past 10 years. Although Novatek's major shareholders purchased their first upstream assets in 1994, Novatek itself only came into existence in 2003, when it was producing around 12 Bcm/year from the East Tarkosalin and Khancheiskoe fields and was operating very much in Gazprom's shadow.[17] Indeed so close was the link between the two companies that shortly after Novatek's IPO in 2005 Gazprom purchased a 19.9 per cent stake in its smaller rival, which at the time was widely viewed as a positive move for Novatek.[18] Gazprom's share has since been reduced to 9.9 per cent, and its influence appears to have declined even further, the most important shareholders at Novatek now being Gennadii Timchenko, widely regarded as having close links to President Putin, and the French oil company Total, which at the time of writing owns a 16 per cent strategic stake that it is aiming to increase to 20 per cent.[19]

The decline in Gazprom's influence at Novatek has not hindered the company's progress, however, as its production has more than doubled from 25 Bcm in 2005 to 57 Bcm in 2012.[20] This growth has largely been based on the development of a single asset, the Iurkharovskoe field, which reached peak output of 34 Bcm/year in 2012. Iurkharovskoe is a gas and condensate field located just north of the Arctic Circle in the Nadym Pur Taz region of West Siberia, and is the basis not only of Novatek's gas output but also of its strategy for condensate production and export.[21] The field is linked to Novatek's newly constructed Purovskii processing plant – which has recently been expanded from its original capacity of 5 mtpa to reach 11 mtpa in early 2014[22] – where the condensate is stabilized and processed before being sent either to the port of Vitino on the White Sea for immediate export or to the company's new fractionation plant at Ust Luga.[23] Here the stable condensate is broken down into various higher-value products before being sold onto global markets. The importance of this condensate production and refining process is that, although only 10 per cent of Novatek's output comes in liquid form, the company generates between 30 per cent and 40 per cent of its revenues from this source (see Figure 13.3), allowing it to be much more competitive in its gas business.

---

[17] *IPO Prospectus*, Novatek, 2005, p.103.

[18] 'Gazprom closed deal to purchase 19.4% of Novatek', Gazprom Press Release, 29 September 2006.

[19] 'Total increased its stake in Novatek', Novatek Press Release, 26 June 2013.

[20] Novatek Management Discussion and Analysis of Financial Statements 2012.

[21] Novatek Annual Report 2012, p. 28.

[22] *Eurobond Prospectus*, Moscow: Novatek, December 2012, p. 2.

[23] Ibid., p. 117.

Non-Gazprom Russian Producers 321

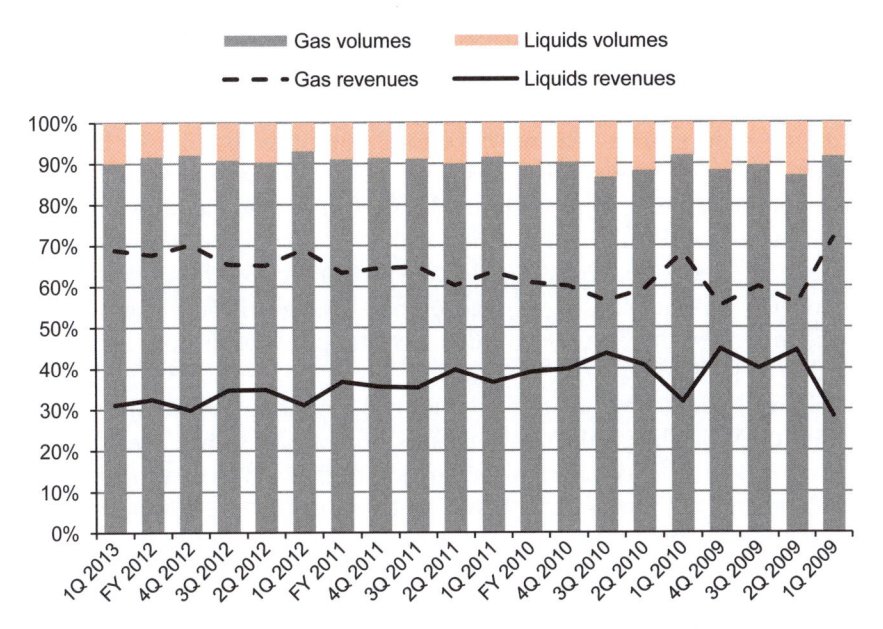

Figure 13.3: Share of Novatek revenues from gas and liquids sales
Source: Novatek IFRS Financial Statements.

Despite its recent growth, Novatek continues to have very ambitious plans for the future, and has announced an overall target to produce more than 110 Bcm of gas and 13 mt of condensate by 2020.[24] It aims to achieve this goal through a combination of organic growth, involving the development of two major new projects, and via the inorganic acquisition of other domestic gas assets, essentially playing a role as one of the consolidators of the independent gas sector. This latter process has already begun through the purchase of an effective 59.8 per cent interest in the Severenergia joint venture, whose other owner is Gazprom Neft,[25] and the purchase of 50 per cent of North Gas in 2013.[26] Both of these assets include 'wet' gas fields, and both were purchased from Gazprom, indicating that the competitive advantage which Novatek has enjoyed with its condensate production over the past few years is set to continue.

---

[24] Novatek Strategy Day presentation, December 2011.

[25] Novatek initially acquired a 25% stake in Severenergia – see 'Novatek, Gazprom Neft JV acquires 51% of Severenergia', *Interfax*, 30 November 2010 – and then increased this to 59.8% after a swap deal with Rosneft that was concluded towards the end of 2013 – see 'Novatek increased its stake in Severenergia to 59.8%', Novatek Press Release 15 January 2014.

[26] 'Novatek increases its stake in North Gas to 50%', Novatek Press Release, 26 June 2013.

The first Severenergia field, Samburgskoe, came into production in 2012 and is set to peak at an output of 8 Bcm/year in the next few years. It will be supplemented by production from three other fields in the joint venture's licence areas that will take total output as high as 35 Bcm by 2020, giving Novatek equity production of 21 Bcm/year.[27] North Gas, which is now a 50:50 joint venture with Gazprom following the purchase of an extra 1 per cent in June 2013 to add to the 49 per cent stake Novatek bought in December 2012, is already producing gas from the North Urengoi field, but plans to expand output through the development of additional reservoirs in order to increase total production from 4 Bcm in 2012 to 9–10 Bcm by 2020.[28]

However, these new acquisitions, although important for the short-term growth of Novatek's production, pale into insignificance when compared to the impact of the company's two major organic growth projects: the development of the Yamal LNG scheme and the exploration and development of the company's licences on the Gydan peninsula. The South Tambei field, which will form the basis of the Yamal LNG project, contains 900 Bcm of 2P reserves and is expected to produce up to 28 Bcm per annum once it reaches peak output, as well as providing 1 mt per annum of condensate to boost the revenues from the field.[29] The key question, though, is at what pace Novatek's ambitious plans to bring the field on stream, and to complete liquefaction and transport facilities, can be achieved. When Novatek announced the final investment decision on the project, in December 2013, it said the first LNG train would be launched in 2017, in contrast to earlier announcements that it was aiming for a start in 2016.[30] The Russian government appears to be very keen for the project to succeed, as part of a plan to develop infrastructure in the Far North, and it has provided significant support both in terms of tax breaks and direct investment in the key facilities. In October 2010 it was announced that the Yamal LNG project would be exempt from the MET gas royalty tax for the first 250 Bcm of gas production and the first 20 mt of condensate output.[31] The project also already benefits from the overall export tax exemption enjoyed by all LNG projects in Russia, meaning that the scheme will effectively only be paying corporation tax at 20 per cent during the initial period of its

---

[27] Novatek Annual Report 2012, p.30.
[28] Novatek presentation, March 2013, 'Harnessing the Energy of the Far North', slide 7.
[29] Ibid., slides 10–20.
[30] 'Final investment decision made on Yamal LNG project', Novatek press release, 18 December 2013.
[31] 'Novatek's Yamal LNG project may be exempt from extraction tax', *Bloomberg*, 11 October 2010.

operations.[32] Furthermore, it has been announced that the state will partly finance the construction of the port and harbour facilities at the Port of Sabetta, and state-owned shipping company Sovcomflot has agreed to construct two LNG tankers for the project, with financing being provided by state bank Vneshekonombank.[33] In addition, the construction of a further 16 ice-class LNG tankers will be undertaken by Daewoo Shipbuilding of Korea, with financing to be provided by third-party shipping companies chosen by Yamal LNG.[34] This once again suggests that Russian state involvement may be provided to assist with the economics of the project.

However, the relatively unique nature of the ice-breaking tankers being constructed for Yamal LNG emphasizes the technical difficulties associated with the project, which could inflate the $27 billion estimated cost and delay first production beyond 2017. The South Tambei field is located in a region of north-west Siberia where the average temperatures are −9 °C and can go as low as −57 °C in the depths of winter, creating permafrost up to a depth of 500 metres and pack ice for 300 days up to a maximum depth of 2 metres.[35] (See maps 11.1 and 11.2.) Although the cold weather can have some benefits, as it reduces the amount of energy required to liquefy gas and so cuts operating costs, it is likely to create considerable problems for construction and logistics. For example, all the facilities will have to be built on stilts because of the permafrost, and the transfer of equipment by river to Sabetta will only be possible between July and October, complicating the development and operating phases of the project.[36] (The site of the planned Yamal LNG terminal at Sabetta is shown in Map 11.2.)

As a result of these complications, Novatek introduced Total as a partner into the project in 2011, and the French IOC has expressed its confidence that the technical issues surrounding the project can be resolved.[37] However, a further complication has been the provision of project financing to undertake the upstream part of the development, which will involve construction of 35 well pads and the drilling of 208 production wells. Banks were reluctant to lend to the project until it became clear how the revenues will be generated; this involved both the signing of LNG sales contracts

---

[32] Novatek Eurobond Prospectus, December 2012, Moscow, p. 140.
[33] 'Novatek signed an agreement on construction of two LNG tankers for Yamal LNG', Novatek Press Release, 20 June 2013, St Petersburg.
[34] 'Yamal LNG names tender winner among shipyards and signs agreement to build LNG tankers', Yamal LNG Press Release, 4 July 2013, Moscow.
[35] Novatek presentation, March 2013, 'Harnessing the Energy of the Far North', slide 11.
[36] Novatek Presentation, 16 April 2013, 'Harnessing the Energy of the Far North', Sberbank Conference – the Russia Forum, Moscow, slides 10–20.
[37] 'Total to buy $4bn stake in Novatek' *Financial Times*, 3 May 2011.

and the related issue of whether the Yamal LNG project would have the right to market its own output, given that up until 2013 Gazprom had a monopoly on all Russian gas exports, including LNG.[38] In 2010 Novatek signed an agency agreement with Gazprom in an attempt to get around this problem,[39] but this was not enough to satisfy banks that do not want to see any break in the link between production and sales. This financing issue has been one of the key practical catalysts for the removal of Gazprom's monopoly on LNG exports that, in late 2013, was resolved in favour of the non-Gazprom players.

In June 2013 Novatek announced that it would be selling a further 20 per cent interest in Yamal LNG to the Chinese company CNPC, who would also purchase up to 3 mtpa of LNG output from the project.[40] This deal was finally confirmed in January 2014, and provided concrete proof that buyers were interested in purchasing LNG from Yamal. It was announced that not only would CNPC take a 20 per cent stake but, with Total, would pay a disproportionate share of the project costs (meaning that Novatek will only pay around 20 per cent of the capital expenditure despite having a 60 per cent stake).[41] Other potential sales agreements have been discussed by Russian Energy Minister Alexander Novak, including deals with BP, and trading entities supplying both European and Asian markets.[42] A second concrete deal, to export gas westwards, was signed in November 2013 with the Spanish utility Gas Natural, for the sale of 2.5 mtpa.[43] This deal practically coincided with the Russian government's announcement that it had approved the lifting of the LNG monopoly. It was stressed that this first deal in Europe does not provide competition to Gazprom, because no Russian gas can reach the Spanish market via pipeline, although of course Novatek's LNG could undermine Gazprom's attempts to develop its own west-facing LNG projects. Nevertheless, the Gas Natural deal does confirm that the Russian Ministry of Energy will take a relatively relaxed stance on the destination for Yamal LNG, providing further comfort to project sponsors and financiers that there will be little, if no, interference in their sales plans. Given this positive news, it seems

---

[38] 'Novatek, Gazprom plan more LNG capacity for Asia', *Reuters*, 10 January 2013.

[39] Novatek Eurobond Prospectus, December 2012, p. 140.

[40] 'Novatek concludes framework agreement on CNPCs entry into Yamal NG', Novatek Press Release, 21 June 2013.

[41] 'Novatek closes sale of 20% interest in Yamal LNG to CNPC', Novatek Press Release, 14 January 2014, Moscow.

[42] 'BP eyes gas from Russia's Yamal LNG project – sources', *Reuters*, 22 May 2013, London.

[43] 'Yamal LNG contract with Spanish Gas Natural does not compete with Gazprom – Novak', *Interfax*, 1 November 2013.

reasonable to assume that Novatek's target of 16.5mt of LNG output from three liquefaction trains can be achieved over the next decade although, given the technical and geographical challenges involved, it is perhaps realistic to expect that it will come on stream a year or two later than planned and not reach peak output until the early 2020s.[44]

Novatek's other major upstream project, located on the Gydan peninsula on the opposite side of the Ob River from the Yamal LNG project, is based around two licences acquired in 2011, and contains two major fields, Geofizicheskoe and Salmanovskoe, with combined reserves of almost 1 Tcm of gas and 45mt of condensate.[45] Geofizicheskoe, the smaller of the two fields, is located closer to the main trunk pipeline system (which starts at Gazprom's Yamburg field, 260km to the south) and so is likely to be developed first, with an estimate of first output in 2016 or 2017.[46] Salmanovskoe would follow a year or two later, with combined production from the two fields estimated to reach 30 Bcm by 2020. The fact that the fields have been designated as potential supply for an LNG scheme has encouraged the Russian government to offer significant tax breaks for their development, including similar MET relief to that offered for the South Tambei field on Yamal.[47] It is unclear whether all the gas from the Gydan fields will go to export markets, with the volume of production suggesting that at least some of it may be destined for domestic consumers, but Novatek has announced that it is currently planning to develop another 3-train LNG plant for the gas output and so it is clear that the field developments will be export-led.[48] As a result, it remains to be seen whether the aggressive timetable set out by Novatek will be met, but it seems clear that government support is encouraging Independent gas supply growth once more.

Overall, Novatek has assets that can sustain production growth over the next ten years, but the rate of growth will depend upon technical and commercial factors which could mean that the company's target of 112 Bcm of output by 2020 may not be met. Figure 13.4 shows a slightly more conservative forecast, with gas output reaching 100 Bcm by 2020, on the assumption that output from Yamal LNG and the Gydan peninsula is somewhat delayed. Nevertheless, Novatek would still have increased gas production by 80 per cent over the 57 Bcm it produced in 2012. It would also

---

[44] Novatek, 21 March 2013, 'Harnessing the potential of the Far North', presentation at HSBC conference, slide 10.
[45] Novatek Eurobond Prospectus, December 2012, 140–1.
[46] Novatek presentation for Strategy Day, 9 December 2011, slide 38.
[47] 'Gas produced at Gydan could get tax, customs breaks if gas liquefied on Yamal – resolution', *Interfax*, 23 December 2013.
[48] 'Government: Russia's Novatek to boost LNG production by 2025', Rigzone, 23 December 2013,

have continued to grow its liquids production, with total condensate output expected to reach 13mt by 2020, an increase of more than 100 per cent.

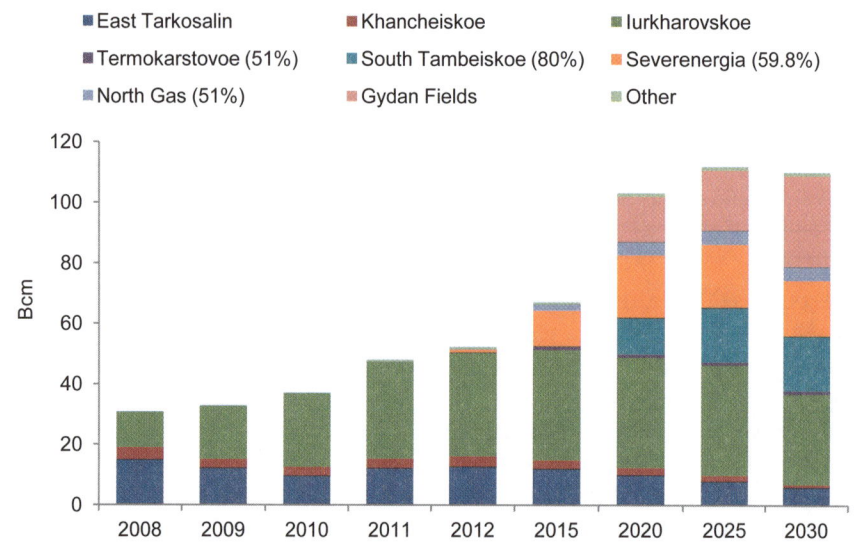

Figure 13.4: Estimate of potential Novatek gas output
Source: Company data and author's estimates.

However, the realization of even this more conservative production profile will depend upon Novatek's ability to find customers for the gas. As discussed above, active negotiations to find buyers for export LNG were already well underway at the time of writing, and Novatek is also one of the most prominent marketers of gas in the Russian domestic market. In Chapter 5 we analysed the changing face of the gas market in Russia, with consumers increasingly having a choice of suppliers at prices related to the regulated price but often below it, leading to increasing competition for customers. This has meant that any company with growth ambitions has been forced to focus its attention either on creating a downstream business that can consume its own gas, or a sales business to sign up buyers. Novatek has chosen the latter option, and in 2013 had a marketing business operating in 35 regions of Russia, providing more than 16 per cent of the gas that was supplied to consumers through the UGSS trunk pipeline system. Its main focus remains the local Yamal Nenets region where its gas is produced, but it also has significant operations in Khanty Mansiisk, Sverdlovsk, Orenburg, Chelyabinsk, Moscow, and St Petersburg. Furthermore, Novatek has been expanding its marketing business through a series of acquisitions, most recently of Gazprom Mezhregiongas Kostroma, the distribution business of Gazprom in the Kostroma region, in order to accelerate its sales reach.[49]

The benefits of this strategy have been seen in both the quantity and quality of the company's customer base. Between the beginning of 2012 and the middle of 2013 Novatek signed a number of significant new contracts that not only increased its sales volumes but also extended the length of its sales profile. Five year deals were signed with Severstal and Uralchem, a 10 year agreement was reached with MMK, an 11.5 year deal with Mechel, and 15 year agreements with E.ON Russia and Fortum, providing Novatek with total sales over the period of all the deals combined of almost 300 Bcm of gas. Significantly, Novatek also signed a three year sales agreement with Mosenergo, a subsidiary of its main competitor, Gazprom, indicating just how aggressive customers have become in their search for low prices.[50] Indeed this last deal also indicates another trend in Novatek's sales portfolio: towards end users (and in particular power generation companies) and away from sales to traders at the wellhead. Figure 13.5 shows how the share of end consumers in Novatek's portfolio has grown over the past six years, while Figure 13.6 shows that power companies make up the largest share of customers, with large industrial users being the next biggest direct consumer of its gas.

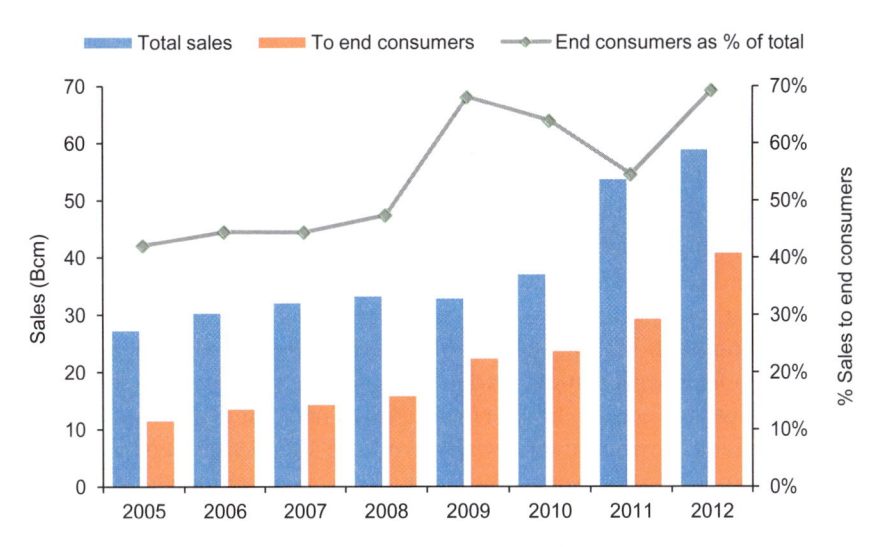

Figure 13.5: Share of end users in Novatek's gas sales
Source: Novatek Annual Reports, 2005–12.

---

[49] Novatek presentation to Sberbank Conference, April 2013, slide 8.
[50] Henderson, J. 'Evolution in the Russian Gas Market', op. cit.

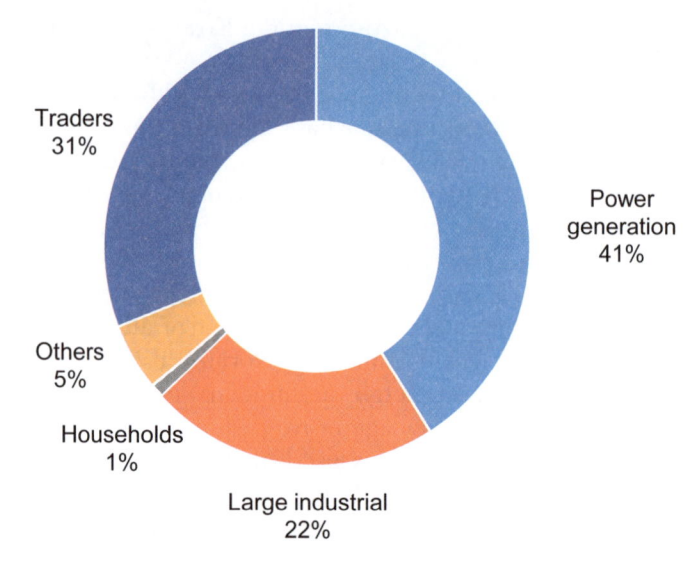

Figure 13.6: Split of Novatek's gas sales in 2012
Source: Novatek Annual Report 2012, p.41.

As Novatek increases its sales effort, it is also having to become more aware of the flexibility that it needs to offer its customers in order to compete with Gazprom and other independent suppliers. Although the details of specific contracts are confidential, it is widely believed that Novatek has been offering prices at a 5–10 per cent discount to Gazprom's regulated price and is also providing much more generous take-or-pay terms (indeed in some cases it has apparently offered complete flexibility to the consumer). In order to provide this service it has also agreed an underground storage deal with Gazprom in order to have gas available in the peak winter months to meet the higher seasonal demand.[51]

A final point about Novatek's marketing business concerns its activities in the European export market, where it signed two deals to supply a total of 2 Bcm/year of gas to the German utility EnBW in 2012.[52] Again, the exact details of the price and contract terms are unknown, but Novatek has revealed that the gas is being sourced from European producers at market prices. This establishes Novatek as a trader and seller of gas in a market in which historically Gazprom has been the sole Russian supplier. Although it is unclear exactly what long-term strategy Novatek is planning in Europe, it certainly seems to be positioning itself for sales of Yamal LNG into the

---

[51] 'Novatek approves RR534bln transport deals with Gazprom', PRIME, 25 April 2013.
[52] 'Russia's Novatek confirms long-term supply deal with EnBW', *Platts*, 14 August 2012.

continent, as evidenced by its October 2013 agreement to sell 2.5mtpa to Gas Natural of Spain,[53] and may even anticipate the ultimate ending of Russia's pipeline export monopoly and the possibility of selling its West Siberian gas into export markets.

Rosneft

Rosneft has had the ambition to develop a significant gas business in Russia since its IPO in 2006.[54] This ambition has been based upon the fact that the company had 992 Bcm of proved gas reserves but only produced 16 Bcm in 2012,[55] meaning that it had a reserve life of over 60 years and obvious potential for significantly increased output and sales. Indeed the company's proved and probable reserves, which totalled 1.36 Tcm, offered a reserve life of 85 years.[56] As a result, in 2008 Rosneft initially announced a target to increase its gas production to 55 Bcm within 10 years,[57] but this goal was consistently moved back as a result of the company's inability to secure access to sufficient domestic customers. The frustration at this constant delay was compounded by the fact that much of the company's gas is located in the gas reservoirs of the Kharampur field, which could be easily and cheaply exploited using much of the infrastructure that has already been used to develop the oil layers of the field.[58] However, Rosneft had been unwilling to commit to the investment required to exploit more than 500 Bcm of gas reserves at the field, which could ultimately produce up to 25 Bcm/year, until it could establish a commercial market for its output.

In 2012, Rosneft finally took a significant step towards achieving its objectives in the gas sector by forming a joint venture with the Itera Group to produce and sell gas owned by both companies.[59] Rosneft initially contributed non-producing assets (its Kynsko-Chasel'skoe Neftegaz subsidiary) plus $173m of cash in order to take a 51 per cent stake in the venture that also included all of Itera's gas assets, but in 2013 it bought out the entire company, so all of its reserves (232 Bcm) and its current output of 13 Bcm/year now belong to Rosneft.[60] Itera's main competitive advantage, however, was not its existing production but its marketing experience, in particular in the

---

[53] 'Yamal LNG and Gas Natural Fenosa sign long-term LNG supply contract', Novatek Press Release, 31 October 2013, Moscow.
[54] Rosneft Annual Report 2006, p.1.
[55] *Annual Report 2012*, Moscow: Rosneft, 2013, p.24.
[56] *Analysts' Databook 2013*, Moscow: Rosneft, 2013, p.6.
[57] *2008 Financial Results Presentation*, Moscow: Rosneft, 2008, p.18.
[58] *Analyst Presentation*, Moscow: Rosneft, 2006, p.65.
[59] 'Rosneft and ITERA Group close deal to create Joint Venture to produce and sell gas', Rosneft Press Release, 6 August 2012, Moscow.
[60] 'Rosneft applies to FAS for green light to buy ITERA', *Interfax*, 4 June 2013, Moscow.

Sverdlovsk region,[61] and Rosneft will hope to exploit this skill set as it looks to market not only the gas controlled by the joint venture but also its other portfolio gas (including that produced at Kharampur). As will be discussed below, progress in achieving this goal has already been made.

Rosneft further enhanced its position in the Russian gas market through the completion of its purchase of TNK-BP for a combined $56 billion from BP and AAR in 2013.[62] As a result, Rosneft has acquired not only Russia's third largest oil producer, but also a company that has significant gas production growth ambitions of its own, largely based around its Rospan subsidiary in West Siberia. TNK-BP produced more than 14 Bcm of gas in 2011 (of which approximately 11 Bcm was associated gas), but plans to grow this output to more than 30 Bcm by 2020, largely thanks to growth in Rospan's output from 3 Bcm to 16 Bcm over the next five years.[63] Rospan is a wet gas field, and so the economics of the gas output are enhanced by associated liquids production (a similar position to Novatek's assets), but production has again been delayed by lack of adequate access to customers over the past decade. However, TNK-BP's active strategy to improve its marketing capability and the increasing availability of pipeline capacity out of West Siberia has meant that a full field development was approved by the company's board in October 2012[64] and production should start to increase from 2013.

Following these two acquisitions, plus the further purchase of Alrosa's gas assets, which brought an extra 200 Bcm of reserves in September 2013,[65] and the acquisition of a 51 per cent stake in Sibneftegaz from Novatek in late 2013,[66] Rosneft now has estimated proved and probable gas reserves in excess of 2 Tcm, and the company estimates that its overall onshore gas resource base could be as high as 6 Tcm.[67] Furthermore, it also estimates that it has an additional 20 Tcm of offshore gas resources, although the majority of these are in the Arctic and are unlikely to be commercial for decades. Of more immediate interest, however, is the fact that Rosneft's annual output was, in 2013, expected to reach 41 Bcm,[68] following the full consolidation of TNK-BP and Itera, making it the third largest gas producer in Russia and placing it in the same league as Novatek.

---

[61] Itera presentation, *Russian Independent Gas Company*, Moscow: ITERA, 2009, p. 24)

[62] 'Rosneft consolidates 100% of TNK-BP', Rosneft Press Release, 21 March 2013.

[63] TNK-BP presentation, *Presentation to Investors*, May 2012, slide 36.

[64] 'TNK-BP Board Meeting Results', TNK-BP press release, 5 October 2012, Cyprus.

[65] 'Rosneft says buys ALROSA gas assets for $1.4 billion', *Reuters*, 23 September 2013.

[66] 'Novatek increased its stake in Severenergia to 59.8%', Novatek Press Release, 15 January 2014.

[67] 'Rosneft wants 20% share of Russian gas market, compared to current 10%', *Interfax*, 23 April 2013, Moscow.

[68] Presentation to Investors by Vlada Rusakova, Head of Rosneft Gas Business, 23 April 2013, slide 4.

One of Rosneft's key corporate objectives to 2020 is to expand this position significantly, and one mark of its ambition has been the hiring of Vlada Rusakova, the former Head of Business Development at Gazprom, to implement its gas strategy.[69] In her first presentation to Rosneft shareholders, Rusakova outlined an aggressive plan to expand production and sales into the domestic market, and the company has since then also announced plans to expand into the export market via sales of LNG from its Sakhalin-1 project. However, it is in the Russian market that it has the most growth potential, having already signed a significant number of contracts with long-term buyers. Rosneft itself signed contracts with E.ON, Fortum, and Inter RAO (the Russian power conglomerate) in 2012, while TNK-BP signed long-term deals with TGKs 5, 7, and 9, and Itera has a longstanding relationship with the Sverdlovsk region where it sells up to 16 Bcm/year.[70] As a result, Rosneft is confident that it has already secured 72 Bcm of firm sales contracts that will be operating by 2016, and its expansion plans include additional sales into the power sector, the chemical industry, and other gas-consuming sectors that could see sales reach 100 Bcm by 2020.[71] Rosneft's planned overall sales split is shown in Figure 13.7. Its sales portfolio, like Novatek's, has quality as well as quantity, with a heavy emphasis on reliable customers who are prepared to pay higher prices.

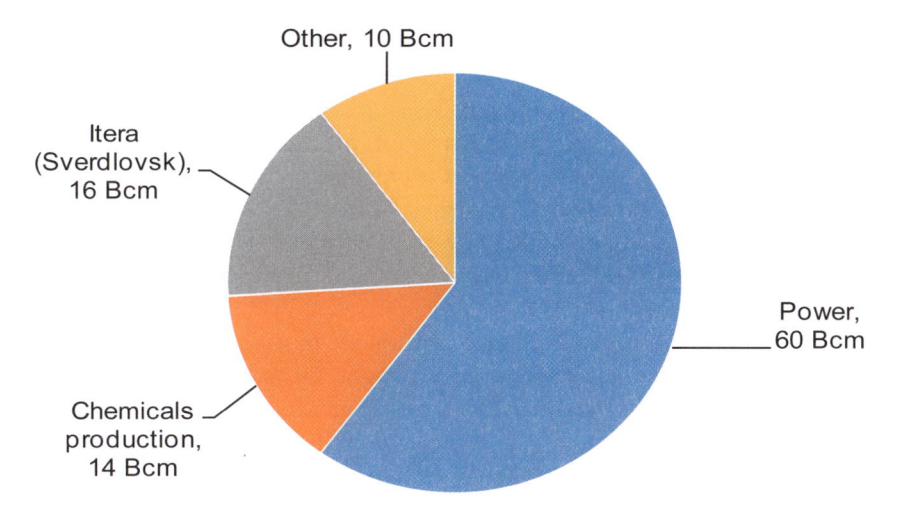

Figure 13.7: Rosneft's planned gas sales portfolio in 2020 (Bcm)
Source: Rosneft presentation to investors, April 2013.

---

[69] 'Gazprom strategist Rusakova appointed Rosneft vice president for Gas Business', *Interfax*, 19 April 2013, Moscow.
[70] Henderson, J. 'Evolution in the Russian Gas Market', op. cit.
[71] Presentation to Investors by Vlada Rusakova, Head of Rosneft Gas Business, 23 April 2013, slide 6.

Although Rosneft undoubtedly has a gas reserve base capable of meeting this potential sales growth, some questions have been asked about its ability to increase production fast enough to satisfy its contracts in the short to medium term. The recent consolidation of 100 per cent of Itera, as well as the acquisitions mentioned above, would seem to have gone some way towards addressing these concerns, and Rusakova has outlined a concrete plan for increasing gas production from the company's existing fields combined with a series of new developments. Figure 13.8 shows the estimated gas production profile for Rosneft as a whole, split by the assets of the companies within the Group as they existed in 2012. Rosneft core assets are expected to produce around 60 Bcm by 2020, TNK-BP fields 34 Bcm, with 100 per cent of Itera bringing around 14 Bcm by the end of the decade, for a potential total of 108 Bcm. Expected output for 2016 is 71 Bcm, close to the currently contracted level in the company's existing sales agreements.

The gas will come from a mixture of sources, including increased production from existing fields as gas flaring is reduced, new associated gas from the giant Vankor field, where new processing facilities are being constructed, and some small contributions from Sakhalin-3 in the Far East and the Kynsko-Chasel'skaya group of fields that had originally been contributed to the Itera JV.[72] However, the bulk of the new gas will come from two major assets: Kharampur, the Rosneft field that has always been seen as the foundation for the company's future gas business and which will produce up to 21 Bcm/year, and Rospan, the main TNK-BP asset. Furthermore, Rosneft's recent purchase of 51 per cent of Sibneftegaz, which means that it now controls 100 per cent of the company as the remaining 49 per cent was already owned by Itera, will also provide a boost to production. Sibneftegaz currently produces 12 Bcm/year but this is expected to rise to 18–20 Bcm over the next five years as new fields are brought online.[73].

Overall, Rosneft appears to have the resource base (including its organic assets and recent acquisitions) and, increasingly, the marketing capability to increase its sales into the Russian domestic gas market rapidly. Rosneft itself has targeted production of 100 Bcm as its ultimate goal,[74] and as Figure 13.8 shows, this appears to be eminently achievable before the end of the decade. Indeed, it would appear that Rosneft could even exceed this goal, although the extra gas may not be destined for the domestic market, as the company has plans to enter the Asian market with gas from its Sakhalin-1 project.

---

[72] 'Gas Business Development', Rosneft presentation, April 2013.
[73] 'FAS clears Rosneft to acquire 51% Sibneftegaz stake', *Interfax*, 24 January 2014, Moscow.
[74] 'Gas Business Development', Rosneft presentation, 23 April 2013, slide 4

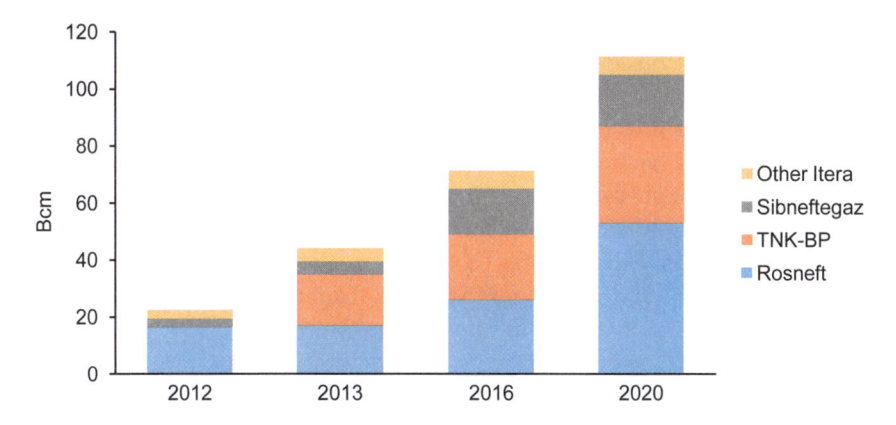

Figure 13.8: Rosneft's potential gas output to 2020
Source: Rosneft presentation to investors, April 2013 plus author's estimates.

Rosneft and its CEO Igor Sechin have been at the forefront of the drive to end Gazprom's LNG export monopoly; in late 2013 this has resulted in legislation approved by the Russian government (see also Chapter 12). Rosneft's Sakhalin-1 project is the first focus of its potential new LNG export undertakings. Indeed the company has already signed contracts with three buyers to cover the 5 mtpa of potential output (1 mtpa will be sold to SODECO, 1.25 mtpa to Marubeni, and 2.75 mtpa to Vitol), with first deliveries scheduled for the first quarter of 2019. It seems likely that these sales could be confirmed, and the overall project approved, once the removal of the LNG monopoly is confirmed early in 2014.[75]

However, although preliminary gas sales contracts have been signed, the actual implementation of the Sakhalin-1 project is at a very early stage. The major partners, Rosneft and ExxonMobil, signed an operating agreement to develop LNG production on Sakhalin in June 2013,[76] but no concrete plans have yet been made for the site of the liquefaction plant. The only significant comment on the potential cost of the project was made in a public conference call between Exxon's President for Development Neil Duffin and Russian President Vladimir Putin, in which a total figure of $15 billion was mentioned, although it was unclear whether this was for a one train (5 mt) or two train (10 mt) project.[77] The Sakhalin Regional

---

[75] 'Rosneft urges Energy Ministry to ensure export liberalisation for three LNG contracts', *Interfax*, 21 June 2013, Moscow.
[76] 'ExxonMobil, Rosneft sign operating agreement to develop LNG production on Sakhalin', *Interfax*, 21 June 2013, St Petersburg.
[77] 'Rosneft, ExxonMobil might invest $15bn in Sakhalin plant', *Interfax*, 4 November 2013, Sakhalin.

Governor Alexander Khoroshavin has stated that a number of potential sites are being considered, five on the Island itself (including one at Prigorodny where the Sakhalin-2 plant operates) and one at De Kastri, where the Sakhalin-1 oil terminal is located.[78] The most logical solution would appear to be to feed Sakhalin-1 gas into an expanded Sakhalin-2 liquefaction plant via the construction of an additional train, but the history of Sakhalin-1 negotiations with Gazprom over gas sales would suggest that this outcome is unlikely. In 2006 ExxonMobil, the operator of Sakhalin-1, reached a preliminary agreement to sell gas from the project to China, but the company needed Gazprom approval to move the gas by pipe to the export market.[79] However, Gazprom was not, at that point, prepared to see its export monopoly broken and insisted that the Sakhalin-1 gas should be sold to it at the wellhead and then transported and exported by itself to China, despite the fact that the Sakhalin-1 PSA gives the project participants the right to market the gas themselves.[80] Pressure has been consistently applied by both Gazprom and the Russian state for the Sakhalin-1 gas to be sold into the domestic market, or at least sold to Gazprom at domestic prices, with the result that no sales agreement has been reached and the majority of the 8 Bcm/year of production to date has been re-injected back into the field.

As a result, it appears unlikely that the Sakhalin-1 partners will ever want to be involved with the Gazprom-operated plant at Prigorodny, but will instead build their own liquefaction facilities. The full stand-alone project will also involve the implementation of a new gas development at the Chaivo field,[81] as the associated gas will continue to be re-injected to maintain oil production, but if the first output target of early 2019 is to be met, then a final investment decision will need to be made in 2014. Negotiations were continuing at the time of writing to try and meet this deadline, but a final decision could be delayed if the high initial cost estimate of $15 billion undermines attempts to raise project financing or the willingness of the buyers to confirm contracts. Nevertheless, the mere fact that Rosneft is pursuing this gas export strategy underlines the serious challenge that it poses to Gazprom in two major markets. If the Sakhalin-1

---

[78] 'Sakhalin-1 gas not considered for Rosneft LNG plant – Governor', *Interfax*, 28 June 2013, Sakhalin.

[79] 'Exxon agrees to sell gas from Sakhalin-1 to China', *Wall Street Journal*, 23 October 2006, New York.

[80] 'Gazprom see Sakhalin-1 gas deal with Exxon', *Wall Street Journal*, 21 June 2011, Washington.

[81] See 'Sakhalin-1' Rosneft website, www.rosneft.com/Upstream/ProductionAndDevelopment/russia_far_east/sakhalin-1/, accessed on 19 November 2013.

project is ultimately successful, it could mark a significant shift in the balance of power in the Russian gas sector.

Lukoil – a more conservative domestic strategy but growing overseas

Russia's second largest oil company, Lukoil, is also implementing plans for significant growth in the gas sector, although its strategy is somewhat different from that of Novatek and Rosneft, not only because domestically it has a closer relationship with Gazprom, but also because it has chosen to invest in gas assets overseas. The company has total proved gas reserves in Russia of just under 500 Bcm,[82] of which the majority are located in the Bolshekhetskaia Depression in West Siberia where the Nakhodkinskoe field accounts for more than 90 per cent of the company's natural, as opposed to associated, gas production. The company's remaining gas reserves are largely located in West Siberia as associated gas, or in the North Caspian, where production from the Iurii Korchagin and Filanovskii fields supplied Lukoil's power assets in southern Russia from 2012.

Lukoil's strategy to sell its Caspian gas to power stations owned by its own subsidiaries is one aspect of a marketing strategy that is somewhat more conservative than those of Novatek and Rosneft, but nevertheless provides a diverse range of potential customers. Up to 2012 the company's approach had been tempered by its strategy of selling most of its gas to Gazprom at the wellhead in West Siberia, and indeed it has an agreement to sell up to 12 Bcm/year on this basis from the Nakhodkinskoe field until 2016.[83] This sales figure is unlikely to be realized, as output from Nakhodkinskoe is not expected to exceed 8 Bcm/year; the development of the Piakiakhinskoe field has been delayed by three years and so peak output of 4–6 Bcm/year will not be reached before around 2018. Nevertheless, output from West Siberia should reach approximately 15 Bcm/year by 2020 as these two fields, and other assets in the region, reach their full potential.

The more innovative approach to gas production and marketing adopted in southern Russia is expected to see Lukoil's North Caspian gas fields supplying around 6 Bcm/year to the company's power assets[84] and a further 2 Bcm/year to a new company-owned petrochemicals complex by 2020.[85] By 2025 gas demand from these assets could have expanded further to around 15 Bcm, and Lukoil has begun the construction of pipelines

---

[82] *Lukoil Databook 2012*, p. 10.

[83] *2012–2021 Strategic Development Programme*, Moscow: Lukoil, 2012a., slide 19.

[84] Alekperov, V. 'Lukoil: Building Value Through Innovation', Presentation: 10 March 2011, New York, Moscow: Lukoil.

[85] 'Building Value Though Innovation' Presentation by Lukoil CEO Vagit Alekperov, 2011,.

from its Caspian fields to meet this level of demand.[86] As a result, Lukoil has essentially created demand for its own gas, first, through buying power assets (that have to date been supplied by Gazprom) and gradually replacing their fuel input with Lukoil gas, and, second, by developing a new downstream petrochemicals business in the south of Russia.

Lukoil's plans in Russia do not just include sales to Gazprom or to its own downstream assets, as it has also begun to offer its gas to third-party consumers; in 2012 it signed a 10-year agreement with E.ON to supply Iaivinskaia GRES with 2.25 Bcm of associated gas piped from Lukoil's nearby oil fields.[87] The company has stated that it is in negotiations for further sales to end users and therefore does appear to be exploring all angles for increasing its domestic gas sales. A target of producing and selling approximately 30 Bcm in Russia by 2020 would seem to be well within reach, with this figure potentially increasing to 40 Bcm by 2025 (see Figure 13.9).

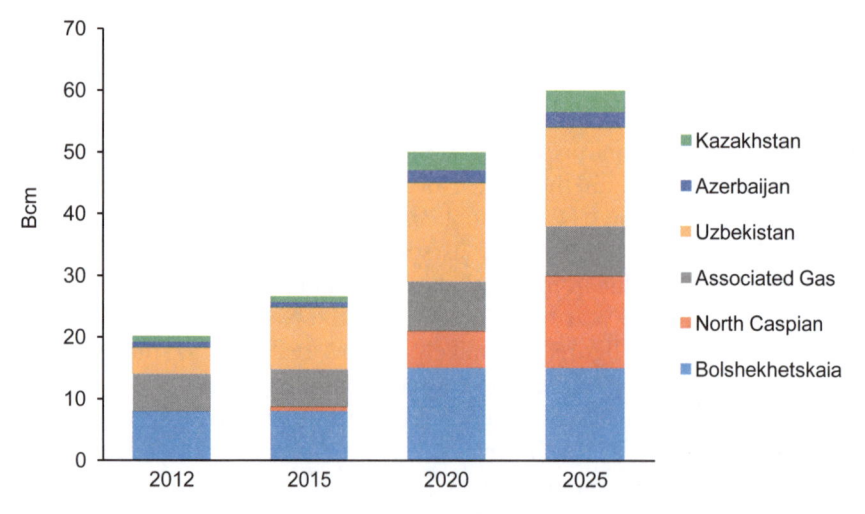

Figure 13.9: Lukoil gas production estimate to 2025
Sources: Lukoil reports and author's estimate.

Figure 13.9 also highlights the other main plank of Lukoil's gas strategy, namely its plans to expand its international business, especially in the Caspian region. A particular focus is Uzbekistan, where Lukoil's investment strategy – which was developed in the mid-2000s, when it was expanding its international oil business and experiencing frustration in the Russian

---

[86] 'Lukoil seeks contractor for onshore section of gas pipeline from North Caspian', *Interfax*, 1 July 2013, Moscow.
[87] 'Lukoil, E.On Russia begin associated gas deliveries to Yaivinskaya power station', *Interfax*, 28 May 2013, Moscow.

domestic gas market – could soon result in the company providing a rival source of supply to Gazprom in the Chinese gas market. Lukoil has two main assets, the Kandym-Khauzak-Shady fields and the South-West Gissar licence, with the former currently producing around 3 Bcm/year and the latter 1 Bcm/year,[88] but full development of both assets could see total production rising to 16 Bcm/year by the end of the decade, with the gas set to be exported via the new Central Asian pipeline to Western China.[89] Further confirmation of Lukoil's aspirations in the Chinese gas market was demonstrated by the signing of an agreement with CNPC in March 2013 to carry out exploration work at five new licences in Uzbekistan.[90] Furthermore, Lukoil's 13 per cent share of gas production from the Karachaganak field in Kazakhstan, which is currently sold into Russia for processing at Orenburg, could also be re-directed towards China from 2015 if current price negotiations with Russia do not produce a suitable outcome.[91]

One final area of potential competition with Gazprom is in western export markets, as Lukoil also owns a 10 per cent interest in the Shah Deniz project in Azerbaijan, which is set to expand gas exports from the country to Europe by 16 Bcm/year following the signing of agreements with the TANAP and TAP pipelines in 2013.[92] Phase 1 of the project is currently selling 9 Bcm/year of gas to the domestic Azeri market and to Turkey, meaning that once Phase 2 has been completed (the estimate of first production is currently 2018–19) Lukoil's equity share of production from the field would reach 2.5 Bcm/year. It is somewhat ironic that, despite not openly competing with Gazprom in Russia, Lukoil is becoming a direct competitor in both the Asian and European export markets.

If all its plans come to fruition, Lukoil could therefore be producing 50 Bcm/year by 2020 and perhaps as much as 60 Bcm/year by 2025, with two-thirds of this gas coming from Russia and the remainder from Central Asia and the Caspian. This level of volumes would equate to the target set by company CEO Vagit Alekperov in 2012, when he stated that 27 per cent of Lukoil's 1.2 billion boe of production in 2021 would come from gas, equivalent to just under 50 Bcm, as the company strives to diversify its

---

[88] *Annual Report 2012*, Moscow: Lukoil, 2013.
[89] Ibid., p.30.
[90] 'LUKOIL, CNPC to jointly develop Uzbek gas deposits', *Interfax*, 29 March 2013, Moscow.
[91] 'Kazakhstan may re-direct Karachaganak gas to China, if no price alignment with Russia', *Interfax*, 7 March 2013, Moscow.
[92] 'Shah Deniz Project Selects TAP as European Gas Pipeline', *Wall Street Journal*, 28 June 2013.

overall portfolio away from oil.[93] It would also mean that Lukoil had become Russia's fourth largest gas company, and potentially the country's largest producer of gas from assets outside Russia.

Other Independent gas production

Among the other Russian oil companies, Surgutneftegaz is the most significant producer, with associated gas output of approximately 12.2 Bcm in 2012. This is mainly sold to local power and industrial companies,[94] although the company is starting to expand the reach of its marketing operations. However, Surgutneftegaz has no specific plans to grow its gas business, and future gas production will remain in line with the company's oil output, which in itself is forecast by company CEO Vladimir Bogdanov to remain at the current level of 61–62 mtpa for the foreseeable future.[95] Surgutneftegaz maintains its focus on keeping associated gas utilization at or above the government target of 95 per cent, but even if it achieves this goal it will find itself slipping down the ranks of independent Russian gas producers due to its lack of a specific dry gas strategy. The lack of such a strategy is reflected in the fact that the company has no real gas reserves estimate, as its resources are entirely dependent upon oil production and therefore do not merit the accreditation of proved reserves.

In contrast, Gazprom Neft has shown significant growth in gas production and sales over the past two years, as it has not only optimized its associated gas output but has brought a number of natural gas projects on stream. Although it is of course debatable whether a 98 per cent subsidiary of Gazprom can be called an independent gas producer, the company is involved in so many different independent gas projects that it is at least worthy of mention in this chapter. Its total proved and probable gas reserves of almost 600 Bcm make it a significant participant in the sector.[96] The company's asset mix comprises some 100 per cent-owned dry gas fields, together with participation in four gas producing joint ventures, the most significant of which, Severenergia, produced its first gas in 2012. However, the most important short-term change has been the development of the 100 per cent-owned Muravlenskoe and Novogodnee fields; this has resulted in overall company gas output jumping from 7 Bcm in 2010 to

---

[93] Alekperov, V. 'Third Decade of Evolution: New Challenges, New Opportunities', Presentation: 14 March 2012b, London, Moscow: Lukoil.

[94] 'Surgutneftegas releases preliminary operating data for 12M 2012', Surgutneftegas Press Release, 15 January 2013.

[95] 'Surgtneftegas chief said policy is not to change, company is for stability', *Interfax*, 10 April 2013, Surgut.

[96] *Gazprom Neft Databook*, Moscow: Gazprom Neft, 2013, p.4.

Non-Gazprom Russian Producers 339

14 Bcm in 2012[97] as the two Cenomanian gas fields are set to contribute up to 9 Bcm/year of new output. However, the majority of the company's future growth will come from the continuing development of the Severenergia JV with Novatek, where production should continue to rise to a peak of 35 Bcm by 2020 (Gazprom Neft's share is about 14 Bcm).[98]

Overall then, Gazprom Neft could be producing as much as 23 Bcm by 2020 (Figure 13.10), with a significant proportion of this being specifically from gas fields rather than associated with oil production. Indeed the company's presence as a gas producer could be further enhanced if plans to transfer Gazprom's 50 per cent stake in North Gas to it are completed;[99] this would make the company a partner with Novatek in another wet gas joint venture and could provide a further 4–5 Bcm of equity gas production. Despite this potential growth, Gazprom Neft's long-term marketing strategy remains unclear, but it must be likely that much of the gas will be sold to Gazprom at the wellhead. An agreement to this effect for 5 Bcm of production, signed in March 2013 – with the price set at the level of regulated industrial consumers in West Siberia – may act as a precedent.[100]

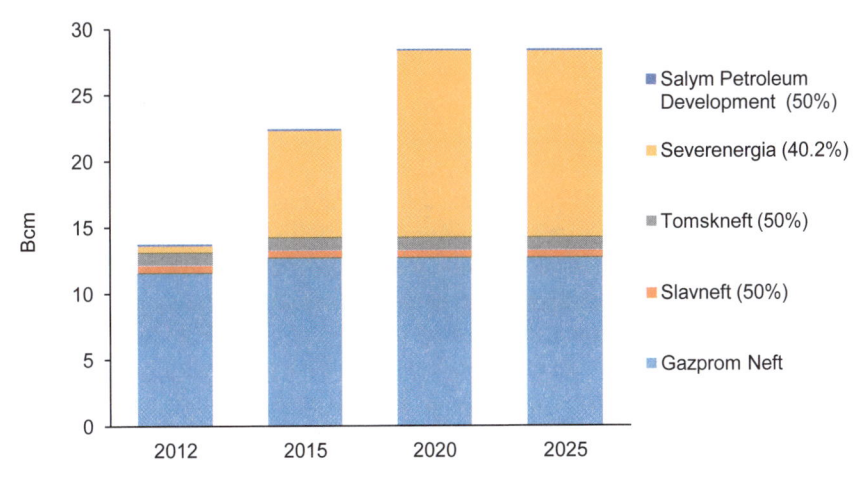

Figure 13.10: Potential Gazprom Neft gas production
Source: Company data and author's estimates.

---

[97] Ibid., p.7.
[98] 'Rosneft farms into Severenergia', *Nefte Compass*, 26 September 2013.
[99] 'GazpromNeft, Novatek to own North Gas on equal footing in future', *Interfax*, 26 February 2013, Moscow.
[100] 'Gazprom Neft, Gazprom sign gas sales contract at $90/mcm', *Interfax*, 25 March 2013, Moscow.

The most significant remaining non-Gazprom producers are the Sakhalin projects in the Far East. As discussed above, Sakhalin-1 produces around 8 Bcm per annum,[101] and although most of this is re-injected, the partnership is in the first stages of planning an LNG export scheme. 80 per cent of the project is owned by foreign companies, led by Exxon,[102] with state oil company Rosneft owning the remaining 20 per cent. Meanwhile Sakhalin-2 – in which Shell owns 27.5 per cent minus one share, Mitsui owns 12.5 per cent, and Mitsubishi owns 10 per cent, while Gazprom owns 50 per cent + one share – produces more than 16 Bcm per annum,[103] all of which is exported as LNG. The potential for an expansion of the LNG project via the addition of a third train is being actively discussed. This additional train could be initially filled with extra Sakhalin-2 gas before being supplied by the new Sakhalin-3 project in the medium to long term.

Much of the remaining non-Gazprom gas is associated with oil production, but there are a few specific domestic gas companies and international players that are worth noting. The Severenergia joint venture mentioned above not only includes Novatek and Gazprom Neft (who are equal partners in a JV that owns a 51 per cent stake) but is also set to make the 49 per cent shareholder Enineftegas JV (a 60:40 JV between ENI and Enel) one of the largest foreign gas producers in Russia, with net output of 18 Bcm/year. As mentioned above, Enel's share of this JV is currently for sale and may be purchased by Rosneft or by the joint venture owned by Novatek and Gazprom Neft,[104] but at the time of writing it remains in foreign hands. Total of France will also be increasing its presence in the domestic gas market via its two joint ventures with Novatek at the Yamal LNG project (20 per cent Total) and the smaller Termokarstovoe field,[105] which could provide combined net production of 6 Bcm/year by 2020. Meanwhile, the German companies Wintershall and E.ON also have significant gas assets in Russia, both being joint partners with Gazprom at the 25 Bcm/year South Russkoe field, while Wintershall also has a 49 per cent interest in the deep gas condensate Achimgaz project with Gazprom.[106] The final major project with foreign participation is the Shtokman field in the Barents Sea, where Total and Statoil remain in negotiations with

---

[101] Production data from *Interfax*, reporting 2011 totals produced by the Fuel and Energy Complex Central Dispatch Department.

[102] The Sakhalin-1 partners include Rosneft (20%), ExxonMobil (30%), SODECO (30%), ONGC (20%).

[103] Production data from *Interfax*, reporting 2011 totals produced by the Fuel and Energy Complex Central Dispatch Department.

[104] 'Novatek eyes Rosneft Severenergia stake', *Interfax*, 3 October 2013.

[105] *Harnessing the Energy of the Far North: Annual Report 2011*, Moscow: Novatek, 2012, p.3.

[106] *Gazprom Bond Prospectus*, Moscow: 2012, pp. 100–102).

Gazprom about the ultimate development schedule for the field, although it now seems unlikely that first production will occur before 2020.[107]

Two final small groups of independent gas companies are also worth noting, if not for their size then at least because they reflect continuing trends in the domestic gas market. The first group consists of regional producers in areas that remain separated from the trunk gas pipeline system. For example, Norilskgazprom and Taimyrgaz jointly supply 3–4 Bcm/year of gas to the Norilsk region, and in particular to the Norilsk Nickel mining complex, while IaTEK (a subsidiary of the Summa Group, formerly called Iakutgazprom) sells 1.5 Bcm/year in the Sakha (Iakutsk) region.[108] However, the lack of potential markets for all three means that their output will almost certainly remain flat for the foreseeable future.[109] More growth may be enjoyed by the second group of small independent producers who are now emerging with a focus on supplying local markets close to the fields which they own. Although the 14 identifiable companies only produced 2.1 Bcm in 2012,[110] they have combined reserves of 500 Bcm which could produce up to 20 Bcm/year in the appropriate market conditions. Many of them, such as JKX and Volga Gas, are located in the south of the country, where their gas can best compete against their larger West Siberian competitors who must pay high transport costs to bring their gas to the market. However, despite this advantage, significant risks remain for smaller companies in the regulatory and business environment that could undermine their progress. Furthermore some, such as the Alrosa subsidiaries mentioned above, are already being taken over by larger players such as Rosneft, but nevertheless the mere fact that a number of small players are emerging in the domestic Russian gas market suggests that the opportunities for competition in the supply chain are real.

Conclusions on Independent production

Overall, then, the future growth in non-Gazprom output in Russia is likely to be driven by two major sources: Novatek and Rosneft (including the recently acquired TNK-BP and Itera), with Lukoil and the remainder of the current Independent producers contributing smaller but still significant increases over the next five to ten years, and with IOCs also having an increasing equity share via their shares in joint ventures with Gazprom (for example, Wintershall at Achimgaz and South Russkoe) or Independent producers (for

---

[107] Ibid., p.104.
[108] Henderson, J. *Non-Gazprom Gas Producers in Russia*, op.cit.; 'YaTEK plans to put 83% of 2012 net profits toward dividends', *Interfax*, 5 March 2013.
[109] Henderson, J. *Non-Gazprom Gas Producers in Russia*, op.cit.
[110] Vorobyov P. 'Lilliputians in the land of giants', *Petroleum Economist*, June 2013, pp. 4–8.

example, Total at Yamal LNG). Combining the possible total output of all these producers, it would seem that the potential for Independent gas production by 2020 is as high as 330 Bcm (see Figure 13.11), and rising to as much as 370 Bcm by 2025 if all the projects identified are completed. It is important to recognize, however, that not all of this gas will be targeted at the Russian domestic market by 2020, as a number of independent export projects are under discussion. Sakhalin-1 and Yamal LNG volumes will be directed at Asia and Europe, and when added to the IOC equity in Sakhalin-2 this will mean that by 2020 approximately 30 Bcm of independent gas could be heading for export markets, leaving 300 Bcm for Russian consumption.

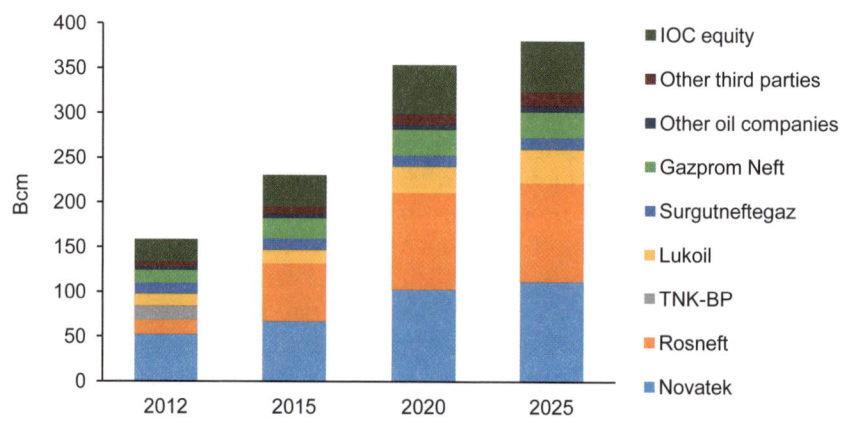

Figure 13.11: Potential non-Gazprom gas production to 2025

Source: Author's estimates based on Company Data (Rosneft data includes Itera and TNK-BP from 2015; Lukoil data includes Russian assets only; IOC equity includes the equity interests of Wintershall, E.On, Eni/Enel, Total, and the IOCs involved in the Sakhalin-1 and Sakhalin-2 projects.)

However, the fact that this potential exists clearly does not mean that it will necessarily be realized, in particular because Gazprom has now brought the Bovanenkovo field on the Yamal peninsula into production, with its potential output of 115 Bcm expected by 2019–21.[111] As a result, it is clear not only that non-Gazprom producers will have to compete aggressively to find their place in the overall market for Russian gas, but also that they will need significant support from the Russian administration, as they will most naturally be displacing the state-owned national gas company. The increase in the market share of Independent producers shown earlier (see Figure 13.2) suggests that this support has been implicitly in place for the past decade, but more overt encouragement for third-party producers has emerged since the financial crisis of 2008–9 as it

---

[111] 'Gazprom puts off production plateau for projects totalling 200 Bcm', *Interfax*, 5 June 2013, Moscow.

became clear that global gas markets were becoming more competitive. For example, in January 2009 the then Prime Minster Vladimir Putin underlined his views that non-Gazprom producers should be given greater access to Gazprom's pipeline system. He stated on German television that:

> ... we see it as our goal to provide our gas producers with more liberal access to Gazprom's pipeline system[112]

He confirmed this intention in May 2009 when he ordered the Federal Anti-Monopoly Commission to investigate accusations that Gazprom was blocking the implementation of several non-Gazprom contracts.[113] At an investor conference later in 2009 he stated that:

> In Russia, like everywhere, we see the negative side of excess monopoly in one area or another, and gas is no different [citing] restraints on the growth of independent producers [as one of the main negative factors, which the government would tackle by] aiming to provide equal access to gas infrastructure.[114]

The political rhetoric was increased in February 2010 when Putin urged Gazprom to take a more active role in developing supplies to industrial customers, warning the company that:

> Gazprom must treat the development of the infrastructure that helps provide the energy sector with gas as responsibly as possible. [...] If the company [Gazprom] itself proves unable to cope with all of these tasks it means we will have to involve other companies.[115]

Not surprisingly other government ministers have backed up this view, with former Energy Minister Sergei Shmatko stating in March 2012 that:

> ... we believe that we need to provide long-term rules of the game on the gas market that support growth of production by independent producers. [We need to offer] the opportunity to prolong previously received access to the gas transport system on a notification basis if there is an extension in relations between supplier and consumer.[116]

---

[112] Interview on German TV network ARD on 15 January 2009.

[113] 'Gazprom blocking contracts of several independent gas producers', *Interfax*, 29 May 2009, Moscow.

[114] 'Putin urges careful approach to gas market liberalisation', *Interfax*, 29 September 2009, Moscow.

[115] Prime Minister Putin quoted at a meeting of power generation companies in February 2010.

[116] 'Rosneft, TNK-BP, Novatek to boost gas output over 30 Bcm in 3 years', *Interfax*, 26 March 2012, Moscow.

In addition to the rhetoric there are a number of other signs that a review of operations in the Russian gas sector is underway. Firstly, a presidential commission for the strategic development of the fuel and energy complex was set up in June 2012 following Putin's re-election as President in order to allow him to continue his direct oversight of the energy, and particularly gas, sector, despite the fact that a parallel structure already exists within the Russian government, centred on the Ministry of Energy.[117] (See also Chapter 1.) Interestingly the presidential commission involves Igor Sechin, a close ally of Putin and the former Deputy Prime Minister with responsibility for Energy who, in 2012, left the government to become CEO of Rosneft. His position as secretary of the presidential energy commission allows him to encourage market development, in order that Rosneft can (in his words) exploit the

> ... new opportunities for monetization of gas and realization of Rosneft's strategic program for bringing significant reserves of gas into development

which were identified when the company finalized its gas JV with Itera in June 2012.[118]

Secondly, President Putin has now started to use the commission to begin a review of the competitive position of the Russian gas sector and its relationship with the Russian economy. At a meeting held in October 2012 he demanded that Gazprom carry out a detailed review of trends in the global gas market, particularly in light of the shale gas revolution in the USA, and to report back to the commission on:

> ... the main principles of its gas export policy ... [and how it can] increase the export potential and competitiveness of Russian energy resources ... because the Russian economy depends upon its effectiveness.[119]

At this same meeting, Putin also addressed the issue of corruption in Gazprom (see Chapter 12), asserting that the company would be investigated and forced to address any allegations. Although these statements themselves do not specifically encourage non-Gazprom players, they nevertheless reinforce the view that Putin is keeping a very close eye on the gas sector.

Thirdly, and perhaps more relevantly for the short-term development of the Russian gas market, Putin and other government ministers have

---

[117] 'Why Dvorkovich and Sechin's Turf War Is Public', *Moscow Times*, 24 July 2012, Moscow.

[118] Quote from Igor Sechin: 'Rosneft, Itera wrap up gas JV creation', *Interfax*, 8 June 2012, Moscow.

[119] President Putin in introductory statement to the Presidential Energy Commission in October 2012, Moscow.

endorsed the re-introduction of an exchange for trading gas, most likely located in St Petersburg.[120] Discussion on this topic has been continuing for a number of years since the dissolution of a trial electronic trading system in 2008,[121] and has been confused by Gazprom's desire to use its own trading platform for physical gas which was established by its subsidiary Mezhregiongaz, rather than accept a separate commodity exchange.[122] However, the Energy Ministry has now taken responsibility for a task force involving Gazprom, St Petersburg International Mercantile Exchange (SPIMEX), Moscow International Commodities and Energy Exchange (MICEX), and representatives from the Russian Gas Association to iron out the final difficulties relating to conflicts with current legislation, ahead of the potential launch of an exchange in 2014.[123] Achievement of this goal would provide a further platform for price competition between suppliers and would also increase access to consumers for independent gas suppliers in Russia.

Finally, even the question of Gazprom's export monopoly appears to be open to discussion in a way that would have been unthinkable even two or three years ago. The issue of the LNG monopoly, which was first raised by Rosneft CEO Igor Sechin at a meeting of the Presidential Energy Commission in February 2013,[124] has now been resolved in favour of the Independents. Legislation has been approved to end the monopoly, and with limited conditions imposed on the potential export sales contracts.[125] The Ministry of Energy had threatened to control the activities of non-Gazprom players in export markets, to ensure that they did not compete

---

[120] 'Putin stresses importance of exchange and electronic trading in determining gas price', *Interfax*, 27 September 2012.

[121] Burgansky, A. *Stand and Deliver*, op.cit.

[122] 'Energy Ministry backs gas trading on exchanges and electronic trading facilities', *Interfax*, 20 November 2012, Moscow.

[123] 'Energy Ministry sets up task force on organised gas trading', *Interfax*, 29 March 2013, Moscow.

[124] Sechin requests LNG liberalisation for shelf, docsn't think Gazprom will suffer', *Interfax*, 13 February 2013, 'Moscow.

[125] 'Russian government committee approves LNG exports' *Reuters*, 29 October 2013,. The new legislation amends Article 3 of the Federal Law on Gas Export and Articles 13 and 24 of the Federal Law on the Fundamentals of Foreign Trade Regulation. Two categories of companies are now allowed to export LNG from Russia. The first are companies developing deposits of federal importance whose licence stated as of 1 January 2013 that they are required to build **LNG** plants or send gas to be liquefied at other **LNG** plants. The second are companies in which the state owns a stake of more than 50% and which develop deposits on the Russian shelf, including in the Black Sea and Sea of Azov, and which produce **LNG** from the gas they produce at those deposits or from gas produced under production-sharing agreements.

with Russia's largest gas exporter, but Minister Novak now seems to see his role more as one of monitoring rather than directing,[126] as Novatek's deal with Spanish utility Gas Natural illustrates. Furthermore, the overall question of Gazprom's export monopoly was raised by Prime Minister Dmitrii Medvedev at Davos in January 2013,[127] and Energy Minister Novak has also raised the question of further liberalization, once the position in the LNG market has been analysed.[128] As a result, although the ending of the pipeline export monopoly still seems very unlikely in the short term, the fact that third parties are being allowed to export LNG without Gazprom's involvement is a clear indication of the increasing role for independent gas producers in Russia, suggesting that in the long term it is no longer inconceivable that they could have an even greater share of both the domestic and export markets.

---

[126] 'Russian government approves bill on LNG export liberalisation', *Interfax*, 30 October 2013.

[127] 'Medvedev allows that Gazprom export monopoly may be ended', *Interfax*, 23 January 2013, Davos.

[128] 'Further steps may be possible if liberalising LNG market proves positive – Novak', *Interfax*, 8 November 2013.

CHAPTER 14

CENTRAL ASIAN AND CASPIAN GAS FOR RUSSIA'S BALANCE

Simon Pirani

Imports of Central Asian gas to the industrial regions of Russia began in Soviet times and continued through the 1990s at relatively low levels of around 3–4 Bcm/year.[1] In the mid-2000s, as Russia struggled to meet the steadily rising demand for its own gas, net Central Asian imports – which by then were running at 51–66 Bcm/year – became an increasingly necessary supplement to its gas balance. In 2008, Gazprom, in the context of efforts to defend its monopsony over Central Asian supplies, agreed to pay import prices linked to the prices of its own gas in Europe. However, in 2009, the market was changed completely by two events: first, the sharp fall in gas demand in Europe, the CIS, and Russia that resulted from the recession; and, second, the completion of the Turkmenistan–China pipeline, which opened up a major new export route for Central Asian gas. Since then, Central Asian exports to and through Russia have fallen to about half their former level. Exports to China are rising steadily, and will continue to do so, while new Azeri gas production is being exported to Georgia and Turkey. By 2012 these changes had led to a situation where the volume of Central Asian and Caspian gas exported to non-Russian destinations exceeded the exports to and through Russia for the first time. An overview of exports from the region is given in Table 14.1.

In this chapter it will be argued that this trend for Central Asian and Caspian exports to switch from Russia to other destinations will continue for the rest of this decade. The context is a general decline in Russia's economic and political ties with the region. Economically, Russian investment has been overtaken by western and Chinese investment in Kazakhstan, particularly in upstream oil and gas, and by Chinese, Korean, and other Asian investment in Turkmenistan and Uzbekistan; in contrast the Russian share of trade flows with Central Asia declined throughout the 2000s.[2] While Kazakhstan has, after some hesitation, joined the Customs

---

[1] Landes, A. *Impasse: Russia Oil and Gas Yearbook 2005* (Renaissance Capital, Moscow, 2005), p.89.

[2] Pirani, S. 'Central Asian and Caspian Gas Production and Constraints on Export', Working Paper NG 69, OIES, December 2012, p. 110; and Zhukov, S.V. and Reznikova, O.B. *Tsentral'naia Aziia i Kitai: ekonomichesko vzaimodeistvie v usloviakh globalizatsii*, Moscow, IMEMO RAN, 2009.

Union with Russia and Belarus, and Armenia decided to do so in late 2013, other Central Asian and Caspian countries have stayed out of it. Politically, Russian influence in the region has also been eroded, not only by China but also by NATO countries (who view Uzbekistan and Kyrgyzstan as potential partners in the 'war on terror').

Table 14.1: Gas exports from the Central Asian and Caspian region

Bcm	2006	2007	2008	2009	2010	2011	2012
Exports to and through Russia	**51.3**	**63.3**	**65.9**	**34.9**	**33.3**	**30.6**	**30.8**
From Turkmenistan	39.3	39.8	42.3	11.8	10.7	11.2	10.9
From Uzbekistan	9	10.3	11.5	13.1	11.4	8	8.7
From Kazakhstan	14.3	15.2	15	10.1	12.4	11.9	11.6
Russian exports to Kazakhstan	*−6.9*	*−2.0*	*−2.9*	*−0.9*	*−2.0*	*−2.0*	*−2.0*
From Azerbaijan	0	0	0	0.8	0.8	1.5	1.6
Russian exports to Azerbaijan	*−4.4*	*0*	*0*	*0*	*0*	*0*	*0*
Exports via other routes	**6.3**	**8.2**	**14.4**	**14.8**	**19.3**	**31.3**	**35.8**
Turkmenistan to China	0	0	0	0	3.5	15.5	20
Turkmenistan to Iran	6.3	6.2	7.1	7	8	8	8
Azerbaijan to Georgia and Turkey	0	2	7.3	7.8	7.8	7.8	7.8

Note.   Total of exports to and through Russia is net of Russian exports to Kazakhstan and Azerbaijan. The intra-Central Asian gas trades are not shown

Sources: Pirani, S. 'Central Asian and Caspian Gas Production and Constraints on Export', Working Paper NG 69, OIES, December 2012, pp. 8–9; Gazprom *Annual Report* 2010 and 2012.

In the gas sector, Russia will continue to yield to China. In particular, Turkmen exports to China will increase still further: the two countries agreed in September 2013 on a programme to raise them to 65 Bcm/year.[3] Russian upstream activity in Central Asia is also changing its function. Historically, it was focused on ensuring supplies for Russia's domestic market, but now the largest Russian-controlled upstream projects, such as Lukoil's in Uzbekistan, are directed towards the Chinese market (see Chapter 13).

---

[3] 'China ups Turkmen supplies', *Platts International Gas report*, 9 September 2013, p. 25.

Russia's declining role as a buyer of Central Asian gas is underlined by the fact that the prices it pays for it are unsustainably high. Having agreed in 2008 to the principle of European netback pricing, Gazprom has been buying gas at Central Asian exporting countries' borders at more than twice the price at which Russian-produced gas is delivered to Russia's main consuming areas. It has continued to buy these volumes mainly for the sake of Russian political interests in Central Asia – but these arrangements could change.[4]

Central Asian gas exports to Russia up to 2009

During the early and mid-2000s, Central Asian gas exports to Russia recovered from the instability of the 1990s, against a background of Russian and European economic growth. Gazprom acted as a monopsonistic buyer, adding the Central Asian volumes to its gas balance and using them to help meet steadily rising demand in Europe, the CIS importing countries, and Russia.[5] Central Asian trade in other products – including oil and oil products, and metals – diversified away from Russia, in keeping with the general economic and political trend in the post-Soviet period. But Russia retained its dominant position in the gas trade, largely because of the constraints imposed by the transport infrastructure.

This network, built in Soviet times, took nearly all of Central Asia's gas to Russia, via (i) the Central Asia–Centre pipeline corridor towards central Russia and Ukraine, (ii) a pipeline to the Orenburg processing plant in the southern Urals, and (iii) the Bukhara–Urals pipeline to the Urals. The only other pipelines from the Central Asian and Caspian countries were those from Turkmenistan to northern Iran (through which 6–7 Bcm/year was exported in the 2000s); and lines built in Soviet times from Azerbaijan to northern Iran and southern Russia, and from Turkmenistan to Afghanistan, that had ceased to be used.

Prior to the 2008–9 recession, four factors shaped the gas trade between Russia and the Central Asian importers, and the prices charged:

---

[4] The arguments are a development of those in: Pirani, S. 'Central Asian and Caspian Gas Production', op.cit. Additional detail on the themes in the chapter is also available there.

[5] More detail on the 1990s and early 2000s, may be found in Henderson, J., Pirani S., and Yafimava, K. 'CIS Gas Pricing: Towards European Netback?' in Stern, J.P. (ed.), *The Pricing of Internationally Traded Gas* (Oxford: OIES/OUP, 2012), pp. 178–223, and in the relevant chapters of Pirani, S. (ed.) *Russian and CIS Gas Markets and their Impact on Europe* (Oxford: OIES/OUP, 2009).

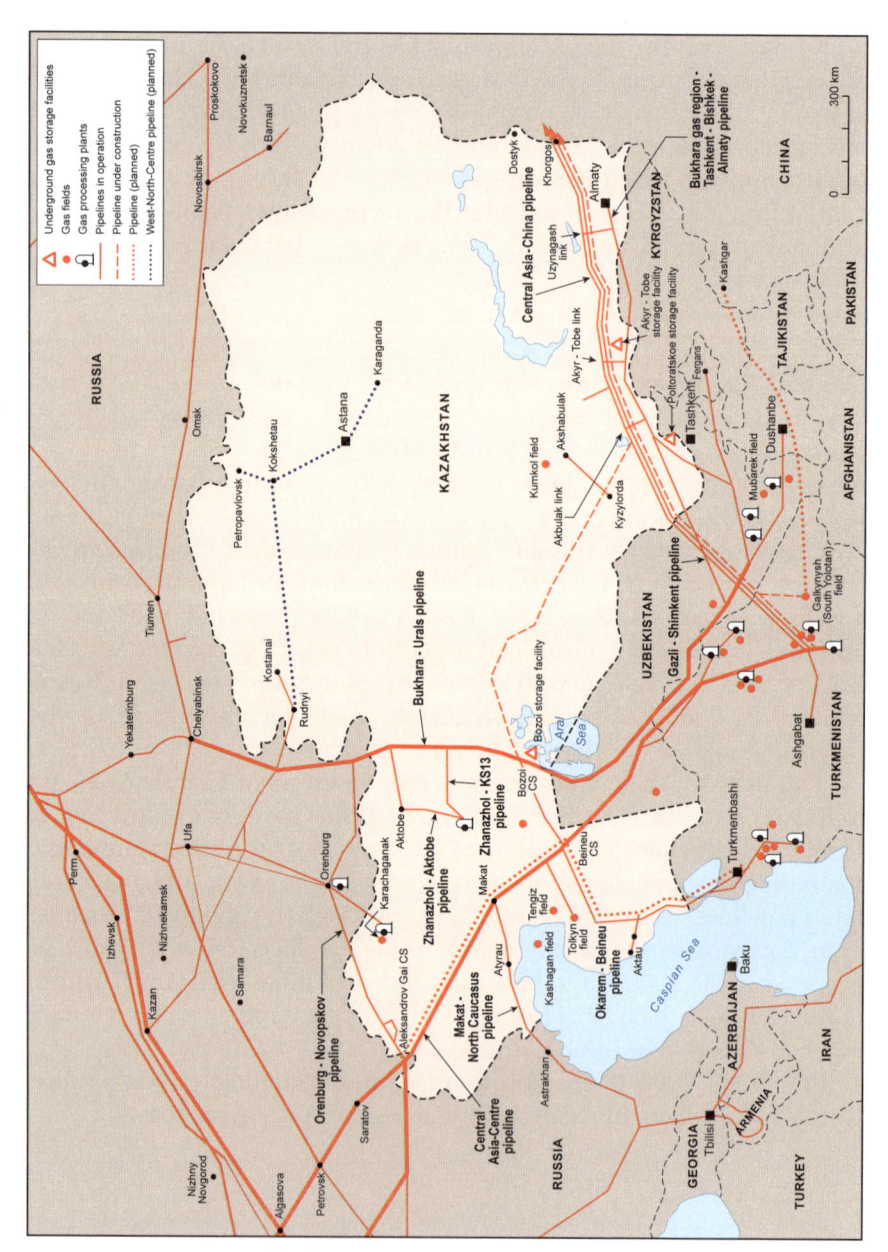

Map 14.1: Central Asian and Caspian gas pipelines
Source: Oxford Institute for Energy Studies.

a) the rising demand for gas in all markets, which Gazprom was struggling to meet, and which made it reliant on Central Asian imports;

b) the high (oil linked) gas prices in Europe, to which the Central Asian producers hoped to achieve some indirect access;

c) political considerations, according to which Russia hoped to use its monopsonistic position in the gas trade to maintain influence in Central Asia when economic links were in general being weakened; and

d) constant tension between Gazprom and various trading companies over who would control Central Asian gas volumes, which – when sold in Ukraine or Europe, priced against Russian gas but free of 30 per cent Russian export duty – yielded a handsome margin.

As a result of all these factors, Gazprom bought all Central Asia's export volumes at prices negotiated politically, but linked indirectly to European netback levels. These prices rose steadily, as shown in Table 14.2.[6]

Table 14.2: Prices of Central Asian gas exports, 2005–9

$/mcm	2005	2006	2007	2008	2009
European border price (est.)*	213.7	285.2	294.1	418.9	307.8
Ukrainian border price**	50–80	95	130	179.5	236.11
Turkmen border price	44–60	65	100	130–150	340***
Uzbek border price	42–64	60–95	100	130–160	190 (est.)
Kazakh border price	47	100 (est.)	100 (est.)	180	180 (est.)

* OIES estimates, based on published German import price and market information.
** 2005–8 prices fixed annually; 2009 average, based on price adjusted quarterly.
*** First quarter only. Westward exports suspended in Q2–4.
Source: information published by companies, press reports.

The way in which Gazprom – almost certainly under the influence of the Russian government, its main shareholder – shared some of the benefits of rising gas prices in Europe with its Central Asian suppliers deserves comment. In May 2006, against a background of rapidly rising European gas prices, the price paid for Kazakh exports rose substantially as a direct result of political intervention: Vladimir Putin, then prime minister, met with Kazakhstan's President Nursultan Nazarbayev and shortly afterwards

---

[6] Pirani, S. 'Central Asian and Caspian Gas Production', op.cit., pp.78–80.

secured agreement with Gazprom to triple the price it was paying.[7] In March 2008, as the oil boom was reaching its apex, and Russia and Ukraine were once again at loggerheads over gas supplies and transit, the chief executives of Gazprom and the three Central Asian producer companies (Turkmengaz, Uzbekneftegaz, and Kazmunaigaz) met in Moscow and agreed that Russia would in future pay 'European prices', in other words, presumably netback from European prices, for Central Asian imports.[8] Three months later, in June 2008, Gazprom CEO Aleksei Miller visited Ashgabat and signed a new agreement with Turkmengaz; reportedly, it provided for purchase prices to rise to a level of $225–295/mcm in the second half of the year, from the $130–150/mcm paid previously.[9]

Gazprom also worked throughout the period up to 2008 to centralize the trading arrangements, and exclude independent traders from them. In 2005–6, Russia severed direct commercial relationships between *Turkmenistan* and Ukraine, under which large volumes had previously been traded. After the Russia–Ukraine gas dispute of January 2006, agreements were made under which all westward Turkmen gas exports were bought at the Turkmen border by Gazprom's international trading division. In 2006–8, almost all of these volumes were then resold to the trading company Rosukrenergo (owned 50 per cent by Gazprom and 50 per cent by the Ukrainian businessmen Dmitry Firtash and Ivan Fursin), which transported them and supplied them to Ukraine and central Europe.[10]

In the case of *Uzbekistan*, whose exports rose steadily from 4.9 Bcm in 2003 to 13.1 Bcm in 2009, volumes were, after 2006, split between Gazprom Export (which bought 7 Bcm and resold them to Rosukrenergo) and ZMB Schweiz, a Swiss-based Gazprom subsidiary, which marketed both volumes produced by Gazprom at two Uzbek projects and small volumes bought from Uzbekneftegaz.[11] The marketing arrangements for gas from

---

[7] Iiulia Sinchuk, 'Gazprom dal Kazakhstanu evropeiskuiu tsenu', *Kommersant*, 22 May 2009. This is an interview with Uzbakbai Karabalin, president of Kazmunaigaz, in which it is suggested that Kazakh export prices had risen to $140/mcm in 2006 from $47/mcm in 2005. Other industry information suggests that the price rise was to $100/mcm (see Table 14.2).

[8] 'Ob itogakh rabochei vstrechi', Gazprom press release, 11 March 2008; 'Sredneaziatskii gaz – po evropeiskim tsenam', *Interfax*, 11 March 2008.

[9] The prices were not disclosed, and are estimated on the basis of press reports. 'Sergei Ivanov proshchupaet uzbekskuiu pochvu', *Kommersant*, 25 August 2008.

[10] Pirani, S. 'Ukraine's Gas Sector', OIES Working Paper NG 21, June 2007, pp. 31–8; Pirani, S.'Ukraine: a gas dependent state' in Pirani (ed.), *Russian and CIS Gas Markets*, op.cit., pp. 99–100.

[11] ZMB Schweiz marketed 6.4 Bcm of Uzbek gas in 2007 and 11.5 Bcm in 2008; some of this may have been repurchased from Rosukrenergo. ZMB (Schweiz) *Annual Report* 2007, p. 70; *Annual Report* 2008, p. 20.

Kazakhstan, whose net exports to and through Russia rose from 7.2 Bcm to 12.1 Bcm between 2006 and 2008, also changed. The 2006 agreements provided for most of these volumes to be traded by Gazprom Export; then in May 2007, Kazrosgaz, a joint venture between Gazprom and Kazmunaigaz, the national oil and gas company, signed a 15-year export contract with Gazprom Germania, for the sale of gas from the Karachaganak project on the Russia–Kazakhstan border.[12]

The economic crisis and opening of Chinese export route

In 2009 Central Asian gas exports to Russia were reduced by about half, as a direct result of the economic crisis and resulting oversupply of gas in Europe, the CIS, and Russia. The exports have remained at this low level since then, and are not expected to increase, because China has established itself as a major alternative source of demand for Central Asian gas. Exports from the Central Asian and Caspian region to all non-Russian destinations (such as China, Iran, and Georgia/Turkey) have already overtaken exports to Russia in volume terms, and China will surely overtake Russia as the largest importer of Central Asian gas in the near future. While Russian purchases from Central Asia and the Caspian have fallen in volume terms, the prices paid have remained high: the concept of European netback agreed between Gazprom and the Central Asian producers in 2008 appears to have been maintained.

Table 14.3 shows purchase prices published by Gazprom, and Gazprom's sales prices in its main markets. In 2009, Gazprom's purchase prices in Central Asia/Caspian were higher than its average sales prices in former Soviet countries (where, in contractual terms, most Central Asian volumes are sold). In 2010–12, the purchase prices were at a discount of $25–45 to sales prices in former Soviet countries, in other words, in the same range as transport costs to Ukraine.

---

[12] 'Eksport gaza' page Kazrosgaz website, www.kazrosgas.org/?p96&version=ru, accessed 1 September 2013. The Kazakhstan–Russia gas trade is complicated by swap arrangements established in Soviet times and/or the 1990s. Small volumes of Kazakh gas are delivered to Gazprom in exchange for (i) volumes it supplies to Kazakh customers in Kostanai region, who are most easily served via a pipeline from Russia, and to customers of Kaztransgaz, Kazakhstan's national gas company, in Georgia, and (ii) Uzbek or Turkmen volumes delivered to customers in south-eastern Kazakhstan. Furthermore, raw gas is delivered from Kazakhstan to the Orenburg processing plant immediately over the border in Russia and returned to Kazakhstan after processing. These arrangements are separate from Kazakh exports to Russia, but have often been covered by the same agreements between companies.

Table 14.3: Gazprom purchase prices in Central Asia & Caspian, and sales prices

$/mcm	2009	2010	2011	2012
Average purchase price of gas at the border of the supplying country	233.6	197.2	244.0	278.6
Average prices of gas sales				
to Russia	60.6	75.6	89.6	92.4
to FSU	202.1	231.7	289.5	305.3
to Europe	296.7	301.8	383.0	385.1

Note. The purchase price is an average for Gazprom purchases from Turkmenistan, Uzbekistan, Kazakhstan, and Azerbaijan.

Source: Gazprom Annual Reports and Management Reports.

The circumstances under which Turkmen exports were reduced in 2009 leaves little doubt that Russia's hand was forced by the economic crisis. Russian purchase prices of Central Asian gas were at an all-time high – $340/mcm, according to President Vladimir Putin[13] – while demand had fallen sharply in all markets, and Gazprom sharply reduced its own production.[14] There can be little doubt that a decision was taken at Russian government level to reduce purchases from Turkmenistan, while maintaining those from Uzbekistan and Kazakhstan, with a view to minimizing the strategic damage done to Russia's position in Central Asia.[15] In April 2009, there was an explosion on the main Central Asia–Centre pipeline, as a result of which shipments of Turkmen gas were suspended. In any case, the damage was soon repaired, but Gazprom announced, in unilateral breach of its purchase contract, that it would not accept any further volumes. It soon demanded that the contract be renegotiated.

Turkmen gas exports to Russia ceased completely between April and December 2009. Sales were resumed on 1 January 2010 after contracts had been amended. The agreement reached provided for Turkmengaz to sell up to 30 Bcm/year to Gazprom, but it was stated by both sides that exports would be 10–11 Bcm/year for the foreseeable future. Price levels were not made public, but Gazprom stated that the price formula was 'indexed to

---

[13] 'Nulevoi potok', *Kommersant*, 11 January 2009; 'Gazprom pereotsenit turkmenskii gaz', *Kommersant*, 23 June 2009.

[14] Gazprom Annual Report 2009, pp. 37–9; 'Proval Gazproma', *Vedomosti*, 3 April 2009; 'Pessimist Gazprom', *Vedomosti*, 17 June 2009; 'Minimum gaza', *Vedomosti*, 30 September 2009.

[15] A full account of these events is in Pirani, S. 'Central Asian and Caspian Gas Production', op.cit., pp. 80–81.

correlate with changes in the prices of oil products' (in other words, reflecting the price levels at which Gazprom was selling into Europe).[16] However, it may be presumed, from the available information, that indexation was at a lower level than in pre-2009 contracts.

Gas export proceeds are the largest item in Turkmenistan's sources of GDP and sources of state revenue, and at many points in post-Soviet Turkmen history the sudden cessation of Russian purchases could have been extremely damaging. Fortunately for the Turkmen economy, the Turkmenistan–China pipeline was completed in 2009, and deliveries of gas began in 2010. They rose from 3.5 Bcm in 2010 to 15.5 Bcm in 2011 and 20 Bcm (estimated) in 2012.[17] At the time of writing, contractual arrangements provide for Turkmen exports to China to rise to 30 Bcm/year, and thereafter to 65 Bcm/year. In September 2013, China and Turkmenistan made a series of agreements providing for pipeline capacity to be raised from its current 40 Bcm/year to 65 Bcm/year, and for China to finance and participate in the second-phase expansion of Turkmenistan's Galkynysh gas field, which will supply most of the gas.[18] Turkmenistan has therefore not only survived its 2009 clash with Russia in financial terms, but has also attained its strategic aim of establishing a major new export destination.

The opening up of the Chinese export route, and the deepening relationship with China, is the most important turning point in the economic history of post-Soviet Turkmenistan. Turkmenistan now has two major gas customers, instead of one, plus its relatively small export trade with Iran. Its future revenues from gas sales, barring a complete collapse in prices, will be far higher than ever before. For this reason, caution should be exercised about Turkmenistan's likely involvement in opening up further export routes, either westwards via a trans Caspian pipeline or south-eastwards via Afghanistan to Pakistan and India. While discussions of a trans Caspian route have receded since the economic crisis, those around a Turkmenistan–Afghanistan–Pakistan–India route have intensified, with active support from the USA, which sees pipeline construction as one means of supporting post-conflict development in Afghanistan.[19] Turkmenistan signed framework

---

[16] 'Zakupki gaza', Gazprom website, www.gazprom.ru/about/production/central-asia/, accessed 1 September 2013.

[17] Pirani, S. 'Central Asian and Caspian Gas Production', op.cit., p. 85.

[18] 'China ups Turkmen supplies', *Platts International Gas Report*, 9 September 2013, p. 25.

[19] 'TAPI Natural Gas Pipeline Project, Phase 3', Asian Development Bank website, www.adb.org/projects/44463-013/details> accessed 25 November 2013; Rusnak, Urban, 'Afghanistan as an energy hub for south Asia' (letter), *Financial Times* 21 March 2013; Tracy, Lynn, US State Department, 'Remarks at the 18th Turkmenistan International Oil and Gas Conference', Ashgabat, 19 November 2013.

sales agreements with Pakistani and Indian companies in May 2012, but pipeline construction remains a distant objective. Turkmenistan, while always interested in additional export diversification, has neither the economic need nor the political capacity to prioritize such projects.[20]

Post-crisis exports to and through Russia

While Turkmen exports to and through Russia were cut by more than two-thirds in 2009, and have since then settled at 10–11 Bcm, exports from Uzbekistan and Kazakhstan to Russia were hardly affected by the 2008–9 economic crisis. Gazprom continued purchases from Uzbekistan under contracts signed in 2002 that covered the years 2003–12. In 2009 and 2010, these purchases were at a similar level to pre-crisis years (13.1 Bcm in 2009 and 11.4 Bcm in 2010), despite the Russian gas balance being oversupplied. In 2011 and 2012, purchases from Uzbekistan fell to 8 Bcm and 8.7 Bcm respectively, but this was because Uzbekistan was unable to deliver any more than this. Uzbekistan faces a gas supply squeeze, due to a gradual fall in output and rising domestic consumption; although the output from fields in natural decline will be replaced by fields now being developed, mainly by Lukoil, this gas will probably be sold to China or to the domestic market.[21] In December 2012, Gazprom and Uzbekneftegaz signed new sales and purchase contracts covering only three years (2013–15). The volumes were not specified, but some media reports suggested that deliveries to Russia would fall to 7.5 Bcm in 2013.[22]

Exports from Kazakhstan continued to be undertaken by Kazrosgaz, the joint venture between Kazmunaigaz and Gazprom. Gas from the Karachaganak oil field is exported under the 2007 contract mentioned above, and Kazrosgaz also exports gas bought by Kazmunaigaz from the operators of the Tengiz oilfield.[23] In August 2013, Kaztransgaz agreed terms with the consortium that operates the giant new Kashagan oil field under which it will buy its gas production, which during the first phase is estimated at 3 Bcm/year.[24]

---

[20] This argument is made in more detail in Pirani, S. 'Central Asian and Caspian Gas Production', op.cit., pp. 99–103.

[21] The Uzbek supply squeeze is discussed in Pirani, S. 'Central Asian and Caspian Gas Production', op.cit., pp. 57–63.

[22] 'Podpisan kontrakt na zakupku uzbekskogo gaza', Gazprom press release, 24 December 2012; 'Rossia v 2013 snizhaet zakupku uzbekskogo gaza', *TsentrAziia*, 26 December 2012 www.centrasia.ru/news2.php?st=1356470100.

[23] Kazmunaigaz *Godovoi Otchet [Annual Report]* 2011, p. 43.

[24] 'Kaztransgaz to buy all gas produced at Kashagan until 2042', *Interfax Natural Gas Daily*, 19 August 2013.

Since 2010, most, if not all, of Gazprom's gas purchases from Central Asia, and the marketing of these volumes, has been undertaken by Gazprom Schweiz, a wholly-owned subsidiary of Gazprom Germania (based in Germany), which is in turn a wholly-owned trading subsidiary of Gazprom Export. The gas is sold in Ukraine, Poland, Serbia, and Montenegro.[25] The trading margin on Central Asian volumes remains high because, although there are additional transport costs to pay, they are priced against volumes from Russia that incur 30 per cent export duty.

Azerbaijan

Azerbaijan's transformation from gas importer to significant exporter has impacted Russia in a number of ways. Azerbaijan had imported Russian gas in Soviet times, and resumed doing so in 2000 as its economy recovered from recession; these imports grew to almost 5 Bcm/year in the mid-2000s. But in 2006, with the start-up of production at the Shah Deniz field, Azerbaijan substituted Russian imports with its own gas, and also replaced Russia as the main provider of gas imports to Georgia (1.5–2 Bcm/year). In 2010, Azerbaijan began exporting gas to southern Russia. This gas, mostly produced from onshore fields, is delivered via the 240 km Gazi–Magomed–Mozdok pipeline that formerly brought Russian gas to Azerbaijan. The compressor station at Siyazan was refurbished in 2010 to handle the reversal of flows. In October 2009, Gazprom agreed with Socar (Azerbaijan's state oil and gas company) to buy 0.5 Bcm/year in 2010–15; at a further meeting between the two companies in January 2010, additional volumes were agreed. In the event, volumes were 0.8 Bcm in 2010, 1.5 Bcm in 2011 and 1.6 Bcm in 2012.[26]

The growth of Azeri exports has had a second and more important consequence for Russia: they have competed with Russian volumes in the Turkish market since 2007, and will do so, albeit on a modest scale, in the

---

[25] Gazprom Schweiz *Annual Report* 2011, pp. 15–20. I have said 'most, if not all' because Gazprom Schweiz reports purchases in Central Asia that are in some years slightly higher, and in some years lower, than the total of Gazprom Group purchases. Gazprom Schweiz purchases gas either directly or in cooperation with Gazprom Export. Physically, the volumes are added to the Russian gas balance in the Russian united transport system. Contractually, they are treated as exports to Ukraine and other destinations; there is no comprehensive information on how much Central Asian gas goes to each destination.

[26] Bowden, J. 'Azerbaijan: from Gas Importer to Exporter', in Pirani, S. (ed.) *Russian and CIS Gas Markets*, op.cit., pp. 222–4; 'Gas export department', Socar website; Socar Annual Report 2010, p. 3; *Foreign Trade of Azerbaijan 2011* (Baku: AzStat, 2011), p. 258. On Georgia, see Chapter 7.

European market from 2019. The exports to Turkey, associated with the start-up of the first phase of the Shah Deniz project, are now running at 6.3 Bcm/year. The second phase of production at Shah Deniz, for which a final investment decision is expected in late 2013, will make available another 6 Bcm/year for Turkey from 2018 and 10 Bcm/year for European customers from 2019.

The gas from Shah Deniz will probably cross Turkey via the new Trans Anatolian Pipeline (TANAP), which is supported by Socar, together with Botas of Turkey (although alternative plans relying on the upgrade of existing infrastructure have also been mooted.) From Turkey's western border much of the gas will be carried to Albania and Italy via the Trans Adriatic Pipeline. These projects will thus open up the 'southern corridor' – bringing Caspian gas to the largest European markets via a non-Russian transit route – for which the European Union has been pushing for many years. However, mainly because of the economic recession and the consequent levelling-off of European gas demand, the corridor is on a much smaller scale than originally envisaged. European hopes were previously focused on the 'Nabucco' project, that was intended to transport greater volumes (at least 30 Bcm/year) via a route more competitively threatening to Russia (via Bulgaria, Romania, and Hungary to Austria).

While the competition offered to Russian gas by Azeri exports to Europe is on a smaller scale than advocates of 'Nabucco' might have wished, it will nevertheless be a factor in the 2020s. Under contracts signed in 2013, Shah Deniz gas will go to Greece (1 Bcm/year), Bulgaria (1 Bcm/year), and Italy and adjacent hubs (8 Bcm/year).[27] This challenge was probably a factor in Russian thinking on the construction of South Stream, although it appears to be of secondary importance when compared to Russia's desire to diversify transit away from Ukraine (see Chapter 4).

Pricing

The purchases by Gazprom of Central Asian gas volumes, priced with reference to Russian exports to Europe, amount to an economic burden on the company. Here it is argued that:

(i) Gazprom's Central Asian purchase prices have been set at a high level for largely political reasons;

---

[27] 'Shah Deniz Major Sales Agreements with European Gas Purchasers Concluded', BP press release, 19 September 2013.

(ii) the differential between the cost of Russian gas and these prices is not likely to narrow substantially, and therefore these purchases will continue at their present levels, and on the same price formation basis, for only as long as there are political reasons to do so; and

(iii) Central Asian export prices to Russia and Central Asian export prices to China may influence each other in future, and there is some evidence that initial exports to China have yielded less of a margin than exports to Russia.

Gazprom's high Central Asian purchase prices

In the 1990s, gas pricing reflected the chaos of the immediate post-Soviet period: intra-Soviet transfers were being replaced by international trading arrangements amidst acute economic crises. Prices were set *ad hoc*, non-payment was ubiquitous, and there was a large element of barter – which continued to be used in the gas trade until the mid-2000s. In the early 2000s, as Russia and other former Soviet states began to recover from recession, Central Asian gas sold to Russia and Ukraine was priced in a range between $36/mcm and $51/mcm. From 2006, when the prices of Russian gas exports to Europe had begun to rise steeply (following oil prices), the Central Asian producers achieved a change in the pricing basis of their exports. Gazprom agreed to set prices with reference to the netback from its export prices in Europe, rather than relative to prices in the Russian market, or to production and transportation costs.

There was little commercial rationale to this decision. Although negotiations on opening up the Chinese route were completed in 2007, Gazprom's Central Asian suppliers had no other *current* major export route for their gas, and would not have until 2010. But concessions on price were very logical in the context of Russia's strategic considerations. Politically, Russia was struggling to keep Central Asian countries within its sphere of influence; and economically it was losing ground to Europe, the USA, and China.

Even in 2009, when recession had led to the supply–demand balance in gas markets tilting sharply against suppliers, although Gazprom reduced purchases from Turkmenistan and renegotiated its contract with Turkmengaz, there is no evidence that it renegotiated prices with Uzbekistan or Kazakhstan. And nor is there evidence of any especially onerous price reduction imposed on Turkmenistan: what little information is available on comparative prices[28] suggests that Turkmen prices were higher than others in early 2009, and that the renegotiation merely brought them into line.

---

[28] Mainly, Ukrainian customs information, such as that cited in Table 14.5 below. But note that this refers to sales prices of Central Asian gas in Ukraine rather than purchase prices in Central Asia.

Political factors remain the main impetus for continuing sales

In the post-crisis period, Gazprom's average purchase price for Central Asia and Caspian gas (in other words, for purchases from Turkmenistan, Uzbekistan, Kazakhstan, and Azerbaijan) has moved in a range between 63 per cent and 73 per cent of the company's average sales price in Europe.[29] This price, apparently set in accordance with the principle of netback from oil-linked prices in Europe, remained 2.5 to 4 times higher than Gazprom's average sales prices in Russia – while Russia's own gas was being produced and transported to the largest domestic markets at, or below, average sales prices (see Chapter 5). Clearly, now that the pre-crisis supply squeeze has been replaced by oversupply, there is no *commercial* rationale for Gazprom to purchase any gas from Central Asia.

Estimates made by researchers at the Moscow School of Management (Skolkovo Energy Centre) show that this gap between Central Asian purchase prices and the costs of production and transportation of Russia's own gas is unlikely to narrow. In other words, purchases of Central Asian gas will continue to be made at a loss, unless the pricing basis is changed. Projections for costs of gas delivered to Moscow in 2020 are shown in Figure 14.1.

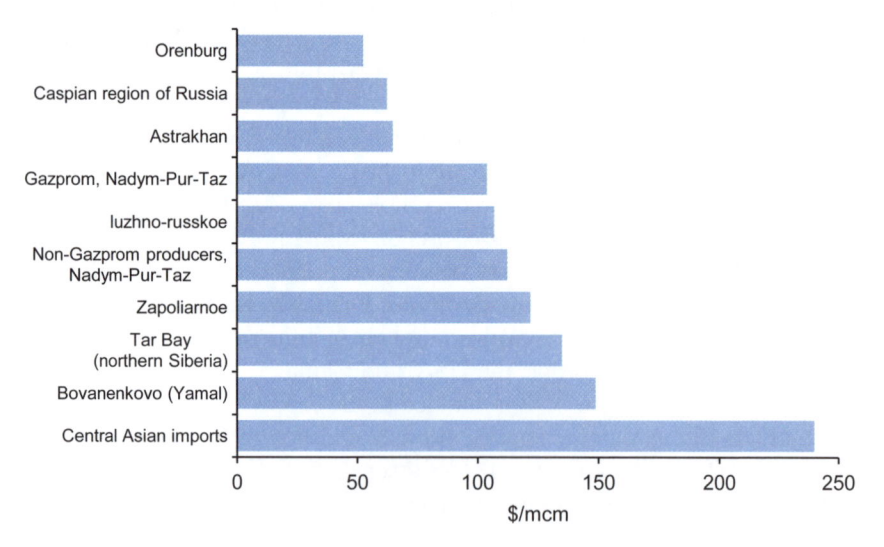

Figure 14.1: Estimates of costs (production + transportation) of gas delivered to Moscow region in 2020, $/mcm

Source: Moscow School of Management, Skolkovo Energy Centre.

---

[29] In 2009, when the price of volumes purchased from Turkmenistan was set under the old contract, Gazprom's average Central Asia/Caspian purchase price was 78.7% of the European price. This ratio moved to 65.3% in 2010, 63.7% in 2011, and 72.3% in 2012. (Calculated from prices in Table 14.3.)

There is a caveat to this. In areas of southern Russia through which pipelines from Central Asia pass, the market dynamics are different. The cost of delivery of Central Asian gas – as opposed to the price currently paid – is lower than in Moscow, and the cost of delivery of Siberian gas (including production and transportation) is higher than in Moscow due to the transport distances. Substantial volumes of gas are produced in southern Russia (42.4 Bcm in 2011), and the cost of production and transportation of this gas would presumably be lower than that of either Central Asian or Siberian gas. Nevertheless, the region will probably still require gas to be brought from elsewhere for the foreseeable future. Table 14.4 lists regions through which the pipelines from Kazakhstan pass, and also shows the extent of Russian non-Siberian production.

Table 14.4: Russian gas demand in areas close to infrastructure links with Central Asia (Bcm, 2012)

Along the Soiuz pipeline (Kazakhstan–Ukraine)	
Volgograd region	6.24
Rostov region	6.47
Along the pipeline from Aleksandrov Gai to Moscow	
Saratov region	5.44
Penza region	2.38
Riazan region	4.68
Linked to pipelines from Orenburg processing plant	
Orenburg region	9.19
Bashkortostan	15.96
Chelyabinsk region	15.53
Nearby to pipeline links from Azerbaijan	
North Caucasus federal district	21.6
Total	**87.5**
Gas production in Central and Southern Russia, 2011	
Southern federal district	16.9
North Caucasus federal district	1.0
Volga federal district	24.5
Total	**42.4**

Source: Rosstat, *Interfax*.

The conclusion from this is that Central Asian supplies could certainly serve this market, but only if commercial rather than political principles were applied, as these volumes would need to be priced against Russian gas rather than on the European netback principle. The question for Central Asian suppliers would then be whether these prices would be sufficiently attractive in comparison to those paid by China.

At present, these observations are merely speculative. The Russian government has apparently decided that high prices, set on a European netback basis, should be paid to Central Asian suppliers for around 30 Bcm/year of gas. However, the sharp reduction of purchases from Turkmenistan in 2009 shows that the political imperative is not immune to economic influence and that, under crisis conditions, it may be re-evaluated.

Possible linkages between export prices to Russia and China

Central Asian gas exports to China began in 2010 and have increased steadily since then. In 2012 they were an estimated 20 Bcm (see Table 14.1 above). In the first half of 2013 customs data indicated that they were 13 Bcm (11.8 Bcm from Turkmenistan and 1.2 Bcm from Uzbekistan), suggesting that they would exceed 25 Bcm for the year. Since these exports began, Chinese gas demand has been strong and rising, while the Russian and European markets have been oversupplied.

Prices are set with reference to oil prices; no more detailed information is available. Purchases are made by PetroChina, a wholly-owned subsidiary of CNPC. For the first few years of the contract, a proportion of the revenue is being taken in repayment of several large loans from Chinese state banks made to Turkmenistan to finance the development of the Galkynysh (South Yolotan) gas field. Customs information indicates that in 2010 Turkmen gas was sold at the Chinese border at a rough average price of $267/mcm ($7.25/MMBtu), at $315/mcm ($8.55/MMBtu) in 2011, $385/mcm (10.50/MMBtu) in 2012, and then falling back to $362/mcm (9.85/MMBtu) in the first half of 2013.[30] To this must be added a transmission tariff for transporting the gas the long distance to large consuming areas in east and south-east China; these, in October 2011, were on average $154/mcm.[31] Since PetroChina sells these volumes at regulated prices, it bears a considerable loss on the sales; in 2012 its total losses on sales of imported gas were $6.8 billion, of which half or more probably related to Turkmen imports.[32] Moreover, the Turkmen volumes

---

[30] Customs information, China, accessed by the author, September 2013.
[31] Pirani, S. 'Central Asian and Caspian Gas Production', op.cit., p.86.
[32] 'China looks local for solution to gas price dilemma', *Reuters*, 29 May 2013. See also Chapter 8.

are more expensive than domestically-produced gas and more expensive than some, but not all, of China's imported LNG.[33]

In short, state-owned Chinese companies are prepared to bear the considerable cost of transporting Turkmen imports across China, in order to guarantee security of supply. While on one hand the cost of imported Turkmen gas delivered to Chinese markets has been higher than most domestically produced gas and other imports, on the other hand it appears that the prices paid by China, netted back to the Turkmen border, are still around $50/mcm *lower* than the prices paid for gas exported to and through Russia. Table 14.5 shows the author's estimates, made on the basis of prices extrapolated from Chinese and Ukrainian customs information, for 2011.

Table 14.5: Central Asian gas exports to China and Ukraine: prices compared (annual averages for 2011)

$/mcm	At border of exporting country	At border of importing country
Turkmen gas for China	241–47	285–90
Turkmen gas for Ukraine	292	355
Uzbek gas for Ukraine	296	296
Kazakh gas for Ukraine	312	351

Assumptions: Transit fees of $3.50/mcm/100 km in Russia (my estimate); $2.50/mcm/100 km in Kazakhstan (Kaztransgaz information); $2.10/mcm/100 km in Uzbekistan (industry information).

Note. Import prices extrapolated from customs information. Netbacks are author's estimate. Estimated prices rounded to nearest dollar.

Source: Pirani, S. 'Central Asian and Caspian Gas Production', op.cit., adapted from Table 23 on page 88.

There are two future possibilities worth highlighting, both probably more relevant to the 2020s than to this decade. The first (which at the time of writing seems unlikely) is that, if the strategic importance of Central Asian gas supplies to China falls – as a result of more domestically-produced gas and other imported gas becoming available, and/or a levelling-off of demand – Central Asian suppliers may come under pressure from China to

---
[33] Industry information; 'China looks to diversify its gas supplies', *Platts* 'The Barrel' blog, 5 August 2013, http://blogs.platts.com/2013/08/05/china-gas/. For a full discussion of gas pricing in China, see Chen, M.X. 'Gas Pricing in China', in Stern, J. (ed.), *The Pricing of Internationally Traded Gas*, op.cit., pp. 310–37.

reduce export prices. The second is that, if Russia concludes an agreement to export its own gas to China (see Chapter 8), then it will have to consider whether or not it wishes to continue paying higher, European-linked prices for Central Asian gas, in order to help support Central Asian export prices to China and, therefore, indirectly support the price of its own exports.

To the extent that Central Asian exports remain competitive in the Chinese market, an element of competition for Central Asian gas may arise between Russia and China, with the possible result that the Chinese price may influence negotiations on the prices of westward exports, and Central Asian countries will increasingly be able to become price makers.[34] The caveat to this is that such an element of competition will depend on the availability of gas for export. Turkmenistan will have very large volumes available and, if and when the construction of its east–west pipeline is completed, ample capacity to redirect them. On the other hand Uzbekistan, and Kazakhstan to a lesser extent, will be constrained by their domestic requirements.

The Changing Russian role in the Central Asian upstream

The presence of Russian companies in the upstream oil and gas sector in the Central Asian and Caspian region has declined during the post-Soviet period. The national oil and gas companies in the region have either undertaken investments themselves or have sought partnerships with American, west European, and Asian companies. In the last five years Turkmenistan has brought Chinese companies into the upstream, in parallel with its commitment to export gas to China; it is developing the Galkynysh gas field, one of the world's largest, with Chinese and Middle Eastern investors, and has also awarded CNPC of China an onshore production sharing agreement.[35] In Kazakhstan: Chinese, European, and American companies dominate the large oil projects and the associated gas from them, which makes up most of the gas balance.[36] In Azerbaijan: European and American companies play a similar role. Only Uzbekistan has significant Russian investment upstream, with Lukoil playing a predominant role.

Lukoil and Gazprom, the two Russian companies that remain the most active in the region's upstream, have pursued markedly different strategies.

---

[34] Henderson, J., Pirani S., and Yafimava, K. 'CIS Gas Pricing: Towards European Netback?' in Stern, J.P. (ed.), *The Pricing of Internationally Traded Gas*, op.cit., p. 218.

[35] Pirani, S. 'Central Asian and Caspian Gas Production', op.cit., pp. 23, 25–26, and 29–30.

[36] See e.g. Luong, P.J. and Weinthal, E. *Oil is Not a Curse: ownership structure and institutions in Soviet successor states* (New York: Cambridge University Press, 2010), pp. 261–5.

Table 14.6: Russian companies upstream in Central Asia and the Caspian

	Share	Annual production (Bcm)	Russian share of production	Comments
Gazprom				
Uzbekistan				
Kokdumalak	12.5%	4.0 (est.)	0.5	
Gissarneftegaz	20%	4.1	0.82	
Shakhpakhty	37.5%	0.2	0.075	
Kazakhstan				
Tsentralnoye (Caspian)	50%	0	0	Pre-exploration
Imashevskoye	50%	0	0	Pre-exploration
Tajikistan				
Three areas	100%	0	0	Under exploration
Total			**1.395**	
Lukoil				
Uzbekistan				
Kandym-Khauzak-Shady	87%	3.8	3.3	35-year PSA from 2004. Plateau output 11 Bcm/year
Aral Sea PSA	26.7%	0	0	Under exploration
South-west Gissar	87%	1.15	1.0	36-year PSA from 2007. Plateau output 5.8 Bcm/year
Kazakhstan				
Karachaganak	13.5%	9	1.2	(Associated gas from oil project)
Tengizchevroil	5.0%	7	0.35	(Associated gas from oil project)
Kumkol	50%	0.222	0.111	(Associated gas from oil project)
Four small projects	25–50%	0.277	0.1	(Associated gas from oil project)
Azerbaijan				
Shah Deniz	10%	7	0.7	
Total			**6.761**	
Rosneft				
Turkmenistan				
Block 23 (Caspian)	100%	0	0	Under exploration

Source: Company websites, author's estimates.

Gazprom has sought to build on its longstanding relationship with the Uzbek and Kazakh state-owned oil companies; it focuses on projects whose output is bought by Gazprom Export and its subsidiary Gazprom Germania. In Uzbekistan, Gazprom reduced its portfolio of upstream assets in a series of transactions in the mid-2000s. Lukoil has, by contrast, invested heavily in Uzbekistan, in projects focused on export to China rather than Russia. It has also built a portfolio of associated gas, produced from large oil projects in Kazakhstan and Azerbaijan in which it has minority stakes. In 2013, press reports indicated that Rosneft had shown interest in investing in future gas production in Azerbaijan, through Total's Absheron field[37] – and such activity would fit logically into its strategy of challenging Gazprom across all markets.

Table 14.6 lists the principal Russian investments in the Central Asian and Caspian region. It shows that Gazprom now produces less than 1.5 Bcm/year of equity gas in the region, while Lukoil's portfolio is more than four times larger, and growing. Lukoil's two major Uzbek projects, Khandym-Khauzak-Shady and South-west Gissar, are projected to raise their aggregate output to 16.8 Bcm/year before the end of the decade. Lukoil hopes to sell most of this gas in China.

Conclusions

Russia's monopsony over Central Asian exports was broken by the opening up of the export route to China. This breakthrough to a second major market was a fundamental change. Central Asian exports to China at the time of writing in mid-2013 comfortably exceed 20 Bcm/year, and by the end of the decade will probably be more than twice, and quite possibly three times, that level. At the same time Azerbaijan, formerly a small market for Russian exports, has itself become an exporter, substituted itself for Russia as the supplier to Georgia, offered competition to Russia in the Turkish market, and has the potential for similar competition in the future (albeit on a small scale) in southern Europe.

Just as the Turkmenistan–China pipeline was approaching completion, the international economic crisis sent gas markets in Europe and Russia into oversupply and triggered Russia's reduction of purchases from Turkmenistan. It is significant that Russia sharply cut the volumes of its purchases from Turkmenistan but apparently made only minor downward adjustments to purchase prices. Since the crisis, Russia has continued to pay for Central Asian gas at prices linked to its own sales prices in Europe.

---

[37] 'Rosneft podpisala soglashenie o sotrudnichestvom s Azerbaidzhanom', *Vedomosti*, 13 August 2013.

It has been argued above that the prime motivation for this is political, in other words prices have been set at a level essentially above the market in order to strengthen Russia's political ties with Kazakhstan, Uzbekistan, and Turkmenistan.

An additional motivation for high prices may relate to Russia's own negotiations with China about future gas exports. If purchase prices in Central Asia remain at relatively high, oil-linked levels, this may strengthen Russia's hand somewhat in its efforts to achieve high, oil-linked prices for East Siberian gas in China. On the other hand, in comparison to the cost of production and delivery of, and actual prices of, Russian gas, Central Asian purchase prices are very high – at least twice as high as those of Russian gas in the largest consuming areas in Russia.

Russia will only continue to buy Central Asian gas on the current pricing basis for as long as political considerations, and possibly the factor mentioned above relating to its negotiations with China, make it strategically worthwhile. This is a decision that will surely be taken at government level rather than by Gazprom. Otherwise the terms of the purchases, which amount to a significant economic burden, would have begun to be changed by the usual commercial methods (renegotiation of prices and volumes, etc.).

It seems likely that, barring a major economic or political shift, Central Asian exports to Russia will remain at their current level, or lower, for the rest of this decade. If Russian policy towards Central Asia changes, the volumes could fall sharply once again and/or prices could be renegotiated. Over the longer term, if market principles were applied, there is sufficient Russian demand – in areas that are closer to Central Asian and Caucasian sources of supply than to Siberia – for exports to continue, or even be increased, but on a different pricing basis. The renegotiation of prices seems to be inevitable, if the Central Asian and Caspian gas trade with Russia is to be revived following China's entry into the market.

CHAPTER 15

SUMMARY: THE LIKELY BALANCE OF RUSSIAN GAS SUPPLY

James Henderson

This chapter summarizes, and highlights common themes in, Chapters 10–14 on the main sources of Russian gas supply. The key shift in relative importance away from Gazprom and towards third-party producers has been the result of the global gas market becoming more competitive. In the period up to 2007/8 Gazprom's position as the dominant player in the Russian gas sector was in no doubt, as it had the role not only of providing most Russian gas supply, but also of co-ordinating additional production from other sources in order to meet the increasing demands being placed upon it in the domestic and export markets. The company imported growing amounts of Central Asian gas and encouraged the Independent producers in Russia to co-operate with it and produce more gas for sale in Russia, in order to allow Gazprom itself to free up its own supply for export to Europe. However, from 2009 the combined impacts of the US shale gas revolution (that caused LNG to be re-directed to Europe and then cheap US coal to be exported across the Atlantic) and the economic crisis in the EU (causing a decline in gas demand) changed the supply–demand balance significantly. From a position in the mid-2000s when it appeared that Russia, and Gazprom in particular, might not have enough gas to meet demand, it now appears that it has a surplus.

In this situation of oversupply, market forces have started to play a key role, even in the regulated environment of the Russian gas sector. Higher-cost options are being de-prioritized where possible and lower-cost producers are winning market share. Unfortunately for Gazprom, and arguably through little fault of its own, this shift to a more competitive and cost-conscious world has been made more difficult by its decision to focus on the development of the vast, but higher-cost, reserves on the Yamal peninsula, with the result that its supply strategy for the foreseeable future appears almost guaranteed to worsen its competitive position.

Central Asia as an example of gas markets driving the Russian gas supply balance

Imports of gas from Central Asia provide a particularly good example of the impact of market forces on Russian gas supply over the past decade. As highlighted in Chapter 14, Central Asian imports played a very small role in the 1990s as Gazprom's own production was more than adequate to meet demand, but the situation started to change in the 2000s as Gazprom's main fields went into decline while demand from both the export and Russian domestic markets continued to grow. Central Asia was a cheap source of gas that could be extensively, although not always exclusively, controlled by Gazprom, as the only available export route for gas from Turkmenistan, Uzbekistan, and Kazakhstan at that time lay north via Russia, limiting the suppliers' market options and bargaining power over price. As a result Gazprom, or other Russian intermediaries who participated in the trade between Central Asia and the western FSU countries, paid a low price (close to Russian domestic levels) and made a significant return by on-selling the gas into the FSU and European export markets.

However, three factors then changed the balance of bargaining power: Gazprom's reduced need for third-party gas after the 2008–9 economic crisis, the higher price demanded by Central Asian countries for their gas, and, most significantly, the emergence of China as an alternative market for Central Asian gas. From 2009, not only could Russia not afford to buy large amounts of Central Asian gas at a price close to, and in some instance higher than, the European export netback, but also Turkmenistan in particular was keen to diversify its gas sales strategy away from its former monopoly buyer towards China. The consequence of this shift has been that Russia has relinquished some political influence in response to severe commercial pressure. One way of looking at this is that Gazprom could no longer afford to buy enough Central Asian gas to maintain Russia's dominant role in the region. It still imports a reduced volume of gas at prices that are much higher than the domestic price in southern Russia, in order to retain some of the old political links, but essentially China has now become the dominant force in the energy economy of Central Asia. China is prepared to provide manpower, financing, and expertise to develop gas (and oil) in the region, to build pipeline infrastructure, and most importantly is prepared to pay a high price for large volumes of gas. Our conclusion is that this trend will continue for the foreseeable future, with Russian imports of Central Asian gas limited to their current levels, perhaps even falling further, as Russia prioritizes utilization of its own sources of gas supply ahead of its geopolitical ambitions in Kazakhstan, Uzbekistan, Turkmenistan, and Azerbaijan.

Having reduced its own purchases from Central Asia, Russia may even be keen to encourage high prices for Central Asian gas in China, as this could help Gazprom in its own negotiations with CNPC. The higher the price of Central Asian gas arriving in Beijing, the stronger the argument for paying a similar price for Russian gas from East Siberia. In this sense, Gazprom has had to respond to market forces by reducing its own imports of Central Asian gas, but may hope to use similar pressure to encourage China to pay a market price for new Russian gas in the east.

Up to 2020, and probably far beyond, the only competitive challenge by the Caspian region's gas producers to Russia in its western markets will come from Azerbaijan, which began exporting gas to Turkey in 2007 and will start to deliver small volumes to Europe in 2019. At present this is viewed by Russia as a relatively small volume threat in south-east Europe.[1]

The rise of the Independents – low cost supply encouraged to meet domestic demand

While imports from Central Asia have been in decline since 2008, production from the Russian Independents has been rising inexorably, encouraged initially by Gazprom's desire to free up its own gas for export, but then by the prospect of higher domestic gas prices and profitable returns from sales to Russian industrial customers. This trend has been exemplified by Novatek, which has almost doubled its output since 2008[2] and has become increasingly competitive in marketing its gas. This culminated in the company offering to sell its output at a discount to Gazprom's regulated price for the first time in 2012, having previously always sold at a premium that limited demand for its gas. This change in strategy catalysed a series of negotiations with power, chemicals, steel, and other industrial companies which led to the signing of a significant number of long-term supply contracts that will underpin continued growth in output to 2020.

However, Novatek is no longer the only major player in the Independent gas sector, with Rosneft emerging as a significant force thanks to a series of acquisitions over the past two years – in particular of Itera and TNK-BP, both of which brought important gas assets.[3] Rosneft has the ambition to produce as much as 100 Bcm of gas by 2020, matching the goal of Novatek at the same date, and to this end it has also been signing a large number of long-term contracts with consumers. There is some concern that it may not have enough gas to meet its contractual obligations, but the company continues to make small acquisitions and to commit to field development plans which should mean that its output can match its sales targets by the end of the decade.

When the ambitions and productive potential of these two companies are added to the other smaller independent gas producers in Russia, then it would appear that the sector as a whole could produce as much as 300 Bcm by 2020, and up to 350 Bcm by 2025 if sufficient demand is available.

---

[1] 'Gazprom doesn't consider TANAP a competitor', AKM News Agency, 7 Mar 2012.
[2] www.novatek.ru/en/business/production/, accessed 28 Nov 2013.
[3] 'Rosneft buys rest of Itera from founder for $2.9bn', *Reuters*, 2 June 2013.

Much of this non-Gazprom gas is relatively low cost as it is either located close to existing infrastructure, or is associated gas from oil production, or is wet gas (having the benefit of condensate sales to improve the economic outcome). As a result, it can be very competitive, and therefore the trend of increasing market share for Independent producers in Russia, at the expense of Gazprom, is likely to continue.

One other important theme in the Independent gas sector is a growing focus on export markets. The existence of Gazprom's export monopoly has frustrated non-Gazprom producers to date, preventing any access for their domestically produced gas to CIS or non-CIS markets. This led Lukoil to adopt a strategy of developing gas assets in the international arena, with its investments in Uzbekistan offering it potential access to the Chinese gas market. However, in 2013 the Russian government provided the first catalyst for non-Gazprom gas exports, by removing, albeit in a limited fashion, Gazprom's monopoly over LNG exports.[4] The new legislation, which allows fields with LNG development embedded in the licence, and offshore discoveries owned by state companies, to export LNG, opens the way for Novatek's Yamal LNG project and Rosneft's Sakhalin-1 scheme to move ahead. Both companies have signed initial sales contracts with Asian (and in Novatek's case also European) customers and seem determined to use this opportunity as a springboard for greater involvement in the gas export market. A number of technical, operational, and commercial issues remain to be resolved, but it now seems very likely that by the end of the decade Gazprom will no longer be the only domestic company exporting gas from Russia.[5]

Gazprom – a swing producer with spare capacity to exploit opportunities if they arise

Gazprom committed itself to the development of the Yamal peninsula gas fields in response to an expectation that demand in Europe would continue on the trend seen in the early 2000s, and without the knowledge that US shale gas would emerge to disturb the equilibrium in the global gas market. Unfortunately, this combination of events has left Russia's largest gas company committed to a project that is gradually making it a relatively high-cost producer in a domestic context. Its response, as one would expect

---

[4] 'Gazprom set to lose export monopoly as Lawmakers pass Bill', *Bloomberg*, 22 Nov 2013.

[5] LNG has been exported from the Sakhalin-2 project by the Sakhalin Energy Company since 2011. Sakhalin Energy is a partnership between Gazprom, Shell, Mitsui, and Mitsubishi, and so it could be argued that three foreign companies have been involved in LNG exports since 2011. However, from a domestic perspective Novatek and Rosneft will be the first non-Gazprom exporters of gas from Russia.

in a market now increasingly driven by the fundamentals of demand and cost of supply, has been to reduce purchases of third-party gas and to cut back its own production in the face of competition from the Independents, as well as slowing the development of its first major Yamal project at Bovanenkovo.[6]

As a result, expectations about the company's gas production in 2020 have been sharply curtailed, from an estimate of 650 Bcm published in 2012 to an estimate of 540 Bcm in February 2013. Indeed company CEO Aleksei Miller openly talks about Gazprom already having the ability to produce 600 Bcm/year,[7] compared to actual output of below 490 Bcm in 2012, implying that the company has, and is likely to continue to have, spare capacity of more than 100 Bcm for the foreseeable future. In effect, Gazprom has become Russia and Europe's swing producer, adapting its output to levels of demand across its western markets and to the availability of alternative supply both domestically and from the global market. When supply is tight, as it appeared to be in Europe in 2013, then its production can rise, but when supply is more abundant, then output is likely to fall while the company continues to pursue its relatively high price oil-linked contract strategy.

Given this apparent production strategy in the west, and the relatively poor outlook for demand for Gazprom's gas in Europe, the FSU, and Russia outlined in Chapters 4–8, it is perhaps no surprise that the company has started to look east for growth markets where it can add to its sales. Russia's Eastern Gas Programme, published in 2007, lays out the strategy for developing Gazprom's very large resource base in East Siberia and the Far East of Russia and transporting the gas through a new trunk pipeline system (the 3,200 km Power of Siberia pipeline) to China and to the Pacific Coast. However, this eastern expansion plan encapsulates many of the main issues that Gazprom is facing. The enormous $80 billion investment required for the whole scheme is a potential source of corruption and vested interest exploitation. The domestic gasification programme that is also one of the goals of the programme, but which is very likely to be loss-making for Gazprom, is an example of politics interfering with the commercial necessities of such an enormous project. The foreign policy implications of establishing a long-term export relationship with China have necessitated the involvement of senior political figures in the negotiations, slowing the process and introducing geopolitical concerns that have complicated the commercial price and volume discussions. These issues, combined with Gazprom's own sense of inertia as it struggles to shake off its Ministry of Gas heritage, have led to delay and uncertainty, so that it remains unclear

---

[6] 'Gazprom cutting Yamal production', *Barents Observer*, 17 October 2013.
[7] 'Miller: No adjustment to Gazprom export plans', *Interfax*, 3 April 2013.

whether the company will be able to exploit its eastern reserves on a timely basis. Indeed, the risk is that it may soon be overtaken by its domestic rivals as they develop their own projects to sell gas into Asia.

These problems bear the worrying hallmarks of Gazprom's Atlantic Basin LNG strategy in the 2000s. At that time, the development of the Shtokman field and the Baltic LNG project seemed set to catapult Gazprom into the global LNG market, with sales targeted at North America and Europe from new projects with international partners. However, market developments, combined with operational and technical issues, caused the postponement of both projects, with Shtokman in particular being consigned to development 'for future generations'.[8] As a result, it now appears that the first Russian LNG project to sell gas into Europe could well be Novatek's Yamal LNG scheme, with Gazprom struggling to recover its position by suggesting the revival of Baltic LNG under the new name of Leningrad LNG. In the east, Gazprom is already the operator of Russia's first LNG project at Sakhalin-2, although this was in reality developed by Shell. However, it is now very unclear whether it will continue to dominate Russia's position in the region, as it seems unable to decide on the optimal balance of pipe and LNG sales from Sakhalin Island, Vladivostok LNG and East Siberia. It may therefore once again find itself beaten to the market by one of its domestic competitors.

However, although this outlook sounds rather bleak, Gazprom's financial position as of 2013 is robust and its production potential provides opportunity as well as risk. If, as the company certainly believes, gas markets in the West recover and alternative sources of global supply disappoint in terms of timing of development, then Russian, and therefore Gazprom, exports can recover quickly, and Gazprom is building significant new export pipeline capacity in anticipation of this outcome. Furthermore, although Gazprom's production is relatively expensive in a Russian context, it can nevertheless compete even with cheap US LNG in Europe, if it is priced competitively. In the east, Gazprom could sign a deal with China in 2014 and catalyse the full development of the Eastern Gas Programme, including Vladivostok LNG, if the company gets its marketing strategy right. The estimated cost of its East Siberian supply suggests that it could find a price to satisfy both Russian and Chinese price requirements.

Unfortunately the outcome of both the western and eastern strategies remains very unclear, and is subject to a number of market forces outside of Gazprom's control. For the first time in its history Gazprom's supply strategy is no longer about finding adequate production to meet demand, but finding adequate demand to meets its supply expectations, and this latter search involves the development of a marketing strategy that is

---

[8] 'Total could quit Shtokman gas field', *The Moscow Times*, 24 September 2013.

competitive in a very uncertain global gas market. Amid such uncertainty, the company is starting to adapt its oil-linked pricing policy to provide prices closer to spot market levels in Europe, but nevertheless the company's relatively reactive, rather than proactive, approach to gas prices, together with its aggressive export pipeline expansion plans, still imply a relatively high risk-strategy which has obvious benefits in a tight gas market but clear downside potential if future global supply expands. Furthermore, in any Gazprom scenario it is also evident that the company will, with Novatek and Rosneft, increasingly become one part of a supply triumvirate in Russian gas production, rather than the utterly dominant force that it has been in the past.

Part 4 consists of the Conclusions (Chapter 16), in which the interaction of changes in gas demand and those on the supply side are discussed.

Part 4
Conclusions

CHAPTER 16

THE CHANGING BALANCE OF THE RUSSIAN GAS MATRIX AND THE ROLE OF THE RUSSIAN STATE

James Henderson and Simon Pirani

Introduction

A central thesis of this book has been that the Russian gas sector can be described using a matrix of supply and demand factors. The main components on the demand side are exports to Europe and the CIS, sales into the domestic Russian market, and the potential for significant expansion into Asia. On the supply side three components of production provide the balancing Russian gas supply portfolio to meet this demand: Gazprom, still the largest producer, the Independents (the non-Gazprom domestic producers), and Central Asian imports. The preceding chapters have described how, in particular since the 2005 publication of Jonathan Stern's book *The Future of Russian Gas and Gazprom*,[1] the balance of this Russian gas matrix has been fundamentally altered. As we consider the future, it is probable that the new trends that have been established – declining Gazprom influence, increasing Independent production in Russia, and a shift in Central Asian exports away from Russia towards China – will continue. These trends have been catalysed by forces in a global gas market that has become increasingly competitive and progressively interconnected, meaning that the Russian gas industry and the Russian government have had to respond to challenges that they have not faced before. Their initial reaction has in itself, we would argue, displayed many market-related responses, even if the players themselves have been reluctant to acknowledge this fact. We summarized the changes on the demand side in Chapter 9, and those on the supply side in Chapter 15. In these conclusions we consider the outcomes to date and the potential consequences for the future. Before doing so, it is important to acknowledge two key issues: firstly, that the impact of market factors that few, if any, commentators, politicians or business executives foresaw, has been enormous, and is likely to continue; and secondly, that Russian domestic factors, and in particular the country's politics, will also play a significant role in the future of the gas industry, irrespective of the economic dynamics at work. We discuss each of these in turn, before returning to our overall conclusions on the future of the Russian gas matrix.

---

[1] Stern, J.P. *The Future of Russian Gas and Gazprom* (Oxford: OIES/OUP, 2005).

The impact of market factors

Europe
Three extraordinary shocks – the economic crisis, the sharp rise in oil prices to almost $150/bbl, and the rise of shale gas in the USA – hit European energy markets in 2006–9. These made a profound impact on Russia's, and Gazprom's, strategic plans, and fundamentally altered the demand and pricing outlook for Russian gas.

The economic crisis and subsequent stagnation that Europe has endured since 2008 was the first of these shocks. The sharp fall in GDP experienced by many countries in the wake of the 2008 banking crisis was reversed to an extent in 2009 and 2010, but since then average GDP growth in the Euro area has been slightly negative, and industrial production has remained more than 10 per cent below the levels seen before the crisis.[2] This crisis produced a sharp interruption in energy demand in 2009, with consumption in the gas markets served by Russia falling much more sharply than the average between 2008 and 2009: in the European Union by 5.9 per cent, in Russia by 6.1 per cent, and in Ukraine by 21.5 per cent.[3]

Since the initial demand shock the recovery has also been slow. Overall in 2012 European energy demand remained 4.5 per cent lower than in 2008, and gas demand remained 8 per cent lower than in 2008.[4] Furthermore Gazprom's exports to Europe fell to a low of 139 Bcm in 2012, 18 per cent lower than the 169 Bcm exported in 2008 (Chapter 3), before recovering to 162 Bcm in 2013.[5]

This sharp fall in Gazprom's exports in 2009–12 was in part a consequence of the second major shock to hit the global energy system in the 2000s: the sharp rise in oil prices to almost $150/bbl. As mentioned above, this was, and continues to be, of particular relevance to Gazprom, because the price of its gas exports is benchmarked against a basket of competing fuels, principally fuel oil and gasoil. Having reached its high of $147/bbl in July 2008, the oil price did initially respond to the global economic crisis – falling back to an average of $66/bbl in 2009 and $84/bbl in 2010 – meaning that Gazprom's export gas price also declined from its 2008 highs (see Figure 16.1). But since 2011 the oil price has rebounded to more than $100 per barrel. This has meant that although Gazprom

---

[2] 'Industrial production up by 1.8% in Euro area', Eurostat News Release, 14 January 2014.
[3] *BP Statistical Review of World Energy 2010*, London: BP. The Russian and Ukrainian consumption figures used are BP's, which are slightly different from the national statistics used in Chapters 5 and 7.
[4] Data from *BP Statistical Review of World Energy 2013*, London: BP, comparing demand in 2012 and 2008
[5] Data from *Gazprom Databook 2013*, the 2013 export figure is preliminary.

started to show some flexibility in its pricing as early as 2009, its gas export prices have nevertheless risen sharply at a time when its customers in Europe have been struggling through a period of economic stagnation and have been offered cheaper energy alternatives. It is perhaps no surprise, then, that during 2009–12 Gazprom's sales to Europe fell faster than the overall average decline in gas demand.

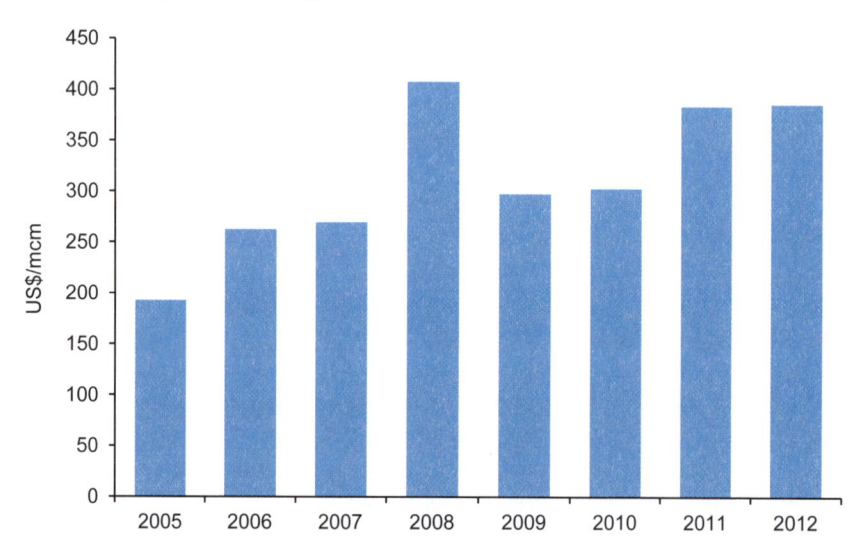

Figure 16.1: Gazprom's gas export price to Europe
Source: Data from *Management Report*, Gazprom (2005–12).

The third unexpected shock was the emergence of shale gas in the USA as a major element in that country's energy mix. The successful implementation of horizontal drilling and hydro-fracking techniques on the vast shale gas resources identified across North America has transformed the gas production outlook.[6] The resulting decline in US gas import needs has had a profound impact on the global gas market, and in particular on Gazprom's major export market, Europe. The initial impact was seen in the re-direction of LNG (that had been developed for supply to the US market) which needed to find a new customer base. The arrival of this extra supply in Europe from 2010 at very competitive prices caused spot

---

[6] For example: Henderson, J.A. 'The Potential Impact of North American LNG Exports', Working Paper NG 68, OIES, October, 2012.; Baker Institute Policy Report, *Shale Gas and US National Security*, No. 49, James A. Baker III Institute for Public Policy of Rice University, October 2011; Foss, M.M. 'Natural Gas Pricing in North America', in Stern, J.P. (ed.), *The Pricing of Internationally Traded Gas* (Oxford: OIES/OUP, 2012), pp. 85–144.

prices to fall and buyers to start to reduce their offtake of more expensive Russian gas, which was being priced against an oil price that was rapidly recovering to a level above $100/bbl. Indeed, European utilities that found themselves stuck in long-term contracts with Russia had no choice but to demand discounts or renegotiate the overall pricing arrangements, in the face of lower hub prices in continental Europe (see Chapter 3).

The effect of increased shale gas production was then compounded by a secondary impact of energy exports from the USA, namely the arrival of cheap US coal in Europe. This coal had been displaced from the US power generation mix due to switching by electricity companies to cheaper US gas.[7] The consequence was that between 2010 and 2012 European coal consumption increased by 6 per cent while gas demand on the continent declined by 9 per cent.[8]

The impact on gas demand of coal imports was exacerbated by two further factors: the collapse in carbon prices in Europe and the political shift towards an energy strategy increasingly favouring renewable energy. The carbon price has fallen from €25 per tonne in 2008 to as little as €2.75 per tonne in mid-2013, meaning that it effectively provided no incentive to prioritize gas over coal in the power generation mix. This has created the anomaly of gas-fired plants in Germany running at an average capacity utilization rate of only 20 per cent as the production cost of electricity from coal-fired stations became on average €20 per Gwh cheaper than production from gas-fired plants.[9] In addition, subsidies provided for the development of renewable energy sources such as solar and wind power have reduced demand for carbon-based power generation. But it has generally been gas rather than coal that has felt the brunt of this shift, due to the price differential between the two fuels.

The FSU
In the FSU countries, which provide a geographical link between European and Russian markets, a commercial link has also taken shape in these post-economic crisis conditions (see Chapter 7). Countries within the Customs Union, most importantly Belarus, are offered gas at prices set on a 'net forward' basis from the Russian domestic gas price, plus transport costs, providing a link between supply, demand, and political strategy in the two countries. In contrast, Ukraine had the price of its Russian gas imports set on the basis of a formula, agreed in 2009, that at times took it higher than a European netback equivalent. This encouraged the Ukrainian authorities

---

[7] 'US coal finds warm embrace overseas', *Wall Street Journal*, 6 February 2013.
[8] Data from *BP Statistical Review of World Energy 2013*, London: BP.
[9] 'German coal-fired plants under threat from renewables', *Argus Media*, 26 November 2013.

to develop a link with the European market in order to buy gas via 'reverse flow'. In 2013 the development of this reverse flow option, the appearance of new importers, and the continuing fall in domestic consumption, further curtailed Russian gas exports to the country, with sales to Naftogaz Ukrainy at less than half the agreed minimum contract level. However, the agreement in December 2013, under which import prices were reduced by more than $130/mcm, and financial support provided to Ukraine from the Russian treasury, has changed the situation dramatically. Gazprom has effectively been forced to cut its gas price, with the likelihood (but not the assurance) of increased sales volumes.

Russia
In Russia itself, the domestic gas market has also been transformed since the economic crisis. Demand growth slowed from 2.3 per cent/year in 2000–6 to less than 0.3 per cent/year in 2007–11, showed no signs of recovery thereafter, and is now projected at 0–1.5 per cent/year into the 2020s (see Chapter 5). Under these conditions of reduced demand growth, elements of genuine competition have taken shape for the first time. Historically Gazprom has been constrained by the need to sell gas at the regulated price set by the Russian government, which until 2009 meant that domestic sales were loss-making. But by 2012, following consistent 15 per cent per annum price rises, Gazprom's problem was reversed. The regulated price had become the high-cost option for consumers who were being offered discounts by the Independents, especially Novatek and Rosneft. As a result Gazprom lost out in a number of major contract negotiations with Russian customers, and its market share began to fall. By 2012, the non-Gazprom producers' sales (150 Bcm) amounted to more than one-third of the gas consumed in Russia (424.5 Bcm) (see Chapter 5). Up to 2020, we expect the non-Gazprom share of the market to continue to rise, albeit less rapidly than in 2009–12.

The impact of Russian political and economic factors

This change in the balance of Russia's domestic market is complicated by the vital role in the political economy of Russia played by gas, which provides close to 50 per cent of total primary energy demand and more than 70 per cent of the fuel input to thermal power generation. While the oil sector historically has been the revenue provider for the Russian budget, the gas sector has a much broader political agenda, including influence over the regions, social support for the population, low energy prices for domestic industries, and provision of largesse for vested interest groups and the political elite (Chapter 1). Gazprom also plays a key role in representing Russia in the global energy economy and is an important actor in a number of significant foreign policy initiatives. However, over the past decade the balance between Gazprom's political and commercial aims has become more difficult to maintain, not only because the global gas market has changed but also because the strategy of Russia's political elite has shifted.

In 2005 Gazprom was by far the most dominant force in the Russian energy economy, but since then the rise of significant domestic competition has been encouraged by the Russian government without any reduction of the burdens on the country's major gas producer. Rosneft, the state-controlled national oil company, has risen in importance under the leadership of Igor Sechin, its influential chief executive, and is now challenging Gazprom in the gas sector. Novatek, again with influential backers close to the Kremlin, has shifted from being a co-operative supporter of Gazprom to being a direct competitor. Both Rosneft and Novatek are also entering, via LNG, Gazprom's previously exclusive export market. Meanwhile Gazprom's own position appears to have been handicapped by the political requirement that it should respond to orders direct from the Kremlin – on issues as wide-ranging as gas exports to Ukraine, the construction of new export pipelines, export negotiations with China, the gasification of East Siberia, pricing strategy in Europe, and diversification into the power sector – without apparent regard for the optimal commercial outcome.

Alongside this political control, Gazprom has also frequently been criticized by the Kremlin for its inefficiency and corruption, despite the fact that many of the economic burdens it carries relate to political objectives set by government and to its role as a subsidy provider to the wider Russian economy. In the coming years, these tensions may intensify for various reasons. The two most apparent are, first, that the government, in view of the anaemic character of Russia's economic recovery, will again turn to gas for such subsidies (implying that the industry will be called upon to sell greater volumes at prices held down by regulation); and, second, that as oil tax incentives are increasingly introduced to encourage development aimed at preventing a potential decline in oil production, so the government will increase the tax burden on gas production and exports in order to balance the impact on budget revenues (implying that gas sector margins could be squeezed if commercial performance does not improve).

Up to the time of writing, such dichotomies between political and commercial objectives have further undermined Gazprom's position. It appears that its political masters are conflicted as to their goals for the company. One of the most important questions asked throughout this book has been whether a balance can be maintained, or whether global gas markets may force a more commercial and less political approach to be taken.

All these interactions between political, economic, and commercial factors may be affected, perhaps drastically affected, by the crisis that erupted in Ukraine in February 2014 (the fall of the Yanukovich government, Russian military action in Crimea, and the resulting tension between western powers and Russia). This is certainly the most serious crisis in Russo-Ukrainian relations, and probably the most serious crisis in Russia's relationship with western Europe, since the fall of the Soviet Union. This book was at an advanced stage of production by the time of these events, and readers should note that the text has not been amended to take account of them.[10]

The inter-linking forces in the Russian gas matrix

The key change to the Russian gas matrix was brought about by the economic crisis of 2008–9 through its impact on Russia's export markets for gas. The outlook suddenly shifted from one of anticipation of continued growth, and concern about whether sufficient supplies would be available to meet rapidly rising gas demand, to the realization that demand had stagnated and that supply competition had increased. In Europe, Gazprom's strategy of pricing its gas relative to a high oil price led to a decline in market share and absolute sales volumes, a trend which continued up to 2012, despite the company's gradual introduction of more flexible pricing terms from 2009. In the largest FSU market, Ukraine, demand for Russian gas fell steeply, while in the Caucasus, Russian supplies were displaced by newly available Azeri gas. The obvious conclusion of this shift in demand patterns was that Gazprom had more gas available to sell in the domestic market than it had anticipated; a complete reversal of the situation in the mid-2000s when the widely held concern was that Gazprom might struggle to supply all of its customers due to rising demand due to a perceived lack of upstream investment.

The supply/demand balance

In the Gazprom-dominated world of the 1990s and early 2000s, such a fall in demand would most likely have meant a Kremlin-supported reduction in Independent gas output and Central Asian imports, in order to allow for increased Gazprom sales. The latter effect did occur, as imports from Turkmenistan in particular were cut back sharply, but Independent output has continued to rise significantly and in 2013 accounted for more than a

---

[10] For an assessment of the issues, by some of this book's authors, see: Pirani S. *et al.* 'What the Ukraine Crisis Means for Gas Markets', Oxford Energy Comment, OIES, March 2014, www.oxfordenergy.org/2014/03/what-the-ukrainian-crisis-means-for-gas-markets/

quarter of overall Russian production. Having been encouraged in the mid-2000s to provide additional domestic supply in order to allow Gazprom to free up gas for exports, and having seen the commercial returns on domestic gas sales improve dramatically thanks to a rising regulated gas price, the Independents were reluctant to cut back supply when Gazprom's export markets went into decline. Furthermore, they have received government support for their expanded role in the Russian market. As a result, Gazprom's ability to match supply and demand across its three traditional markets has been significantly reduced, with the competitive pressure in the domestic market being increased by the significant political influence enjoyed by its main rivals.

Gazprom's decision, taken in 2006–7, to invest in the development of the gas resources on the Yamal peninsula (Chapter 11) has added greatly to its difficulties in balancing supply and demand. Although with hindsight the multi-billion dollar investment in new fields and pipeline infrastructure appears to have been a huge commercial mistake, at the time it was encouraged by the Russian government and all western observers, and appeared commercially logical given the generally anticipated growth in demand for Russian gas. The outcome, however, has been that Gazprom has developed the Bovanenkovo field, with an initial production potential of up to 115 Bcm/year, at a time when there is no obvious need for the gas, committing itself to a supply portfolio that, in a Russian context at least, is also relatively high cost when compared to those of its domestic rivals.

This inter-linkage between demand reduction in the three main markets for Russian gas, the investment in new sources of supply at a time of stagnating consumption, and the reduction in Gazprom's ability to manage the balance between supply and demand for its own gas over the past five years, has led to a potential gas bubble in Russia. Figure 16.2 shows the situation in 2012 and a projection for 2020, with Russia having a 100+ Bcm/year surplus in both time periods.

The 2012 bar reflects Gazprom senior managers' claim that, if necessary, the company could produce up to 600 Bcm/year. When this potential figure is added to 2012 Independent production, it can be seen that Russia had the capacity to produce more than 750 Bcm/year. This was about 100 Bcm above 2012 demand for Russian gas. (These conditions persisted in 2013, as is self-evident from the fact that Gazprom produced substantially less than 500 Bcm in that year.)

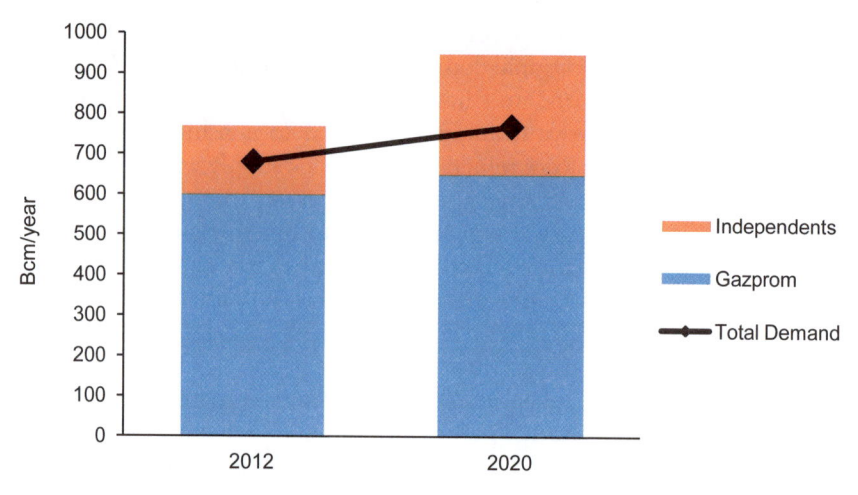

Figure 16.2: The potential oversupply of Russian gas
Source: Oxford Institute for Energy Studies.

Looking out to 2020, Gazprom forecast as recently as 2012 that it would produce as much as 650 Bcm/year of equity gas at that date.[11] When this is added to possible Independent output of 300 Bcm/year at the same time (see Chapter 13), Russian gas supply could theoretically reach 950 Bcm/year, perhaps more than 150 Bcm/year above the demand level we estimate for the end of the decade. These figures are clearly speculative, and in reality actual supply would of course match demand (through constraining production below design capacity). But some important conclusions may be drawn from the fact that Russia, and in particular Gazprom, is acquiring the potential to act as a swing producer in the global gas market.

Post 2008 Gazprom could be described as having become a buffer, or a shock absorber, for the global gas market, adjusting its own supply in response to demand in an attempt to support relatively higher prices, especially in Europe.[12] The most obvious evidence of this is the fact that the company in 2013 reduced its 2020 production estimate from 650 Bcm to 540 Bcm,[13] effectively slowing its development plans for the Yamal peninsula resources and pushing back the timing for peak production on a number of fields by up to three years (see Chapter 11). As mentioned above, imports of gas from Central Asia have also been cut back sharply, and

---

[11] 'Scale Does Matter', Gazprom Investor Day presentation, February 2012, slide 9.

[12] Rogers, H., 'The Impact of a Globalising Market on Future European Gas Supply and Pricing: The Importance of Asian Demand and North American Supply', Working Paper NG 59, OIES, January 2012.

[13] 'Gas for the Future', Gazprom Investor Day presentation, February 2013, slide 9.

although we would expect political objectives in the region to encourage a continuation of some modest import volumes we do not anticipate a return to the levels of the mid-2000s.

The markets

This strategy of adjusting supply is effectively a reactive response to the change in global gas markets, and has been justified by Gazprom as providing short-term support for higher gas prices in a slack gas market, before a tightening of global gas supply or a recovery of demand causes its sales volumes to rebound. Indeed the evidence of 2013 would tend to support this theory, at least in the short term, as Gazprom's sales to Europe have increased to 163 Bcm, from 139 Bcm in 2012. With few major new gas supply projects expected to come on stream before 2016,[14] and with the possibility that many of the new LNG projects under construction and expected to reach FID could suffer delays and cost over-runs, it is possible that Gazprom's strategy could pay dividends, at least up to 2018. Certainly the company's plans to expand significantly its gas export transport capacity, via the construction of new pipelines such as South Stream, would suggest that Russia's export strategy is at least partly based on this belief – despite the potential regulatory problems for new infrastructure created by the EU's Third Energy Package (Chapter 4).

However, creation of 'surplus' production and transport capacity in anticipation of a failure by competing supply (LNG projects in particular) to start up on time, is clearly a high-risk strategy. If Gazprom is wrong, either about the direction of demand in Europe and the FSU, or about the timing of development of alternative sources of gas (or broader energy) supply, then it could be wasting vast sums on ill-timed investment projects. Gazprom is apparently proceeding with such projects – in relation to transport, under government instruction that transit via Ukraine is minimized and possibly abandoned – but also appears to have understood the risks, and is adopting a more flexible and competitive strategy both in terms of price and volumes across all its existing, and potential new, markets. This emerging new trend reflects the fact that Gazprom does not need to pursue a high-price strategy as its Yamal gas is able to reach EU borders at a cost of around $7.50/MMBtu (at a 10 per cent internal rate of return including export tax). This is significantly below the European hub prices of $10 to $11/MMBtu pertaining in 2013, and below the delivered cost of much new imported supply, including LNG from the USA.

In Europe, as noted above, Gazprom started to demonstrate some pricing flexibility in its negotiations with customers as early as 2009, but it

---

[14] IEA Medium Term Gas Report 2013, p.139.

was only in 2012–13 that the company's prices responded to market forces. In the face of enforced renegotiations and arbitration cases, Gazprom offered a combination of base price reductions and rebates in its oil-linked contracts with customers in competitive markets that, by 2013, effectively aligned its sales prices with spot prices on European hubs. It is possible that during 2014 or 2015 the European Commission's competition investigation into Gazprom's commercial practices in Eastern Europe could prompt the company to make a more significant shift towards hub-based pricing. Gazprom has made it clear that it will be prepared to compete with the most immediate new source of competition, LNG exports from the USA, that are expected to be available from 2016. A move towards a more market-based pricing strategy would certainly allow it to do this, given the amount of extra gas it can make available for export. Indeed, even if Gazprom does not follow a full hub-based trading strategy, the combination of a more flexible pricing mechanism and the ability to control the marginal supply of gas into Europe could provide it with increasing market power on the continent, and the ability to influence and perhaps control European hub prices, unless, or perhaps until, substantial volumes of additional LNG begin to flow to Europe. However, the risk it faces is that if it holds prices too high it encourages the development of competing supplies and ultimately a loss of market share.

The future evolution of Russian gas exports to FSU countries also seems likely to be increasingly influenced by the conditions in adjacent markets, in particular Europe. The agreement made between Russia and Ukraine in December 2013 bears this out. It may put an end, at least temporarily, to Ukrainian purchases of gas at European prices via the 'reverse flow' route, as Russian gas now appears to be priced on a more accurate European netback basis, eliminating the incentive for 'reverse flow' arbitrage. It remains to be seen how these relationships will unfold as prices move in Europe, and whether the price of gas in Ukraine could ultimately have any impact further west in the EU. But the overall conclusion remains that European prices have forced Russia and Gazprom to adopt a more market-based approach to pricing exports to Ukraine.[15]

The dynamics of the Russian gas market are also continuing to evolve, and here too Gazprom has the opportunity to adopt a more proactive strategy. Legislation is being introduced in 2014 to allow Gazprom to offer gas at a discount to the regulated price (as well as at a premium should it wish), meaning that a de facto 'free market' for gas would be introduced, albeit with an indicative regulated price still in place. In effect this will mean that Gazprom will have more, but not complete, freedom both at

---

[15] This book was at an advanced stage of production when the Ukraine crisis broke out in February 2014. See footnote 10 above.

home and in its export markets to choose whether to maximize price or volume in its sales strategy and also to create a more optimal balance between the various markets.

It is even possible to conceive of a time when volumes and prices in the domestic market, which already impact prices in some FSU countries, could also have an impact in Europe through re-export of supplies from Russia by FSU countries. The price differentials and Gazprom's ongoing control of the export monopoly make this unlikely in the short term, but interaction between all of Russia's western markets is certainly becoming a more viable long-term concept.

Asia as an indicator of the future

Gazprom appears to be gradually adopting a more proactive commercial strategy in its traditional markets, but perhaps an even more significant indicator of the future of the Russian gas sector, and Gazprom's role in it, may be found in the development of Asia, and in particular China, as a new export opportunity (see Chapters 8 and 12). Negotiations for pipeline exports to China have been continuing for more than a decade, as Gazprom has sought to diversify its sales options and possibly even to link its exports to Europe and Asia, thereby placing itself and Russia at the heart of the global gas industry. To date China has resisted these moves, ostensibly on the grounds of price but also due to a perceived reluctance to become reliant on assets owned by a politically sensitive neighbour over which it has no control. Both sides appear to have strong reasons to conclude an agreement, with Russia keen to develop a new gas supply region in East Siberia and to increase and diversify export sales, and China needing to supply its rapidly expanding gas demand. However, the failure to agree a deal has now raised the question on the Russian side as to whether Gazprom is best equipped to lead the country's eastern expansion – with the emergence of Novatek and in particular Rosneft as possible alternative implementers of an Asian strategy that could be led by LNG developments. There would appear to be three potential outcomes to the current stand-off, and the ultimate result could reveal much about the future of the Russian gas sector over the next decade.

The first, and perhaps most likely, result is that during 2014 Gazprom and CNPC finally manage to agree an export deal. Gazprom CEO Aleksei Miller continues to assert that this will happen[16] and indeed the logic of linking Russia's huge gas resources with China's growing demand has seemed irrefutable for at least the past decade, despite the lack of agreement between the two countries. We believe that a gas price can be found that

---

[16] 'Miller: China gas contract might be signed before Feb', *Interfax*, 17 December 2013.

would allow all parties to be satisfied commercially, especially if Gazprom finally concedes some equity in the upstream part of the project. Indeed this concession would be a further demonstration of the increasing flexibility already shown by Gazprom in its existing markets. Construction of the Power of Siberia pipeline would then underpin both pipeline exports to China and LNG exports from Vladivostok, backed up by existing, and perhaps expanded sales from Sakhalin. Rosneft and Novatek could still develop their Sakhalin-1 and Yamal LNG projects (Chapter 13), but Gazprom would be at the forefront both of Russia's eastern development and its LNG strategy.

The second, and completely contrary, possible result of the negotiations with China is that no deal is struck and Russia fails to find a significant place in the Asian gas market. Gazprom and CNPC may find that they cannot agree a price which both companies can justify commercially, and China may decide to wait for alternative gas supplies to become available, either from regional pipeline or global LNG suppliers or from its own indigenous resources. Russia could still find a small role in Asia as an LNG producer, with Gazprom's role diluted as its Sakhalin-2 and Vladivostok projects would have a smaller relative position and would compete for customers with the Rosneft and Novatek schemes. Gazprom would remain heavily reliant on its west-facing markets, facing the domestic and global competitive threats discussed above, while Russia would be only a minor player in the world's fastest growing gas market.

A final, and more radical, third outcome could see Gazprom's position significantly undermined by its failure to do a deal in China. Essentially the Russian government could decide, prompted by Gazprom's domestic competitors, that the opportunity for piped export sales to Asia is too important to miss, and that if Gazprom cannot complete the deal then an alternative company should. This alternative would most likely be Rosneft, given its successful history with oil exports to China and its huge domestic political influence. Rosneft could then become the leader of Russia's overall eastern energy strategy, with Gazprom and Novatek in supporting roles. Although we do not regard this outcome as likely, the ending of Gazprom's monopoly over LNG exports in late 2013 indicates that Russia's export strategy may start to evolve. It is no longer inconceivable that Gazprom's overall export monopoly could at some point be challenged. Asia is becoming the first test of its ability to respond to the challenge, as its domestic competitors are already doing deals with customers in the region.

The outcome of the Asia negotiations is crucial on two levels. Firstly, for Russia as a whole the signing of a deal with China in 2014 would mean that pipeline gas could be exported before 2020, with the peak level of 38 Bcm/year being reached shortly thereafter. Failure to conclude a deal, on the other hand, could see a major delay in the arrival of significant Russian gas in Asia, in particular piped gas that would generate export tax for the Russian budget. Secondly, failure to sign a deal could also have significant implications for Gazprom. The initiative could be handed to others, most likely Rosneft, with Gazprom left to develop its LNG assets based only on the Sakhalin-2 and Sakhalin-3 licences. If, however, it can conclude a deal with CNPC for gas sales to China, then it can confirm the entire Eastern Gas Programme, with the Power of Siberia pipeline and the Vladivostok LNG plant moving ahead and underpinning Gazprom's position as leader of Russia's Asian market initiative.

The future of Russian gas and Gazprom

In each of the markets on the demand side of the matrix, Gazprom faces strategic decisions: how to meet the challenge of hub pricing in Europe; how to deal with falling demand and potential supply diversification in the FSU; how to manage the new, competitive conditions in the Russian market; and how to manage the opening-up of Asian export markets. We expect that the company's major decisions will continue either to be taken, or heavily influenced, by government, and will be considered alongside other factors unrelated, or only loosely related, to gas market fundamentals. These factors will include: in Europe, Russia's larger energy strategy; in the FSU, the management of political relationships with Ukraine and other countries; and in Russia, the calls on gas sector revenues by the state and by activities such as infrastructure projects, and the political pressures constraining electricity and gas price reform.

On the supply side of the Russian gas matrix, one element – Central Asia – has already been reduced to a minor role. There is no apparent prospect that volumes purchased from Central Asia could increase before 2020, and they could fall further; any future revival of the Central Asian trade would depend on a shift in price formation towards a link with the Russian market. The big question mark in terms of supply up to 2020, therefore, is the speed at which Gazprom's Yamal fields increase their output, how effectively this gas competes with non-Gazprom production, and under what market rules. This competition between Gazprom and non-Gazprom producers will again be played out in the drive to open up Asian markets for much larger volumes of Russian gas in the decade after 2020. In both cases, government will be decisive.

While the government's decision-making power is increasingly concentrated in the hands of the president, he is influenced by a range of interest groups in the state sector and in business circles. (We think it most unlikely that this system of government will change fundamentally before 2020 – although, if it did, that could alter the situation dramatically.) In the case of the gas sector, such influence is wielded not only by Gazprom but also by its competitors Rosneft and Novatek. All these major gas industry players may find themselves under increasing financial pressure from government's increased need for more revenue from the sector. But government decisions about the gas sector are increasingly being taken under the pressure of market forces. At stake is the future character and role of Gazprom. In a positive scenario, Gazprom could use its market power in all three of its existing major markets to create a strong position from which to maximize profits and revenues for itself and for the Russian budget. Arguably it is already making progress towards this goal in Europe, where it has started to adapt to market price and competition, and the FSU, where in Ukraine it is now pricing its gas on a netback equivalent basis with Europe, albeit as a result of political negotiation as much as commercial logic. In the domestic Russian market too, Gazprom could change its strategy in 2014 as it is now permitted to sell gas at below, as well as above, the regulated price, making competition with the Independents possible. If in 2014 Gazprom could also succeed in opening up the Asian market, by means of a pipeline deal with China and two world-scale LNG export terminals on the Pacific coast, then it could create a robust foundation for itself, and as a result for Russia, in the global gas market. Alternatively, in a worst case scenario for Gazprom, it could fail to expand in Asia, and a reactive strategy that encouraged supply competition could make it a declining force in the domestic and western export markets. Clearly there are other outcomes, in which elements of these two extremes might be combined. However, the mere fact that a debate about the future of Gazprom is relevant, and that the potential decline of Gazprom in the face of domestic and international competition can even be suggested, underlines how much markets have driven, and are continuing to drive, change in the Russian gas sector.

APPENDICES

Appendix 1. Statistical issues

Russian gas consumption levels in 2000–11 are shown in Table 5.1, and sectoral trends are highlighted in some tables in Chapter 6. Statistics from Rosstat, the national statistical agency, have been used to compile these tables. The sectoral breakdown of consumption is compiled according to the All-Russian Classifier of Types of Economic Activity. Researchers consider that the sectoral and regional breakdowns provided by Rosstat make the statistics the best available. They are not published, and become available only after considerable delay.

A second set of statistics is compiled by the Central Dispatching Unit of the Fuel and Energy Complex (TsDU TEK), an agency of the Energy Ministry. This gives a much more detailed regional breakdown – monthly statistics for each of the 83 regions in the Russian Federation – although these do not include any sectoral breakdown. These statistics are not published, although *Interfax* has in recent years published annual summaries.

A third set of statistics is compiled by Gazprom. This covers only gas distributed through the United Gas Supply System. Gazprom uses its own methodology to work out sectoral breakdowns, which does not conform to the standards used by statistical agencies.

Finally there are statistics published by the IEA, derived from questionnaires completed by Rosstat, following IEA methodology. The IEA publishes a sectoral breakdown, using different categories from Rosstat's. The IEA's statistics give a more detailed sectoral breakdown than Rosstat's and illustrate more clearly, for example, volumes of manufactured gases in the iron and steel sector. The IEA uses a problematic categorization of power stations and CHP, which in the Russian context is confusing.[1]

Table A1.1 below shows the total volumes of Russian gas consumption as recorded by different agencies.

Comments on the sectoral breakdown used by Rosstat in Table 5.1.

Heat. The row 'Power stations' appears to include gas purchased both by wholesale generating companies for large thermal power stations, and territorial generating companies for CHP. At least one-third of that total

---

[1] See Pirani, S. 'Elusive Potential: Gas Consumption in the CIS and the Quest for Efficiency', Working Paper NG 53, OIES, July, 2011, pp. 32, 33–5, and 39–44.

(at least 56–64 Bcm/year in recent years) may therefore be considered to be gas consumed for heat production. In addition, the row 'Centralized boilers' (68–70 Bcm/year in recent years) is entirely for heat production. Smaller amounts of gas used to produce heat are concealed in the rows 'Manufacture and processing sectors', 'Other economic sectors' and 'Residential use'.

The reasons why it is difficult to measure the volume of gas used for heat production include: (i) there are different ways of assessing the proportion of fuel in CHPs that should be attributed to the production of electricity and heat respectively; (ii) many statistics (including Rosstat's) do not separate out gas consumed for heat production in CHPs integrated into other industrial plants; and (iii) there is no clear way of counting gas consumed by individual heat sources.

Industry. The rows 'Manufacture and processing sectors', 'Gas processing plants' and 'Feedstock' all reflect gas consumption by industry. All volumes categorized as 'Feedstock' are consumed by chemical plants, mostly fertilizer producers. The category 'Mining and extractive sector' may also include some industrial plants.

Other economic sectors (25.5 Bcm in 2011). This probably comprises commercial and public-sector organizations supplied via low-pressure networks, as opposed to large industrial plants supplied directly from trunk pipelines, so some of these volumes are also for industry.

Residential use (49.7 Bcm in 2011) and *Technical use and losses* (52 Bcm in 2011) are the other substantial sectors. Construction, Agriculture, and Transport and communications together consume comparatively little, for example, only 3.5 Bcm in 2011.

Table A1.1: Statistical measurements of Russian gas consumption (Bcm)

	2000	2001	2002	2003	2004	2005	2006	2007	2008	2009	2010	2011	2012
IEA	394.92	402.72	403.15	424.13	429.15	432.88	444.07	453.17	453.43	433.77	473.86	484.44	479.01
Minenergo									458.55	430.50	495.40	496.22	459.64
Gazprom publications						444.4	458.9	467.1	462.5	432.2	460.3	473	465.4
Rosstat	398.5	406.4	412.5	426.4	434.9	440.7	456.3	466.7	462.0	432.6	462.1	472.8	n/a
CDU Tek										354.45	365.20	365.20	362.56

Sources: CDU Tek, from *Interfax*; Gazprom Total internal gas consumption, *Gazprom in Figures 2007–11*, p. 62; earlier years from earlier editions; IEA is from *Energy Statistics of non-OECD countries*, various editions and *Natural Gas Information* (2013), converted by the author from Terajoules, the unit of measurement used by the IEA. See Pirani, S. 'Elusive Potential: Gas Consumption in the CIS and the Quest for Efficiency', Working Paper NG 53, OIES, July, 2011 for methodology.

Appendix 2. Russian domestic gas consumption by region

Table A2.1. summarizes gas consumption, broken down into the eight Federal Districts. The last two columns show the growth of consumption by volume, between 2003 and 2007, and between 2007 and 2011.

Table A2.1: Russian gas consumption: regional breakdown

| | \multicolumn{12}{c}{Consumption (Bcm)} | \multicolumn{2}{c}{Growth (%)} |
	2000	2001	2002	2003	2004	2005	2006	2007	2008	2009	2010	2011	2003–7	2007–11
Central Federal District: total	**93.0**	**98.4**	**98.3**	**100.2**	**99.8**	**102.2**	**104.0**	**103.7**	**103.0**	**99.8**	**105.1**	**106.8**	**3.5**	**3.1**
Power stations	41.3	44.1	43.4	43.9	43.2	45.1	46.2	47.2	46.7	43.4	46.8	47.8	3.3	0.6
Boilers	22.9	24.3	24.3	24.9	25.0	24.8	24.7	23.9	23.1	22.8	23.1	22.2	-1.0	-1.7
Industry, agri., construction, etc.	17.7	18.5	18.9	19.2	19.5	20.0	20.7	20.5	20.2	20.0	21.0	22.6	1.3	2.1
Residential	11.2	11.5	11.6	12.2	12.1	12.3	12.5	12.1	13.0	13.5	14.2	14.1	-0.1	2
North-West Federal District: total	**30.5**	**31.5**	**32.6**	**33.0**	**35.2**	**35.6**	**36.9**	**38.4**	**38.0**	**36.7**	**39.2**	**40.2**	**5.4**	**1.8**
Power stations	12.2	12.5	13.7	13.7	15.0	15.2	16.1	16.9	17.3	16.7	18.1	19.4	3.2	2.5
Boilers	8.5	8.5	8.5	8.5	8.5	8.4	8.2	8.2	7.7	7.9	8.2	7.3	-0.3	-0.9
Industry, agri., construction, etc.	8.7	9.2	9.0	9.4	10.4	10.6	11.2	11.9	11.6	10.6	11.4	12.1	2.5	0.2
Residential	1.2	1.3	1.4	1.4	1.4	1.4	1.4	1.4	1.4	1.4	1.5	1.5	0	0.1
Southern Federal District: total	**28.2**	**29.4**	**30.0**	**30.9**	**31.0**	**31.9**	**32.0**	**32.0**	**32.7**	**29.7**	**33.0**	**33.1**	**1.1**	**1.1**
Power stations	7.0	7.5	7.5	7.8	8.0	8.2	7.9	8.0	8.2	7.1	8.2	8.4	0.2	0.4
Boilers	4.9	4.9	5.2	5.2	5.0	5.0	4.8	4.9	4.9	4.7	4.5	4.6	-0.3	-0.3
Industry, agri., construction, etc.	9.0	9.6	9.7	10.0	10.2	10.8	11.2	11.1	11.3	9.5	11.6	10.6	1.1	-0.5
Residential	7.4	7.4	7.6	7.9	7.8	7.9	8.1	7.9	8.4	8.4	8.6	9.5	0	1.6
North Caucasus Fed. Dist.: total	**19.5**	**20.2**	**20.2**	**20.6**	**20.4**	**20.8**	**21.8**	**22.2**	**22.1**	**20.9**	**20.4**	**21.6**	**1.6**	**-0.6**
Power stations	4.8	4.9	4.7	4.5	4.4	4.5	4.9	4.9	5.3	4.9	5.1	5.3	0.4	0.4
Boilers	2.2	2.1	2.2	2.3	2.1	2.1	2.1	2.1	2.1	1.9	1.8	1.9	-0.2	-0.2
Industry, agri., construction, etc.	5.6	6.4	6.5	6.6	6.5	6.9	7.3	7.7	6.4	5.9	6.1	6.4	1.1	-1.3
Residential	6.9	6.8	6.9	7.2	7.3	7.3	7.5	7.5	8.3	8.2	7.4	7.9	0.3	0.4

Table A2.1: Russian gas consumption: regional breakdown *(continued)*

| | \multicolumn{12}{c}{Consumption (Bcm)} | \multicolumn{2}{c}{Growth (%)} |
	2000	2001	2002	2003	2004	2005	2006	2007	2008	2009	2010	2011	2003-7	2007-11
Volga Federal District: total	**101.3**	**102.3**	**102.6**	**104.5**	**107.5**	**107.6**	**110.5**	**112.7**	**110.2**	**101.9**	**110.0**	**112.5**	**8.2**	**-0.2**
Power stations	47.1	47.7	47.1	47.7	49.8	50.1	51.3	52.9	52.6	46.7	52.0	53.0	5.2	0.1
Boilers	17.2	17.4	17.7	17.6	17.6	17.8	18.3	18.3	16.5	16.3	16.6	16.6	0.7	-1.7
Industry, agri., construction, etc.	25.0	24.7	24.5	25.7	26.6	26.7	27.8	28.4	28.1	25.7	28.2	29.2	2.7	0.8
Residential	12.1	12.5	13.2	13.6	13.4	13.0	13.2	13.1	12.9	13.1	13.3	13.6	-0.5	-0.5
Urals Federal District: total	**61.9**	**61.4**	**64.0**	**67.0**	**70.9**	**74.3**	**79.4**	**84.1**	**80.2**	**78.2**	**80.7**	**82.7**	**17.1**	**-1.4**
Power stations	32.1	31.1	33.2	34.5	36.5	38.4	40.1	42.8	43.6	42.5	44.1	45.9	8.3	3.1
Boilers	11.3	11.1	10.9	11.0	11.3	10.9	11.7	11.3	10.6	10.7	10.8	10.2	0.3	-1.1
Industry, agri., construction, etc.	17.4	17.9	18.5	20.4	21.6	23.4	25.8	28.2	24.1	23.1	23.7	24.5	7.8	-3.7
Residential	1.1	1.2	1.4	1.5	1.5	1.6	1.8	1.7	1.9	1.9	2.0	2.1	0.2	0.4
Siberian Federal District: total	**13.1**	**14.5**	**14.0**	**15.4**	**15.7**	**15.5**	**16.2**	**15.9**	**15.6**	**15.7**	**17.1**	**16.7**	**0.5**	**0.8**
Power stations	5.5	6.4	5.6	6.1	6.3	6.1	5.9	5.4	5.6	5.7	6.3	6.1	-0.7	0.7
Boilers	2.7	3.1	3.1	3.4	3.4	3.5	3.8	3.5	3.5	3.5	3.8	3.5	0.1	0
Industry, agri., construction, etc.	4.7	5.0	5.2	5.8	5.7	5.6	6.2	6.7	6.1	6.0	6.4	6.4	0.9	-0.3
Residential	0.1	0.1	0.1	0.1	0.2	0.2	0.3	0.3	0.4	0.5	0.6	0.6	0.2	0.3
Far Eastern Federal District: total	**3.3**	**3.3**	**3.3**	**3.5**	**3.4**	**3.4**	**3.7**	**8.4**	**9.4**	**13.7**	**13.1**	**13.8**	**4.9**	**5.3**
Technical and other	0.2	0.2	0.2	0.2	0.2	0.2	0.2	4.4	5.0	8.7	7.1	6.5	4.2	2.1
Power stations	1.7	1.7	1.7	1.6	1.6	1.7	2.0	2.2	2.5	2.5	3.1	3.6	0.6	1.4
Boilers	0.9	0.8	0.8	0.8	0.8	0.7	0.8	0.8	0.8	1.0	1.0	1.0	0.0	0.2
Industry, agri., construction, etc.	0.5	0.5	0.6	0.7	0.7	0.6	0.6	0.8	0.9	1.3	1.6	2.5	0.1	1.7
Residential	0.1	0.1	0.2	0.2	0.2	0.2	0.2	0.2	0.2	0.2	0.3	0.3	0.1	0.0

Table A2.1: Russian gas consumption: regional breakdown *(continued)*

| | \multicolumn{12}{c|}{*Consumption (Bcm)*} | \multicolumn{2}{c}{*Growth (%)*} |
	2000	2001	2002	2003	2004	2005	2006	2007	2008	2009	2010	2011	2003–7	2007–11
Russian Federation: total	**398.5**	**406.4**	**412.5**	**426.4**	**434.9**	**440.7**	**456.3**	**466.7**	**462.0**	**432.6**	**462.1**	**472.8**	**40.3**	**6.1**
Technical and other	47.7	45.7	47.7	51.5	51.1	49.6	52.0	53.6	55.9	44.8	50.7	51.9	2.1	-1.7
Power stations	151.6	156.0	157.0	159.8	164.8	169.3	174.3	180.4	181.7	169.4	183.6	189.5	20.6	9.1
Boilers	70.4	72.1	72.6	73.7	73.6	73.3	74.3	73.0	69.3	68.8	69.8	67.3	-0.7	-5.7
Industry, agri., construction, etc.	88.6	91.7	92.9	97.4	101.4	104.7	110.7	115.4	108.8	102.3	110.2	114.3	18	-1.1
Residential	40.1	41.0	42.3	44.0	43.9	43.9	45.0	44.3	46.4	47.2	47.8	49.7	0.3	5.4

Note: This table abbreviated by the author. The row 'Industry, agriculture, construction, etc.' comprises all fuel use outside the power and heat sector, minus residential consumption, plus non-fuel use (i.e. feedstock for the chemical industry). The row 'technical and other' is pipeline fuel gas, and some small volumes used in the Far Eastern region upstream and in LNG production. Due to an anomaly, most fuel gas is recorded by Rosstat in the Russian statistics but not in any of the regional statistics. All regions except the Far East record zero technical gas use.

Source: Rosstat.

Appendix 3. Data on gas transmission and storage in Russia

The age of pipelines in the Russian high-pressure Gas Transmission Network

Table A3.1: Gazprom's high-pressure Gas Transmission Network (GTS), 31 December 2012

Years since construction	Length (km)	% of total
Up to 10 years	22,200	13.2
11–20 years	20,400	12.1
21–30 years	61,700	36.6
31–40 years	36,800	21.9
41–50 years	18,800	11.2
Over 50 years	8,400	5.0
Total	**168,300**	**100**

Source: Gazprom Annual Report 2012, p.55.

Programme of repairs

Gazprom has carried out significant in-line and electronic inspections (19,000 km and 20,000 km annual average, respectively). The company replaced 432 km of pipe in 2011 and 132 km in 2012 and increased the level of capital repairs (it repaired 2,500 km of trunk pipelines). These measures appear to have improved the reliability of the network, and Gazprom has reported that the level of faults and interruptions is falling. There were 16 faults reported in 2012, compared to 32 in 2002, with on average one fault per 10,000 km in 2012, 2.5 times below the level of 2002.[2] However, it is not clear whether these measures have increased the level of productivity of the network, which in 2002 stood at around 9 per cent below its design capacity.[3]

Storage

Of Gazprom's 25 storage facilities in Russia, 17 are in depleted gas fields and 8 are in aquifers. Average daily deliverability is 535.9 mmcm/day and

---

[2] Gazprom press conference, GTS, 2013.
[3] Stern, J.P. *The Future of Russian Gas and Gazprom* (Oxford: OIES/OUP, 2005), p. 36.

maximum deliverability is 671.1 mmcm/day. Gazprom plans to increase the latter to 727.8 mmcm/day by the end of 2013.[4]

Three storages with total active capacity of 5.6 Bcm are under construction, with a view to commissioning in 2013–14: Kaliningradskoe, Volgogradskoe, and Bednodemianovskoe. Gazprom's 2012 investment programme provides for capital investments of RR17.3 billion for gas storage.

Appendix 4. Proposed legal framework for third-party access to the Russian gas transport system

The Federal Antimonopoly Service (FAS) has made several attempts to introduce new legislation to replace Government Resolution No. 858, which sets out the legal framework for third-party access to the Russian gas transport system. The most recent draft resolution, proposed by the FAS in April 2012,[5] remains under discussion in government.

The Draft Resolution laid down rules on non-discriminatory access to high-pressure pipelines as follows.

It defines 'spare capacity' as technically possible capacity existing at any section of pipeline and at any point in time, minus volumes to be transported:

a) for state and municipal purposes;

b) for citizens' communal and social needs;

c) under transportation contracts concluded before the new Resolution entered into force (importantly, supplies under extensions of these contracts would have lower priority for access compared to new contracts, see below);

d) for supplies under Russia's intergovernmental agreements.

The draft Resolution suggested a 'sunset period' – marked by expiry dates of existing transportation contracts (both Gazprom's and third parties'), apart from those under which gas is transported for purposes stated in (a), (b) and (d) – after the expiry of which all capacity in the UGSS would be allocated to Gazprom and third parties on equal terms.

---

[4] *Gazprom in Figures 2008–12 Factbook*; Gazprom Magazine, August 2013.

[5] Draft Resolution 'Rules On *Non-Discriminatory Access to High-Pressure Gas Pipelines in the Russian Federation*, available on the Russian Ministry of Economic Development website,
www.economy.gov.ru/wps/wcm/connect/ec61c7004ad85dc2a080abaf3367c32c/pp.pdf?MOD=AJPERES&CACHEID=ec61c7004ad85dc2a080abaf3367c32c.

The draft Resolution envisages the following capacity allocation mechanism: long-term capacity to be allocated to each shipper proportionally to volumes requested by all shippers (for the first year and beyond), while also taking into account the plans for system development (for the period after the first year). Short-term capacity is to be allocated proportionally to requests, but only once long-term capacity has been allocated.[6]

Should spare capacity be insufficient for granting all the requests, then capacity is to be allocated – in part and proportionally to requests – in the following order in respect of:

1. supplies for domestic consumption;

2. in respect of shippers extending their existing contracts, concluded prior to the new Resolution coming into force (in respect of volumes not higher than actual supplies under these contracts);

3. supplies outside Russia;

4. other supplies including for transit across the territories of Customs Union member states.

Every request for capacity made in line with the rules outlined by the draft Resolution must be considered and the request must be: granted fully, granted partially, granted on an interruptible basis (provided the shipper's request contains this agreement), or refused. Transportation is to be carried out on the basis of a contract between a transportation organization and a shipper. This contract must contain *inter alia* the following information: terms and conditions of gas injection, transportation, and offtake; volumes for which transportation service is requested; points of entry to and exit from the high-pressure pipeline; payment conditions; and a dispute settlement mechanism.

The draft Resolution, if adopted, will have to be implemented under the existing legal framework, in which the principle of UGSS indivisibility is firmly embedded. Although the draft Resolution aims to reinforce regulated access to networks (where previously there was a mixture of negotiated and regulated access), its reach will remain limited as long as the principle of UGSS indivisibility remains in place. If it does, Gazprom – as the UGSS owner – will continue to remain in charge of provision and oversight of UGSS access, both to itself and to third parties.

---

[6] In line with the draft Resolution, transportation service provided for a period up to 1 year is considered short-term, and for period above 1 year – long-term.

Appendix 5. Gas transportation tariff methodology

The tariff methodology adopted in 2005 for transportation of gas in Russia[7] introduced a two-part tariff, which consisted of:

a) a fee charged for usage of the UGSS which is set depending on the zones[8] where the gas entered and exited the system (measured in currency units per one mcm), and

b) a fee charged for transportation (measured in currency units per mcm/100km).[9]

According to this methodology: a zone of entry into the system consists of:

- all entry (inflow) points to high-pressure pipelines from gas fields;
- gas processing plants, storages, and other sources;
- all cross-border points (in case of transport of gas from storages outside Russia).

A zone of exit from the system consists of:

- all exit (outflow) points from high-pressure pipelines (several zones of exit can be created in those regions where unidirectional transportation distance is above 400km);
- all connections points of Russia-based storage facilities;
- all cross-border points on the sections adjacent to a Russian region.

The Resolution on tariffs for third parties adopted by the FTS in June 2012 lists zones of entry from:[10]

---

[7] Methodology for calculation of tariffs charged for transportation of gas through high-pressure pipeline, approved by FTS Order No 388-e/1, 23 August 2005, see FTS Information Bulletin No. 11 (385), 26 March 2010. Various amendments were made to by FTS Orders No 245-e/2 (7 November 2006), No 286-e/4 (25 October 2007), and No 174-e/6 (17 September 2008). Further details in Appendix E. The FTS was acting under the government resolution no. 1021 (2000), which mandated the FTS (together with the Ministry of Economic Development and Trade, MEDT) to develop methodological principles of tariff setting. It also mandated the FTS (together with the MEDT and Gazprom) to develop a corresponding methodology.

[8] Currently there are 34 entry and 95 exit zones.

[9] FTS Information Bulletin, No 16 (246), 2 May 2007.

[10] FTS Resolution No 143-e/1, 'On approving the tariffs for transportation of gas though Gazprom's high-pressure pipelines, which are part of the UGSS, charged to third parties', 8 June 2012.

- gas producing fields/regions (Yamburg, Purovskaia, Urengoi, Pangody, Tarkosale, Gubkinskaia, Vyngapur, Taezhnaia, Nizhnevartovskii, Vertkos, Vyktul, Orenburg, Astrakhan', Lokosovo, Iuzhno-Balyksk, Zapoliarnoe);

- compressor stations in producing areas (Nadym, Vyngapurovskaia, Iagelnaia, Iuzhno-Balykskaia);

- balancing points (Iuzhno-Balykskaia);

- storage facilities (Kanchurinsko-Musinskiy, Severo-Stavropolskoe, Kushevskoe, Gatchinskoe, Nevskoe, Kaliuzhskoe, Kasymovskoe/ Uviazovskoe, Karashurskoe, Punginskoe, Sovkhoznoe, Saratovskoe, Krasnodarskoe, Shelkovskoe, Samarskie).

The Resolution lists the following zones of exit from:

65 Russian regions (including the cities of Moscow and St Petersburg);

- compressor stations in producing areas (the same as for entry, see above);

- storages (the same as for entry, see above);

- gas metering stations installed on Russia's international borders (Imatra (Russia–Finland), Izborsk (Russia–Estonia), Smolensk (Russia–Belarus), Sudzha (Russia–Ukraine), Valuiki (Russia–Ukraine), Serebrianka (Russia–Ukraine), Pisarevka (Russia–Ukraine), Sokhranovka (Russia–Ukraine), Platovo (Russia–Ukraine), Beregovaya (Russia–Turkey) Novo-Filia (Russia–Azerbajan), Chmi (Russia–Georgia)).

Under this methodology, one-part tariffs are still applied if no compression of gas occurs during transportation or if the share of cost of gas (power) used by the system for technical purposes in the total cost of transportation is not above 1 per cent. A one-part tariff can also be set for regional gas supply systems.

The one-part tariff, and the fee charged for transportation as part of the two-part tariff, are both set based on the volume of gas transported (in other words, for transportation of one mcm) if the gas is transported through the same route for all shippers;[11] otherwise both are set based on distance (that is, for transportation of one mcm for 100 km).

The fee charged for usage of the system as part of the two-part tariff itself consists of two parts: constant (does not depend on distance but

---

[11] Such tariff setting is also allowed for regional supply systems.

reflects the cost of gas distribution[12]) and variable. The variable part is calculated on the basis of a matrix of weighted average transportation distances between all zones of entry and exit, with special coefficients applied to ensure that the cost of gas (including the cost of transportation) is directly dependent on the distance of transportation. (If transportation takes place within the same zone – this is usually the case when gas is transported over short distances – the fee for usage is calculated on the basis of actual distance travelled by the gas). A special model optimizes transportation flows from certain entry zones to certain exit zones by minimizing a total volume of transportation in the UGSS as a whole.[13]

The fee for usage of the system is set for most of the pairs of entry and exit zones (currency units per mcm). The range of usage fee varies depending on the entry–exit pair; for example, as of July 2012, for gas entering at Yamburg and exiting at various Russian regions, the fee is in the range of RR532–RR1912 ($18–64) per mcm; exiting at various Russian storages, the fee is in the range of RR1035–RR1899 ($34–63) per mcm.[14] For those pairs of zones where a usage fee is not defined, a set fee is provided within the range RR51/mcm (if distance is 10 km) and RR1995/mcm (if distance is 4680 km). As of July 2012, the fee for transportation of gas is set at RR12.02/mcm/100 km for transportation in Russia and other Customs Union countries, and at RR13.31/mcm/100 km for transportation outside (this only applies to Gazprom as it has a *de jure* monopoly on gas exports)

Here we calculate the transportation tariff that would be charged, as of January 2011, for transportation of gas across 2400 km, the average transportation distance for gas to a domestic Russian customer, on the basis of FTS order No 497-e/2.[15] Given that the fee for transportation is RR11.23/mcm/100km, and the usage fee is RR1105.65 (assuming that the usage fee for a given entry–exit pair is not defined), the total charge for transporting a thousand cubic metres of gas across 2400 km is RR1375.17 (11.23×24+1105.65) (approximately $46). Another way of defining a tariff could be to add the fee for transportation and a usage fee for some entry–exit zone, the length of pipelines within which is approximately 2400km (for example, Yamburg–Tatarstan). Then the tariff would be RR1407.38 (11.23×24+1137.86) (approximately $47.90).[16]

---

[12] Not charged if gas does not go through gas distribution station at the end of its route of transportation (e.g. transportation to balancing points where gas is sold at the electronic trading platform, transportation to storages); it is also not charged if a gas distribution station does not belong to Gazprom.
[13] FTS Information Bulletin, No 11 (385), 26 March 2010.
[14] FTS Order No 143-e/1.
[15] FTS Order No 497-e/2, 30 December 2010.
[16] RR/$ assumed at 29.39 (average over 2011).

Appendix 6. Statistical information on the power sector

A breakdown of fuel shares for thermal power generation by Federal district – highlighting the predominance of gas in European Russia and of coal in Siberia and the Far East – is shown in Table A6.1. The same information appears in Map 6.1, in the text.

Table A6.1: Territorial structure of fuel consumption for thermal power generation

%	Gas	Coal	Oil prod.	Other
Central	93.41	5.63	0.86	0.11
North-west	84.24	9.38	6.37	0
Southern	82.42	17.02	0.56	0
North Caucasus	99.67	0	0.33	0
Volga	97.87	0.63	1.14	0.36
Urals	76.25	23.54	0.22	0
Siberian	7.68	91.75	0.57	0
Far East	24.07	72.96	2.96	0

Source: APBE Agentsvo po prognozirovaniu balansov v elektroenergetike, *Funktsionirovanie i razvitie elektroenergetiki v Rossiiskoi federatsii v 2011 godu*, [Agency for Forecasting Energy Balances, *Electricity in Russia: functioning and development, 2011*], p. 62.

Tables A6.2 and A6.3 present information, gathered by the Agency for Forecasting Energy Balances, on relative prices of gas and coal. Table A6.2 shows how prices have changed since 2000; Table A6.3 gives more detailed information on prices in 2011.

Table A6.2: Gas and coal prices for power stations, 2000–11, average prices in Federal Districts, rubles per unit of standard fuel

Federal District	2000 Gas	2000 Coal	2003 Gas	2003 Coal	2006 Gas	2006 Coal	2009 Gas	2009 Coal	2010 Gas	2010 Coal	2011 Gas	2011 Coal
Central	326	430	752	970	1403	1677	2049	2285	2565	2288	2961	2829
North-west					1469	1547	1995	1875	2464	1748	2854	2056
Southern					1470	1478	1804	2158	2540	1882	2936	2068
Volga					1308	2250	1871	2238	2297	1828	2601	2389
Urals	321	479	726	741	987	980	1485	1365	1843	1541	2060	1714
Siberia					824	817	966	1055	2201	1232	2537	1388
Far East					1452	1083	2101	1622	2181	2207	2396	2529

Note.   1 unit of standard fuel = 1.1454 mcm of gas or 0.6 tonnes of coal.

Sources: Figures for 2000 and 2003, compiled before federal districts were established, are from APBE, *Funktsionirovanie i razvitie 2005*, p. 57. Those for 2006–11 are calculated by the author from APBE, *Funkshionirovanie i razvitie*, various issues.

Table A6.3: Gas and coal prices for power stations, 2011

Federal district	Shares of power sector fuel supply for region (%) Gas	Shares of power sector fuel supply for region (%) Coal	Population centre	Gas price as proportion of coal price	Regulated gas price (RR/mcm), 2011 Min	Regulated gas price (RR/mcm), 2011 Max
Central	93.4	5.6	Moscow	1.22	3071	3378
			Tula	0.97	3064	3370
			Ryazan	1.01	3006	3307
North-west	84.2	9.4	Vologodskaia	1.12	2763	3039
			Arkhangelsk	1.44	2569	2826
			Komi republic	1.64	2430	2673
Southern	82.4	17	Rostov	1.52	3133	3446
Volga	97.9	0.6	Kirovskaia	1.03	2698	2968
Urals	76.2	23.5	Sverdlovsk	1.4	2588	2847
			Chelyabinsk	1.42	2637	2901
			Kurganskaia	1.6	2455	2701
Siberia	7.7	91.7	Omsk	1.55	2620	2882
			Tomsk	1.44	2644	2908
			Novosibirsk	1.54	2710	2981
			Altai krai	1.72	2862	3148
			Kemerovo	2.09	2872	3159
Far East	24.1	73	Sakha (Yakutia)	1.15	n/a	
			Khabarovsk	1.14	n/a	
			Sakhalin	0.67	n/a	
			Primorskii krai	1.14	n/a	

Source: APBE, *Funktsionirovanie i razvitie 2011*, p. 70; Prilozhenie 2 k prikazu FST no. 412, 10 December 2010, Predel'nyie minimal'nye i predel'nyie maksimal'nye optovye tseny na gaz.

Tables A6.4, A6.5, and A6.6 focus on investment in new power generation capacity. Table A6.4 shows total new capacity commissioned, and capacity decommissioned, in 2005–11. Table A6.5 shows estimates of new power generation capacity commissioning made by the Energy Ministry for 2013–19 and by Alfa Bank for 2012–20. Table A6.6, based on information reported by companies and compiled by Alfa Bank, lists plans to commission new capacity in 2012–18.

Table A6.4: Total new electricity generating capacity commissioned, 2005–11

	2005	2006	2007	2008	2009	2010	2011
Newly commissioned, MW	2866.5	1479.4	2087.9	1321.8	1364.3	3232.5	4598.3
Decommissioned, MW	−260.2	−392.2	−413.7	−1074.3	−286.4	−1027.5	−1507.2
Net increase	**2606.3**	**1087.2**	**1674.2**	**290.0**	**1077.9**	**2205.0**	**3091.1**
Plant replaced and other, MW*	217.4	410.6	1750.7	412.4	69.0	1149.2	186.1
Total capacity at year end, GW	**219.2**	**212.01**	**215.4**	**216.1**	**216.9**	**214.9**	**218.1**

* The row 'Plant replaced and other' includes new plant installed at already operating power stations and other similar changes.

Note: The volume of total capacity was recalculated by the APBE in 2006 to exclude capacity sited in closed zones, and again in 2011 to exclude networks separate from the main grid. The volumes given cannot therefore be directly compared with each other.

Source: APBE, *Funktsionirovanie i razvitie elektroenergetiki v RF*, various issues.

Table A6.5: Timing of new power generation capacity installation

MW	2012	2013	2014	2015	2016	2017	2018	2019	2020	Total
Energy Ministry estimates, 2013										
New nuclear capacity		0	3179	2369	1170	2250	1150	1150		2994 net
Nuclear decommissioning		*−1000*	*−2000*	*−1000*	*0*	*−1417*	*−1417*	*−1440*		
New thermal capacity		4166	5582	5972	1671	200	437	0		7316 net
Thermal decommissioning		*−1264*	*−2478*	*−2537*	*−2164*	*−470*	*−1237*	*−561*		
New hydro and renewable capacity		2425	355	169	182	345	0	0		3476
Alfa Bank estimates, 2012										
New nuclear capacity	70	1180	3126	1180	2326	1070				5238 net
Nuclear decommissioning					*−417*	*−417*	*−1440*	*−440*	*−1000*	
New commissions under CSAs										
European Russia/Urals	2600	2370	4150	4540	980	200				14,840
Siberia	500	0	500	1610						2610

Sources: Ministry of Energy of the Russian Federation, Skhema i programma razvitiia Edinoi energeticheskoi sistemy Rossii na 2013–2019 gody, pages 31 and 39; *Russian Power Generators: No Catalysts, Negatives Mostly Priced In*, Alfa Bank, pp.18–19.

Table A6.6: Commissioning plans by generating companies, 2012–18

Company/ownership	Plant	Type of capacity/fuel	Federal district	Installed capacity (MW)	Year of commissioning
RusHydro	Gotsatlinskaia HPP	Hydro	South	100	2013
	Zelenchukskaia HPP	Hydro	South	140	2013
	Zagorskaia pumped storages	Hydro	Centre	840	2014
	BEMO	Hydro	Siberia	2997	2015
	Zaramagskie HPPs	Hydro	South	352	2016
	Nizhne-Bureyskaia HPP	Hydro	Far East	320	2016
	Ust'-Srednekanskaia HPP	Hydro	Far East	570	2018
RAO ES of the East (owned by RusHydro)		Gas + coal	Far East	n/a	2016
InterRAO	Ivanovskie CCGTs	CCGT Gas	Central	325	2012
	Urengyskaia GRES	CCGT gas	Urals	450	2012
	Khanarorskaia GRES	Coal-fired turbine	Siberia	225	2012
	Nizhnevartovskaia GRES	CCGT gas	Urals	410	2013
	Gusinoozerskaia GRES	Coal-fired turbine	Siberia	210	2013
	Dzhubginskaia TES	Gas	South	180	2013
	Yuzhnouralskaia GRES	CCGT gas	Urals	400	2014
	Yuzhnouralskaia GRES	CCGT gas	Urals	400	2014
	Cherepetskaia GRES	Coal-fired turbine	Central	214	2014
	Verkhnetagilskaia GRES	CCGT gas	Urals	420	2015
	Permskaia GRES	CCGT Gas	Urals	800	2016
OGK-2 (owned by Gazprom)	Troitskaia GRES	Coal-fired turbine	Urals	660	2014
	Serovskaia GRES	CCGT gas	Urals	420	2014
	Novocherkasskaia GRES	Coal-fired turbine	South	330	2014
	Cherepovetskaia GRES	CCGT gas	North W	420	2014
	Stavropolskaia GRES	CCGT gas	South	420	2016
E.ON Russia	Berezovskaia GRES	Coal-fired turbine	Siberia	800	2014
TGK-1 (owned by Gazprom/Fortum)	Pravoberezhnaia CHP-5	CCGT gas	North W	450	2013
	Cascade 1, Vulksinskie GES	Hydro	North W	240	2014
	Tsentralnaia CHP	CCGT gas	North W	100	2016
Mosenergo (owned by Gazprom)	CHP 9	CCGT gas	Central	61.5	2013
	CHP 20	CCGT gas	Central	420	2014
	CHP 12	CCGT gas	Central	420	2015
	CHP 16	CCGT gas	Central	420	2015

Note. This does not include new nuclear capacity.

Source: *Russian Power Generators: No Catalysts, Negatives Mostly Priced In*, Alfa Bank, 15 October 2012, p.31.

Appendix 7. Reserves, production and costs at Gazprom's major production associations

Table A7.1: Remaining proven and probable reserves at Gazprom's major production associations (31 December 2012, Bcm)

Production association	Main gas fields	Proven reserves	Probable reserves
Western Siberia (Urals Federal District)			
Gazprom Dobycha Urengoi	Urengoiskoe, Severo-Urengoiskoe, En-Iakhinskskoe, Pestsovoe	2612	399
Gazprom Dobycha Yamburg	Yamburgskoe, Zapoliarnoe, Tazovskoe, Severo-Parusovoe	3528	523
Gazprom Dobycha Nadym	Medvezh'e, Iamsoveiskoe, Iubilyeynoe, Kharaseveiskoye, Bovanenkovskoe	7074	1256
Gazprom Dobycha Noiabr'sk	Komsomol'skoe, Eti-Purovskoe, Zapadno-Tarkosalinskoe	533	51
Severneftegazprom	Iuzhno-Russkoe	669	7
OAO Gazprom	Zapadno/Severo Tambeiskoe, Kruzenshternskoye, Maliginskoe, Tasiiskoe, Antipaiutinskoe, Tota-Iakhinskoe, Semakovskoe	3615	840
Purgaz*	Gubkinskoe	188	13
Southern Russia (Southern Federal District)			
Gazprom Dobycha Astrakhan	Astrakhanskoe, Zapadno-Astrakhanskoe	268	18
South Urals (Privolzhsky Federal District)			
Gazprom Dobycha Orenburg	Orenburgskoe	254	103
Other	Tomsk and Irkutsk	325	316
Total		**19,065**	**3525**

* Gazprom share 51 per cent.

Note: Only 4 per cent of proven and 12 per cent of probable reserves are located outside western Siberia.

Sources: Production associations and fields from *Gazprom in Figures 2008–12*, pp. 15–16; reserves data from DeGolyer and McNaughton, Loan Notes 2013, Appendix B, B3.

Table A7.2: Production and costs of production at Gazprom's major production associations

Production association	Production (Bcm)			Cost of production ($/mcm)		
	2011	2010	2009	2011	2010	2009
Gazprom Dobycha Nadym	54.2	56.2	51.8	530	552	746
Gazprom Dobycha Noiabr'sk	58.1	61.1	53.9	460	328	347
Gazprom Dobycha Urengoi	111.6	107	106.4	791	665	646
Gazprom Dobycha Yamburg	203.3	203.3	177.8	570	446	471
Total	**427.2**	**427.6**	**389.9**			

Source: 'Gazprom's average production costs on Yamal top $20 per thousand cubic metres in 2011', *Interfax*, 26 September–3 October 2012, p. 50.

BIBLIOGRAPHY

Books, articles, reports, and presentations

Agentsvo po prognozirovaniu balansov v elektroenergetike, *Funktsionirovanie i razvitie elektroenergetiki v Rossiiskoi federatsii v 2011 godu* [Agency for Forecasting Energy Balances, *Electricity in Russia: functioning and development, 2011*].

Ahrend, R. and Tompson, B., 'Realising the Oil Supply Potential of the CIS: The Impact of Institutions and Policies', OECD Working Paper, No ECO/WKP(2006)12, 2006.

Alekperov, V., 'Third Decade of Evolution: New Challenges, New Opportunities', Presentation: 14 March 2012b, London, Moscow: Lukoil.

Alekperov, V., 'Lukoil: Building Value Through Innovation', Presentation: 10 March 2011, New York, Moscow: Lukoil.

Alfa Bank, *Russian Power Generators: No Catalysts, Negatives Mostly Priced In*, 15 October, 2012.

Allsopp, C. and Stern, J., 'The Future of Gas: what are the analytical issues related to pricing', in Stern, J.P. (ed.), *The Pricing of Internationally Traded Gas* (Oxford: OIES/OUP, 2012), pp. 10–39.

Aslund, A., 'Gazprom's demise could topple Putin', Peterson Institute for International Economics, 2013. www.piie.com/publications/opeds/oped.cfm?ResearchID=2419.

Aslund, A., 'Why Gazprom resembles a crime syndicate', Peterson Institute for International Economics, 2012. www.piie.com/publications/opeds/oped.cfm?ResearchID=2056.

Baev, P. and Overland, I., 'The South Stream versus Nabucco Pipeline Race', *International Affairs*, 86:5, 2010, pp. 1075–90.

Baker Institute Policy Report, *Shale Gas and US National Security*, No. 49, James A. Baker III Institute for Public Policy of Rice University, October 2011.

Belyi A., 'Trends of Russia's Gas Sector Regulation', Fourth Annual Conference on Competition and Regulation in Network Industries. Brussels, 25 November 2011.

Boute, A., 'The Russian Electricity Market Reform: Towards the Re-regulation of the Liberalized Market', in Sioshansi, F.P. (ed.), *Evolution of Global Electricity Markets: New Paradigms, new challenges, new approaches*, London: Elsevier Academic Press, pp. 461–96, July 2013.

Bowden, J., 'Azerbaijan: from Gas Importer to Exporter', in Pirani, S. (ed.), *Russian and CIS Gas Markets and their Impact on Europe* (Oxford: OIES/OUP, 2009).

BP Energy Outlook 2030, London: BP, 2013.

BP Statistical Review of World Energy 2010, London: BP , 2010.

BP Statistical Review of World Energy 2013, London: BP , 2013.

Bruce, C. and Yafimava, K. 'Moldova's Gas Sector', in Pirani, S. (ed.), *Russian and CIS Gas Markets and Their Impact on Europe* (Oxford: OIES/OUP, 2009).

Burgansky, A., *Stand and Deliver: Oil and Gas Yearbook 2010* (Moscow: Renaissance Capital, 2010).

Cedigaz, *Natural Gas in the World: 2012 Edition*, Rueil Malmaison, 2012.

CEER Blueprint on Incremental Capacity, Council of European Energy Regulators, Brussels: CEER, 23 May 2013, www.energy-regulators.eu/portal/page/ portal/EER_HOME/EER_PUBLICATIONS/CEER_PAPERS/Gas/ Tab3/C13-GIF-06-03%20_CEER_blueprint_on_incremental_capacity_ final_0.pdf.

Center for Strategic Studies and Reforms, Chisinau-Tiraspol, *Moldova trade diagnostic study*, 2003.

Chen, M.X., 'Gas Pricing in China', in Stern, J. (ed.), *The Pricing of Internationally Traded Gas* (Oxford: OIES/OUP, 2012), pp. 310–37

Chibis, A., 'Modernizatsiia kommunal'noi infrastruktury v usloviiakh tarifnykh ogranichenii', presentation by A. Chibis of the municipal sector working group of the government, July 2013.

Chyong, C.K., Noel, P., and Reiner, D.M., 'The Economics of the Nord Stream Pipeline System', University of Cambridge, EPRG Working Paper 1026, 2010, www.eprg.group.cam.ac.uk/wp-content/uploads/2010/09/Binder14. pdf.

Colton, T.J. and Holmes, S., *The State After Communism: Governance in the New Russia* (London: Rowan & Littlefield, 2006).

Covi, G., 'A case study of an advanced Dutch disease: The Russian oil', IMEF Ca Foscari University of Venice, May, 2013. http://mpra.ub.uni-muenchen. de/46670/1/MPRA_paper_46670.pdf.

Darbouche, H., El-Katiri, L., and Fattouh, B. 'East Mediterranean Gas: what kind of a game-changer?', Working Paper NG71, OIES, 19 December 2012, www.oxfordenergy.org/2012/12/east-mediterranean-gas-what-kind-of-a-game-changer/.

DeGolyer and McNaughton, Loan Notes 2013

Deloitte, *Tax and Legal Guide to the Russian Oil & Gas Sector*, 2012.

Deutsche Bank, *Russian Gencos: spot price dynamics key to 2013 performance*, 21 February 2013.

Deutsche Bank, *Electric Utilities: incorporating tariff freeze into valuations*, October 2013.

Dienes, L., 'Natural gas in the context of Russia's energy system', *Demokratizatsiia*, Fall 2007, pp. 408–29.

Dülger, F., Lopcu, K., Burgaç, A., and Balli, E., 'Is Natural Resource-Rich Russia Suffering from the Dutch Disease?', International Conference on Eurasian Economies 2012.

Energeticheskii Institut im. G. Krzhizhanovskogo, *Razrabotka programmy modernizatsii elektroenergetiki Rossii na period do 2020 goda* [G. Krzhizhanovskii Energy Institute, *Development of the Programme of Electricity Modernization in Russia up to 2020*] (Moscow 2011).

European Commission, 'The EU Energy Policy: Engaging with Partners beyond Our Borders', Communication from the Commission to the European Parliament, the Council, the European Economic and Social Committee and the Committee of the Regions, on security of energy supply and international cooperation –COM(2011) 539 final, Brussels, 7 September, 2011.

European Commission, *External Energy Relations – from principles to action*, Communication from the Commission to the European Council, COM (2006) 590 final, Brussels 12 October, 2006.

Evrokhim press release, 'EuroChem announces fertiliser project in Louisiana', July 2013.

Evrokhim, *Annual Report and Accounts 2012*.

Fattouh, B. and Stern, J.P. (eds.), *Natural Gas Markets in the Middle East and North Africa* (Oxford: OIES, 2011).

Forbes Magazine, 'Ekskliuzivnoe interviu Rema Viakhireva: Putin kogda uslishal, chto ua ukhozhu, tak obradoval'sia', *Russian Forbes Magazine*, 11 September, 2012.

Foreign Trade of Azerbaijan 2011 (Baku: AzStat, 2011).

Foss, M.M., 'Natural Gas Pricing in North America', in Stern, J.P. (ed.), *The Pricing of Internationally Traded Gas* (Oxford: OIES/OUP, 2012), pp. 85–144.

Gao, F., 'Will there be a shale gas revolution in China by 2020?', Working Paper NG 61, OIES, April, 2012.

Gas Exporting Countries Forum, Moscow Declaration of the GECF, The Second Gas Summit of the Heads of State and Government of GECF Countries, 1 July, Moscow, 2013. www.gecf.org/moscow-declaration--second-gas-summit.

Gazprom Annual Reports, IFRS Consolidated Financial Statements, Management Reports, Management's Discussion and Analysis of Financial Condition and Results of Operations, for the years 1998–2012. These can be accessed at: www.gazprom.com/investors/reports/2012/.

Gazprom Bovanenkovskoe Field Brochure, www.gazprom.com/about/production/projects/deposits/bm/.

Gazprom Databook 2013.

Gazprom in Figures 2003–07, www.gazprom.com/investors/reports/2007/.

Gazprom in Figures 2007–11.

Gazprom Investor Day Presentation, 'Gas for the Future', 12 February 2013, www.gazprom.com/investors/events-presentations/investor-day-feb-2013/.

Gazprom Loan Notes 2012: Preliminary Financial Terms, Open Joint Stock Company Gazprom, 2012.

Gazprom Loan Notes 2013: Preliminary Base Prospectus, Open Joint Stock Company Gazprom, 12 July 2013.

Gazprom *Management Report*, Gazprom (2005–12).

Gazprom Neft Databook (Moscow: Gazprom Neft, 2013).

Gazprom Neft. *Full Year Results 2010* (Moscow: Gazprom Neft, 2011).

Gazprom Press Conference 'Perekhod na rynochnie printsipy sotrudnichestva s respublikami bivshego SSSR', 20 June 2006. Gazprom website.

Gazprom Press Conference 2009 Production: *Press-konferentsiia na temu: 'Razvitie mineral'no-syrevoi bazy. Dobycha gaza. Razvitie GTS'*, 16 July 2009. www.gazprom.ru/f/posts/02/094829/shifr_rus_09.06.16.pdf.

Gazprom Press Conference presentation 'Mineral and Raw Material Base Development', May 2013, www.gazprom.com/f/posts/59/719889/presentation-press-conf-2013-05-21-en.pdf.

Gazprom Press Conference Transcript 2013, *Razvitie mineral'no-syrevoi bazy. Dobycha gaza. Razvitie GTC*, 21 May 2013 (in Russian only) www.gazprom.ru/f/posts/83/172307/development-resources-stenogram-2013-05-21-rus.pdf.

Gazprom Press Conference: Gazprom's Financial and Economic Policy, 27 June 2013, www.gazprom.com/press/conference/2013/economic-policy/.

Gazprom, Stenogramme of press conference by K. Seleznev, 'Postavki gaza na vnutrennyi rynok' ['Supply of gas to the domestic market'], 28 May 2013.

Gazprom, Yamal Megaproject Brochure, www.gazprom.com/f/posts/25/697739/book_my_eng_1.pdf.

Gazprom Bond Prospectus, Moscow: Gazprom, 2012.

Gény, F., 'Can Shale Gas be a Game-Changer for European Gas Markets?', Working Paper NG46, OIES, 1 December, 2010, www.oxfordenergy.org/2010/12/can-unconventional-gas-be-a-game-changer-in-european-gas-markets/.

Giamourdis, A. and Paleoyannis, S., 'Security of Gas Supply in South Eastern Europe: potential contribution of planned pipelines', LNG and storage', Working Paper NG52, OIES, July 2011. www.oxfordenergy.org/2011/07/security-of-gas-supply-in-south-eastern-europe-potential-contribution-of-planned-pipelines-lng-and-storage/.

GIIGNL, 'The LNG Industry in 2012', Brussels: GIIGNL, 2012, www.giignl.org/sites/default/files/publication/giignl_the_lng_industry_2012.pdf.

Goldman, M.I., *Oilopoly: Putin, Power and the Rise of the New Russia* (London: OneWorld Books, 2008).

Gordon, D. and Sautin, Y., 'Opportunities and challenges confronting Russian oil', Carnegie Endowment for International Peace, 2013. http://carnegieendowment.org/2013/05/28/opportunities-and-challenges-confronting-russian-oil/g6x5.

Gore, O., Viljainen, S., Makkonen, M., and Kuleshov. D. 'Russian electricity market reform: deregulation or reregulation?', *Energy Policy*, 41 (2012), 676–85.

Gorska, A., 'Changes on the Ukrainian Gas Market', Centre for Eastern Studies, 21 July 2010.

Grant, C., 'Stifling progress in China and Russia', Centre for European Reform, 2012, accessed at www.cer.org.uk/in-the-press/stifling-progress-russia-and-china.

Grigas, A., 'The Gas Relationship between the Baltic States and Russia: politics and commercial realities', Working Paper NG 67, OIES, October, 2012. www.oxfordenergy.org/2012/10/the-gas-relationship-between-the-baltic-states-and-russia-politics-and-commercial-realities/.

Grigas, A., *The Politics of Energy and Memory between the Baltic States and Russia* (Burlington: Ashgate, 2013).

Gurvich E., 'Dolgosrochnie perspektivy Rossiiskoi ekonomiki', *Ekonomicheskaia politika*, 3, 2013.

Gurvich E. and Prilipskiy I., 'Kak obespechit' vneshniuiu ustoichovost' rossiiskoi ekonomiki', *Voprosy Ekonomiki*, 9, 2013. http://www.eeg.ru/files/lib/230913.pdf?PHPSESSID=810ba946012738af91b2195255128a2d.

Gustafson, T., *Wheel of Fortune: the battle for oil and power in Russia* (Harvard University Press, 2012).

Haase, N., *European Gas Market Liberalisation* (Groningen: Energy Delta Institute, 2009).

Heather, P., 'Continental European Gas Hubs: are they fit for purpose?' Working Paper NG63, OIES, 14 June, 2012. www.oxfordenergy.org/2012/06/continental-european-gas-hubs-are-they-fit-for-purpose/.

Heather, P. and Henderson, J., 'Lessons from the February 2012 European gas "crisis"', Energy Comment, OIES, April, 2012. www.oxfordenergy.org/2012/04/lessons-from-the-february-2012-european-gas-%e2%80%9ccrisis%e2%80%9d/.

Hedlund, S., 'Russia as a neighbourhood energy bully', *Russian Analytical Digest*, No. 100, July 2011, pp. 2–5.

Helm, D., *The Russian dimension and Europe's external energy policy*, September, 2007. www.dieterhelm.co.uk/node/655.

Henderson, J., 'Domestic Gas Prices in Russia – Towards Export Netback?', Working Paper NG57, OIES, November 2011.

Henderson, J., 'Evolution in the Russian Gas Market: The Competition for Customers', Working Paper NG 73, OIES, January, 2013. www.oxfordenergy.org/wpcms/wp-content/uploads/2013/01/NG_73.pdf.

Henderson, J., 'The Asian Key to Russian Gas' (Trusted Sources, London, 18 September, 2013).

Henderson, J., 'The Strategic Implications of Russia's Eastern Oil Resources', Working Paper WPM 41, OIES, April, 2011.

Henderson, J., 'Tight Oil Developments in Russia', Working Paper WPM 52, OIES, 8 October 2013.

Henderson, J., *Non-Gazprom Gas Producers in Russia* (Oxford: OIES, 2010).

Henderson, J., Pirani S., and Yafimava, K., 'CIS Gas Pricing: Towards European Netback?' in Stern, J.P. (ed.), *The Pricing of Internationally Traded Gas* (Oxford: OIES/OUP, 2012), pp. 178–223.

Henderson, J.A., 'The Potential Impact of North American LNG Exports', Working Paper NG 68, OIES, October, 2012.

Henderson, J.A., 'The Pricing Debate over Russian Gas Exports to China', Working Paper NG 56, OIES, October, 2011.

Higashi, N., *Natural Gas in China: Market Evolution and Strategy* (Paris: IEA, 2009).

Hogselius, P., *Red Gas: Russia and the Origins of European Energy Dependence* (Basingstoke: Palgrave Macmillan, 2013).

Honoré A. 'Economic Recession and Natural Gas Demand in Europe: what happened in 2008–2010?', Working Paper NG47, OIES, 25 January 2011 www.oxfordenergy.org/2011/01/economic-recession-and-natural-gas-demand-in-europe-what-happened-in-2008-2010/.

Honoré A., *The Outlook for Natural Gas Demand in Europe* (Oxford: OIES/OUP, 2014, forthcoming).

Honoré, A., *European Natural Gas Demand, Supply and Pricing: Cycles, Seasons and the Impact of LNG Price Arbitrage* (Oxford: OIES/OUP, 2010).

IEA *Gas: Medium Term Market Report 2013: market trends and projections to 2018* (Paris: IEA/OECD, 2013).

IEA Monthly Gas Survey, January 2013 (Paris: IEA/OECD, 2013).

IEA *Natural Gas Information* (2012 Edition) (Paris: IEA/OECD, 2012).

IEA *Optimising Russian Natural Gas: Reform and Climate Policy* (Paris: IEA/OECD, 2006).

IEA *Russia Energy Survey 2002* (Paris: IEA/OECD, 2002).

IEA *Russia Energy Survey 2005* (Paris: IEA/OECD, 2005).

IEA *Russian Electricity Reform 2013 Update: Laying an Efficient and Competitive Foundation for Innovation and Modernisation* (Paris: IEA/OECD, 2013).

IEA *World Energy Outlook 2011* (Paris: IEA/OECD, 2012).

IEA *World Energy Outlook 2011: Are We Entering a Golden Age of Gas?* (Paris: IEA/OECD, 2011).

IEA *World Energy Outlook 2012* (Paris: IEA/OECD, 2012).

IEA *World Energy Outlook 2013* (Paris: IEA/OECD, 2013).

IMF, *Article IV Report on Ukraine, 2012*.

Institute of Energy and Finance, *Sovremennie tsenarii razvitia mirovoi energetiki: resul'taty issledovanii 2009–12 gg*, Institut Energetiki i Finansov, [*Current scenarios for the development of world energy: research results, 2009-12*, Moscow: Institute of Energy and Finance], 2012.

International Gas Union, *Wholesale Gas Price Survey 2013 Edition – a global review of price formation mechanisms 2005–2012*, June 2013. www.igu.org/gas-knowhow/publications/igu-publications/Wholesale%20Gas%20Price%20Survey%20-%202013%20Edition.pdf.

ITERA *Russian Independent Gas Company*, Moscow: ITERA, 2009.

JRC/EU *Unconventional Gas: Potential Energy Market Impacts in the European Union*, JRC Scientific and Policy Reports, European Commission Joint Research Centre, Luxembourg: European Union, 2012.

Kinnander, E., 'The Iran–Turkey gas relationship: politically successful, commercially problematic?', Working Paper NG 38, OIES, January, 2010. www.oxfordenergy.org/2010/01/the-turkish-iranian-gas-relationship-politically-successful-commercially-problematic/.

Konoplyanik, A.A., Pravovye aspekty protsedury nediskriminatsionnovo konkurentnogo dostupa k svobodnym moshchnostiam transportirovki (DEKH, TAG i ECG)', *Neftegaz, Energetika i Zakonodatel'stvo*, 8, 2009, pp. 142–56.

Kushkina, K., *Golden Age of Gas In China* (Moscow: The US Russia Foundation, 2012).

Landes, A., 'Gazprom – Ready for Lift Off', Renaissance Capital, Moscow, 2001.

Landes, A., *Gazprom: The Dawning of a New Valuation Era*, Renaissance Capital, Moscow, 2002.

Landes, A., *Impasse: Russia Oil and Gas Yearbook 2005* (Renaissance Capital, Moscow, 2005).

Locatelli, C., 'EU Gas Liberalization as a Driver of Gazprom's Strategy', Paris: IFRI/Russia-NIS Centre, February, 2008.

Lough, J., 'Russia's Energy Diplomacy', Chatham House Briefing Paper, 2011. http://www.chathamhouse.org/publications/papers/view/171229.

Lukoil, 'Global Trends In Oil And Gas Markets To 2025', Lukoil, Moscow. 2013.

Lukoil, *2012–2021 Strategic Development Programme*, Moscow: Lukoil, 2012.

Lukoil, *Annual Report 2012*, Moscow: Lukoil, 2013.

Lukoil, *Databook 2012*, Moscow: Lukoil, 2012.

Lukoil, *Fact Book 2012*, Moscow: Lukoil, 2012.

Luong, P.J. and Weinthal, E., *Oil is Not a Curse: ownership structure and institutions in Soviet successor states* (New York: Cambridge University Press, 2010).

Makarov A., Mitrova T., and Malakhov V., (2013). 'Prognoz mirovoi energetiki i sledstviia dlia Rossii', *Voprosy prognozirovaniia*, 2013. No.5.

Makarov, A., Grigoriev, L., and Mitrova, T. (eds.), 'Global and Russian Energy Outlook up to 2040', ERI RAS – ACRF, 2013.

Makarov, A.A., Mitrova, T.A. et al., *Vliianie rosta tsen na gaz i elektroenergiiu na razvitie ekonomiki Rossii* [*The influence of gas and electricity price rises on the development of the Russian economy*], Institute of Energy Research of the Russian Academy of Sciences, Moscow, 2013.

Malakhov V.A., 'Otsenka zavisimosti VVP i sprosa na energonositeli ot udorozhaniia topliva i energii na vnutrennem i vneshnem rinke', ['Assessment of dependence of the GDP and demand for energy resources on the growth of fuel and energy prices in internal and international markets'] *TEK Rossii*, No.1, 2012, pp. 16–24.

Milov, V., 'The Risk of Russian Energy Supply Disruptions', *Energy Policy*, 2006. http://www.energypolicy.ru/files/Paris-May16-2006-2.ppt.

Mitrova, T., *Vnutrennyi spros na gaz: perekhodnyi vozrast* [*Domestic demand for gas: a time of transition*], Skolkovo Business School presentation, February 2013.

Mitrova, T., Pirani S., and Stern, J., 'Russia, the CIS and Europe: gas trade and transit', in Pirani, S. (ed.), *Russian and CIS Gas Markets and Their Impact on Europe* (Oxford: OIES/OUP, 2009).

Nemtsov B. and Milov, V., *The Nemtsov White Paper, Part II, Putin and Gazprom*, 2008. http://larussophobe.wordpress.com/2008/09/28/the-nemtsov-white-paper-part-ii-gazprom-the-full-text/.

Noel, P., 'Beyond Dependence: how to deal with Russian gas', European Council on Foreign Relations, Policy Brief, November 2008.

Noel, P., Findlater, S., and Chyong, C.K., 'The Cost of Improving Supply Security in the Baltic States', Electricity Policy Research Group, University of Cambridge, EPRG Working Paper 1203, 2012, www.eprg.group.cam.ac.uk/wp-content/uploads/2012/01/EPRG-WP-1203_-Complete.pdf.

Novak A., 'Prioritety gosudarstvennoi politiki v rossiiskoi neftegazovoi otrasli', Natsional'nyi Neftegazovyi Kongress, 19 March 2013.

Novatek, *Eurobond Prospectus*, Moscow: Novatek, December 2012.

Novatek, *Focus on Growth*, Moscow: Novatek, 2011.

Novatek, *Harnessing the Energy of the Far North: Annual Report 2011*, Moscow: Novatek, 2012.

Novatek, *Investor Presentation: Harnessing the Energy of the Far North*, Moscow: Novatek, 2013.

Novatek, *IPO Prospectus*, Moscow: Novatek, 2005.

Novatek, *Sberbank Conference Presentation: Harnessing the Energy of the Far North*, Moscow: Novatek, 2013.

Novikov, S., 'Tarifnoe regulirovanie v 2013 g', [Tariff regulation in 2013], presentation by S. Novikov of the RTS, July 2013.

O'Sullivan, S., *China Gas: The Price Is Becoming Right* (London: Trusted Sources, 2013).

O'Sullivan, S., and Henderson, J., 'UPDATE China Gas – End of the shale dream, but not of the growth plan', Trusted Sources, 8 October 2013, at http://www.trustedsources.co.uk/em-energy/china/ accessed on 11 November 2013.

OECD, *Economic Survey: Russian Federation* (Paris: OECD, 2004).

Paik, K.-W., 'Northeast Asian Countries' Oil and Gas Relations with Russia', presented at Clingendael Energy Conference 'The Geopolitics of Energy in Eurasia: Russia as an Energy Lynch Pin', 22–23 January Clingendael International Energy Programme/Clingendael Asia Studies, (The Hague: Clingendael Institute, 2008).

Paik, K.-W., *Sino-Russian Oil and Gas Co-operation* (Oxford: OIES/OUP, 2012).

Paik, K.-W., 'Pipeline gas introduction to the Korean peninsula', Chatham House Report, January 2005.

Pirani S., *Change in Putin's Russia: power, money and people* (London: Pluto Press, 2009).

Pirani, S. (ed.), *Russian and CIS Gas Markets and their Impact on Europe* (Oxford: OIES/OUP, 2009).

Pirani, S., 'Central Asian and Caspian Gas Production and Constraints on Export', Working Paper NG 69, OIES, December 2012.

Pirani, S., 'Consumers as players in the Russian gas sector', Energy Comment, OIES, January, 2013.

Pirani, S., 'Elusive Potential: Gas Consumption in the CIS and the Quest for Efficiency', Working Paper NG 53, OIES, July, 2011. www.oxfordenergy.org/wpcms/wp-content/uploads/2011/08/NG-531.pdf.

Pirani, S., 'Ukraine's Gas Sector', OIES Working Paper NG 21, June 2007.

Pirani, S., Stern, J.P., and Yafimava, K., 'The April 2010 Russo-Ukrainian gas agreement and its implications for Europe', Working Paper NG 42, OIES, June, 2010. www.oxfordenergy.org/wpcms/wp-content/uploads/2011/05/NG_42.pdf.

Pirani, S., Stern, J.P., and Yafimava, K., 'The Russo-Ukrainian gas dispute of January 2009: a comprehensive assessment', Working Paper NG27, OIES, February 2009. www.oxfordenergy.org/2009/02/the-russo-ukrainian-gas-dispute-of-january-2009-a-comprehensive-assessment/.

Pirani S., Henderson, J., Honoré, A., Rogers, H., and Yafimava, K., 'What the Ukraine Crisis Means for Gas Markets', Oxford Energy Comment, OIES, March 2014, www.oxfordenergy.org/2014/03/what-the-ukrainian-crisis-means-for-gas-markets/

Pittman. R., 'Restructuring the Russian electricity sector: recreating California?', *Energy Policy*, 35 (2007), pp. 1872–83.

Pomelnikova, Maria et al., *Russia Country Report*, no. 2 (Raiffeisen Research, October, 2013).

Porokhova, N., 'Gas and Power Markets in 2012 and New Challenges', Gazprombank presentation, March 2013.

Pozsgai, P., 'The Evolution of EU security of gas supply policy', *EDI Quarterly*, Vol. 4, No.3, 2012, pp. 35–7.

Puzakov, V. and Polivanov, V., 'Russia', in Euroheat & Power, *District Heating and Cooling 2013 Survey*, Brussels, 2013.

Riley A., 'Commission v. Gazprom: the antitrust clash of the decade?' CEPS Policy Brief No. 285, 2012. www.ceps.be/book/commission-v-gazprom-antitrust-clash-decade.

Riley, A., 'The Coming Russian Gas Deficit: consequences and solutions', Centre for European Policy Studies, Policy Brief 116, Brussels, October 2006.

Rogers, H., 'The Impact of a Globalising Market on Future European Gas Supply and Pricing: The Importance of Asian Demand and North American Supply', Working Paper NG 59, OIES, January 2012.

Rosneft, *2008 Financial Results Presentation*, Moscow: Rosneft, 2008.

Rosneft, *Analyst Presentation*, Moscow: Rosneft, 2006.

Rosneft, *Analysts' Databook 2011*, Moscow: Rosneft, 2012.

Rosneft, *Analysts' Databook 2013*, Moscow: Rosneft, 2013.

Rosneft, *Non-Deal Roadshow*, Moscow: Rosneft, 2012.

Rosneft, Presentation to Investors by Vlada Rusakova: Gas Business Development, Moscow: Rosneft, 2013.

Rosneft, *Annual Report 2012*, Moscow: Rosneft, 2013.

Sartori, N., 'The European Commission vs. Gazprom: an issue of fair competition or a foreign policy quarrel?', IAI Working Papers 13 March 2013, www.iai.it/pdf/DocIAI/iaiwp1303.pdf.

Sberbank, *Russian Oil and Gas – the unconventional issue*, October 2012.

Sberbank, *Russian Gencos: what goes around comes around*, August 2013.

Seliverstov, S.S. and Gudkov, I.V., 'The Development of Electricity and Gas Networks in Russia' in Roggenkamp, M.M., Barrera-Hernandez, L., Zillman, D.N., and del Guayo, I. (eds.) *Energy Networks and the Law* (Oxford: OUP, 2012), p. 400.

Shapot D.V. and Malakhov V.A., Vliianie vneshnikh tsen na otsenku perspektiv razvitia ekonomiki Rossii', *Voprosi Ekonomiki* ['The influence of external prices on the assessment of the Russian economy's development prospects', *Questions of Economics*] 4. 2012.

Simmons, D. and Murray, I., 'Russian Gas: will there be enough investment?' *Russian Analytical Digest*, 27 July 2007, pp. 2–5.

Smirnova, O., 'Anti-monopoly regulation of gas markets', presentation made at the round table 'Formation of domestic gas market: taxes, tariffs, prices, and investments', Forum 'Russia's Gas – 2012', 20 November 2012.

Société Générale, 'European Gas and Power Drivers', SocGen Cross Asset Research, 29 October 2013.

Stern, J.P. (ed.), *The Pricing of Internationally Traded Gas* (Oxford: OIES/OUP, 2012).

Stern, J.P., 'Continental European long-term gas contracts: is a transition away from oil product-linked pricing inevitable and imminent?', Working Paper NG34, OIES, 1 September 2009. www.oxfordenergy.org/2009/09/continental-european-long-term-gas-contracts-is-a-transition-away-from-oil-product-linked-pricing-inevitable-and-imminent/.

Stern, J.P., 'Future Gas Production in Russia: is the concern about lack of investment justified?', Working Paper NG 35, OIES, 1 October 2009.

Stern, J.P., 'Is there a rationale for the continuing link to oil product prices in Continental European long term gas contracts?', Working Paper NG19, OIES, 1 April, 2007. www.oxfordenergy.org/2007/04/is-there-a-rationale-for-the-continuing-link-to-oil-product-prices-in-continental-european-long-term-gas-contracts/.

Stern, J.P., 'The Russian Gas Balance to 2015: difficult years ahead', in Pirani, S. (ed.), *Russian and CIS Gas Markets and Their Impact on Europe* (Oxford: OIES/OUP, 2009), pp. 54–92.

Stern, J.P., 'The Russian Natural Gas "Bubble": Consequences for European Gas Markets', RIIA Chatham House, 1995.

Stern, J.P., 'The Russian-Ukrainian Gas Crisis of January 2006', Working Paper, OIES, January 2006.

Stern, J.P. and Bradshaw, M.K., 'Russian and Central Asian Gas Supply for Asia', in J. Stern (ed.), *Natural Gas in Asia* (Oxford: OIES/OUP, 2008), pp. 220–78.

Stern, J.P. and Rogers, H., 'The Transition to Hub-Based Gas Pricing in Continental Europe', in Stern, J.P. (ed.), *The Pricing of Internationally Traded Gas* (Oxford: OIES/OUP, 2012), pp. 145–77.

Stern, J.P. and Rogers, H., 'The Transition to Hub-Based Pricing in Continental Europe: A Response to Sergei Komlev of Gazprom Export', OIES, 12 February 2013, www.oxfordenergy.org/2013/02/the-transition-to-hub-based-pricing-in-continental-europe-a-response-to-sergei-komlev-of-gazprom-export/.

Stern, J.P., *Competition and Liberalization in European Gas Markets: a diversity of models*, London: RIIA, 1998.

Stern, J.P., *The Future of Russian Gas and Gazprom* (Oxford: OIES/OUP, 2005).

The Coordinators of the EU–Russia Energy Dialogue, *Roadmap: EU–Russia Energy Cooperation Roadmap Until 2050*, The Coordinators of the EU–Russia Energy Dialogue, March, http://ec.europa.eu/energy/international/russia/doc/2013_03_eu_russia_roadmap_2050_signed.pdf.

Ulchenko, N., 'What is so special about Russian-Turkish economic relations', *Russian Analytical Digest*, No. 125, 25 March, 2013, pp. 5–9.

UNDP Human Development Report for Russia, 2013, Russian edition (*Ustoichivoe razvitie: vyzovy Rio*).

US EIA, *International Energy Outlook 2013* (Washington: US Energy Intelligence Agency, 2013).

US EIA, China Report, April 2013.

van der Marel, E., 'Beyond Dutch Disease: When Deteriorating Rule of Law affects Russian Trade in High-Tech Goods and Services with Advanced Economies', London School of Economics, July 2012.

Vorobyov, P., 'Lilliputians in the land of giants', *Petroleum Economist*, London, 24 May, 2013.

Westphal, K., 'Security of Gas Supply: four political challenges under the spotlight', SWC Comments 2012/C, June, Stiftung Wissenschaft und Politik, www.swp-berlin.org/en/publications/swp-comments-en/swp-aktuelle-details/article/security_of_gas_supply.html.

Whist B.S., 'Nord Stream: Not Just A Pipeline', FNI Report 15/2008, Fridtjof Nansen Institute, Oslo.

World Bank, *Energy Efficiency in Russia: Untapped Reserves*, (Washington: World Bank, 2008).

World Bank, *Implementation, Completion and Results Report*, IBRD-46010, 10 December 2008.

World Bank, *Russian Economic Report* no. 30: *Structural Challenges to Growth Become Binding*, Washington, September 2013.

Yafimava, K., 'Belarus: the domestic gas market and relations with Russia', in Pirani, S. (ed.), *Russian and CIS Gas Markets and Their Impact on Europe* (Oxford: OIES/OUP, 2009).

Yafimava, K., 'The 2007 Russia–Belarus gas agreement', Oxford Energy Comment, OIES, January 2007, www.oxfordenergy.org/2007/01/the-2007-russia-belarus-gas-agreement/.

Yafimava, K., 'The EU Third Package for Gas and the Gas Target Model: major contentious issues inside and outside the EU', Working Paper NG 75, OIES, April 2013. www.oxfordenergy.org/wpcms/wp-content/uploads/2013/04/NG-75.pdf.

Yafimava, K., 'The June 2010 Russian-Belarusian gas transit dispute: a surprise that was to be expected', Working Paper NG43, OIES, 1 July, 2010. www.oxfordenergy.org/2010/07/the-june-2010-russian-belarusian-gas-transit-dispute-a-surprise-that-was-to-be-expected/.

Yafimava, K., 'The Third Energy Package and the Gas Target Model: contentious issues inside and outside the EU', Working Paper NG 75, OIES 2013.

Yafimava, K., *The Transit Dimension of EU Energy Security: Russian Gas Transit Across Ukraine, Belarus, and Moldova* (Oxford: OIES/OUP, 2011).

Zhukov, S.V. and Reznikova, O.B., *Tsentral'naia Aziia i Kitai: ekonomichesko vzaimodeistvie v usloviakh globalizatsii* [*Central Asia and China: economic cooperation in the context of globalisation*], Moscow, IMEMO RAN, 2009.

Government publications

'Ekonomia i berezhlivost' – *glavnye faktory ekonomicheskoi bezopasnosti gosudarstva'* [*'Economy and thrift: the main factors of the state's economic security*], Directive no. 3 14.06.2007 [Belarus].

'Gosudarstvennaia kompleksnaia programma modernizatsii osnovnykh proizvodstvennykh fondov... na period do 2011' [*Integrated state programme for the modernisation of principle production assets*], ukaz no. 575 ot 15.11.2007 goda [Belarus].

APBE Agentsvo po prognozirovaniu balansov v elektroenergetike, *Funktsionirovanie i razvitie elektroenergetiki v Rossiiskoi federatsii v 2011 godu*, [Agency for Forecasting Energy Balances, *Electricity in Russia: functioning and development, 2011*].

AzStat, Foreign Trade of Azerbaijan 2011 (Baku, 2011).

Draft Resolution 'Rules On *Non-Discriminatory Access to High-Pressure Gas Pipelines in the Russian Federation*, available on the Russian Ministry of Economic Development website, http://www.economy.gov.ru/wps/wcm/connect/ ec61c7004ad85dc2a080abaf3367c32c/ pp.pdf?MOD=AJPERES&CACHEID=ec61c7004ad85dc2a080abaf3367c32c.

Federal'nyi zakon no. 103-F3 'o vnesenii izmenenii v federal'nyi zakon 'o kontsessionnykh soglasheniiakh'', [Federal law 'on the amendment of the law 'on concession agreements'] (7 May 2013).

Federal'nyi zakon no. 190-F3 'o teplosnabzhenii', [Federal law 190-F3 'on heat supply'] (27 July 2010).

Federal'nyi zakon no. 291-F3 'o vnesenii izmenenii v otdel'nyi zakonodatel'nyie akty RF v chasti sovershenstvovaniia regulirovaniia tarifov', [Federal law 291-F3 'on the amendment of legislation in order to improve the regulation of tariffs'] (30 December 2012).

FTS Information Bulletin No. 11 (385), 26 March 2010.

FTS Information Bulletin, No 16 (246), 2 May 2007.

FTS Order No 143-e/1.

FTS Order No 497-e/2, 30 December 2010.

FTS Resolution No 143-e/1, 'On approving the tariffs for transportation of gas though Gazprom's high-pressure pipelines, which are part of the UGSS, charged to third parties', 8 June 2012.

Government resolution, *On gas prices*, N 122, 27 December 2012 [Belarus].

Ministerstvo energetiki Rossiiskoi Federatsii, 'Energeticheskaia strategiia Rossii na period do 2020 goda'[Ministry of Energy of the Russian Federation, 'Energy Strategy of Russia to 2020'], (Moscow, May 2003).

Ministerstvo energetiki Rossiiskoi Federatsii, 'Skhema i programma razvitiia Edinoi energeticheskoi sistemy Rossii na 2013–2019 gody', [Ministry of Energy of the Russian Federation, 'Scheme and programme for the development of the united energy system of Russia for 2013–2019'] (June 2013).

Ministerstvo energetiki Rossiiskoi Federatsii, Doklad ministra energetiki RF S.I. Shmatko 'O proekte general'noi skhemy razvitiia gazovoi otrasli na period do 2030', [Ministry of Energy of the Russian Federation, Report by Minister of Energy S.I. Shmatko 'on the draft general scheme for the development of the gas industry to 2030'], 12 October, 2010.

Ministry of Energy of the Russian Federation, Moscow, 'Energy Strategy of Russia for the Period up to 2030', 2010.

Ministverstvo ekonomicheskogo razvitiia Rossiiskoi Federatsii, 'Prognoz dolgosrochnogo sotsial'no-ekonomicheskogo razvitiia Rossiiskoi Federatsii na period do 2030 goda', [Ministry of Economic Development of the Russian Federation, 'Projection of long-term social-economic development in the Russian Federation up to 2030'] (March 2013).

Ministverstvo ekonomicheskogo razvitiia Rossiiskoi Federatsii, 'Prognoz sotsial'no-ekonomicheskogo razvitiia RF na 2014 god i na planovyi period 2015 i 2016 godov', [Ministry of Economic Development of the Russian Federation, 'Projection of social-economic development in the Russian Federation for 2014 and in the planning period 2015–2016'].

Prognoz dolgosrochnogo sotsial'no-ekonomicheskogo razvitiia RF pa period do 2030 goda, Ministerstvo ekonomicheskogo razvitiia, [*Long-term projections of economic development to 2030*, Ministry of Economic Development], Moscow, March 2013.

Zakon Ukrainy Pro zasadi funktsionuvannia rinku prirodnogo gazu [the Law of Ukraine on the Functioning of the Natural Gas Market] (law no. 2467-VI of 8 July 2010), http://zakon1.rada.gov.ua/cgi-bin/laws/main.cgi?nreg=2467-17.

Journals and newspapers

Argumenty i fakty Ukraina

Ekonomychna Pravda

Ekspert

Energobiznes

Forbes Ukraina

Gas Matters

Interfax Natural Gas Daily

Interfax Russia and CIS Oil and Gas Weekly

Interfax *Russia's Gas Industry (2012)*, Moscow: Interfax, 2012.

ITAR-TASS

Kommersant'

Kommersant'-Ukrainy

Obozrevatel'

Oil and Gas Journal

Platts European Gas Daily

Platts European Gas Daily Monthly Averages

Platts International Gas Report

Promyshlennye i otopitel'nye kotel'nye i mini-TETs

Vedomosti

Vzgliad

Zerkalo Nedeli

Online resources

Asbarez Armenian News

Assossiatsiia po prognozirovanii balansov v elektroenergetike, [Association for Forecasting Energy Balances], www.e-apbe.ru/.

Business Caucasus

East European Gas Analysis, http://www.eegas.com/ukr-ns-2012-2020e.htm.

EU information portal (europa.eu)

Fond Sodeistvia Reformirovaniiu ZhKKh, [Cooperative Fund for Municipal Services Reform], www.fondgkh.ru/.

Gazprom in Figures 2008–12, http://reports2.equitystory.com/cgi-bin/show.ssp?companyName=gazprom&report_id=if-2012&language=English.

Gazprom website

Interfaks-Ukraina

Interfax Belarus news agency website (www.interfax.by)

ITAR-TASS news agency http://itar-tass.com

Management Report 2012, Gazprom.

Platts The Barrel blog

Russian Energy Monthly, Eastern Bloc Research, www.easternblocenergy.com.

UNIAN news agency

www.oilcapital.ru

INDEX

Absheron field 366
Achimgaz project 340, 341
Agreement on Partnership and Cooperation (1994) 82
Alekperov, Vagit 337
Algeria 55
Alrosa 330
Altai 217, 218, 271
Ananenkov, Alexander 265, 268, 307
annual contract quantity (ACQ) 59, 60
Anti-Monopoly Service 130
Arctic offshore 11
Armenia 348
   gas exports to 181–3, 206, 207
   gas prices 206, 207
   Gazprom gas storage 141
Armrosgazprom 206
Asian gas market 217–45
   and LNG 41, 239–43
   importance of 23, 25, 216
   Russian export strategy 231, 238–9, 246–9
   and Russian gas matrix 387–8
Asia–Pacific Economic Cooperation Summit 21
associated gas 2, 29, 254, 258, 315, 335–6, 364–6, 371
Australia, Chinese investment in 226
Austria 51, 53, 61, 77, 83, 90, 358
avoidance pipelines 70–81
Azerbaijan 55, 98, 99, 181, 336, 337, 349, 364, 369, 370
   gas exports 347–8, 357–8, 366
      to Georgia 206, 207, 347, 348, 357
      to Russia 22, 95, 357–8, 360
      to Turkey 347, 348, 357–8
   gas imports from Russia 182, 206–7, 214, 357
   Russian upstream investment 365–6
   Shah Deniz field 98, 181, 206, 337, 357–8, 365

Balkans 22, 23, 53, 62
Balkan corridor 204, 213
Baltic LNG 280, 282, 373
Baltic States 94–5
barter, and gas trade 359
base price (P0) reduction 64
base-load plants 164
Bay of Bengal fields 225

Bazhenov tight oil formation 275
Behre Dolbear 14
Belarus 23, 197–201, 215
   and Beltransgaz sale 197, 199, 200–1, 211–12
   and Customs Union 30, 206, 348
   domestic gas sector 200–1
   energy balance 199–200
   gas consumption 197–8, 199–200, 201
   gas contracts
      January 2007 197–9, 211
      2012–14 199, 211
   gas imports from Russia 181–3, 197–201, 211, 379
   gas prices 198, 199, 200–1, 211, 379
   gas security 199–200, 201
   and gas transit 70, 73, 80, 201, 207, 211–12, 303
   Gazprom gas storage 141
   and Russia 22, 23, 197, 201, 211
Belgium 51, 52
Belgorod 18
Belousov, Andrei 127
Beltopgaz 200–1
Beltransgaz 197, 199, 200–1, 211–12
Beregovo 72
Bering Sea, gas reserves 217
biomass, and power generation 16
Black Sea naval base 23, 185, 209–10
Blagoveshchensk 297
Blue Stream 30, 75–7, 80, 95, 303
Bogdanchikov, Sergei 309
Bogdanov, Vladimir 338
boilers, centralized 156, 168, 169, 171
Bolshekhetskaia Depression 335
Bosnia and Herzegovina 51, 61
Botas 59, 96, 358
Bovanenkovo field 43, 124, 256, 263–5, 267–8, 269, 271, 273, 283–4, 342, 360, 372, 383
Bovanenkovo–Ukhta pipeline system 138, 272
Bovanenkovskoe zone 265
BP 243, 279, 324, 377
   and TNK-BP 31, 33, 124, 126, 218, 220, 296, 297, 330–3, 342, 370
Briansk 18
Bukhara–Urals pipeline 349
Bulgaria, 75, 204, 213, 358
   Gazprom exports to 51, 61, 96, 205
Burma, and China 219, 223, 225, 227

430

Canada, Chinese investment in 226
capacity delivery agreement (CDA) 165–6, 168
carbon price, decline of, in Europe 54, 379
Caspian gas 98, 347–67
Caspian region, pipelines 349–50
Caucasus, gas imports 206–7
cement manufacture, gas consumption 173, 176
Central Asian gas 347–67
    and China 219, 223, 369–70
    gas export prices 316, 351–2, 358–64
        to China 362–4
        to Ukraine 363
    gas exports 347–8
        through Russia 347–8
        to China 347, 353, 353–6, 362–4
        to Russia 349–53, 356–7, 362–4, 368–70
    gas imports 41, 253–5, 257, 315
    and global economic crisis 353
    political factors 351–2, 359–62
    Russian market, impact on 368–70
    upstream investment, Russian 364–6
Central Asia, pipelines 349, 350
Central Asia–Centre pipeline 224, 349, 354
Central Dispatching Unit of the Fuel and Energy Complex 391
Chaianda field 229, 232, 234, 238, 239, 297–8, 300, 301
Chaivo field 334
Cheliabinsk 33
chemical industry 173–5, 180
Chernomorneftegaz 191
Chernomyrdin, Viktor 2, 217, 252
Chevron 191
Chikanskoe field 298
China
    12th Five Year Plan 222
    and Burma 219, 223, 225, 227
    and Central Asian investment 347
    coal bed methane 224
    gas consumption 216
    gas demand 223–4
    gas, domestic production 223, 224
    gas imports 221, 223, 224–5
        from Central Asia 219, 223, 225, 347, 353, 362–4, 369–70
    gas pricing, domestic 221–3, 242
    gas pricing, imports 218–20, 222, 223, 237, 243–4, 362–4
    gas supply, diversification of 219, 223, 244
    as geopolitical threat to Russia 229–31
    LNG imports 219–20, 223, 225–9, 276
    Russia, gas negotiations with 217–29, 236–9, 387–8
    security of supply 244, 263
    shale gas resources 238
    transmission tariffs 362
    and Turkmenistan 219, 220, 223, 224–5, 348
    unconventional resources 224, 244
    upstream investments, international 223, 226
    wellhead price 221, 222
Chukchi Sea, gas reserves 217
CIS countries 1, 2, 36, 97, 181–3, 214–15, 246, 247, 248
    and gas transit 70, 183, 207–14
    and geopolitics 22, 248
    Russian gas exports to 22, 33, 39, 41–3, 69, 97, 181–3, 214–15
    and transit-avoidance pipelines 209, 213–14, 215
    see also Armenia, Belarus, Georgia, Moldova, and Ukraine
Climate Doctrine Action Plan 25–6
CNOOC 226
CNPC 217, 219, 224, 243, 244, 362, 364, 369
    and Australian LNG assets 226
    and Canadian shale assets 226
    and East Siberian gas 243
    and Gazprom 218, 227, 237–8, 245, 271, 312, 387, 388
    and Lukoil 337
    and Mozambique assets 226
    and Yamal LNG 241–3, 304, 324
coal, and power generation 16, 18, 19, 158–9, 165, 405–6
coal bed methane 224, 274
combined cycle gas turbine plant (CCGT) 166, 167, 168
combined heat and power (CHP) plants 155, 168, 171, 391–2
contracts, long-term, with European countries 53, 56–9, 60
'contractual mismatch' 84
Copenhagen Accord 25
corporate income tax 29
credit ratings 14
Croatia 51, 52
Customs Union 23, 30, 206, 211, 379
Czech Republic 51, 52, 61

De Kastri 334
Denmark 61
destination clause 57, 68
DF Group 188
DG COMP *see* European Commission's Competition Directorate
district heating 110, 112, 155–6, 168–72, 179–80
Dnipr-Donetsk basin 191

Dobycha 306
'Doing Business' rating 14
Druzhba (Brotherhood) pipeline 189
dry gas 43–4, 273
DTEK 188, 195
Duffin, Neil 333

E.ON 65, 331, 336, 340
E.ON Ruhrgas 73, 78
E.ON Russia 126, 166, 327
East Siberian Sea, gas reserves 217
East Siberia-Pacific Ocean (ESPO) pipeline 231, 232
East Tarkosalin field 320, 326
Eastern Gas Programme 25, 26, 219, 229–31, 294, 303, 372
Eastern Gas Strategy 228
East–West pipeline, China 218
economic diversification and energy pricing 134–5
EDF 75
Edison 59, 65, 66
Egypt 55
Ekaterinburg 171
electricity demand 160–61, 167–8, 179
electricity generation 12, 16, 163, 404–9
electricity pricing 162–3, 168
Electronic Trading System (ETS) 121
Enel 340
energy balance 97
  Belarus 199
  European countries 53, 54, 103, 105, 107
  Russia 16–19, 25, 169, 180, 247, 302
Energy Charter Treaty 83, 208
Energy Community 190, 195
Energy Community Treaty (EnCT) 205, 213
energy companies, state-owned 162, 163
energy consumption, primary, role of gas 16–17
energy intensity, reduction of 25
energy policy framework 25–6
Energy Strategy to 2030 18–19, 25
Enhanced Oil Recovery (EOR), potential for 11
ENI 59, 75, 340
Enineftegas JV 340
entry/exit (EE) system, gas transportation 84
Erdgas Import Salzburg 65
Estonia 51, 91, 94, 95
Eural Trans Gas 33, 69, 255
Europe
  Azerbaijan, gas imports from 354, 358, 366, 370
  diversification away from Russian gas 91–2, 97–9
  and economic recession 247, 280, 283, 315, 353, 368, 377–8

energy balance 53, 54, 103, 105, 107
energy regulation reform 57
gas contracts, long-term 53, 58–64, 68, 70, 93–4
gas demand 40, 52–6, 78, 104–5, 107
gas diversification 96–9
gas price mechanisms 9, 56–7, 58, 62, 63–8, 93–7, 189, 248, 351, 360, 382, 389
gas pricing
  country variations 61, 62
  international arbitration 64–6, 94
  'political' 92–7, 99, 101, 104, 107
gas security of supply 43, 70, 90–92, 96–7, 99–100, 103–4, 207, 213–14, 303
  Gazprom storage in 90–91
geopolitics of gas 22, 23, 91, 99–101, 104, 107
LNG imports 58, 105, 245, 256, 280, 324, 373
market changes 56–8, 377–9
Russian gas exports to 1, 3, 4, 41, 50–56, 58–64, 107, 120, 288–9, 316, 385
  pipeline capacity 44, 79–81
  strategy 25, 46, 101–7, 216, 231, 245, 246–9, 377–9, 385–6
and Russian gas matrix 3, 39, 377–9
and shale gas 106, 378–9
utility structures 56–8
European Commission, response to 2009 Ukrainian crisis 207–8
European Commission Competition Directorate (DG COMP) 67–8, 92, 94
European Union (EU)
  accession countries, and Russia 95, 97, 103
  carbon reduction targets 54, 107
  Emissions Trading Scheme 54
  Energy Charter Treaty 83, 208, 213
  energy security 83, 92, 208
  First Gas Directive (1998) 83
  Gas Target Model 83–8
  and gas transportation 77, 83–6, 102, 106, 146
  Gazprom, investigation into 67–70
  household spending on gas 20
  liberalization process 40, 82–3, 88–9, 248
  membership, and gas pricing 97
  network codes 84, 103
  regulations and export to Europe 82–107, 304
  and Russia 22, 36, 67, 82, 88, 93, 103–4
  Second Gas Directive (2003) 83
  Second Strategic Energy Review (2008) 208
  and 'Southern Corridor' 97–9, 358
  Third Energy package 57, 83–8, 102, 103
  and Ukraine 1, 187, 189, 191, 207

Index 433

EU–Russia Energy Dialogue 82, 208
EU–Russia Energy Cooperation Roadmap to 2050 82
EU–Russia Gas Advisory Council 88
EU–Russia Partnership and Cooperation Agreement 103
EU–Ukraine Energy Dialogue 208
Europol Gaz 72
Eustream 189
Evrokhim 173–5, 180
exploration, and the Subsoil law 26–7
export duty *see* export tax
export markets, influence on supply strategy 246–9
export tax 11, 15, 28, 30, 36, 118, 187, 206, 209, 231, 274, 305, 322, 351, 357, 389
ExxonMobil 34, 191, 236, 241, 242, 333–4, 340

Far East Federal District onshore , gas reserves 217
Far East, gasification programme 24, 177, 178, 180
Federal Antimonopoly Service (FAS) 24, 143, 145, 161, 343, 399
Federal Energy Commission (FEC) 24, 146
Federal Tariff Service (FTS) 24, 118, 122, 128, 146, 171
fertilizer production, gas consumption 173–4
Filanovskii field 335
Finance Ministry 25
Finland 51, 61, 94
Firtash, Dmitry 188, 352
France 51, 61
FSU, Gazprom gas revenue share 288
FSU, and Russian gas matrix 379–80
fuel mix, power sector 157–9, 164–5
Fukushima nuclear disaster, and LNG supply 56
Fursin, Ivan 352

GAIL 276
Galkynysh (South Yolotan) gas field 225, 355, 362, 364
gas, as 'cash source of last resort' 21
gas, as domestic political tool 19–22
gas, as instrument of foreign policy 22–3
gas, role in energy balance 16–19
gas, role in Russian economy 9, 15–23, 380–81
gas, statistical issues 391–3
gas, volume measurements xvi
gas condensate xvi, 29, 132, 174–5, 263, 274, 288, 310, 319, 320, 371
gas consumption, Russian 109–17
 district heating 155–6, 168–72
 industry 172–6

power sector 155–68
residential use 176–8
system losses 178
technical use 178
gas demand, from Europe 52–6, 247
gas demand, Russian 17, 41, 247–8
 forecasts 113–17
 and declining population 176
Gas Exchange 120–21, 129–30, 135, 345
gas market globalization 249
Gas Natural 324, 346
gas pricing
 arbitration, international 64–6, 94
 barter 359
 base price (P0) reduction 64
 gas-to-gas 248
 hub-based 40, 42, 57, 58, 63, 66, 385–6
 market-based 385–6
 netback 37, 56–7, 62, 93, 97, 118–19, 131, 185, 187, 223, 237, 255–6, 353, 359
 'net-forward' 199
 oil-linked 42, 57, 63–5, 237, 247
 'political' 44, 62, 92–7, 99, 101, 104, 107, 134, 351–2, 360–62, 380–81
 price formation changes 248
 pricing strategy 374
 shale gas, impact of 57–8, 378–9
 spot pricing 58, 189, 239–40, 276–7
gas pricing, domestic
 China 221–3, 242
 Russian 17–20, 108, 117–36, 149, 152–3, 255–6
  regulated 32–3, 115, 117–23, 149, 152–4, 177, 316, 317, 380, 386
  liberalization of 135–6
  netback parity 28, 30, 34, 108, 118–23, 128, 131, 135, 153, 256, 316
  and tax revenue 131–3
gas pricing, exports from Russia
 Armenia 206, 207
 Belarus 183, 198, 199, 200–1, 211, 379
 Baltic States 94–5
 China 218–20, 222, 223, 237, 243–4, 246, 362–4
 CIS countries 183
 Europe 9, 56–7, 58, 62, 63–8, 93–7, 189, 246, 351, 360, 377–8, 389
 Moldova 202–3, 204–5
 Ukraine 183–9, 193, 209–10, 211, 214, 215, 248, 351, 363
gas pricing, imports to Russia 37
 Central Asia 316, 351–2, 358–64
 Turkmenistan 316, 351, 354, 367
 Uzbekistan 348, 352, 354, 356, 362, 369
gas production *see* Gazprom and entries on individual countries

gas statistics 391–2
gas storage 90–91, 140–41, 398–9
Gas Supply Law 318
Gas Target Model (EU) 84
gas taxation *see* taxation
gas transit disputes 207–14
 see also transit
gas transmission 84, 85, 136–40, 398–9
 see also transmission
Gas Transportation System (GTS) 137–9
 third-party access 142–4
gas transportation *see* gas transmission,
gas 'weapon' 101
gasification programmes 24, 176–8, 180, 229, 372
gas-to-gas pricing 248
gas-to-power demand 168
Gasunie 73
Gaz-Energo-Alians 143
Gazi–Magomed–Mozdok pipeline 357
Gazprom, company finances
 capital expenditure 291-4
 cash flow 287, 294–5
 cost efficiency 291
 customer debt 128–9, 134, 169–70
 EBITDA 289, 291, 292, 295
 financial position 32, 46–7, 287–96, 311–12, 373
 gas revenues 287–8
 investments 47, 88–91, 293–4, 312, 315
 liquids revenue 288–9, 310
 margins 291
 non-gas revenue 288–9
 oil power assets 260, 309, 311
 revenues, total 289–90
Gazprom, company strategy
 Asian strategy 216, 256–7, 316, 387–8
 Eastern Gas Programme 25, 26, 219, 229–31, 294, 303, 372
 Eastern LNG strategy 234–6, 238, 239–40, 244, 312
 Eastern Siberia supply plans 296–301
 European export strategy 25, 35, 42, 46, 101–7, 216, 231, 245, 246–9, 368, 377–9, 385–6
 Far East supply plans 296–300
 future strategy 46, 286–313, 389–90
 investment plans 35, 47, 88–91, 259, 260, 262–3, 265, 293–4
 LNG strategy 228, 275–82
 supply strategy 3, 258, 283–5, 286, 373–4, 382–5, 389
 trading house strategy 89
 transit diversification policy 209–14
Gazprom, company structure 31, 136, 138, 300–6

corruption, alleged 307–9, 344
governance 307–9
and government's role 45–7
history of 1, 2, 31–3, 252–6, 258–9, 301–2, 314–15
management reform 307
ownership 32, 300–6
reform of 306–11
Sochi Olympics 21, 115
statistics 391
vested interests 22
Gazprom and Armenia 141, 182, 206
Gazprom and Asia 216, 219, 239, 245
 Eastern Gas Programme 229–31
 Eastern LNG strategy 234–6, 238, 239–40, 244, 312
 strategy 256–7, 316, 387–9
Gazprom and Azerbaijan 99, 182, 357–8
Gazprom and Baltic States 94–5
 Lithuania price arbitration 66
Gazprom and Belarus 141, 182, 197, 199–201, 211–12
 Beltransgaz 199, 200–1, 211
Gazprom and Central Asia 347–67
 gas imports from 44, 349–52, 369–70
 Kazakhstan 181, 353, 356
 price 44, 347, 349, 351–4, 358–64
 Turkmenistan 354–5
 upstream investment 364–6
 Uzbekistan 316, 352, 356, 359, 364–6, 369
Gazprom and China 236–9, 243, 217–18, 220, 232, 236
 CNPC 218, 227, 237–8, 245, 271, 312, 387, 388
Gazprom and CIS markets 181–215
 transit 207–15
Gazprom and Europe
 downstream investments 88–9, 94
 EU competition investigation 67–70
 EU liberalization, response to 83, 84, 89
 EU transportation changes 84–8
 European export strategy 25, 35, 42, 46, 101–7, 216, 231, 245, 246–9, 377–9, 385–6
 European pricing policy 62–6, 96–7, 377–8
 European supply contracts 85–6
 gas exports to Europe 3, 42, 50–52, 55, 56, 59–62, 80–81, 368, 377–8
 security of supply to Europe 91–2
Gazprom, exports 2, 316
 export strategy 42, 246–9, 273–4
 gas export monopoly 32, 33–4, 46, 68–9, 252–3, 345–6
 government approval 24

Index

LNG export monopoly 240-42
  end of 34, 69, 104, 240, 242, 243, 282, 304, 312–13, 333, 345–6, 371
'reverse flow', opposition to 189–90
take-or-pay volumes 59, 60
see also Europe and individual country entries
Gazprom gas storage 90–91, 141, 210, 398–9
Gazprom and Georgia 182, 206
Gazprom and Independent producers 31, 124–5, 127, 128–31, 253, 316–19, 370–71, 380
Gazprom and LNG
  Asian LNG sales 244–5
  Atlantic Basin LNG strategy 277, 280, 373
  LNG export monopoly 240-42
    end of 69, 104, 240, 242, 243, 282, 304, 312–13, 333, 345–6, 371
  LNG spot sales 239–40, 276–7
  LNG strategy 228, 275–82
  Pechora LNG 280–811
  Shtokman field 21, 35, 73, 82, 263, 269, 270, 271, 275–9, 282, 294, 340, 373
  Vladivostok LNG 232–4, 238, 239, 244, 299–300, 304, 312, 373, 389
Gazprom and Lukoil 335–7
Gazprom and Aleksei Miller 81, 83, 219, 227, 228–9, 232, 237, 238, 280, 302, 303, 306, 307, 308, 352, 357, 372, 387
Gazprom and Moldova 182, 202, 204–5, 212–13
  Moldovagaz 204
  Transdniestrian debt 202, 204, 205, 212
Gazprom and Novatek 32–3, 125, 292, 317, 320, 321, 322, 324, 328
Gazprom and pipelines 21, 106–7, 398
  capacity 80–81
  GTS 138–9, 398
  pipeline system, control of 252–3
  third-party pipeline access 31, 108, 144, 252–3, 318-19, 343
  transit, CIS countries 207–15
  transit avoidance 43, 70, 77–9, 90, 100–3, 209–14
  transmission costs 290
  transmission operators 138
  transmission tariffs, internal 144, 147, 148–9, 150–53
  UGSS 31, 34, 37, 136–40
    third-party access 141–8, 154, 326, 399–400
    third-party tariffs 139–40, 146–8
  see also Nord Stream, Power of Siberia, South Stream, Yamal–Europe

Gazprom and pricing policies
  domestic pricing 32, 117–31, 149, 152–3, 255–6, 317
  European pricing policy 62–6, 96–7, 368, 377–8, 385–6
  Gas Exchange 121, 129–30, 345
  hub-based pricing 42, 63, 385–6
  market-based pricing 385–6
  netback prices 255
  oil-linked pricing 42, 63–5, 237, 247
  political pricing 44, 62, 351–2
    in Europe 62, 91, 91–7
  price arbitration, international 64–6
  price formation changes 248
  pricing strategy 374
  see also gas pricing
Gazprom production
  Bovanenkovo field 43, 124, 256, 263–5, 267–8, 269, 271, 273, 283–4, 342, 360, 372, 383
  Chaianda field 297–8
  eastern Siberia 232–3
  Kovykta purchase 219, 220
  oil diversification 309–11
  petrochemicals 311
  production 3, 41, 43, 120, 254, 256, 258, 262–4, 272, 411
  production associations 410–11
  production costs 290, 411
  production forecasts 263, 264, 269–70, 271, 301, 372, 383–4
  reserves 262, 410
  Sakhalin 228, 233–6, 298, 334–5, 340
  as swing producer 371–3, 384
  West Siberia fields 258–61
  wet gas 272, 273–4, 275
  Yamal 123, 263–72, 283, 286, 293–4, 311–12, 315, 371–2, 383
Gazprom, and Vladimir Putin 68, 72, 232, 236, 245, 300, 302–3, 307, 308, 309, 312–13, 343
  and corruption allegations 307–8, 344
  LNG export monopoly 242, 291, 304, 312–13
Gazprom and Rosneft 32–3, 302, 304, 309–10, 333, 334
Gazprom, and Russian government
  domestic gas prices 123–31
  domestic market competition 119–21
  export approval 24
  geopolitics 99–101
  as government revenue source 21, 295–6, 305, 381
  government involvement 31–2, 34–5, 45–7, 302, 303–4, 300–6, 372, 3812
  Mineral Extraction Tax 29, 132–3

'political' pricing 62, 91–7, 183, 351–2, 362
Presidential Energy Commission 25, 240, 242, 344–5
share of Russia's GDP 118
Russian budget revenue 305
transit avoidance 72, 79
Gazprom and Russian market
  demand estimates 3, 113–14, 270
  demand reduction 32–3
  district heating 169–70, 172, 180
  gas supply balance 254–5, 371–4
  gasification programme 20, 24, 177, 178, 180
  household tariffs 134–5
  market position 33, 117, 252–4
  power diversification 309–11
  power sector supply 164
  pricing regulation 32, 117–31, 255–6
  rural gas supply 20
  sales by customer segment 129
Gazprom supply 252–7, 268–75
  East Siberia and Far East 296–300, 301, 303
  strategy 258, 262–3, 283–5, 373–4, 389
  supply/demand balance 382–4
  third-party supply 315, 316
Gazprom and Turkey 95–7, 100
Gazprom and Ukraine 43, 81, 181, 182, 185, 187, 191, 194, 196, 210–11
  'reverse flow', opposition to 189–90
  transit 207, 209–10
Gazprom Export 33, 352, 353
Gazprom Gazoraspredelenie 306
Gazprom Germania 89, 357
Gazprom Global LNG 239–40, 276
Gazprom Marketing Trading (GMT) 89, 276, 282
Gazprom Mezhregiongas Kostroma 327
Gazprom Neft 274, 275, 290, 309, 310–11, 321, 338–9, 342
Gazprom Pererabotka 306
Gazprom PKhG 306
Gazprom sbyt Ukrainy
Gazprom Schweiz 357
Gazpromenergoholding 163, 171, 311
GDF Suez 73
GDP, Russian, 6, 7, 9, 11–12, 116
General Schemes (Master Plans) 25
Geofizicheskoe field 325
Georgia 181–3, 206–7, 353
geothermal energy, and power generation 16
Germany 51, 60, 61, 90
Giprospetsgaz 281
global economic crisis, impact of 11, 40
global energy market, impact on Russian economy 10

Greece 51, 52, 61
Griazovets–Vyborg pipeline 138
Guangdong, gas pricing project 222
Guangxi, gas pricing project 222
Gunvor 32
Gydan Peninsula 322, 326

helium 297
Heritage Foundation 14
high-tech industry 6–7
household tariffs 134–5
  subsidy 19–20, 134
hub pricing 40, 57, 58, 63, 66
Hungary 51, 61, 189
hydro-electric power 16, 157–9, 167, 168, 409

IatTEK 341
IEA International Energy Agency
International Energy Agency (IEA) statistics 391
Independent gas producers 124–7, 253–6, 257, 314–46, 368, 370–71
  competing with Gazprom 316
  as gas exporters 371
  gas production 354–5, 316, 342, 383–4
  government support 342–3
  market share 119, 318
  and MET 29, 132–3
  pipeline access 318–19
  prices 118, 119, 124–5, 126
  regional 341
  selling at discount to Gazprom 370, 380
  supply 124–7, 314–29
  wet-gas production 319
India 239
industry
  capital investment 12, 13
  gas consumption 110, 112, 172–6
  gas prices 19–20, 28, 135
  production 12–13
INPEX 242
Inter RAO 163, 166, 331
interconnection points (IP), gas transportation 84, 85
inter-fuel competition 17, 18–19
investment, foreign 13–14, 27
Iran 353, 355
Irkutsk Centre 229
iron and steelmaking, gas consumption 173, 175
Italy 51, 60, 61
Itera 3, 33, 119, 255, 319, 329–30, 331, 332
ITGI pipeline 98
Iurii Korchagin field 335
Iurkharovskoe field 320, 326

Iuzhno-Russkoe field 78, 260
Iuzivska field 191, 195

Japan 233, 239, 240, 242, 276
Japan Far East Gas Consortium 232
Jiang Jiemin 227

Kaliningrad, subsidized gas prices 20
Kamchatka peninsula 298
Kamennomysskoe More field 268, 284
Kandym-Khauzak-Shady fields 337, 365
Karachaganak oil field 353, 356, 365
Kashagan oil field 356
Kashgar 225
Kazakhstan 181, 224, 225, 226, 316, 336
  and Customs Union 347–8
  gas exports 2, 348
    to China 363
    to Russia 348, 351, 356, 359
  investment in 347
  Russian upstream investment 365–6
  trading arrangements 353
Kazmunaigaz 352, 353, 356
Kazrosgaz 353, 356
Khancheiskoe field 320, 326
Khar'iaga project 27
Kharampur field 329,330, 332
Khorgos 225
Khoroshavin, Alexander 334
Kirinskoe subsea field 298, 299, 300, 301
Klepach, Andrei 123
KOGAS 276
Komlev, Sergey 63
Korea 233, 239, 347
Korovinskoe field 281
Kostroma 33
Kovykta field 217, 218, 219, 220, 229, 232, 238, 239, 243, 296–8, 301
Krasnoiarsk Centre 229
Krasnoiarsk region 298
Kruglov, Andrey 237, 279
Kumzhinskoe field 281
Kuwait 240
Kuzbass Basin 274
Kynsko-Chasel'skaya group of fields 332
Kynsko-Chasel'skoe Neftegaz 329
Kyoto Protocol 22
Kyrgyzstan 225

Laptev Sea, gas reserves 217
Latvia 51, 94, 95
Law on Gas Exports 69
Law on Gas Supply 136, 141
Law on Trunk Pipelines, draft 139, 399–400
Leningrad LNG 373
Libya 55

Lithuania 51, 66, 94, 95
LNG
  and Asian market 227–9, 239–45
  and Baltic States 95
  China regasification plants 219–20, 225
  European market 105, 107, 256
  excess supply 40
  exports 30, 34, 41, 69
  gas market, impact on 249
  Gazprom 228, 240–42, 275–82
    end of export monopoly 69, 104, 240, 242, 243, 282, 304, 312–13, 333, 345–6, 371
  shale gas, impact of 368
  spot sales 239–40, 276–7
  supply to Europe 56, 58
  tankers, ice-breaking 323
  US exports 105
  see also Gazprom and LNG, Sakhlin-1 LNG, Sakhalin-2 LNG, Shtokman, Vladivostok LNG, Yamal LNG
Lukoil 31, 33, 119, 335–7, 341
  and Azerbaijan 336, 337
  and Chinese gas market 337
  downstream investment 336
  gas production 336–8, 342
  gas supply contracts 124–5, 126
  marketing strategies 335–6
  and North Caspian 335–6
  power assets 335, 336
  third-party sales 336
  and Uzbekistan 336–7, 349, 364–6, 371
  and West Siberia 335, 336

Macedonia, Gazprom exports to 51, 61
manufacturing industry 6, 12, 173
Markelov, Vitaly 300
Market Council 161
Marubeni 242
Medvedev, Aleksandr 234
Medvedev, Dmitrii 25, 88, 122, 127, 346
Medvezh'e field 254, 258–9, 315
merit order, power generation 164
Mezhregiongaz 120, 121
Mikhelson, Leonid 216
Miller, Aleksei 81, 83, 219, 227, 228–9, 232, 237, 238, 280, 302, 303, 306, 307, 308, 352, 357, 372, 387
Milov, Vladimir 307
Mineral Extraction Tax (MET) 11, 15, 21, 28–30, 36, 131–3, 305, 322, 325
mining, gas consumption 173
Ministry of Economic Development 24
Ministry of Energy 24, 70, 167, 345
Ministry of Gas 2
Ministry of Natural Resources 24

'missing gas' incident 140, 210
'missing transportation link' 86
Mitsubishi 340
Mitsui 340
Moldova 18
   and European Commission 205
   gas consumption 202, 203, 205
   gas imports 181–3, 202–6
      prices 202–3, 204–5
   and gas transit 70, 203, 204–6, 212
   Gazprom supply contracts 204–5
   South Stream impact 204, 213
   see also Transdniestria
Moldovagaz 204
Moldova–Romania interconnector 204, 206
Molodtsov, Kirill 130
monopolies, regional 33
Moscow International Commodities and Currency Exchange (MICEX) 129
Moscow United Energy Company (MOEK) 171
Mosenergo 124, 166, 171, 327
Movsisian, Armen 206
Mozambique 226
municipal services 170–72, 176–7
Muravlenskoe field 338
Murmansk region 21
Murmansk, LNG export terminal 73
Mytishi district 170

Nabucco pipeline 98, 358
Nadra Ukrainy 191
Nadym Pur Taz (NPT) region 258, 259, 270, 271, 272, 320
Naftogaz 210
Naftogaz Ukrainy 184, 185, 186, 187–9, 189 191, 194, 195
Nakhodkinskoe field 335
NATO 97, 348
Nazarbayev, Nursultan 351
Nemtsov, Boris 307
Nenets Autonomous District 280, 281
Neste 73
netback pricing 37, 56–7, 62, 93, 97, 118–19, 131, 185, 187, 223, 237, 255–6, 353, 359
netback parity, Russian domestic pricing 28, 30, 34, 108, 118–23, 128, 131, 135, 153, 316
'net-forward' pricing 199
Netherlands 51, 61, 90
Nevinnomysskii Azot plant 174
non-associated gas 43–4
non-Gazprom producers 33, 41, 314–46, 370–71
non-hydrocarbon energy 19

Nord Stream 73–5, 78, 79, 80, 81, 87, 103, 104, 201, 209, 212, 292, 303
North Africa, gas exports 55, 60, 105
North Caucasus 18, 20
North European Pipeline 73
North Gas 321, 322, 326, 339
North Kamennomysskoe field 268, 284
North Korea 23
North Transgas 73
North Urengoi field 322
Northern Lights 212, 268
Northern Sea Route 21
Norway 55, 60, 63
Novak, Aleksandr 228, 234, 240, 281, 300, 324, 346
Novatek 3, 21, 32, 216, 257, 319–29, 330, 341, 370
   and CNPC 243, 303, 324
   and condensate 320, 321
   domestic gas sales 119, 317, 326–8
   EBITDA 291, 292
   gas prices 124–5, 317, 380
   gas production 33, 119, 255, 320, 321, 322, 325–6, 342
   gas supply 33, 114, 124, 126, 315, 327
   gas trader 328–9
   and gasification, regional 20
   and Gazprom 32, 34, 46, 255, 320, 374, 381
   and Gazprom Neft 310, 321, 339
   and Gydan peninsula 322, 325, 326
   government assistance 322–3, 325
   LNG exports 34, 35, 69, 240–42, 245, 282, 346, 381
   MET rates 132
   and North Gas 322, 359
   pipeline access 108, 143, 149, 164
   and power sector supply 164
   Severenergia, joint venture 321, 322, 339, 340
   Spanish exports 324, 329
   Total share 320, 323, 324, 340
   wet gas 319, 273
   Yamal LNG 34, 239, 241–3, 280, 304, 322–5, 388
Novogodnee field 338
Novomoskovskiy Azot fertilizer plant 174
Novoportovskoe oilfield 310
nuclear power 16, 18, 19, 157–9, 166, 167, 168, 408

OAO Gazprom 2
Ob–Bovanenkovo railway 267
Ob–Taz Bay fields 265, 266, 268, 284
oil exports 6–8, 15
oil-linked pricing 58, 62, 63–5, 66, 248

oil MET 15
oil prices, world, impact on Russian economy 9–10
oil production 6, 10–11, 258
Okhotsk, Sea of 242
'On Gas Exports' (2006) 33–4
OPAL pipeline 71, 75, 87, 102, 106
Orel 18, 171
Orenburg 272, 326, 360, 361
Orenburg processing plant 337, 349, 350, 353, 361
Ostchem Holding 186, 188, 195
over-supply, Russia 124, 125

peak-power plants 164
Pechora LNG plant 280–81
PetroCanada 276, 279
PetroChina 221–2, 362
PGNiG 62, 65
Piakiakhinskoe field 335
pipeline leaks 178
pipeline regulation 136–53
pipelines *see* Gazprom and pipelines, Nord Stream, Power of Siberia, South Stream, transit, transmission, Yamal–Europe
Podremont Orenburg 306
point to point (PP) system, gas transportation, 84
Poland 51, 52, 61, 62, 189
political gas pricing 44, 62, 92–7, 99, 101, 104, 107, 134, 351–2, 360–62, 380–81
political pressure, and gas market 248, 380–81
Power of Siberia pipeline 35, 230, 232–3, 234, 237, 238, 241, 244, 293, 296, 297, 300, 303–4, 312, 372, 388, 389
power sector 179
 coal prices 405–6
 fuel mix 157–9
 gas consumption 110, 112, 155–68
 gas prices 405–6
 infrastructure investment 161–2, 165–7, 407–9
 oil consumption 16, 158–9
 reform of 161–5
 statistical information 404–9
President, Russian, and executive authority, energy sector 24, 27, 45, 300, 308, 344–5, 389
Presidential Energy Commission 25, 240, 242, 344–5
price *see* gas pricing
Prigorodny 334
Prirazlomnoe oil field 310
'Prodi Initiative' 82
Production Sharing Agreement (PSA) 27, 30–31
profit-based taxation 11

'Projects of Common Interest' (PCI) 87
Purovskii processing plant 320
Putin, Vladimir 9, 14, 32, 70, 291, 307, 344
 and Central Asian gas prices 351–2, 354
 and China 218, 220, 227, 228, 236–7, 238, 245, 312
 and domestic gas prices 37, 120, 123–4, 127, 131, 172, 255-6
 and Eastern Gas Programme 228–9, 231
 and EU competition investigation, response to 68, 70
 and executive authority 24, 27, 45, 300, 308, 344–5, 389
 and Gas Exchange 344–5
 and gas industry review 344
 and Gazprom 68, 72, 245, 300, 302–3, 307, 308, 309, 312–13, 343
  and corruption allegations 307–8, 344
  LNG export monopoly 242, 291, 304, 312–13
 and 'netback parity' 120, 121, 316
 and non-Gazprom producers 343
 and North American shale gas 227
 and Novatek 32, 312, 320
 and oil-indexed pricing 63–4
 and Power of Siberia pipeline 232, 303–4
 and Presidential Energy Commission 25, 240, 242, 344–5
 and Sakhalin-1 333
 and third-party access 318, 343
 and transit avoidance pipelines 72, 79, 104
 and Ukraine 103, 104, 184, 185

RAB (regulated asset base) 171
Real-Gaz 143
recession, impact on gas market 247
regulated gas prices 149, 152, 177, 380
 changing policy 153–4
 discount on 386
 and Independent suppliers 32–3, 124, 316, 317
regulated heat and heat transportation tariffs 171
Regulation 715 83–4
Regulation on Gas Security, EU 92
renewable energy 19, 25, 379
residential gas consumption 110, 112, 176–8
Resolution No. 858 141, 142, 143, 145, 399
'reverse flow' 92, 187, 189–90, 196, 204, 214–15, 380, 386
Romania 51, 61, 213
Rosatom 163
Rosgazifikatsiia, Gazprom shares 302–3
Rosneft 108, 119, 257, 304, 309–10, 341, 388
 Alrosa gas assets 330
 and Far East gas strategy 242, 388

gas output portfolio 332–3
gas sales portfolio 331–2
gas supply 125, 126, 329–35
and Itera 33, 329–30, 331, 332
as Independent gas producer 332, 370
LNG exports 240–42, 245, 333–5
production 329, 330–31, 332, 342
regulated transmission tariffs 149
reserves 329, 330, 332
and rural gas supply 20
Sakhalin-1 LNG 34, 331
Sakhalin-2 236
and Sibneftegaz 330, 332
TNK-BP purchase 33, 330
Turkmenistan upstream investment 365
see also Sechin, Igor
Rosneftegaz and Gazprom shares 302
Rospan field 330, 332
Rosstat 391
Rosukrenergo 188, 352
Rotenberg brothers 22
royalty tax *see* Mineral Extraction Tax
Rusakova, Vlada 265, 331, 332
RusHydro 163, 166, 409
Russia
  and Azerbaijan 357
  and the Balkans 22, 23, 53, 62
  and Belarus 22, 23, 197–201, 211
  and Central Asia 347–9, 351–62, 364–6, 368–70
  chemical industry 173–5, 180
  and China 217–29, 231, 236–9, 364, 367, 387–8
  and CIS countries 22–3, 41, 97, 181–3
  coal, and power generation 16, 18, 19, 158–9, 165, 405–6
  coal bed methane 224, 274
  coal prices, domestic 17–18
  combined heat and power (CHP) plants 155, 168, 171, 391–2
  Customs Union 23, 30, 206, 211, 379
  demography 115
  district heating 155–6, 168–72, 179–80
  economic diversification and energy pricing 134–5
  economic recovery 113–14
  economy 6–14, 113–14
  electricity
    demand 160–61, 167–8, 179
    generation 12, 16, 163, 404–9
    pricing 162–3, 168
  energy balance 16–19, 25, 169, 180, 247, 302
  energy companies, state-owned 162, 163
  energy consumption, primary, role of gas 16–17

energy intensity, reduction of 25
energy policy framework 25–6
and EU 22, 22, 36, 67–70, 82, 88, 93, 103–4
EU–Russia Energy Dialogue 82, 208
EU–Russia Energy Cooperation Roadmap to 2050 82
EU–Russia Gas Advisory Council 88
EU–Russia Partnership and Cooperation Agreement 103
and Europe 1, 3, 4, 25, 41, 46, 50–56, 101–7, 216, 231, 245, 246–9, 288–9, 377–9, 385–6
fertilizer production, gas consumption 173–4
gas, as domestic political tool 19–22
gas, as instrument of foreign policy 22–3, 99–101, 229–31
gas, role in Russian economy 7–9, 15–23, 295–6, 305, 380–81
gas balance 254, 368–74
gas consumption 109–17, 394–7
  district heating 155–6, 168–72
  industry 172–6
  power sector 110, 155–68
  residential use 110, 112, 176–8
  system losses 178
gas demand 17, 41, 113–17, 153–80, 247–8, 361–2
gas market reform 108–9, 115, 117–36
gas prices, domestic 17–20, 32–4, 108, 117–36, 149, 152–4, 255–6, 316, 317, 380, 386
  see also gas pricing
gas supply 252–7, 268–75, 296–300, 368–74
gasification programme 20, 24, 177, 178, 180
GDP 6, 7, 9, 11–12, 116, 118
geothermal energy, and power generation 16
global economic crisis, impact of 11, 40
high-tech industry 6–7
hydro-electric power 16, 157–9, 167, 168, 409
Independent gas producers 124–7, 253–6, 257, 314–46, 368, 370–71
industry 12–13, 19–20, 28, 110, 112, 135, 172–6
inter-fuel competition 17, 18–19
investment, foreign 13–14, 27
iron and steelmaking, gas consumption 173, 175
manufacturing industry 6, 12, 173
merit order, power generation 164
mining, gas consumption 173

and Moldova 182, 202, 204–5, 212–13
monopolies, regional 33
municipal services 170–72, 176–7
non-hydrocarbon energy 19
nuclear power 16, 18, 19, 157–9, 166, 167, 168, 408
oil industry 6–11, 15, 258
'political' gas pricing 44, 62, 92–7, 99, 101, 104, 107, 134, 351–2, 360–62, 380–81
political pressure, and gas market 248, 380–81
power sector 16, 110, 112, 155–68, 179, 309–11, 404–9
renewable energy 19, 25, 379
state regulation 23–5
taxation 1, 10, 11, 15, 18–19, 21, 25, 28–31, 34, 36, 38, 47, 290, 295, 312, 381
    see also export tax, Mineral Extraction Tax
territorial generating companies (TGKs) 126, 155, 156, 163, 168, 171, 331, 409
thermal power 157–9, 165–7, 404–9
and Ukraine 1, 23, 81, 90, 103, 106, 184–5, 188, 190–91, 196, 207–10, 303, 381–2, 390
unconventional gas resources 274
wholesale generating companies (OGKs) 155, 156, 163
see also gas pricing, Gazprom, Putin, Vladimir, and entries on individual companies, countries, and gas fields
Russian elite, and Gazprom 307–8
Russian Energy Strategy to 2030 231, 269, 276, 296
Russian Federal Property Agency, Gazprom shares 303
Russian Gas Association 129
Russian gas matrix 40–47, 376–90
RWE 62, 65–6, 189

Sabetta 323
Saiano-Shushenskaia hydro power plant 161
St Petersburg International Mercantile Exchange (SPIMEX) 121, 129, 345
Sakha-Iakutiia 230
Sakhalin 21, 217, 219, 228, 232–6, 298
Sakhalin-1 27, 235, 236, 334, 340, 342, 371
Sakhalin-1 LNG 240, 241, 242, 331, 333–4
Sakhalin-2 27, 216, 235, 298, 300, 301, 340
Sakhalin-2 LNG 233, 234, 236, 239, 241, 258, 298, 299, 373
Sakhalin-3 234, 235, 298, 299, 300, 301, 332, 333, 340
Sakhalin Centre 229
Sakhalin Energy 233–4

Sakhalin–Khabarovsk–Vladivostok (SKV) pipeline 21, 234, 298–9, 303, 308
Salmanovskoe field 325
Samburgskoe field 322
Sechin, Igor 32, 121, 216, 291, 309, 381
    and Gazprom LNG export monopoly 240, 242, 304, 312–13, 333, 345
    and Presidential Commission 25, 240, 344
Second Gas Directive (2003) 83
Second Strategic Energy Review 208
security of supply, winter weather 90
Serbia 51, 61, 90, 91, 357
Severenergia 310, 321, 322, 326, 338, 339, 340
Severneft Urengoi 174
Shah Deniz field 98, 181, 206, 337, 357–8, 365
shale oil 9, 10, 274–5
shale gas 9, 10, 191, 274
    and China 223, 224, 226, 228, 238, 244
    and Europe 55, 105, 106
    US 'revolution' 40, 54, 57–8, 105, 227, 246, 256, 275, 377, 378
        Russian impact 227, 247, 249, 263, 279, 283, 344, 368, 371, 378–9
Shatalov, Sergei 133
Shell 191, 195, 226, 234, 241, 274, 299, 301, 340, 371, 373
Shmatko Sergei 178, 343
Shtokman field 21, 35, 73, 82, 263, 269, 270, 271, 275–9, 282, 294, 340, 373
Siberia, gasification programmes 24, 177, 178, 180
Siberian Federal District, gas reserves 217
Sibneft 309, 310
Sibneftegaz 330, 332, 333
Sibur 175
Sidanco 218, 243
Single Economic Space (SES) gas sector agreement 199, 211
Sinopec 226
Slovakia 51, 61, 72, 78, 91, 189, 212
Slovenia 51, 61, 77
Sobinskoe field 229
Socar 357, 358
Sochi Olympics 21, 115
SODECO 242
South Korea 23, 217, 240
South Russkoe field 340, 341
South Stream pipeline 23, 43, 62, 75–81, 86–7, 96, 97, 99, 102–3, 106, 204, 205, 209, 210, 213, 293, 303, 304, 312, 358
South Tambei field 322, 323, 326
'Southern Corridor' 55, 77, 97–9, 138–9, 358
South-West Gissar licence 337, 365, 366
Sovcomflot 323

Spain, LNG imports 324, 329
'spare capacity', gas transportation 142
spot prices in Europe 58
SRTO–Torzhok pipeline 138
Statoil 278, 340
Strategic Co-operation Agreement 218
Stroitransgaz 287, 307
Subsoil law 26–7
Subsurface Use Tax 30
Surgutneftegaz 31, 338, 342
SurgutNG gas supply contract 126
Sverdlovsk 33
Sverdlovsk region 330, 331
Switzerland 51, 53, 61

Taiwan 233, 276, 277
Tajikistan 225, 241, 365
take-or-pay (ToP) volumes 59, 60
Tambeiskoe zone 265, 267, 268
tax break 21, 37, 133, 322, 325
tax revenue, and gas pricing 37, 131–3, 163
taxation 1, 10, 11, 15, 18–19, 21, 25, 28–31, 34, 36, 38, 47, 290, 295, 312, 381
  see also export tax, Mineral Extraction Tax
Termokarstovoe field 340
territorial generating companies (TGKs) 126, 155, 156, 163, 168, 171, 331, 409
Thailand 240, 277
thermal power 157–9, 165–7, 404–9
Third Energy Package, EU 42, 83–8, 103
Third Gas Directive, EU 83–4, 88
third parties, transmission tariffs 146–7, 401–3
third-party access (TPA) 57, 87, 125, 141–7, 318–19, 399–400
third-party production *see* Independents
tight gas 274
Timchenko, Gennadii 22, 32, 287, 307, 320
TNK-BP 31, 33, 124, 126, 218, 219, 220, 297, 330–33, 341–2, 370
Total 241, 242, 278, 320, 323, 324, 340, 342, 366
trading house system, demise of 88–90
Trans-Adriatic Pipeline (TAP) 98, 358
Trans-Anatolian Pipeline (TANAP) 98, 358
Trans-Austria Gasleitung (TAG) pipeline 83
trans Caspian pipeline 355
Transdniestria 202–6, 212–13
Trans-European Networks (TENS) initiative 82
Transgaz companies, Gazprom 148–51, 306
transit-avoidance pipelines 43, 77–9, 100–3, 104, 209
transit avoidance strategy 100, 209, 213, 303
transit countries 70

transit disputes 207–14
transit diversification 70, 73, 183, 209–14, 215, 248, 303
Transit Protocol 83
transit security 73, 208, 213
transit tariff *see* transmission tariffs
transmission tariffs 37, 85, 106, 125, 135–53, 154, 199, 203, 362
  entry–exit 146
  Gazprom 148–9
  methodology 148–9, 401–3
  regulated 149–52, 200
  third party 146–8
  zonal 146
Transparency International 14
transport tariffs *see* transmission tariffs
transportation contracts, European 85
Turkey 30, 54, 95–7, 337, 347–8, 353, 357–8
  and Azerbaijan gas 347–8, 350, 353, 358, 370
  and Iranian gas 99–100
  political gas pricing 62, 94, 95–6
  and Russian gas 51, 52, 53, 60, 61, 75, 79, 204, 213
  as transit country 70, 98
Turkmen gas exports 352
Turkmen–China pipeline 255
Turkmengaz 352, 354
Turkmenistan 2, 44, 98, 191, 255, 316, 347, 349, 355–6, 369
  and China 44, 219, 220, 223, 224–5, 348, 355, 362–4
  Chinese upstream investment 347, 364
  gas exports 348, 355, 363
  gas prices to Russia 316, 351, 354, 367
  reserves 44
  and Russia 2, 352, 354–6, 359–60, 362, 366, 369, 382
  Russian upstream investment 365
Turkmenistan–Afghanistan–Pakistan–India pipeline 23, 355–6
Turkmenistan–China pipeline 347, 355, 366
Ukraine 1, 2, 174, 214–15
  2009 contract 185, 187
  2010 deal 209–10
  2010 gas law 194–5
  Black Sea naval base 183, 185, 209, 210
  crisis (2006) 69, 184, 352
  crisis (2009) 43,73, 75. 91–2, 100, 103–4, 184–5, 196, 207–10
  crisis (2014) 1, 183, 382
  December 2013 renegotiation 184, 187, 188, 189, 191, 215, 380, 386
  and Europe 186, 188, 189, 190, 195, 196–7, 207–8, 380

gas consumption 192–4
　declining 116–17, 153, 181, 192–4, 196, 214, 377
gas debts 184
gas hub potential 190, 197
gas imports 181–92, 379–80
　Central Asia 3, 181, 351, 357, 359, 363
　government policy 187–8
　Russia 41, 42–3, 181, 182, 183–9, 209–11, 215, 379–80, 386, 390
gas market law (2010) 190, 194–5
gas market reform 184, 190–92, 194–7
gas prices
　households 193
　imports 183–9, 193, 209–10, 211, 214, 215, 248, 351, 363
　industry 193
gas production, domestic 184, 191–2, 214
gas storage 90, 140, 197, 208, 210
and gas transit 70, 75, 79–81, 103, 184–5, 191
　risk 73, 92, 95, 100, 197, 207
GDP 116, 192
LNG regasification plant 191
Naftogaz Ukrainy 184, 185, 186, 187–9, 189, 191, 194, 195
oil-linked prices and spot prices 189
pipelines 71, 72, 80
'reverse flow' 186, 189–92, 191, 197, 214–15, 380, 386
and Russia, relations with 1, 23, 81, 90, 103, 106, 184–5, 188, 190–91, 196, 207–10, 303, 381–2, 390
supply diversification 188, 189–92, 214–15, 247
take-or-pay obligation 210–11
and transit avoidance 36, 73, 77–9, 99, 100, 106, 185, 209, 212, 215, 385
transportation network 140, 187, 209, 210
Ukrgazvydobuvannya 191
Ukrgazvydobuvannya fields 192
Ukrnafta 191
unconventional gas resources, Russia 274
Ungheni–Iasi pipeline 204
Unified Gas Supply System (UGSS) 31, 34, 37, 136–40
　and gas storage 140–41
　and third-party access 141–5, 154, 326, 399–400
　third-party tariffs 139–40, 146–8
United Energy Systems of Russia 161
United Kingdom 51, 52, 61
United States of America 20, 39, 174–5, 180, 246
　and coal exports 54, 368, 379

and geopolitics 23, 355, 359
and LNG 40, 226, 275–7, 280, 282, 385, 386
'shale gas revolution' 40, 54, 57–8, 105, 227, 246, 256, 275, 377, 378
urea, prilled, gas costs 174
Urengoi field 254, 258–9, 261, 267, 273, 315, 322, 411
Ust Luga 280, 320
Uzbekistan 44, 224, 225, 226, 316, 348, 356
　gas exports 2, 348, 363, 364
　　to China 225, 362
　　to Russia 348, 352, 354, 356, 362, 369
　gas prices to Russia 316, 351, 354, 359, 360, 367
　and Gazprom 316, 352, 356, 359, 364–6, 369
　and Lukoil 336, 337, 348, 365–6, 371
　Russian upstream investment 336, 364–6
Uzbekneftegaz 352, 356

Vankor field 332
VAT 28–9
Velke Kapusany 72
Verkhne-Salym field 275
Vetek 188, 195
Viakhirev, Rem 302, 306, 307
'virtual reverse flow' 190
virtual trading point (VTP) 84
Vitol 242
Vladimir 171
Vladivostok LNG 35, 232–3, 234, 238, 239, 241, 312, 373, 389
　feed-gas supply 233–4, 244, 299–300, 304
volume-oriented strategy 35

Warnig, Matthias 307
West Siberia fields 218, 258–61
West–East pipeline 224
wet gas 43–4, 263, 272, 273–4, 275, 310, 319, 321, 330, 371
wholesale generating companies (OGKs) 155, 156, 163
WIEH 89
Wingas 89
winter weather, and security of supply 90
Wintershall 73, 75, 78, 89, 273, 340, 341
World Bank 14

Xi Jinping 236
Xinjiang 217, 224

Yakutia 21
Yamal LNG 21, 33, 239, 241, 257, 265, 280, 322, 371, 373, 388

and ending of Gazprom LNG export monopoly 34, 69, 282, 324, 328, 371
financing problems 323–4
government assistance 322–3
partners 243–4, 304, 324, 340, 342
technical challenges 323
'Yamal Megaproject' 265, 270, 283, 286, 293, 294, 311
Yamal peninsula fields 43, 123, 258, 261, 265–9, 283–5, 342, 368, 383
   development costs 47, 254, 256, 267, 272, 293–4, 311–12, 315, 368
   development timing 247, 263–4, 272, 283, 368, 371–2, 383
   infrastructure requirements 268
   and MET 29, 322
   permafrost 267
   production 267–8, 269–70, 271, 283, 284, 384, 389
   reserves 267
   Southern region 267, 268
   see also Bovanenkovo field
Yamal–Europe pipeline 43, 70–73, 82, 104, 211, 212, 293
Yamal–Europe-2 72, 79, 80, 81, 86, 103, 212
Yamal Nenets region 199, 261, 280, 326
Yamburg field 132–3, 254, 258–9, 267, 268, 273, 315
Yanukovich, Viktor 183, 184, 185, 187, 194, 382
Yeltsin era 302
Yukos 309–10
Yurkharov field, MET rates 132-3
Yushchenko, Viktor 185

Zapoliarnoe field 254, 259, 260, 264, 273, 315
ZMB Schweiz 352